Hybrid Polymer Composite Materials

Related titles

Ceramic Nanocomposites (ISBN: 978-0-85709-338-7)

Environmentally-friendly Polymer Nanocomposites (ISBN: 978-0-85709-777-4)

Physical Properties and Applic. of Polymer Nanocomposites (ISBN: 978-1-84569-672-6)

Woodhead Publishing Series in Composites Science and Engineering

Hybrid Polymer Composite Materials

Applications

Edited by

Vijay Kumar Thakur

Manju Kumari Thakur

Asokan Pappu

ELSEVIER

WP
WOODHEAD PUBLISHING
An imprint of Elsevier

Woodhead Publishing is an imprint of Elsevier
The Officers' Mess Business Centre, Royston Road, Duxford, CB22 4QH, United Kingdom
50 Hampshire Street, 5th Floor, Cambridge, MA 02139, United States
The Boulevard, Langford Lane, Kidlington, OX5 1GB, United Kingdom

British Library Cataloguing-in-Publication Data
A catalogue record for this book is available from the British Library

Library of Congress Cataloging-in-Publication Data
A catalog record for this book is available from the Library of Congress

ISBN: 978-0-08-100785-3 (print)
ISBN: 978-0-08-100786-0 (online)

For information on all Woodhead Publishing publications
visit our website at https://www.elsevier.com/books-and-journals

Publisher: Matthew Deans
Acquisition Editor: Gwen Jones
Editorial Project Manager: Charlotte Rowley
Production Project Manager: Debasish Ghosh
Cover Designer: Greg Harris

Typeset by MPS Limited, Chennai, India

Contents

List of Contributors ix

1 Conducting polymer–graphite binary and hybrid composites: structure, properties, and applications **1**
B.T. S. Ramanujam and Pratheep K. Annamalai
1.1 Introduction 2
1.2 Processing of polymer–graphite binary and hybrid composites 7
1.3 Structure development and properties enhancement in conducting polymer composites 13
1.4 Applications 19
1.5 Conclusions and future perspectives 28
Acknowledgments 29
References 29

2 Hybrid polymer composites for structural applications **35**
Hai Nguyen, Wael Zatar and Hiroshi Mutsuyoshi
2.1 Introduction 35
2.2 Hybrid FRP composites for structural applications 37
2.3 Hybrid FRP–concrete composites structural members 44
2.4 Conclusions 49
References 50

3 Hybrid polymer composites for electromagnetic absorption in electronic industry **53**
Charalampos A. Stergiou, Marina Y. Koledintseva and Konstantin N. Rozanov
3.1 Introduction 54
3.2 Shielding effectiveness definitions 60
3.3 Applications 64
3.4 Specular absorber for far field operation 76
3.5 Material's design 84
3.6 Summary 95
Acknowledgments 96
References 96

4 **Hybrid polysaccharide-based systems for biomedical applications** **107**
Paula I.P. Soares, Coro Echeverria, Ana C. Baptista, Carlos F.C. João, Susete N. Fernandes, Ana P.C. Almeida, Jorge C. Silva, Maria H. Godinho and João P. Borges
4.1 Introduction 107
4.2 Biosensors and actuators 110
4.3 Theranostic systems 117
4.4 Tissue engineering 123
4.5 Conclusions and future perspectives 134
Acknowledgments 135
References 135

5 **Hybrid multifunctional composites—recent applications** **151**
Naheed Saba, Mohammad Jawaid, Mohamed Thariq Hameed Sultan and Othman Alothman
5.1 Introduction 151
5.2 Multifunctional materials 153
5.3 Multifunctional composites 154
5.4 Applications of multifunctional materials and multifunctional composites 161
5.5 Conclusion 163
Acknowledgment 164
References 164

6 **Conducting polymer-based thermoelectric composites: principles, processing, and applications** **169**
Temesgen A. Yemata, Qun Ye, Hui Zhou, Aung K.K. Kyaw, Wee S. Chin and Jianwei Xu
6.1 Introduction 169
6.2 Preparation and processing of TE composites 172
6.3 Conductive polymers for TE materials 175
6.4 Polymer–inorganic hybrid TE materials 175
6.5 Carbon-based nanostructures/polymer nanocomposites TE materials 179
6.6 Conclusions and outlook 186
Acknowledgment 191
References 191

7 **Using recycled polymers for the preparation of polymer nanocomposites: properties and applications** **197**
Shadpour Mallakpour and Vajiheh Behranvand
7.1 Introduction 197
7.2 NCs obtained from recycled polymers 199

7.3 Summary 217
7.4 Conclusions and challenges 222
Acknowledgments 222
References 222

8 Recent developments in the synthesis of hybrid polymer/clay nanocomposites: properties and applications 227
Shadpour Mallakpour and Shima Rashidimoghadam
8.1 Introduction 227
8.2 Preparation of PCNs and their applications 234
8.3 Conclusions 259
Acknowledgments 259
References 260

9 Graphene-based materials and their potential applications: a theoretical study 267
Hong-ping Zhang and Youhong Tang
9.1 Introduction 267
9.2 Properties of different graphene, including pristine graphene, defective graphene, graphene oxide, and metal-doped graphene 268
9.3 Theoretical study methodologies 268
9.4 Graphene-based materials with applications 269
9.5 Conclusion and remarks 282
References 283

10 Synthesis and applications of cellulose nanohybrid materials 289
Nathaniel T. Garland, Eric S. McLamore, Carmen Gomes, Ethan A. Marrow, Michael A. Daniele, Scott Walper, Igor L. Medintz and Jonathan C. Claussen
10.1 Introduction 289
10.2 Cellulose 291
10.3 Regenerated cellulose 291
10.4 Nanocrystalline cellulose 293
10.5 Functionalization 295
10.6 Applications 295
10.7 Conclusion 310
References 311

11 Sonochemical preparation of hybrid polymer nanocomposites: properties and applications 321
Bharat A. Bhanvase and Shirish H. Sonawane
11.1 Introduction 321
11.2 Properties of polymer nanocomposite: role of ultrasound 322
11.3 Sonochemical preparation of polymer nanocomposite: case studies 327

11.4 Future perspectives 338
11.5 Conclusion 339
References 339

12 Biomedical applications of hybrid polymer composite materials 343
Burhan Ates, Suleyman Koytepe, Sevgi Balcioglu, Ahmet Ulu, and Canbolat Gurses
12.1 Introduction 343
12.2 Tissue engineering applications 347
12.3 Drug delivery 373
12.4 Biosensor applications 384
12.5 Other biomedical applications 399
12.6 Conclusion 400
References 401

13 Liquid crystalline DNA: a smart polymer with a variety of applications ranging from photonics to plasmonics 409
Luciano De Sio, Ferdinanda Annesi, Tiziana Placido, Roberto Comparelli, Alfredo Pane, Maria L. Curri, Cesare Umeton and Roberto Bartolino
13.1 Deoxyribonucleic acid 409
13.2 Plasmonic nanoparticles 413
13.3 Liquid crystalline DNA organization in polymeric template 413
13.4 Gold nanorods organized in liquid crystalline DNA 417
13.5 Conclusion 419
Acknowledgments 420
References 420

Index 423

List of Contributors

Ana P.C. Almeida i3N/CENIMAT, Department of Materials Science, Faculty of Science and Technology, Universidade NOVA de Lisboa, Caparica, Portugal

Othman Alothman Aerospace Manufacturing Research Centre, Level 7, Tower Block, Faculty of Engineering, Universiti Putra Malaysia, Serdang, Selangor Darul Ehsan, Malaysia

Pratheep K. Annamalai Australian Institute for Bioengineering and Nanotechnology, The University of Queensland, Brisbane, QLD, Australia

Ferdinanda Annesi CNR-Lab. Licryl, Institute NANOTEC, Arcavacata di Rende, Italy

Burhan Ates Department of Chemistry, Inonu University, Malatya, Turkey

Sevgi Balcioglu Department of Chemistry, Inonu University, Malatya, Turkey

Ana C. Baptista i3N/CENIMAT, Department of Materials Science, Faculty of Science and Technology, Universidade NOVA de Lisboa, Caparica, Portugal

Roberto Bartolino CNR-Lab. Licryl, Institute NANOTEC, Arcavacata di Rende, Italy; Department of Physics, Centre of Excellence for the Study of Innovative Functional Materials, University of Calabria, Arcavacata di Rende, Italy; Interdisciplinary Institute B, Segre of the National Academy of Lincei, Rome, Italy

Vajiheh Behranvand Organic Polymer Chemistry Research Laboratory, Department of Chemistry, Isfahan University of Technology, Isfahan, Islamic Republic of Iran

Bharat A. Bhanvase Department of Chemical Engineering, Laxminarayan Institute of Technology, Rashtrasant Tukadoji Maharaj Nagpur University, Nagpur, Maharashtra, India

João P. Borges i3N/CENIMAT, Department of Materials Science, Faculty of Science and Technology, Universidade NOVA de Lisboa, Caparica, Portugal

Wee S. Chin Department of Chemistry, National University of Singapore, Singapore

Jonathan C. Claussen Department of Mechanical Engineering, Iowa State University, Ames, IA, United States

Roberto Comparelli CNR-IPCF, Institute for Physical and Chemical Processes, Bari, Italy

Maria L. Curri CNR-IPCF, Institute for Physical and Chemical Processes, Bari, Italy

Michael A. Daniele Joint Department of Biomedical Engineering, UNC-Chapel Hill/NC State University, Raleigh, NC, United States; Department of Electrical & Computer Engineering, NC State University, Raleigh, NC, United States

Luciano De Sio Beam Engineering for Advanced Measurements Company, Orlando, FL, United States; CNR-Lab. Licryl, Institute NANOTEC, Arcavacata di Rende, Italy

Coro Echeverria i3N/CENIMAT, Department of Materials Science, Faculty of Science and Technology, Universidade NOVA de Lisboa, Caparica, Portugal

Susete N. Fernandes i3N/CENIMAT, Department of Materials Science, Faculty of Science and Technology, Universidade NOVA de Lisboa, Caparica, Portugal

Nathaniel T. Garland Department of Mechanical Engineering, Iowa State University, Ames, IA, United States

Maria H. Godinho i3N/CENIMAT, Department of Materials Science, Faculty of Science and Technology, Universidade NOVA de Lisboa, Caparica, Portugal

Carmen Gomes Department of Biological and Agricultural Engineering, Texas A&M, College Station, TX, United States

Canbolat Gurses Department of Molecular Biology and Genetics, Inonu University, Malatya, Turkey

Mohammad Jawaid Laboratory of Biocomposite Technology, Institute of Tropical Forestry and Forest Products (INTROP), Universiti Putra Malaysia, Serdang, Selangor, Malaysia; Aerospace Manufacturing Research Centre, Level 7, Tower Block, Faculty of Engineering, Universiti Putra Malaysia, Serdang, Selangor Darul Ehsan, Malaysia; Department of Chemical Engineering, College of Engineering, King Saud University, Riyadh, Saudi Arabia

Carlos F.C. João i3N/CENIMAT, Department of Materials Science, Faculty of Science and Technology, Universidade NOVA de Lisboa, Caparica, Portugal

Marina Y. Koledintseva Electromagnetic Compatibility Design Engineering, Oracle, Santa Clara, CA, United States

Suleyman Koytepe Department of Chemistry, Inonu University, Malatya, Turkey

Aung K.K. Kyaw Institute of Materials Research and Engineering, Agency for Science, Technology and Research (A*STAR), Singapore, Singapore

Shadpour Mallakpour Organic Polymer Chemistry Research Laboratory, Department of Chemistry, Isfahan University of Technology, Isfahan, Islamic Republic of Iran; Nanotechnology and Advanced Materials Institute, Isfahan University of Technology, Isfahan, Islamic Republic of Iran; Center of Excellence in Sensors and Green Chemistry, Department of Chemistry, Isfahan University of Technology, Isfahan, Islamic Republic of Iran

Ethan A. Marrow Joint Department of Biomedical Engineering, UNC-Chapel Hill/NC State University, Raleigh, NC, United States

Eric S. McLamore Agricultural and Biological Engineering Department, Institute of Food and Agricultural Sciences, University of Florida, Gainesville, FL, United States

Igor L. Medintz Center for Biomolecular Science and Engineering, U.S. Naval Research Laboratory, Washington, DC, United States

Hiroshi Mutsuyoshi Department of Civil & Environmental Engineering, Saitama University, Saitama, Japan

Hai Nguyen College of Information Technology & Engineering, Marshall University, Huntington, WV, United States

Alfredo Pane CNR-Lab. Licryl, Institute NANOTEC, Arcavacata di Rende, Italy

Tiziana Placido University of Bari, Chemistry Department, Bari, Italy; CNR-IPCF, Institute for Physical and Chemical Processes, Bari, Italy

B.T.S. Ramanujam School of Science and Engineering, Navrachana University, Vadodara, Gujarat, India

Shima Rashidimoghadam Organic Polymer Chemistry Research Laboratory, Department of Chemistry, Isfahan University of Technology, Isfahan, Islamic Republic of Iran

Konstantin N. Rozanov Institute for Theoretical and Applied Electromagnetics, RAS, Moscow, Russia

Naheed Saba Laboratory of Biocomposite Technology, Institute of Tropical Forestry and Forest Products (INTROP), Universiti Putra Malaysia, Serdang, Selangor, Malaysia

Jorge C. Silva i3N/CENIMAT, Physics Department, Faculty of Science and Technology, Universidade NOVA de Lisboa, Caparica, Portugal

Paula I.P. Soares i3N/CENIMAT, Department of Materials Science, Faculty of Science and Technology, Universidade NOVA de Lisboa, Caparica, Portugal

Shirish H. Sonawane Department of Chemical Engineering, National Institute of Technology, Warangal, Telangana, India

Charalampos A. Stergiou Laboratory of Inorganic Materials, Centre for Research and Technology-Hellas, Thessaloniki, Greece

Mohamed Thariq Hameed Sultan Aerospace Manufacturing Research Centre, Level 7, Tower Block, Faculty of Engineering, Universiti Putra Malaysia, Serdang, Selangor Darul Ehsan, Malaysia; Department of Aerospace Engineering, Faculty of Engineering, Universiti Putra Malaysia, Serdang, Selangor Darul Ehsan, Malaysia

Youhong Tang Centre for NanoScale Science and Technology and School of Computer Science, Engineering and Mathematics, Flinders University, Adelaide, SA, Australia

Ahmet Ulu Department of Chemistry, Inonu University, Malatya, Turkey

Cesare Umeton CNR-Lab. Licryl, Institute NANOTEC, Arcavacata di Rende, Italy; Department of Physics, Centre of Excellence for the Study of Innovative Functional Materials, University of Calabria, Arcavacata di Rende, Italy

Scott Walper Center for Biomolecular Science and Engineering, U.S. Naval Research Laboratory, Washington, DC, United States

Jianwei Xu Institute of Materials Research and Engineering, Agency for Science, Technology and Research (A*STAR), Singapore, Singapore; Department of Chemistry, National University of Singapore, Singapore

Qun Ye Institute of Materials Research and Engineering, Agency for Science, Technology and Research (A*STAR), Singapore, Singapore

Temesgen A. Yemata Institute of Materials Research and Engineering, Agency for Science, Technology and Research (A*STAR), Singapore, Singapore; Department of Chemistry, National University of Singapore, Singapore

Wael Zatar College of Information Technology & Engineering, Marshall University, Huntington, WV, United States

Hong-ping Zhang School of Materials Science and Engineering, Southwest University of Science and Technology, Mianyang, China

Hui Zhou Institute of Materials Research and Engineering, Agency for Science, Technology and Research (A*STAR), Singapore, Singapore

Conducting polymer–graphite binary and hybrid composites: structure, properties, and applications

1

B.T. S. Ramanujam[1] *and Pratheep K. Annamalai*[2]
[1]School of Science and Engineering, Navrachana University, Vadodara, Gujarat, India,
[2]Australian Institute for Bioengineering and Nanotechnology, The University of Queensland, Brisbane, QLD, Australia

Chapter Outline

1.1 Introduction 2
1.1.1 Conducting polymers and classification 2
1.1.2 Electrical conductivity variation in binary conducting composites 4
1.1.3 Percolation theory 5
1.1.4 Factors affecting electrical percolation threshold 6
1.2 Processing of polymer–graphite binary and hybrid composites 7
1.2.1 Graphite and graphite derived high aspect ratio fillers 7
1.2.2 Carbon nanofiber 11
1.2.3 Processing methods 11
1.3 Structure development and properties enhancement in conducting polymer composites 13
1.3.1 Electrical properties 14
1.3.2 Mechanical properties 16
1.3.3 Thermal properties 17
1.3.4 Structure development in thermoplastic–graphite composites 18
1.4 Applications 19
1.4.1 Alternate energy technology 19
1.4.2 Electromagnetic interference shielding devices 22
1.4.3 Solar cells 24
1.4.4 Anticorrosion coatings 25
1.4.5 Supercapacitors 26
1.4.6 Other applications 27
1.5 Conclusions and future perspectives 28
Acknowledgments 29
References 29

Hybrid Polymer Composite Materials: Applications. DOI: http://dx.doi.org/10.1016/B978-0-08-100785-3.00001-2

1.1 Introduction

The electrically conducting polymeric materials and their composites with low density, low cost, good physical, thermal, and mechanical properties have been explored for many applications such as fuel cell bipolar plates, electromagnetic interference (EMI) shielding devices, and others. As driven by the demand for light weight intelligent materials and the recent advancements in organic electronics, nanotechnology and electrochemistry, the global market for these electrically active polymeric materials is skyrocketing and is projected to reach US$ 5.2b by 2022 (http://www.strategyr.com/MarketResearch/Conductive_Polymers_Market_Trends.asp). These materials can be developed using organic polymers with enhanced conducting properties or dispersing conducting phases at certain quantity in insulating polymer matrices. By exploiting the different properties of two different conducting particles/dispersed phases in polymer synergistically, the hybrid polymer composite materials can also be synthesized. In general, the development of polymer composite materials is highly preferred, not only because they can be synthesized via industrially viable processes with less environmental impact, but also their properties can be tailored/engineered by controlling the dimension, shape, and morphology of dispersed phases and their interaction with the polymers. In last three decades, advancements in the nanotechnology have also offered us unprecedented potential to have control over the properties and performance of these hybrid composite materials. Typically, to enhance the conducting properties of the polymer matrices, micro/nanoparticles of conducting carbon black (CB), graphite and others fillers are incorporated. This chapter focuses on recent developments in the area of electrically conducting binary and hybrid composite materials.

1.1.1 Conducting polymers and classification

Until the discovery of iodine doped polyacetylene, whose conductivity was in the order of 10^3 S cm^{-1}, polymers were regarded as insulators. After the discovery of transpolyacetylene by Alan G MacDiarmid, Hideki Shirakawa and Alan J Heeger for which they were awarded Nobel prize in the year 2000, many conducting polymers have been synthesized such as polyaniline (PANI), polypyrrole (PPy), and polythiophene (PTh) (Shirakawa et al., 1977; Nalwa, 2004). The conductivity of polymeric materials with range of conductivities, as shown in Fig. 1.1, can also be achieved by suitable dopants.

In general, electrically conducting polymers can be classified into following types:

(a) Inherently conducting polymers (ICPs), (b) polymer charge transfer complexes, (c) organometallic polymer conductors, and (d) conducting polymer composites (CPCs).

a. Inherently conducting polymers

Quasi infinite π electron system extending over many monomers and conjugation are the characteristics of ICPs. The motion of charged defects within the conjugated

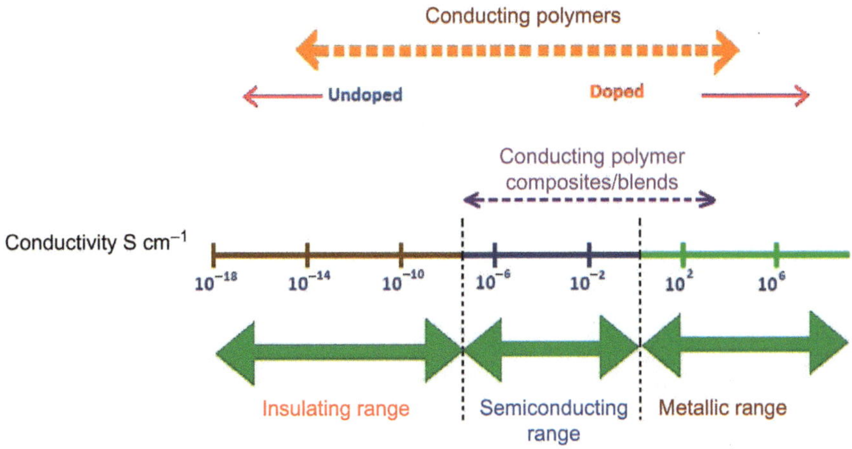

Figure 1.1 Range of electrical conductivities of different materials.

framework forms the basis of conduction in this class of conducting polymers. PANI, PTh, and PPy are examples for this type.

b. Polymer charge transfer complexes

Addition of acceptor molecules in insulating polymers results in the formation of polymer charge transfer complexes. Mulliken (1950, 1951) and Dewar and Rogers (1962) have proposed the theory of charge transfer complexes. The donor–acceptor interaction promotes orbital overlap resulting in enhanced electron delocalization.

c. Organometallic polymer conductors

Organometallic polymers comprise of transition metal ions and highly conjugated ligands in the back bone which eventually results in conducting polymers (Nalwa, 1990). The electron delocalization is enhanced by the overlap of metal d-orbital and π orbitals of the organic structure. Poly(metallophthalocyanines), poly(ferrocenylene), and others are examples for this class of conducting polymers. Many organometallic polymers are intrinsic conductors and in others, metallic like conductivity is achieved by molecular doping process.

d. Conducting polymer composites

This class of conducting polymers can be synthesized by dispersion of electrically conducting fillers in an insulating polymer matrix (either thermoplastics or thermosets) or ICPs so that appreciable level of electrical conductivity can be achieved depending up on the end applications. Also, it is possible to employ polymer blends in which the conducting phases can be dispersed. Usually conducting CB (Pinto et al., 1999; Yacubowicz et al., 1990), graphite (Tchmutin et al., 2003; Saini et al., 2009), metal powders (Baker et al., 1988; Chiang and Chiang, 1992), carbon nanotubes (CNTs) (Winey et al., 2007; Bryning et al., 2005), graphene (Potts et al., 2011; Verdejo et al., 2011; Stankovich et al., 2006), expanded graphite (ExGr) (Zheng et al., 2004; Usuki et al., 1993; Chen et al., 2003), carbon nanofiber (CNF) (Chatterjee et al., 2007; Higgins and Brittain, 2005; Cortes et al., 2003), and others are being used as conducting fillers.

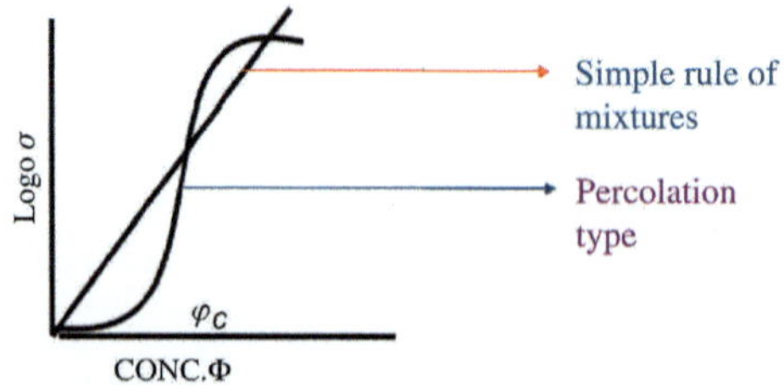

Figure 1.2 Electrical conductivity variation in binary polymer composites.

1.1.2 Electrical conductivity variation in binary conducting composites

The electrical conductivity variation in most of the conducting binary composites does not follow rule of mixtures instead nonlinear variation of electrical conductivity is observed as depicted in Fig. 1.2. The nonlinear variation of electrical conductivity in many binary composites has been explained through percolation theory. There are three distinct regions which can be identified in Fig. 1.2. The binary composites behave as insulators. The insulating nature of the composites is reflected through very low value of electrical conductivity up to a particular loading of filler in the polymer matrix up on increasing the concentration of fillers (below φ_c). At a critical concentration of fillers (φ_c), the electrical conductivity increases to many orders, known as electrical percolation threshold which essentially represents insulator to semiconductor transition of the composite. Above critical concentration, the electrical conductivity is saturated. The nonlinear variation of electrical conductivity is advantageous as the loading of fillers is lesser than what it would have been if the variation of electrical conductivity follows rule of mixtures to achieve higher value of conductivity. This aids in easy processing of CPCs when mass production technique such as injection molding is employed. The electrical conductivity of CPCs is affected by the degree of filling and the proximity of the conducting particles.

The composite remains as insulator when the conducting particles are isolated. This is reflected below the electrical percolation threshold. At critical concentration (φ_c) of fillers, contact between them is established, and hence, the conductivity of the composite starts increasing. The close proximity of conducting particle in the polymer matrix causes current flow across the gap. Above φ_c, saturation in contacts between filler particles results in saturation in electrical conductivity. The charge transport across the insulating gap can be by hopping (Dyre and Schrøder, 2000) or tunneling (Blythe and Bloor, 2005) mechanisms. When the volume fraction of fillers is above the percolation threshold, composites conduct through network formation between the particles. There could be band type conduction. In general hopping conduction can be clearly delineated through ac conductivity studies. If the hopping conduction mechanism is operative, the conductivity will be dependent on the frequency of the applied voltage and the ac conductivity will be higher than that of dc conductivity. The latter is independent of frequency of the applied voltage. Usually, at low concentration of fillers in a polymer matrix nearer to the electrical percolation threshold, hopping conduction can be clearly identified.

1.1.3 Percolation theory

Percolation model is mainly used to account for the variation of properties of materials by taking into consideration the formation of network among its constituents. There are two types of percolation namely site percolation and bond percolation. For the former case, points are defined in the lattice, and there will be finite probability for the point to exist in every lattice site. In the latter case, bonds are defined between neighboring sites in the lattice. In both cases, clusters which are nothing but the structure of connected points will be formed with a finite probability "p". Eventually, cluster that has a path spans the whole system will be formed. The first such model was used in 1940 to explain polymerization phenomenon that resulted in gelation. In many binary conducting composites, the electrical conductivity variation and the electrical percolation threshold are theoretically predicted using percolation theory though this model cannot be used to describe all binary or hybrid composites. One of the drawbacks of statistical percolation is that lattice has to be defined. In truly random situation, it is very difficult to define a lattice. When the properties of individual components in composites vary significantly, the percolation theory shall be applicable. The classical percolation equations are given below.

$$\sigma_{\text{eff}} = \sigma_2 \left(\frac{\varphi - \varphi_c}{1 - \varphi_c} \right)^t \qquad \varphi > \varphi_c \tag{1.1}$$

$$\sigma_{\text{eff}} = \sigma_1 \left(\frac{\varphi_c - \varphi}{\varphi_c} \right)^{-s} \qquad \varphi < \varphi_c \tag{1.2}$$

where σ_{eff} is the effective conductivity of the composite consisting of a conductor with conductivity σ_2 and an insulator essentially the polymer with a conductivity σ_1, φ is the volume fraction of filler, φ_c is the percolation threshold. "t" and "s" are critical exponents characterizing the conductivity in the conducting and the insulating phase, depend on the dimensionality of the network formation. The values of the exponents when three-dimensional network formation exists are $t = 2$ and $s = 0.87$ (Nan, 2002) for lattice percolation. The percolation threshold for different binary composites is shown in Table 1.1.

The variation of electrical conductivity with respect to the concentration of high abrasion resistance CB in noncrystallizable chloroprene rubber has been reported to follow scaling law above the percolation threshold as reported elsewhere (All and Abo-Hashem, 1997). Similarly, there are many reports in the literature on the nonlinear variation of electrical conductivity observed in various conducting polymer binary and hybrid composites. (Chiu and Chen, 2015; Noël et al., 2014; Meier

Table 1.1 Example for percolation threshold for few binary composites

Host polymer matrix	Conducting phase	Electrical percolation threshold	References
High density polyethylene (HDPE)	Single walled carbon nanotube (SWCNT)	4 wt%	Zhang et al. (2006)
Unsaturated polyester	Graphite nanosheets	0.64 vol%	Lu et al. (2006)[a]
Epoxy	Multiwalled carbon nanotube (MWCNT)	<0.02 wt%[b]	Faiella et al. (2012)
Polyvinylidene fluoride (PVDF)	MWCNT PPy	0.3% 10 wt%	Silva et al. (2013)

[a]They have also reported critical exponent close to two.
[b]Depends on curing temperature and sonication time.

et al., 2011; Zhang et al., 2015; Ghasemi and Sundararaj, 2012; Zeng et al., 2011; Vasileiou et al., 2015).

1.1.3.1 Other theories/models

There are many models propounded to account for the variation of electrical conductivity of CPCs. Mamunya et al. (1997) have developed thermodynamic model for which surface energy of filler, polymer, melt viscosity of the polymer are taken into consideration. At all points above electrical percolation threshold, the conductivity variation of the composite could be well accounted. However, this model explains CB filled polymer composite systems well but not to other fillers. Similarly, structure oriented model proposed by Nielsen (Clingermann et al., 2002) takes into consideration the filler orientation as well as the aspect ratio of filler to account for the electrical conductivity variation. Nielsen model is used to describe electrical conductivity, thermal conductivity and the modulus of metal–polymer systems well. Geometrical models have also been proposed to account for the electrical conductivity variation in polymer composites (Youngs, 2003). The conduction mechanisms in CPCs are not universal. Various conduction mechanisms can be grouped into four major types such as uniform model, uniform channel model, non-tunneling barrier model and tunneling barrier model as reported elsewhere (Celzard et al., 1997).

1.1.4 Factors affecting electrical percolation threshold

The electrical percolation threshold as discussed in Section 1.3 is an important parameter which should be minimized as much as possible without affecting the electrical conductivity. This will help in easy processing of the CPCs when mass production techniques such as injection molding will be employed to make products. The electrical percolation threshold in binary composites is dependent on filler particle size, particle size distribution of fillers, dispersion of fillers in a given polymer matrix, processing routes employed, aspect ratio of fillers, polymer characteristics, and others. Lower particle size, narrow particle size distribution, high aspect ratio of fillers which will enhance interparticular contacts at low loading and enhanced dispersion of fillers without agglomeration are preferable to reduce the electrical percolation threshold.

1.2 Processing of polymer–graphite binary and hybrid composites

Though there are many conducting fillers such as metal particles, fibers, and others, in this article only carbon based especially graphite and graphite derived fillers and their effect on electrical percolation threshold and various properties will be addressed.

1.2.1 Graphite and graphite derived high aspect ratio fillers

1.2.1.1 Graphite

One of the allotropes of carbon, namely graphite is being used as conducting filler in many insulating polymer matrices such as polypropylene (PP), polystyrene (PS), epoxy, and others due to very high in-plane conductivity of the order of 10^4 S cm^{-1} (Chung, 2002).

Graphite has layered structure with carbon atoms arranged in hexagonal pattern in AB stacking sequence as shown in Fig. 1.3. The unit cell parameters are $a = 2.46$ Å, $b = 2.46$ Å, and $c = 6.71$ Å. The crystal structure corresponds to that of $P6_3/mmc$ space group (Chung, 2002). There are four atoms denoted by M, M', N, and N'. The atoms M, N are on one layer, and M', N' are on other layer displaced by half the c-axis spacing. Thus, the interlayer spacing is 3.35 Å. The carbon atoms in graphite are sp^2 hybridized and hence in-plane conductivity is very high. Hence from electronic structure perspective, it is regarded as semimetal (Chung, 2002).

Graphite can be broadly classified into natural and synthetic. It can form intercalation compounds in which foreign molecules are inserted in the interlayer spacing without disturbing the carbon layers. The foreign molecule, known as intercalant, can increase the interspacing of graphite layers. Majority of graphite intercalation compounds (GIC) belong to the class of ionic intercalation compounds. The bonding between carbon and the intercalant has certain ionic character or polar character in this category. The stage of the compound defined as the number of carbon layers between intercalated layers which can be ascertained through other characterization technique such as X-ray diffraction. Graphite–halogen, graphite–alkali metal, graphite–acid compounds are examples of ionic intercalation compounds. Out of

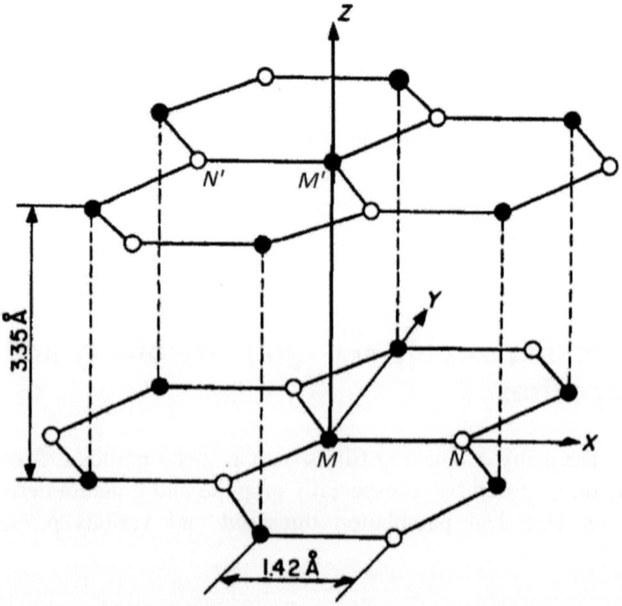

Figure 1.3 Crystal structure of graphite.

them, graphite-acid compounds are very interesting and frequently used for synthesizing ExGr. The intercalation process is pivotal in order to synthesize high aspect ratio graphite nanosheets (GNS) and in few cases graphene, which is a monolayer of graphite with one carbon atom thick. Conventionally, high purity graphene can be produced through chemical vapor deposition process (Wang et al., 2015). It is very important to synthesize high aspect ratio nanofillers as the increased surface area will reduce the electrical percolation threshold to a greater extent.

1.2.1.2 Expanded graphite

By proper intercalation of acids such as sulfuric acid, it is possible to increase the interlayer spacing of natural graphite, where by expansion along *c*-axis of graphite unit cell results in GICs or expandable graphite (Duquesne et al., 2003). When GICs are heated at high temperature around 1000°C for few seconds (Chen et al., 2003) or exposed to microwave irradiation (Zhang et al., 2016; Sykam and Kar, 2014; Ramanujam et al., 2010), volume expansion of few hundred times is observed. The resultant porous structured graphite is known as expanded graphite (ExGr). The schematic of the synthesis procedure is represented in Fig. 2.2. In a simple procedure, mixture of concentrated sulfuric acid and concentrated nitric acid in 4:1 volume ratio is mixed with natural flake graphite and the mixture is stirred continuously for 16 h, and the acid treated graphite particles are washed with distilled water till the pH becomes neutral and then dried as reported by Chen et al. (2003). Up on subsequent high temperature heat treatment, expansion of graphite

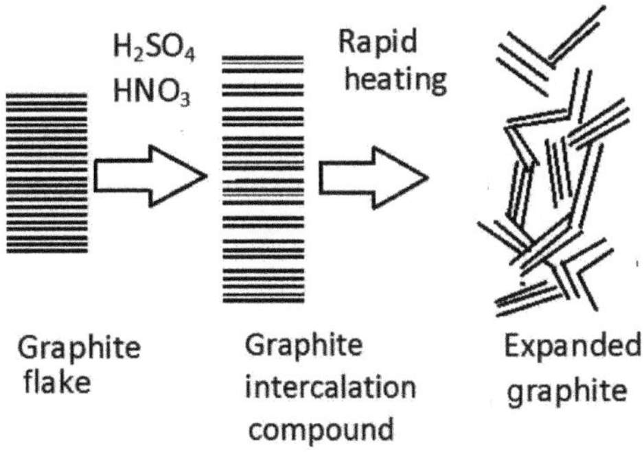

Figure 1.4 Schematic representation of formation of expanded graphite.

Figure 1.5 SEM picture of expanded graphite.

layers due to vaporization of the intercalant results in the formation of ExGr with varying pore sizes. The schematic of formation of ExGr is shown in Fig. 1.4. The scanning electron microscope image of commercial ExGr particle is shown in Fig. 1.5. Ramanujam et al. (2010) have studied charge transport and impedance characteristics of polyether sulfone-sonicated ExGr composites. Chung (Chung, 1987) has written an excellent review article on the exfoliation of graphite. The porous structure is used to adsorb oil from oil–water mixture.

Up on ultrasonication, the pore structure is destroyed and GNS are formed which are used as nanofillers in polymer matrix as shown in Fig. 1.6.

The high aspect ratio GNS when mixed in polymer matrices both thermoplastic and thermoset, very low electrical percolation threshold results. Since the density of ExGr is less than that of natural graphite, it is used for making fuel cell bipolar plates.

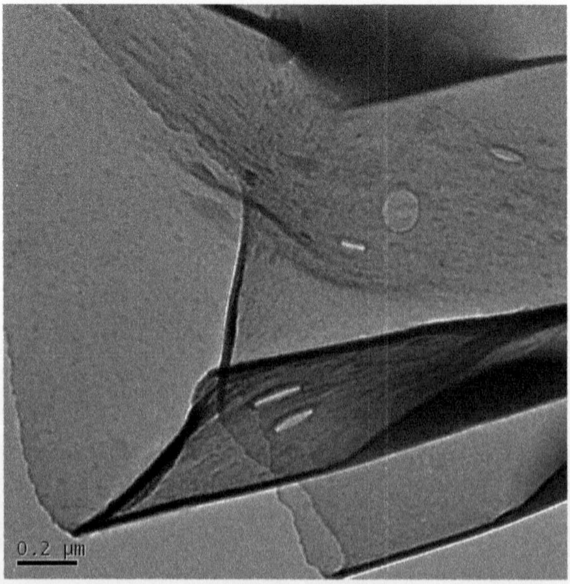

Figure 1.6 TEM picture of sonicated expanded graphite.

1.2.1.3 Graphite oxide to graphene

Graphene is being focused for its exceptional properties such as very high Young's modulus (~1100 GPa) (Lee et al., 2008), very high mobility of charge carriers (200,000 $cm^2\ V^{-1}\ s^{-1}$) (Bolotin et al., 2008), and others. There are four different ways of producing graphene (Park and Ruoff, 2009).

1. Chemical vapor deposition
2. Micromechanical exfoliation of graphite
3. Epitaxial growth on insulating substrates
4. Colloidal suspensions

The production of large quantities of graphene is still a challenge. Chemical routes are more promising for the production of graphene in scalable quantities. Underlying idea behind the production of graphene through chemical routes is to first oxidize graphite completely to form insulating graphite oxide followed by reduction using reducing agents such as hydrazine and others. Graphite oxide can be synthesized by Brodie, Staudenmaier, and Hummer's methods (Park and Ruoff, 2009). When graphite is treated with strong acids in presence of oxidizing agents can result in graphite oxide. The level of oxidation can be varied on the reaction conditions, precursor graphite used, and others. Graphite oxide is hydrophilic and water molecules readily intercalate resulting in graphene oxide (GO). The interlayer distance in GO can reversibly increase with increase in humidity. The zeta potential measurement proves that GO sheets bear negative charges when dispersed in water. The electrostatic repulsion between the sheet results in better dispersion in water

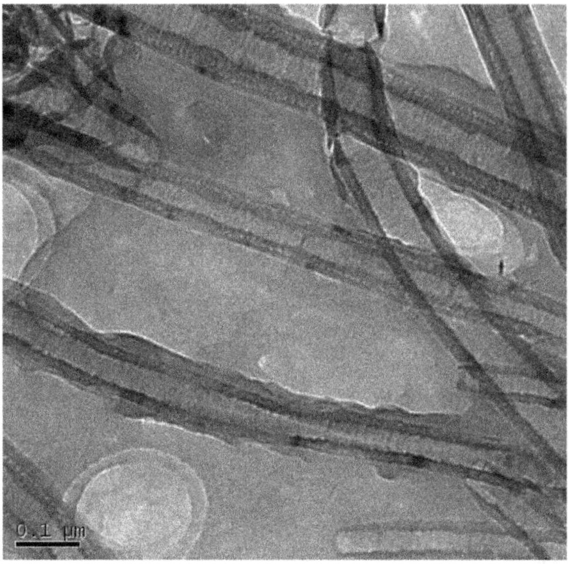

Figure 1.7 TEM picture of commercial VGCNF.

and hence colloidal suspension can be formed. Thermal methods (McAllister et al., 2007) and ultraviolet-assisted methods (Williams et al., 2008) have produced electrically conducting chemically modified graphene.

1.2.2 Carbon nanofiber

CNFs are one of the high aspect ratio nanofillers based on carbon and act as excellent replacement for CNTs as the cost of CNTs is very high though CNTs exhibit higher electrical and thermal conductivities. Vapor grown CNFs (VGCNFs) are produced by catalytic chemical vapor deposition of hydrocarbon such as natural gas, and others, or carbon monoxide on the surface of a metal or metal alloy catalyst. The catalyst can be deposited on a substrate or directly fed with the gas phase. Fiber thickness depends on the operating conditions, catalyst size, and others (Mordkovich, 2003). The production technique and post treatment method decides the properties of VGCNFs (Al-saleh and Sundararaj, 2009). Transmission electron microscope image of commercial VGCNF (Pyrograph products, USA) is shown in Fig. 1.7.

1.2.3 Processing methods

The electrical, thermal, and mechanical properties of the both binary and hybrid polymer composites critically depend on the dispersion of fillers in the polymer matrix which depends on the synthesis method adopted. Hence, different ways of synthesizing CPCs are highlighted below.

1.2.3.1 Powder mixing route

In this route, the filler and the polymer in powder form are mixed together in a mixer to make sure that the dispersion of filler is uniform. This route is simple and mainly used when micron sized fillers are mixed with polymer matrix especially thermoplastics which exists in powder form at room temperature. This method may have the limitation in terms of agglomeration of fillers.

1.2.3.2 Solution blending route

In solution blending route, required amount of polymer according to weight fraction calculation is dissolved in fixed amount of solvent, and then, the nanofillers or micron fillers are added to the dissolved polymer preferably thermoplastics and the mixing is continued for few hours. Finally, solvent is evaporated to form composite film or composite powder. The composite powder is then crushed and pelletized for further property measurements and characterization. This method is preferred for facilitating the intercalation of polymer chains into graphite layers. Under solvated conditions, the polymer may intercalate/crawl into the interlayer and try to exfoliate them. Depending on polymer polarity the solvents can be chosen. As this method can enhance the dispersion of conducting filler particles, the electrical percolation threshold can be reduced and enhancement in mechanical and thermal properties of CPCs can be obtained.

Ramanujam and Radhakrishnan (2015) have investigated the effect of processing route on the electrical behavior of polyethersulfone (PES)–graphite–CB hybrid composites. The frequency dependent conductance study clearly proved that the electrical conductance of solution blended PES-7 wt% graphite-2 wt% CB hybrid composites is four orders higher than that of powder mixed composites as depicted in Fig. 1.8.

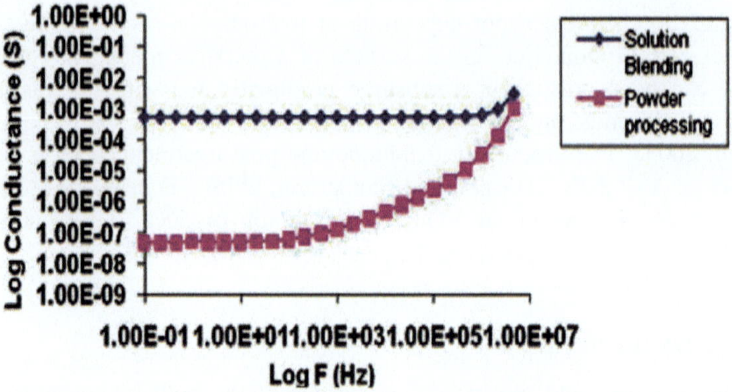

Figure 1.8 Effect of processing route on the frequency dependent conductance of PES-7 wt % graphite-2 wt% CB.
Reproduced with permission from Ramanujam and Radhakrishnan (2015).

The enhancement in the electrical conductance is attributed to the reduction in the particle size of graphite in solution blending route compared to powder processing route. He et al. (2015) have reported higher ac conductivity almost six-order enhancement in exfoliated graphite nanoplatelets (xGNPs)–syndiotactic PS (sPS) composites prepared by solution blending route compared to pure sPS. This is attributed to better dispersion of xGNPs in the polymer matrix. He et al. (2014) have studied dielectric properties of chemically functionalized xGNPs-sPS synthesized by solution blending route. They have reported three order enhancement in the ac conductivity at 1000 Hz for the composites compared to pure sPS. This is once again attributed to the better dispersion of xGNPs in the polymer matrix due to solution blending method.

1.2.3.3 In-situ polymerization route

In in-situ polymerization route, the conducting particles are first swollen within the liquid monomer and then with suitable initiator, polymerized either by heat or radiation. Many nanocomposites such as polymethyl methacrylate (PMMA)–ExGr (Chen et al., 2003), isotactic PP–nanographite (Cromer et al., 2015), ethylene vinyl acetate (EVA)–PANI (Rahaman et al., 2012), phenolic foam-multiwalled CNT (MWCNT) (Li et al., 2016), PS–graphene (Yu et al., 2007) have been synthesized by in-situ polymerization routes. Very low electrical percolation threshold, enhanced mechanical properties have been observed in many nanocomposites (Kuilla et al., 2010).

1.2.3.4 Melt compounding

Melt compounding technique of synthesizing polymer composites is advantageous since solvents need not be used. A thermoplastic polymer is mechanically mixed with conducting fillers and conventional injection molding or extrusion technique can be employed. Depending up on the type of fillers and polymer, intercalation of polymer has also been realized in many polymer composite systems. Variety of polymer composites has been synthesized through melt compounding technique (Kalaitzidou et al., 2007a,b; Zhao et al., 2007; Kim et al., 2009; Zhang et al., 2010). This process is being used extensively in the industry.

1.3 Structure development and properties enhancement in conducting polymer composites

The conducting or electrically active behavior of CPCs and other properties depend not only on the types and dimensions of conducting particles, but also the quality of dispersion and the structural-morphology developed in the host polymer matrices. Hence, it is imperative to understand the relationship between the processing, structure, electrical, mechanical, and thermal properties of the final CPCs.

1.3.1 Electrical properties

In CPCs, the interparticular distance between conducting fillers in a given polymer matrix decides the resultant conductivity apart from the matrix characteristics such as crystallinity and others. Compared to three-dimensional conducting fillers, two-dimensional fillers such as GNS, graphene and one-dimensional nanofillers like CNF and CNT can result in low electrical percolation threshold. This is due to the high aspect ratios of low-dimensional fillers. Appropriate selection of polymer matrix and by proper functionalization of fillers, it is possible to enhance the dispersion of fillers in the polymer matrix. The electrical percolation threshold, processing method for few polymer–filler system is given in Table 1.2.

It has been reported that by selectively making graphite nanoplatelets (GNPs) to be located on the electrospun thermoplastic polyurethane nanofibers resulted in enhancement in the electrical conductivity. High conductivity would not have resulted if the GNPs are distributed inside the nanofibers (Gao et al., 2013). Segregated 3D graphene network in styrene–butadiene-rubber (SBR)-enhanced electrical conductivity (Lin et al., 2016). Similarly, electrical properties of various thermoplastics such as polycarbonate (PC), PMMA, PS into which VGCNF has been mixed by different technique is reported. Excellent review on vapor grown CNF–polymer composites has been published in the literature (Al-saleh and Sundararaj, 2009). The processing routes and the particle–particle contact between fillers, dispersion, polymer characteristics, and others have been cited as the reasons for the reduction in the electrical percolation threshold. Das et al. (2012) have investigated the effect of dispersion of ExGr, GNPs, CNTs in SBR and their effect on the electrical, mechanical properties.

In similar lines, the electrical properties of conducting polymer hybrid composites have also been investigated. Radhakrishnan et al. (2007) have reported addition of small quantities of CB (5 wt%) in PES–graphite composites prepared by solution blending method, resulted in increase in electrical conductivity. High temperature sintering of PES–graphite composites at 240°C for few hours resulted in reduction in crystallite size of graphite. The anisotropy in electrical resistance has been reported to be less in solution mixed PES–graphite than the composites prepared by powder mixing route. Ramanujam and Radhakrishnan (2010) have investigated the electrical properties of powder mixed polyphenylene sulfide (PPS)–graphite–ExGr hybrid composites. The percolation threshold has been reported to be 0.25 wt% ExGr in that system. The authors have also studied charge transport and impedance analysis. The charge transport is by hopping mechanism at lower loading of ExGr (0.75 wt%) in PPS-7 wt% graphite. Ramanujam et al. (2015) have investigated the electrical percolation threshold when second conducting fillers such as CB, ExGr, and CNF are mixed in PP-7 wt% graphite. The barrier for the charge transport is reduced when the aspect ratio of second conducting fillers is higher. Thus in both binary and hybrid CPCs, the electrical conductivity can be enhanced by proper choice of filler. For application-oriented development of CPCs, not only the electrical properties but also mechanical and thermal properties need to be improved.

Table 1.2 Electrical percolation thresholds in different polymer composites

Host matrix	Conducting phase	Processing route	Electrical percolation threshold	Reference
Polyethersulfone (PES)	Natural graphite	Solution blending	5–10 wt%	Ramanujam and Radhakrishnan (2015)
	Carbon black (CB)	Solution blending	10 wt%	
Polystyrene	Graphene	Solution mixing and compression molding	0.33 vol%	Qi et al. (2011)
Polystyrene (PS) and PLA	Graphene	Solution mixing and hot pressing	0.075 vol%	Qi et al. (2011)
Polyamide 12 (PA12)	Graphene	Melt compounding	0.03 vol%	Yan et al. (2012)
Epoxy	Graphene foam	Impregnation of epoxy in Graphene foam and curing	0.05 wt%	Jia et al. (2014)
Thermoplastic polyurethane (PU)	Thermally reduced graphite oxide	Solution blending	0.5 wt%	Kim et al. (2010)
Polystyrene (PS)	Graphite nanosheets	Compression molded and heat treatment	0.057 vol%	Yang et al. (2014)
Epoxy	Expanded graphite	Casting and curing	3 wt%	Gao et al. (2013)
Polyimide	Carbon nanofiber (CNF) and Functionalized CNF	In-situ polymerization	<1 vol%	Lin et al. (2016)

1.3.2 Mechanical properties

Carbon-based fillers are extensively focused not only as electrically conducting fillers but also reinforcing agents in thermoplastic or thermo set matrices in order to improve mechanical properties such as tensile modulus, flexural modulus, stress at break, and others. Thermo mechanical characterization of carbon-based fillers reinforced polymers has also been studied. The improvement in mechanical properties depend up on the concentration of fillers as above certain loading, agglomeration causes poor mechanical properties. Hence, it is a challenge to retain higher value of electrical conductivity and improve mechanical properties of the filled system. Carbon-based nanofillers offer advantage compared to conventional micron-sized fillers as at lower loading better enhancement in mechanical properties can be realized. In many cases, depending up on the choice of polymer matrix, fillers have to be functionalized in order to ensure better dispersion, and hence, enhancement in mechanical properties for the composite system can be obtained. Epoxy-based composites reinforced by three-dimensional interconnected graphene layers (3DGS) prepared by resin transfer molding have been shown to exhibit 120.9%, 148.3% enhancement in tensile and compressive strengths at 0.2 wt% loading of 3DGS (Ni et al., 2015). The dispersion level of the 3DGS has been the reason for enhanced mechanical properties. The dynamical mechanical analysis study of the same system has shown appreciable enhancement in storage modulus due to better reinforcement. Ha et al. (2012) have investigated the effect of chemical modification of graphene on the mechanical, electrical, and thermal properties of polyimide (PI)-based composites. It has been shown that 5 wt% GO reinforced PI composite exhibits 12 fold and 18 fold increase in tensile strength and tensile modulus of PI due to higher loading capacity of GO/PI composites than reduced GO/PI composites. Enhanced mechanical properties such as young's modulus, ultimate tensile strength, fracture toughness, fracture energy, material resistance to fatigue crack propagation have been reported for 0.1 wt% graphene–epoxy nanocomposite which are much higher than MWCNT-loaded epoxy nanocomposites with the same loading of MWCNT as that of graphene. The superior mechanical properties are attributed to high surface area of graphene as well as matrix adhesion/interlocking arising from wrinkled surface of graphene (Rafiee et al., 2009). Similarly, increased surface area of graphene has been attributed to enhanced mechanical properties of spirocyclic phosphazene epoxy–graphene composites (Feng et al., 2013). Kim et al. (2010) have also discussed chemically or thermally reduced graphite oxide (TRGO) in different polymer matrices either prepared by melt compounding method or solvent-based mixing and the associated mechanical, electrical, and gas barrier properties. The dispersion and high aspect ratio of graphene are the key factors for enhancement in mechanical properties. Hatui et al. (2015) have investigated the effect of ExGr and polyether imide (PEI)-*co*-silicon rubber on the thermomechanical properties of composites based on poly(ether imide) and liquid crystalline polymer (LCP). The storage modulus has been found to increase when modified ExGr and PEI-*co*-silicon rubber in PEI, LCP systems. When 2 wt% of TRGO is mixed with natural rubber, stiffness of the composite has been reported to have increased. The graphite oxide is prepared by Brodie's method. This is due to highly interconnected graphene layers which promote rubber–TRGO interactions (Aguilar-Bolados et al., 2016). The surface modification of GNPs by organosilane has been found to enhance mechanical properties such as wear resistance and storage modulus of UHMWPE–GNP nanocomposites due to better dispersion of the filler in the polymer matrix (Liu et al., 2016). Composite fiber with graphene as the filler when prepared by melt spinning and electrospinning techniques exhibit better mechanical properties. In this regard excellent review has been published by Ji et al. (2016). Not only graphene, but also one-dimensional filler such as CNF reinforced composites exhibit enhanced mechanical properties. The filler dispersion and distribution while maintaining the aspect ratio of nanofillers in a polymer matrix, filler adhesion are the key factors for enhancing mechanical properties of both binary and hybrid composites. Uniform stress distribution will be realized when dispersion of filler is good in a polymer matrix. The mechanical properties of 3-glycidoxypropyltrimethoxysilane functionalized CNF in epoxy matrix exhibits higher storage modulus compared to neat epoxy matrix. This is attributed to the better dispersion and adhesion of CNFs (Nie and Hübert, 2012). Improved mechanical properties have also been reported for 3D PI fiber felt-CNF–epoxy composites. The wear resistance of the composites has been shown to increase by 18 times when compared to pure epoxy matrix (Zhu et al., 2016). Mechanical properties of composites of CNF-thermoplastic or thermosets matrix have been reviewed by Al-Saleh and Sundararaj (2011). From the foregoing discussion, it is clear that to improve mechanical properties of composite systems, one has to focus on how to improve filler dispersion, adhesion between matrix, and the filler. Designing a strategy for improving those factors will eventually help in the tailor making of composites with appreciable level of mechanical properties.

1.3.3 Thermal properties

Thermal properties such as thermal conductivity of CPCs need a thorough focus for the development of CPC-based heat sink, thermal interface materials, and others. Since in CPCs, conducting fillers are incorporated in the insulating polymer matrix, not only the electrical conductivity, but also thermal conductivity can be enhanced. Unlike the case of electrical conductivity, thermal conductivity is a slowly varying parameter. High thermal conductivity is preferred in order to dissipate heat fast. Phonon conduction is the main mechanism of thermal conduction in insulating polymers. Lattice imperfections such as voids, impurities, dislocations, and others result in phonon scattering which will reduce thermal conductivity of the electrically insulating materials. The thermal conductivity in metals is dominated by free electrons than phonon. Thermally, CPCs offer advantages over metals in terms of good corrosion resistance, reduction in density, easy process ability. For improving thermal conductivity of the polymer composites conducting fillers such as metal powders, carbon-based material such as graphite are being used as fillers due to higher thermal conductivity of those fillers. In order to use polymer composites as thermal interface materials, the thermal conductivity has to be in the range of

1–30 W m^{-1} K^{-1} (Heiser and King, 2004). Thermal conductivity of many binary composites such as polyethylene (PE)–graphite (Agari and Uno, 1986), PP–VGCF (Biercuk et al., 2002), PI–CNF (Ghose et al., 2006) have been reported in the literature. Recently, hybrid filler systems in order to improve thermal conductivity of polymer composites have been investigated. The bulk thermal conductivity of PC-20 wt% GNPs composites has been reported to be 1.13 W m^{-1} K^{-1}. When MWCNT is added in small quantities, the thermal conductivity has been observed to be synergistically enhanced to 1.39 W m^{-1} K^{-1} as reported (Yu et al., 2016). Addition of multigraphene flakes and graphene foam in polydimethyl siloxane resulted in enhancement in the thermal conductivity (Zhao et al., 2016). Graphene at the interface of poly(ε-caprolactone) (PCL)/poly (lactic acid) (PLA) blend improves thermal conductivity remarkably. This is due to occupation of graphene at the interface of cocontinuous structure formed by the blend (Huang et al., 2016). Several models have also been proposed to account for thermal conductivity especially for polymer composite systems. The rule of mixture or the parallel model and the series model are the basic models. In the former model, interparticle contact is the major factor contributing to thermal conductivity of the composite. The series model assumes no contact between the filler particles. Nielsen model takes into consideration the shape, orientation of the filler particles in the polymer matrix (Nielsen and Landel, 1994). Although many models have been proposed to account for thermal conductivity of polymer composite systems, there is no universal model which predicts the thermal conductivity of the filled polymer composites.

1.3.4 Structure development in thermoplastic–graphite composites

When semi crystalline polymers are reinforced with various types of organic/inorganic reinforcements like fibers, different morphology and crystallinity of the polymer matrix can be induced. There can be new phases or preferred orientation of particular plane depending up on the processing routes employed. Fibers may act as heterogeneous nucleating agents and they can aid nucleation along the interface. These nuclei may hinder lateral extension and growth in one direction especially perpendicular to the fiber surface and result in columnar crystalline layers. Thus, the formation of transcrystalline layers is not fully understood, and there are no clear cut rules defined. The formation of transcrytsalline morphology is specific to polymer matrix-filler system. Fiber topography, surface coating of the fiber, processing conditions such as cooling rate and others of the composites are the factors which influence the transcrystalline morphology.

Saujanya and Radhakrishnan (2001) have explained transcrystals formation in polyethylene-terephthalate (PET)-fiber-filled PP composites through lattice mismatch theory. From the lattice parameters of α phase of PP and with that of PET, the lattice mismatch between *b*-axis of PP and twice *c*-axis of PET is less than 3% while that between *c*-axis of PP and *b*-axis of PET is less than 8.5%. They further reported that the *bc* planes of PP crystals get aligned initially along the fiber axis of PET fibers and subsequently these crystals grow with preferential *b*-plane orientation. Thus, the alignment of *b*-axis of PP cystallites gives rise to change in intensity of 0 4 0 reflection in X-ray diffraction pattern. It is further reported when the lattice mismatch factor is less than 10%, epitaxial growth of one phase over the other can be expected. According to lattice mismatch theory, the mismatch factor δ is defined as

$$\delta = \frac{|pl_s - ql_g|}{|l_s|} \times 100 \tag{1.3}$$

where l is the lattice parameter along any axis; p and q are integers. The subscripts "s" and "g" represent substrate and growing media. Xia et al. (2004) have reported for PP–CNT composites prepared by melt mixing route that when the intensity ratio of α-PP (1 1 0) reflection to (0 4 0) reflection is greater than 1.5, *a*-axis of PP will be found along the surface. When the same is less than 1.3, a strong *b*-plane orientation will be observed.

Strong *b*-plane orientation is reported in PP–ExGr system when the composites are synthesized by melt crystallization route at 180°C (Ramanujam and Radhakrishnan, 2016). The X-ray diffraction pattern showed a strong *b*-plane orientation with the addition of ExGr particles in PP, which has been explained through lattice mismatch theory as explained above between monoclinic alpha phase of PP and hexagonal structure of graphite. However for melt crystallized PPS-ExGr composites, *b*-plane orientation is not observed. Hence, it is clear that the structural change is specific to polymer–filler system. Melt crystallized PP–ExGr/PPS–ExGr composites exhibit higher electrical percolation threshold due to increase in interparticular distance of filler particles due to transcrystals formation or due to reduction in the viscosity of the polymer. Hence, the processing route and the processing conditions should be properly optimized to tailor make the composites.

1.4 Applications

1.4.1 Alternate energy technology

Because of the depletion of fossil fuels, recent research is much focused on alternate energy technology. Fuel cells are electrochemical energy conversion devices and green sources of energy. Out of six different types of fuel cells, proton exchange membrane fuel cells (PEMFCs) are operated at low temperature (80°C). In a typical PEMFC, hydrogen is supplied at the anode, oxygen is supplied at the cathode, and the two electrodes are connected by a proton exchange membrane which acts as a barrier for electrons and allows only protons to pass through as depicted in Fig. 1.9.

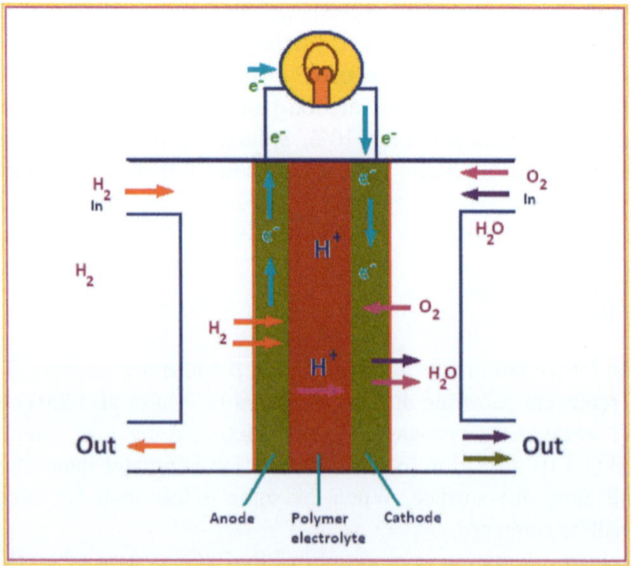

Figure 1.9 Structure of a PEM fuel cell.

Typical electrochemical reactions occurring at the respective electrodes are as follows:

$$\text{Anode}: \quad H_2 \rightarrow 2H^+ + 2e^-$$
$$\text{Cathode}: \quad \frac{1}{2}O_2 + 2H^+ + 2e^- \rightarrow H_2O$$

$$\text{Overall}: \quad H_2 + \frac{1}{2}O_2 \rightarrow H_2O + \text{Electrical Energy} + \text{Heat Energy}$$

Since the bye product is only water, fuel cells are sources of clean energy.

A PEMFC has three major components: (1) membrane electrode assembly (MEA) which contains anode, cathode and the polymer membrane, (2) catalyst layer (platinum based), and (3) hardware (backing layer, bipolar plates, and end plates). Out of three components, bipolar plates which are essentially current collectors and account for 80% of stack weight and 45% of stack cost (Tsuchiya and Kobayashi, 2004). They are designed for distributing reactants uniformly over active areas, removal of heat, and others. These plates must be inexpensive, light weight and easily manufactured by mass production techniques. The materials investigated to be used in fuel-cell bipolar plate application can be classified as (1) non porous graphite/electrographite, (2) metals both coated and noncoated, and (3) polymer–carbon and polymer metal composites. Mehta and Cooper (2003) have reported the properties which should be met by any bipolar plate material.

Graphite, whether natural or synthetic, is used as bipolar plate material. However, graphite plates exhibit poor mechanical characteristics, and the cost of machining is also high. Stainless steel has also dragged considerable attention, but the major concern with that material is the extent of corrosion and the contact resistance of the surface passivation film. In order to avoid corrosion, metallic bipolar plates are coated with protective layers. But coating results in increased cost of the bipolar plate and also the density of metal bipolar plates is higher. Compared to metals and graphite, polymer composites are being focused extensively due to light weight, easy processing, and reduced cost. Both thermoplastic and thermoset resins are being used with conducting carbonaceous fillers for bipolar plate applications. Epoxy (Du and Jana, 2007), phenolic (Dhakate et al., 2008), vinyl ester (Kuan et al., 2004) thermoset resins are employed extensively for bipolar plate application. Thermoplastics such as PP, polyvinyldiene fluoride (PVDF), PPS, PMMA, and others have been used as thermoplastic matrices with graphite, CB, CNF, CNT, ExGr. Usually, very high loading of conducting micron sized fillers (>50 wt%) are incorporated in various thermoplastic matrices in order to achieve the required electrical conductivity as reported elsewhere (Mehta and Cooper, 2003). Higher loading of micron-sized fillers may improve electrical conductivity of the composite but deteriorates mechanical properties. Hence, hybrid composites are developed to meet the required level of properties. The synergistic effect, due to combination of various carbonaceous fillers, has gained tremendous attention. Radhakrishnan et al. (2007) have reported that addition of small quantity of CB (5 wt%) particles even in PES-50 wt% graphite resulted in decrease in resistance. The same is true in the case of PPS–graphite-based hybrid composites with CB as the second conducting filler. In composite bipolar plates, there should be enough filler contacts so that conductive networks can be realized. Poor adhesion between the filler and the matrix, agglomeration of fillers, and poor distribution of fillers in the matrix cause electrical percolation threshold to be higher and the mechanical properties to be inferior. Coupling agents have been employed to improve matrix-filler binding (Brovko and Friedrich, 2002). The degree of dispersion of fillers and hence the various properties will be affected depending on the types of processing routes employed. The composite bipolar plates have been manufactured through the well-known injection and compression molding. Mass production of bipolar plates can be done through injection molding. In recent years, new technique known as wet-lay process has been reported to make composite bipolar plate sheets with PPS–graphite–glass or carbon fibers. The process involves making slurries and then compression molding to make composite sheets with high in-plane conductivity of the order of 200–300 S cm^{-1} (Huang et al., 2005). Not only electrical properties but also mechanical properties are enhanced to a greater extent when wet lay process has been employed. Lee et al. (2015) have investigated electrical resistance, flexural strength and gas permeability of PP/PE/carbon fiber/MWCNT composite bipolar plate prepared by compression molding technique. The authors have claimed that the flexural strength of PP sol and PP/PE sol specimen exceeded the US DOE value of 25 MPa. Kang et al. (2016) have prepared ultra-thin composite bipolar plate (0.6 mm thick) with epoxy–carbon fiber prepreg and reported in-plane conductivity to be 172 S cm^{-1} and through plane conductivity to be 38 S cm^{-1} with serpentine flow channel. Hot rolling process has been employed for the making of carbon fiber/epoxy composite bipolar plate satisfying the target values mentioned by US DOE for different properties (Kim et al., 2015). A review of composite and metallic bipolar plates from material selection to fabrication has been published elsewhere (Taherian, 2014). In spite of identifying correct process, materials, achieving desired property levels, still cost reduction as well as filler levels have to be brought down.

1.4.2 Electromagnetic interference shielding devices

EMI is a serious concern as it can affect the performance of various electronic devices as well as human health. It is a kind of pollution and EMI should be minimized. For the past two decades, EMI shielding has been attempted through various strategies and accordingly novel materials starting from carbon materials, conducting polymers, CPCs have been synthesized (Saini et al., 2009). No single material could meet the requirements of EMI shielding materials from the perspectives of cost, reduction in density, easy fabrication, shielding efficiency, and others. Though metals are good EMI shielding materials, the density of metals is very high. Conducting polymer nanocomposites have shown promise towards EMI shielding and hence attracted worldwide attention. The term shield refers to an enclosure which protects the electronic products and hence acts as a barrier to the transmission of electromagnetic (EM) waves. Hence, in a way, it is a process of achieving attenuation of EM radiation. The EM radiation blocking efficiency of a shield usually comprises of three parts: (1) reflection, (2) absorption, and (3) multiple reflection. The schematic of EMI shielding is shown in Fig. 1.10. The shielding efficiency can be expressed as (Choudhary et al., 2012).

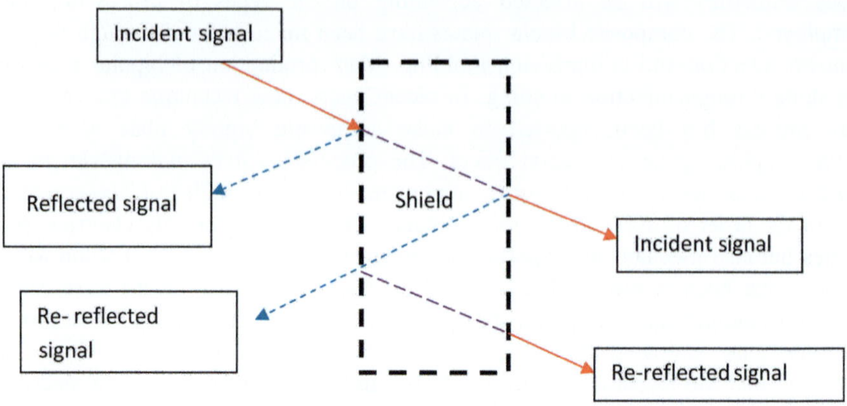

Figure 1.10 EMI shielding representation.

$$\mathbf{SE_{Total}} = \mathrm{SE}_R + \mathrm{SE}_A + \mathrm{SE}_M = 10\,\mathrm{Log}_{10}\left(\frac{P_T}{P_I}\right) = 20\,\mathrm{Log}_{10}\left(\frac{E_T}{E_I}\right) = 20\,\mathrm{Log}_{10}\left(\frac{H_T}{H_I}\right) \tag{1.4}$$

where SE refers to shielding efficiency, the subscripts R, A, and M refer to reflection, absorption and multiple reflection phenomenon. P_I, E_I, and H_I refer to power of incident EM waves, electric component and magnetic component of incident EM radiation. Similarly, P_T, E_T, and H_T refer to power of transmitted EM waves, electric component, and magnetic component of transmitted EM radiation from the shield.

The theoretical expressions for reflection loss, absorption loss, and multiple reflection loss for plane wave radiation (far field condition) have been reported (Choudhary et al., 2012).

$$\mathrm{SE}_R = -10\,\mathrm{Log}_{10}\left(\frac{\sigma}{16\varepsilon\omega\mu_r}\right), \quad \mathrm{SE}_A = -8.68t\left(\sigma\pi f\mu_r\right)^{1/2} \tag{1.5}$$

Thus, SE_R is a function of ratio of electrical conductivity (σ) and relative permeability of the shield material. Shielding efficiency is directly related to the square root of the product of electrical conductivity and the permeability of the shield. Thus, good absorbing materials should have high electrical conductivity, permeability, and sufficient thickness (t). "f" is the frequency of the EM radiation. In case of thick shields, the shielding due to multiple reflections can be neglected. Thus, the requirements of shielding materials are, they should have mobile charge carriers as well as electric or magnetic dipoles. The shielding in metal is due to reflection of EM radiation. However, they suffer from high reflectivity, corrosion, high density, and others. Absorption of EM radiation requires electric or magnetic dipoles in the shield. In the process of replacing metals, CPCs with very low percolation thresholds have been formulated. Nanomaterials such as graphene, CNTs, and others have been employed for the synthesis of nanocomposites. PVDF-functionalized graphene nanocomposites have been synthesized and 12 order enhancement in the electrical conductivity when compared to PVDF matrix has been reported. For 5 wt% functionalized graphene loading in PVDF, an EMI shielding efficiency of −20 dB in X-band has been reported (Eswaraiah et al., 2011). ICP such as PANI-coated PET film exhibits high shielding efficiency. 15%–53% shielding efficiency has been reported when the optical transmittance varied between 58% and 78% (Kim et al., 2010). For Radar absorption, stealth technology, the shielding of EM radiation should be through absorption mechanism. To improve absorption of EM radiation, materials like $BaTiO_3$, TiO_2, and magnetic materials such as γ-Fe_2O_3, Fe_3O_4 have been incorporated into the polymer matrices. Abbas et al. (2006) have studied the shielding efficiency of polyurethane/$BaTiO_3$/PANI/conducting carbon system. PP/PE blend with 5 vol% graphene nanoplatelets:CNT hybrid nanofiller has been synthesized by melt blending. The shielding efficiency has been reported to be increasing with the higher loading of CNT (Al-Saleh, 2016). Although better shielding efficiency is achieved, still the problem of agglomeration of nanofillers and their dispersion in various polymer matrices pose great challenge. Similarly, shielding efficiency of nanocomposites containing only dielectric or magnetic inclusions often results in narrow band action. Hence, still there exists a great scope to identify novel materials so that they can act as effective EMI shielding devices.

1.4.3 Solar cells

Solar cells are the most promising green sources of energy. Commercially, silicon-based solar cells have been used due to the higher efficiency compared to other types. However manufacturing cost of silicon-based solar cells is very high. Though thin film solar cells based on especially CdTe have been fabricated, the toxicity of cadmium poses problem. Hence, organic materials have been focused due to easy processing, high throughput. The absorption coefficient of organic semiconductors is usually very high, and hence, low cost thin solar cells can be fabricated (Zhokhavets et al., 2006). Further, the efficiency of organic solar cells increases with increase in temperature, whereas most of the inorganic solar cells lose efficiency with increase in temperature. Conjugated polymers, different carbon materials such as amorphous carbon, fullerenes, graphene, and CNTs are focused for the development of both organic and organic–inorganic hybrid solar cells (Wright and Uddin, 2012). Transparent electrically conducting electrodes are one of the components of solar cell to extract energy. Conventionally, indium tin oxide (ITO) and fluorine doped tin oxides are being used. Limited resources of tin oxide and the mechanical rigidity made researchers to probe into alternatives for flexible solar cells. CPCs have been focused in that regard. Transparent graphene thin films can be prepared and their work function can also be engineered. Usually, the work function of transparent electrodes lies in the range of 4.5–5.2 eV (Wan et al., 2011). Composites of graphene and conducting polymers have also been used as transparent conducting electrodes. In this regard, poly(3,4-ethylenedioxythiophene) (PEDOT):PS sulfonate (PSS) composite electrodes have been fabricated and introducing PEDOT:PSS contact between graphene layers improves flexibility.

Out of different types of solar cells, dye-sensitized solar cells (DSSC) are important due to easy fabrication of the cell. Typical DSSC comprises of photoanode usually made of nanocrystalline titania, an electrolyte that provide redox couple (I_3^-/I^-) and a counter electrode (O'Regan and Grätzel, 1991). Platinum-coated fluorine doped tin oxide is usually used as counter electrode due to excellent catalytic activity towards reducing I_3^-. Platinum corrodes in electrolyte containing I^- and forms PtI_4 (Murakami and Grätzel, 2008). Hence, platinum metal catalyst is replaced with many other materials such as titanium nitride, carbon derivatives, and others. Carbon derivatives such as CB are very attractive due to low cost, provide high catalytic activity and offer chemical stability. Since carbon powders cannot retain their shapes, they must be mixed with binders such as polymers. Chen et al. (2013) have reported the use of CB/polymer composite counter electrode comprising of PANI, PPy, and others. Novel stain less steel (SS) counter electrodes coated with a composite of CB and 3D networked polymers to replace conventional Pt counter electrodes on FTO glass has been reported (Kang et al., 2016). The authors claim that the photovoltaic performance for the device using CB coated SS counter electrodes remained nearly constant for 30 days of storage at 65°C, whereas the reference Pt/FTO reported to have exhibited 50% photovoltaic performance. In the recent years, thin film plastic solar cells using polymer–fullerene (Shaheen et al., 2001) and polymer–polymer bulk hetero junctions as absorbers (Granström et al., 1998) have been reported. The typical structure of these cells consists of composite of two materials sandwiched between two electrodes. Easy processibility is the advantage of this type of solar cells. Though organic solar cells have advantages, still the photo conversion efficiency achieved is far behind the silicon-based solar cells. Hence, lot of scope exists to probe further to take them towards commercialization.

1.4.4 Anticorrosion coatings

Corrosion causes severe damage to metals, alloys, and composites. The electrochemical theory of corrosion of metals in presence of water involves two half-cell reactions at a neutral pH namely oxidation of metals and reduction of water. Because of this destructive phenomenon metal loses luster, strength, and others. Mild steel which is mostly used as structural material is severely affected by corrosion. Conventionally chromate-based conversion coatings were used. As chromate ions are toxic, in search of alternate anticorrosion coating materials, conducting polymers and their composites fill the gap. Anodic protection is the mechanism of corrosion protection by conducting polymers by raising its potential to passive region (Spinks et al., 2002). The dual aspects of conducting polymers imparting barrier property in the insulating state and the ability to store charges at the metal/coating interface makes them unique as anticorrosion coatings. The stored charges can be effectively used to form passive layer on the metal and hence gives corrosion protection. PANI and PPy have been focused more for anticorrosion coatings application. PPy has excellent adherence to the metal substrate whether it is chemically synthesized or electrochemically coated onto the metal. The mechanical integrity and thermal stability of bare conducting polymer coatings are lost in harsh climatic conditions and in presence of diffusive chloride ions also. Thus CPCs coatings have gained popularity. Addition of nanoinorganic materials, such as SiO_2 or TiO_2, significantly improved corrosion resistance of the base metal. More recently conducting polymer–carbon nanostructure composite coatings gained popularity especially PANI-graphene coatings. Chang et al. (2012) have prepared PANI-graphene composite. They prepared 4-aminobenzoyl group grafted graphene like sheets in PANI through chemical oxidation polymerization of aniline. The possible improvement of anticorrosion property with the addition of graphene is that these high aspect ratio filler offers barrier for the transport of oxygen and water as depicted in Fig. 1.11.

PANI/PMMA microfiber coating in which PANI dosage of 25 wt% has been reported to show enhanced anticorrosion behavior (Zhao et al., 2016). This is attributed to compact morphology of microfiber as electrospinning technique has been

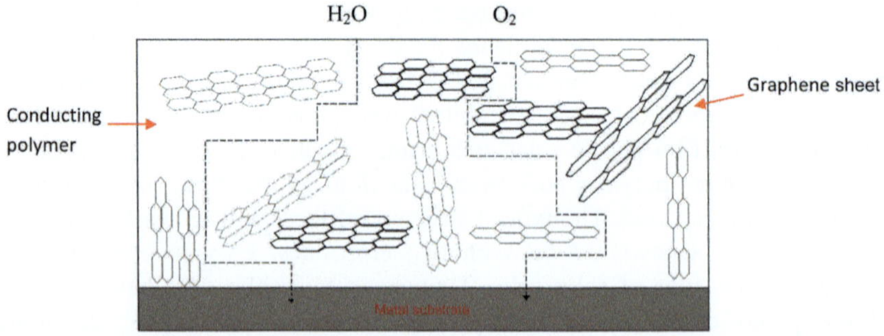

Figure 1.11 Tortuous path effect on gas diffusion in graphene filled conducting polymers.

employed. Jafari et al. (2016) have reported enhanced anticorrosion behavior of PANI-graphene nanocomposite film which has been electrochemically deposited on copper electrode. The effect of graphene on the corrosion inhibition behavior of hybrid epoxy ester–siloxane–urea polymer has been studied by Okafor et al. (2015). The coating resistance on the aluminum alloy has been investigated through direct current polarization and electrochemical impedance spectroscopy. The corrosion protection mechanism has been attributed to restriction of polymer chain motion due to high aspect ratio graphene nanosheets. Thus, conducting polymers and their composites act as efficient anticorrosion coatings. The anticorrosion property of conducting polymer coatings depends on how they are coated on to the substrates and condition of corrosion experiment. More efforts are required still to elucidate the mechanism of corrosion protection by conducting polymers though they are more promising materials for these applications.

1.4.5 Supercapacitors

Electrochemical supercapacitors have gained prime focus due to exceptional charge storage behavior, high power densities at relatively high energy density and long cycle life. These characteristics have already been utilized in power electronics, smart grid, and others. They are also characterized by fast discharge rates and low maintenance. Super capacitors have been classified into two major types: (1) electrical double layer capacitors (EDLCs) and (2) pseudocapacitors. In EDLC type, the interface between electrode and the electrolyte results in double layer capacitance. The rapid responses of the interface to change in electrode potential and high reversibility are exploited in EDLCs. In the pseudocapacitor type, Faradaic electrochemical redox mechanism of charge storage is operative. Materials with high surface area such as activated carbon are targeted for EDLC type where as for pseudocapacitors, oxides such as ruthenium dioxide and manganese dioxide are being focused. In the hybrid supercapacitor type composite material which can exhibit both EDLC and psuedocapacitance behavior are used. For supercapacitors to be commercially viable for large scale applications, high energy density and low cost of fabrications need to be met. High energy density can be obtained by increasing the electrode capacitance or electrolyte voltage window. Hence, the use of high capacitance electrode material and nonaqueous electrolytes with larger window of electrochemical stability will ensure good supercapacitor characteristics. Various electronic conducting carbon materials such as activated carbon, graphene, CNTs, and carbon aerogel are being focused to improve the electrode capacitance. Wang et al. (2009) have reported capacitance as high as 205 F g^{-1} with graphene synthesized by chemical reduction of GO by hydrazine. Electrochemical tests were performed with aqueous electrolyte over a 1 V window. Graphene/PANI composites have been made by electro polymerization of aniline onto graphene paper and have been shown to exhibit capacitance of 230 F g^{-1} in sulfuric acid over 1 V window (Wang et al., 2009). For PANI/graphene composites synthesized by in-situ polymerization route, higher capacitance of about 530 F g^{-1} at 200 mA g^{-1} has been reported (Wang et al., 2009). Apart from transition metal oxides, transition metal nitrides, sulfide and carbides have also been focused. Choi et al. (2006), Choi and Kumta (2005), and Choi and Kumta (2006) have investigated super capacitance when various nitrides in aqueous media were tested. Not only electrode materials but also development of nonaqueous electrolyte for supercapacitor application is receiving great focus. This is due to the fact that energy density has squared dependence of voltage window of the cell. Aqueous electrolytes are susceptible to gas evolution due to decomposition at high voltages. To overcome this, ionic liquids such as tetraethyl ammonium tetrafluoro borate in acetonitrile is used. Use of asymmetric cells using a capacitor type electrode in conjunction with battery type electrode is the new direction in the supercapacitor field to realize higher energy density. Though composite materials show promising features, still there exists a great scope to improve stability at high rates and improve cyclability.

1.4.6 Other applications

Conducting polymers and their composites are being used as scaffold materials. These materials are graft or porous substrates that interact with cells to form a precise functional tissue. The idea of tissue engineering is to guide the host cells to regenerate tissue, cultivation of cells on a biomaterial scaffold. The extra-cellular matrix which can be nanocomposite material, not for only providing mechanical support for embedded cells but also interacts with cells and promotes and regulates adhesion, proliferation, and others. The scaffold must present high degree of porosity, pore interconnection, proper pore size, and others. In the biomedical field, nanofibers synthesized by electrospinning technique are used as suitable substrates for controlled release and drug delivery, scaffolds, wound dressing (Beachleya and Wena, 2010). The biocompatibility of PANI has been realized by surface modification with bioactive peptide. It has been reported that PANI films functionalized with the bioactive-laminin-derived adhesion peptide enhanced PC12 cell attachment and differentiation (Wei et al., 2004). PPy has also been used as biomaterials. An excellent review has been published on the progress in the field of conducting polymers for tissue engineering applications (Bendrea et al., 2011).

Not only biomaterials applications, CPCs have also been used to fabricate super hydrophobic materials, thermal interface materials, heat sinks, sensors, and others. Graphene-based materials have been realized as smart materials with high efficient removal of inorganic and organic pollutants from aqueous solutions (Wang et al., 2013). The applications of conducting polymers, and their composites are ever expanding as in this chapter, bird's eye view of their applications have been presented.

1.5 Conclusions and future perspectives

The development of CPCs or nanocomposites for different applications such as fuel cell bipolar plates, EMI shielding devices requires proper choice of filler, and their dispersion in the chosen polymer matrix. Incorporating high aspect ratio nanofillers especially carbon nanofillers such as GNS, graphene, CNF, and others in thermoplastics or thermosets will reduce electrical percolation threshold and in a way improve mechanical properties at lower filler loading itself. In order to reduce the cost of the resultant composite, hybrid composites can be focused in which the nanofillers will be added in lesser quantities along with micron-sized cheap fillers. The enhancement in properties could only be realized if the dispersion of nanofillers can be enhanced either by adopting proper synthesis routes as well as functionalization of the nanofillers according to the polymer matrix. The electrical behavior of conducting polymer nanocomposites is not only dependent on the filler size, shape, aspect ratio, processing routes, dispersion but also on the characteristics of the polymer matrix. Hence, to achieve target values of electrical, mechanical and thermal properties for a given application, polymer characteristics such as crystallinity must be taken into consideration. In filled system, new structures can be induced by the fillers depending up on the processing routes employed. The filler particles can act as nucleating agents and a particular phase of the polymer can be nucleated. The interparticular distance between conducting filler particles could be changed in different processing routes. Graphene, GNS which are derived from graphite play an important role in improving electrical conductivity, mechanical properties at lower loading itself. The easy processing, light weight of the end product, and reduction in cost make CPCs or nanocomposites more interesting materials.

Though the applications in which CPCs can be used are plenty but still lots of work needs to be carried out especially in the areas of plastics solar cells, fuel cells, and theoretical modeling. The solar cell efficiency needs to be improved nearer to that of silicon-based solar cells. Similarly for fuel cell bipolar plate application, over all filler loading must be reduced. The dispersion of fillers in a given polymer matrix poses a great challenge. Further, there is a need to develop theoretical model which can be used to predict the electrical percolation threshold for range of polymer–filler system including multicomponent CPCs.

Acknowledgments

Authors thank Dr. S. Radhakrishnan, Dean Research, MIT-Pune, India for fruitful discussions and the Provost, Navrachana University, Vadodara, Gujarat, India for constant encouragement and support.

Abbreviations

CB	carbon black
CNF	carbon nanofiber
EMI	electromagnetic Interference
EVA	ethylene vinylacetate
ExGr	expanded graphite
GIC	graphite intercalation compounds
GNS	graphite nanosheets
GO	graphene oxide
HDPE	high density polyethylene
I	current
LCP	liquid crystalline polymer
MWCNT	multiwalled carbon nanotube
PA12	polyamide 12
PANI	polyaniline
PCL	poly(ε-caprolactone)
PEDOT	poly(3,4-ethylenedioxythiophene)
PEI	polyether imide
PES	polyether sulfone
PET	polyethylene terephthalate
PLA	polylactic acid
PMMA	polymethyl methacrylate
PP	polypropylene
PPS	polyphenylene sulfide
PPy	polypyrolle
PS	polystyrene
PSS	poystyrene sulfonate
PTh	polythiophene
PVDF	polyvinylidene fluoride
SWCNT	single-walled carbon nanotube
TRGO	thermally reduced graphite oxide
UHMWPE	ultrahigh molecular weight polyethylene
V	voltage
VGCF	vapor grown carbon fiber
VGCNF	vapor grown carbon nanofiber

References

Abbas, S.M., Chandra, M., Verma, A., Chatterjee, R., Goel, T.C., 2006. Composites A. 37, 2148.

Agari, Y., Uno, T., 1986. J. Appl. Polym. Sci. 32, 5705.

Aguilar-Bolados, H., Lopez-Manchado, M.A., Brasero, J., Aviles, F., Yazdani-Pedram, M., 2016. Composites B: Eng. 87 (15), 350.

All, M.H., Abo-Hashem, A., 1997. J. Mater. Process. Technol. 68, 163.
Al-Saleh, M.-H., 2016. Synth. Met. 217, 322.
Al-saleh, M.H., Sundararaj, U., 2009. Carbon. 472, 2.
Al-Saleh, M.-H., Sundararaj, U., 2011. Composites A: Appl. Sci. Manuf. 42, 2126.
Baker, Z.Q., Abdelazeez, M.K., Zihlif, A.M., 1988. J. Mater. Sci. 23, 2995.
Beachleya, V., Wena, X., 2010. Prog. Polym. Sci. 35, 868.
Bendrea, A.-D., Cianga, L., Cianga, L., 2011. J. Biomater. Appl. 26, 3.
Biercuk, M.J., Llaguno, M.C., Radosavljevic, M., Hyun, J.K., Johnson, A.T., Fischer, J.E., 2002. Appl. Phys. Lett. 80, 2767.
Blythe, T., Bloor, D., 2005. Electrical Properties of Polymers. Cambridge University Press, New York.
Bolotin, K.I., et al., 2008. Solid State Commun. 146, 351.
Brovko, Rosso O.P., Friedrich, K., 2002. J. Mater. Sci. Lett. 21, 305.
Bryning, M.B., Islam, M.F., Kikkawa, J.M., Yodh, A.G., 2005. Adv. Mater. 17 (9), 1186.
Celzard, A., McRae, E., Furdin, G., Mareche, J.F., 1997. J. Phys.: Condens. Matter. 9, 2225.
Chang, C.H., Huang, T.C., Peng, C.W., Yeh, T.C., Lu, H.I., Hung, W.I., et al., 2012. Carbon. 50, 5044.
Chatterjee, A., Alam, K., Klein, P., 2007. Mater. Manuf. Process. 22 (1), 62.
Chen, G., Weng, W., Wu, D., Wu, C., 2003. Eur. Polym. J. 39, 2329.
Chen, G., Wu, D., Weng, W., Wu, C., 2003. Carbon. 41, 579.
Chen, G.-H., Wu, C.-L., Wang, W.-G., Wu, D.-J., Yan, W.-L., 2003. Polymer. 44, 1781.
Chen, P.W., et al., 2013. RSC Adv. 3, 5871.
Chiang, W.Y., Chiang, Y.S., 1992. J. Appl. Polym. Sci. 46, 673.
Chiu, F.-C., Chen, Y.-J., 2015. Composites A: Appl. Sci. Manuf. 68, 62.
Choi, D., Blomgren, G.E., Kumta, P.N., 2006. Adv. Mater. 18, 1178.
Choi, D., Kumta, P.N., 2006. J. Electrochem. Soc. 153, A2298.
Choi, D.W., Kumta, P.N., 2005. Electrochem. Solid-State Lett. 8, A418.
Choudhary V., Dhawan S.K., Saini P. Polymer based nanocomposites for electromagnetic interference (EMI) shielding. In EM Shielding – Theory and Development of New Materials, 2012: Eds. Maciej Jaroszewski and Jan Ziaja. pp. 1. (ISBN: 978-81-308-0499-6).
Chung, D.D.L., 2002. J. Mater. Sci. 37, 1475.
Chung, D.D.L., 1987. J. Mater. Sci. 22, 4190.
Clingermann, M.L., King, J.A., Schulz, K.H., Meyers, J.D., 2002. J. Appl. Polym. Sci. 8, 1341.
Cortes, P., Lozano, K., Barrera, E.V., Bonilla-Rios, J., 2003. J. Appl. Polym. Sci. 89 (9), 2527.
Cromer, B.M., Scheel, S., Luinstra, G.A., Bryan Coughlin, E., Lesser, A.J., 2015. Polymer. 80 (2), 275.
Das, A., Kasaliwal, G.R., Jurk, R., Boldt, R., Fischer, D., Stöckelhuber, K.W., et al., 2012. Compos. Sci. Technol. 72 (16), 1961.
Dewar, M.J.S., Rogers, H., 1962. J. Am. Chem. Soc. 84, 395.
Dhakate, S.R., Sharma, S., Borah, M., Mathur, R.B., Dhami, T.L., 2008. Int. J. Hydrogen Energy. 33 (23), 7146.
Du, L., Jana, S.C., 2007. J. Power Sources. 172 (2), 734.
Duquesne, S., Bras, M.L., Bourbigot, S., Delobel, R., Vezin, H., Camino, G., et al., 2003. Fire Mater. 27, 103.
Dyre, J.C., Schrøder, T.B., 2000. Rev. Modern Phys. 72 (3), 873.
H. S. Nalwa). 2004;2:153.

Eswaraiah, V., Sankaranarayanan, V., Ramaprabhu, S., 2011. Macromol. Mater. Eng. 296, 894.
Faiella, G., Antonucci, V., Buschhorn, S.T., Luis, A.S.A., Prado, L.A.S.A., Schulte, K., et al., 2012. Composites A: Appl. Sci. Manuf. 43 (9), 1441.
Feng, H., Wang, X., Wu, D., 2013. Ind. Eng. Chem. Res. 52 (30), 10160.
Gao, J., Hu, M., Dong, Y., Li, R.K.Y., 2013. ACS Appl. Mater. Interfaces. 5, 7758.
Ghasemi, H., Sundararaj, U., 2012. Synth. Met. 162 (13–14), 1177.
Ghose, S., Working, D.C., Connell, J.W., Smith Jr., J.G., Watson, K.A., Delozier, D.M., et al., 2006. High Perform. Polym. 18, 961.
Granström, M., Petritsch, K., Arias, A.C., Lux, A., Andersson, M.R., Friend, R.H., 1998. Nature. 395, 257.
Ha, H.W., Choudhury, A., Kamal, T., Kim, D.-H., Park, S.-Y., 2012. ACS Appl. Mater. Interfaces. 4 (9), 14623.
Hatui, G., Malas, A., Bhattacharya, P., Dhibar, S., Kundu, M.K., Das, C.K., 2015. J. Alloys CompD. 619 (15), 709.
He, F., Lam, K.-H., Fan, J., Chan, L.H., 2014. Carbon. 80, 496.
He, F.-A., Wu, H.-J., Yang, X.-L., Lam, K.-H., Fan, J.-T., Chan, L.-W.H., 2015. Polym. Test. 42, 45.
Heiser, J.A., King, J.A., 2004. Polym. Compos. 25 (2), 186.
Higgins, B.A., Brittain, W.J., 2005. Eur. Polym. J. 41 (5), 889.
http://www.strategyr.com/MarketResearch/Conductive_Polymers_Market_Trends.asp. April 2016.
Huang, J., Baird, D.G., McGrath, J.E., 2005. J. Power Sources. 150, 110.
Huang, J., Zhu, Y., Xu, L., Chen, J., Jiang, W., Nie, X., 2016. Compos. Sci. Technol. 129 (6), 160.
Jafari, Y., Ghoreishi, S.M., Shabani-Nooshabadi, M., 2016. Synth. Met. 217, 220.
Ji, X., Xu, Y., Zhang, W., Cui, L., Liu, J., 2016. Composites A: Appl. Sci. Manuf. 87, 29.
Jia, J., Sun, X., Lin, X., Shen, X., Mai, Y.-W., Kim, J.-K., 2014. ACS Nano. 8 (6), 5774.
Kalaitzidou, K., Fukushima, H., Drzal, L.T., 2007a. Carbon. 45, 1446.
Kalaitzidou, K., Fukushima, H., Drzal, L.T., 2007b. Compos. Sci. Technol. 67, 2045.
Kang, G., Choi, J., Park, T., 2016. Sci. Rep. 6, 1.
Kang K., Park S., Jo A., Lee K., Ju H. Int. J. Hydrogen Energy, In Press, Corrected Proof, Available online 25 May 2016. Available from: http://dx.doi.org/10.1016/j.ijhydene.2016.05.027.
Kim, B.R., Lee, H.K., Kim, E., Lee, S.-H., 2010. Synth. Met. 160, 1838.
Kim, H., Abdala, A.A., Macosko, C.W., 2010. Macromolecules. 43 (16), 6515.
Kim, H., Miura, Y., Macosko, C.W., 2010. Chem. Mater. 22, 3441.
Kim, M., Choe, J., Lim, J.W., Lee, D.G., 2015. Compos. Struct. 132, 1122.
Kim, S., Do, I., Drzal, L.T., 2009. Polym. Compos. 31, 755.
Kuan, H.-C., Ma, C.-C.M., Chen, K.H., Chen, S.-M., 2004. J. Power Sources. 134 (1), 7.
Kuilla, T., Bhadra, S., Yao, D., Kim, N.H., Bose, S., Lee, J.H., 2010. Prog. Polym. Sci. 35, 1350.
Lee, C., Wei, X., Kysar, J.W., Hone, J., 2008. Science. 321, 385.
Lee, H.E., Han, S.H., Song, S.A., Kim, S.S., 2015. Compos. Struct. 134, 44.
Li, Q., Chen, L., Li, X., Zhang, J., Zhang, X., Zheng, K., et al., 2016. Composites A: Appl. Sci. Manuf. 82, 214.
Lin, Y., Liu, S., Peng, J., Liu, L., 2016. Compos. Sci. Technol. 131 (2), 40.
Liu, T., Eyler, A., Zhong, W.-H., 2016. Mater. Lett. 177 (15), 17.
Lu, W., Lin, H., Wu, W., Chen, G., 2006. Polymer. 47, 4440.
Mamunya, E.T., Davidenko, V.V., Lebedev, E.V., 1997. Compos. Interf. 4, 169.
McAllister, M.J., et al., 2007. Chem. Mater. 19, 4396.

Mehta, V., Cooper, J.S., 2003. J. Power Sources. 114, 32.
Meier, J.G., Crespo, C., Pelegay, J.L., Castell, P., Sainz, R., Maser, W.K., et al., 2011. Polymer. 52 (8), 1788.
Mordkovich, V.Z., 2003. Theor. Found. Chem. Eng. 37 (5), 429.
Mulliken, R.S., 1950. J. Am. Chem. Soc. 72, 600.
Mulliken, R.S., 1951. J. Chem. Phys. 19, 514.
Murakami, T.N., Grätzel, M., 2008. Inorg. Chim. Acta. 361, 572.
Nalwa, H.S., 1990. Appl. Organometallic Chem. 4 (2), 91.
Nan, C.W., 2002. Prog. Mater. Sci. 37 (1), 1.
Ni, Y., Lei Chen, L., Teng, K., Shi, J., Qian, X., Xu, Z., et al., 2015. ACS Appl. Mater. Interfaces. 7, 11583.
Nie, Y., Hübert, T., 2012. Composites A: Appl. Sci. Manuf. 43 (8), 1357.
Nielsen, E., Landel, R.F., 1994. Mechanical Properties of Polymers and Composites. 2nd edition Marcel Dekker Inc, New York.
Noël, A., Faucheu, J., Chenal, J.-M., Viricelle, J.-P., Bourgeat-Lami, E., 2014. Polymer. 55 (20), 5140.
O'Regan, B., Grätzel, M., 1991. Nature. 353, 737.
Okafor, P.A., Singh-Beemat, J., Iroh, J.O., 2015. Prog. Org. Coat. 88, 237.
Park, S., Ruoff, R.S., 2009. Nat. Nanotechnol. 4, 217.
Pinto, G., Cipriano, L.G., Ana, J.M., 1999. Polym. Compos. 20, 804.
Potts, J.R., Dreyer, D.R., Bielawski, C.W., Ruoff, R.S., 2011. Polymer. 52 (1), 5.
Qi, X.-Y., Yan, D., Jiang, Z., Cao, Y.-K., Yu, Z.-Z., Yavari, F., et al., 2011. ACS Appl. Mater. Interfaces. 3, 3130.
Radhakrishnan, S., Ramanujam, B.T.S., Adhikari, A., Sivaram, S., 2007. J. Power Sources. 163, 702.
Rafiee, M.A., Rafiee, J., Wang, Z., Song, H., Yu, Z.-Z., Koratkar, N., 2009. ACS Nano. 3 (12), 3884.
Rahaman, M., Chaki, T.K., Khastgir, D., 2012. Compos. Sci. Technol. 72 (13), 1575.
Ramanujam, B.T.S., Mahale, R.Y., Radhakrishnan, S., 2010. Compos. Sci. Technol. 70, 2111.
Ramanujam, B.T.S., Radhakrishnan, S., 2016. Conducting thermoplastic-graphite nanosheets composites: Understanding structure and properties. In: Michel, Siluvai (Ed.), Recent advances in Functional Materials for Device Applications. Vishnu Prints Media, Chennai, p. 48. (ISBN:978-93-85374-34-0).
Ramanujam, B.T.S., Radhakrishnan, S., 2010. Int. J. Plast. Technol. 14, 37.
Ramanujam, B.T.S., Radhakrishnan, S., Deshpande, S.D., 2015. J. Thermoplast. Compos. Mater0892705715614063 (Before Print).
Ramanujam, B.T.S., Radhakrishnan, S.J., 2015. Thermoplast. Compos. Mater. 28 (6), 835.
Saini, P., Choudhari, V., Sood, K.N., Dhawan, S.K., 2009. J. Appl. Polym. Sci. 113, 3146.
Saini, P., Choudhary, V., Singh, B.P., Mathur, R.B., Dhawan, S.K., 2009. Mater. Chem. Phys. 113, 919.
Saujanya, C., Radhakrishnan, S., 2001. Polymer. 42 (10), 4537.
Shaheen, S.E., Brabec, C.J., Padinger, F., Fromherz, T., Hummelen, J.C., Sariciftci, N.S., et al., 2001. Appl. Phys. Lett. 78 (6), 84.
Shirakawa, H., Louis, E.J., MacDiarmid, A.J., Chiang, C.K., Heeger, A.J., 1977. J. Chem. Soc. Chem. Commun.578–580.
Silva, L.B.D., Marini, J., Gelves, G., Sundararaj, U., Gregório Jr., R., Bretas, R.E.S., 2013. Eu. Polym. J. 49, 3318.

Spinks, G.M., Dominis, A.J., Wallace, G.G., Tallman, D.E., 2002. J. Solid State Electrochem. 6, 85.
Stankovich, S., Dikin, D.A., Dommett, G.H.B., Kohlhaas, K.M., Zimney, E.J., Stach, E.A., et al., 2006. Nature. 442, 282.
Sykam, N., Kar, K.K., 2014. Mater. Lett. 117, 150.
Taherian, R., 2014. J. Power Sources. 265 (1), 370.
Tchmutin, I.A., Ponomarenko, A.T., Krinichnaya, E.P., Kozub, G.I., Efimov, O.N., 2003. Carbon. 41, 1391.
Tsuchiya, H., Kobayashi, O., 2004. Int. J. Hydrogen Energy. 29, 985.
Usuki, A., Kojima, Y., Kawasumi, M., Okada, A., Fukushima, Y., Kurauchi, T., et al., 1993. J. Mater. Res. 8, 1179.
Vasileiou, A.A., Kontopoulou, M., Gui, H., Docoslis, A., 2015. ACS Appl. Mater. Interfaces. 7 (3), 1624.
Verdejo, R., Bernal, M.M., Romasanta, L.J., Lopez-Manchado, M.A., 2011. J. Mater. Chem. 21 (10), 3301.
Wan, X., Long, G., Huang, L., Chen, Y., 2011. Adv. Mater. 23, 5342.
Wang, D.W., Li, F., Zhao, J.P., Ren, W.C., Chen, Z.G., Tan, J., et al., 2009. ACS Nano. 3, 1745.
Wang, H., Yuan, X., Wu, Y., Huang, H., Peng, X., Zeng, G., et al., 2013. Adv. Colloid Interface Sci. 19, 195–196.
Wang, H.L., Hao, Q.L., Yang, X.J., Lu, L.D., Wang, X., 2009. Electrochem. Commun. 11, 1158.
Wang, Y., Shi, Z.Q., Huang, Y., Ma, Y.F., Wang, C.Y., Chen, M.M., et al., 2009. J. Phys. Chem. C. 113, 13103.
Wang, Y., Xiuwen, X., Zhiqiang, T., Zhao, L., Bo, Y., Yongfeng, L., et al., 2015. Electrochim. Acta. 173 (10), 715.
Wei, Y., Lelkes, P.I., MacDiarmid, A.G., Guterman, E., Cheng, S., Palouian, K., 2004. Electroactive polymers and nanostructured materials for neural tissue engineering. In: Zhou, Q.F., Cheng, S.Z.D. (Eds.), Contemporary Topics in Advanced Polymer Science and Technology. Peking University Press, Beijing, p. 430.
Williams, G., Serger, B., Kamat, P.V., 2008. ACS Nano. 2, 1487.
Winey, K.I., Kasiwagi, T., Mu, M., 2007. Mater. Res. Bull. 32 (4), 348.
Wright, M., Uddin, A., 2012. Sol. Energy Mater. Sol. Cells. 107, 87.
Xia, H., Wang, Q., Li, K., Hu, G.-H., 2004. J. Appl. Polym. Sci. 93, 378.
Yacubowicz, J., Narkis, M., Benguigui, L., 1990. Polym. Eng. Sci. 30, 459.
Yan, D., Zhang, H.-B., Jia, Y., Hu, J., Qi, X.-Y., Zhang, Z., et al., 2012. ACS Appl. Mater. Interfaces. 4 (9), 4740.
Yang, L., Wang, Z., Ji, Y., Wang, J., Xue, G., 2014. Macromolecules. 47 (5), 1749.
Youngs, I.J., 2003. J. Phys. D: Appl. Phys. 36, 738.
Yu, J., Choi, H.K., Kim, H.S., Kim, S.Y., 2016. Composites A: Appl. Sci. Manuf. 88, 79.
Yu, J., Lu, K., Sourty, E., Grossiord, N., Koning, C.E., Loos, J., 2007. Carbon. 45, 2897.
Zeng, X., Xu, X., Shenai, P.M., Kovalev, E., Baudot, C., Mathews, N., et al., 2011. J. Phys. Chem. C. 115 (44), 21685.
Zhang, F., Zhao, Q., Yan, X., Li, H., Zhang, P., Wang, L., et al., 2016. Food Chemistry Part A. 197, 943.
Zhang, H., Bilotti, E., Tu, W., Lew, C.Y., Peijs, T., 2015. Eur. Polym. J.(68), 128.
Zhang, H.B., ZhengWG, Yan, Q., Yang, Y., Wang, J., Lu, Z.H., et al., 2010. Polymer. 51, 1191.
Zhang, Q., Rastogi, S., Chen, D., Lippits, D., Lemstra, P.J., 2006. Carbon. 44, 778.

Zhao, Y., Zhang, Z., Yu, L., Tang, Q., 2016. Synth. Metals. 212, 84.
Zhao, Y.F., Xiao, M., Wang, S.J., Ge, X.C., Meng, Y.Z., 2007. Compos. Sci. Technol. 67, 2528.
Zhao, Y.-H., Zhang, Y.-F., Bai, S.-L., 2016. Composites A: Appl. Sci. Manuf. 85, 148.
Zheng, G., Wu, J., Wang, W., Pan, C., 2004. Carbon. 42, 2839.
Zhokhavets, U., Erb, T., Gobsch, G., Al-Ibrahim, M., Ambacher, O., 2006. Chem. Phys. Lett. 418, 347.
Zhu, Y., Wang, H., Yan, L., Wang, R., Zhu, Y., 2016. Wear. 101, 356–357.

Hybrid polymer composites for structural applications

2

Hai Nguyen[1], *Wael Zatar*[1] *and Hiroshi Mutsuyoshi*[2]
[1]College of Information Technology & Engineering, Marshall University, Huntington, WV, United States, [2]Department of Civil & Environmental Engineering, Saitama University, Saitama, Japan

Chapter Outline

2.1 Introduction 35
2.2 Hybrid FRP composites for structural applications 37
2.2.1 Aircraft structures 37
2.2.2 Automotive components 37
2.2.3 Civil infrastructures 38
2.2.4 Energy sector components 40
2.2.5 Marine transportation 41
2.2.6 Smart structures 42
2.2.7 Sporting goods 43
2.2.8 Telecommunication components 43
2.3 Hybrid FRP–concrete composites structural members 44
2.3.1 Hybrid FRP–concrete girder/deck systems 44
2.3.2 Hybrid FRP–concrete tubular columns/piles 47
2.4 Conclusions 49
References 50

2.1 Introduction

The term "hybrid" is generally understood as an object made from two or more different elements. Hybrid composite materials (a.k.a. hybrid composites or hybrid materials) are defined as composites consisting of at least two dissimilar materials embedded in either thermosetting or thermoplastic resin matrix. Mechanical properties of hybrid composites based on natural fibers (e.g., flax, jute, hemp, kapok, and sisal) and synthetic fibers (e.g., basalt, carbon, glass, and Kevlar) have been systematically reviewed by the authors in Chapter 4, Mechanical Properties of Hybrid Polymer Composite Natural fibers have found to have extensive applications in various fields such as automobile, biomedical, building and civil engineering, railway,

Hybrid Polymer Composite Materials: Applications. DOI: http://dx.doi.org/10.1016/B978-0-08-100785-3.00002-4

furniture, and paper/pulp industry. Few different applications of natural fibers are shown in Fig. 2.1. Natural fibers are cheap, biodegradable, and environment friendly, but they have some drawbacks such as relatively poor fiber-matrix bonding, poor fire resistance, and low impact strength and durability. The properties of natural-fiber composites can be enhanced by hybridizing with synthetic fibers. Glass fibers are by far most commonly used man-made fibers on earth. Glass fibers are often combined with natural fibers to make hybrid composites due to their good mechanical properties such as high-specific strength/stiffness, low cost, low density, good fire and chemical resistance, good electric insulation, and insensitive to moisture. Though hybrid synthetic/natural-fiber composites exhibit improved mechanical properties compared to all-natural-fiber composites, their applications are quite limited (e.g., secondary components of vehicles such as floor/door panels, parcel shelfs, and seat coverings). Hybridization of high strength, fully synthetic fibers including carbon/glass and carbon/Kevlar composites is generally used for high-performance structures such as commercial airplanes and navy vessels. This chapter reviews recent structural applications of hybrid synthetic-fiber composites used for aircrafts, automobiles, civil infrastructures, energy sector components, marine and smart structures, sport goods, and telecommunication components. In addition,

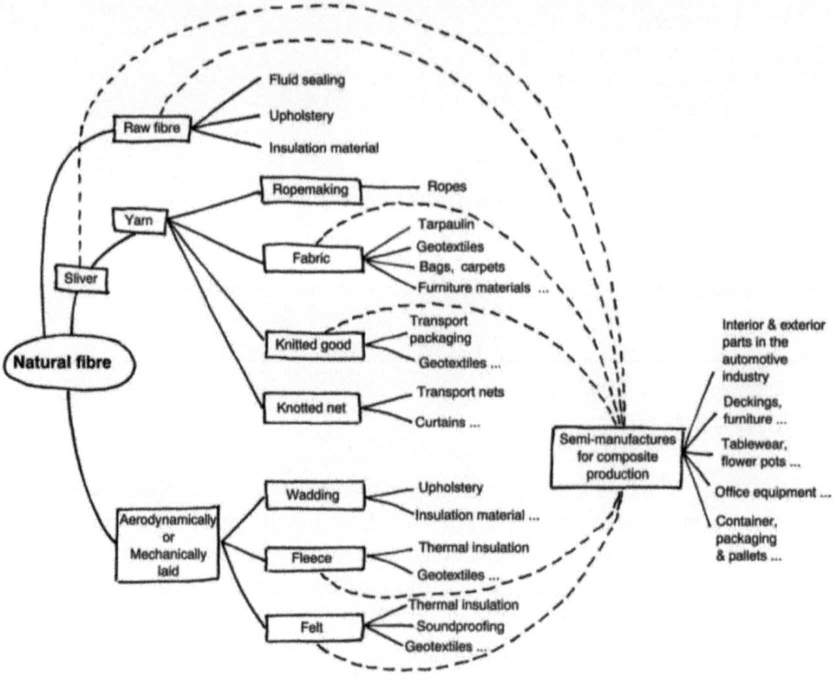

Figure 2.1 Different applications of natural fibers.
Source: Stevens, C. Industrial applications of natural fibres: structure, properties and technical applications. Edited by J. Müssig. Vol. 10. John Wiley & Sons, 2010.

applications of fiber-reinforced polymer (FRP) composites in civil engineering will be discussed with the main focus on hybrid FRP–concrete structural components such as deck slabs, beams/girders, and columns/piles.

2.2 Hybrid FRP composites for structural applications

2.2.1 Aircraft structures

In the pioneering days, aircraft structures (e.g., Havilland mosquito aircraft DH98 and de Havilland Albatross DH91) were manufactured from natural composites of ply wood, balsa, or untwisted flax fibers. Aluminum alloys took over in the 1930s and have dominated the aircraft industry to the recent time (Soutis, 2005). Fiber–metal laminates (FMLs) are a new class of hybrid composite structures used for advanced aerospace structural applications. FMLs include thin sheets of metal alloy bonded together with FRP plies. They have been increasingly used for high-performance and lightweight structures in aircraft industry. The hybrid fiber–metal composites offer the advantages of both metallic and fiber-reinforced materials. Metals have isotropic characteristics, high bearing, and impact strength and are easy to repair and manufacture, while the hybrid composites have high-specific strength and stiffness and excellent fatigue resistance (Sinmazçelik et al., 2011). The most common FMLs are aramid-reinforced aluminum laminate (ARAL), carbon-reinforced aluminum laminate (CARAL), glass-reinforced aluminum laminate (GLARE), and titanium-based and magnesium-based FMLs. ARAL and GLARE materials are commercially available in four and six, respectively, different standard grades (i.e. ARAL 1-4 and GLARE 1-6), depending on metal type/thickness and fiber thickness/orientation. ARAL (incorporated aramid fibers, aluminum 7475-T761, and epoxy resin) was successfully developed at Delft University of Technology, The Netherlands in 1978 (Botelho et al., 2006). ARAL has been used for the lower skin panels of the former Fokker 27 aircraft and the cargo door of the Boeing C-17 (Sinmazçelik et al., 2011). GLARE has been used in two sections of Airbus A380's upper skin fuselage (Fig. 2.2) and is selected for the Boeing 777 impact resistant bulk cargo floor (Sinmazçelik et al., 2011). CARAL was developed to improve the poor compressive strength of ARAL. The combination of high strength and stiffness and good impact resistance makes CARAL a great choice for aerospace applications. Other applications of CARAL are impact absorbers for helicopter struts and aircraft seats (Sinmazçelik et al., 2011). In addition, engineers have concentrated on replacing the secondary structures of civil aircrafts with nonmetal fiber-reinforced composites, where the reinforcements are either carbon, glass, Kevlar, or hybrids thereof. Typical examples of the extensive application of composites in this manner are the Boeing 757, 767, and 777, and the Airbus A310, A320, A330, and A340 airliners (Soutis, 2005).

2.2.2 Automotive components

Several researchers have reported the application of natural and synthetic fiber-reinforced composites in automotive industry (Marsh, 2003; Bledzki et al., 2006;

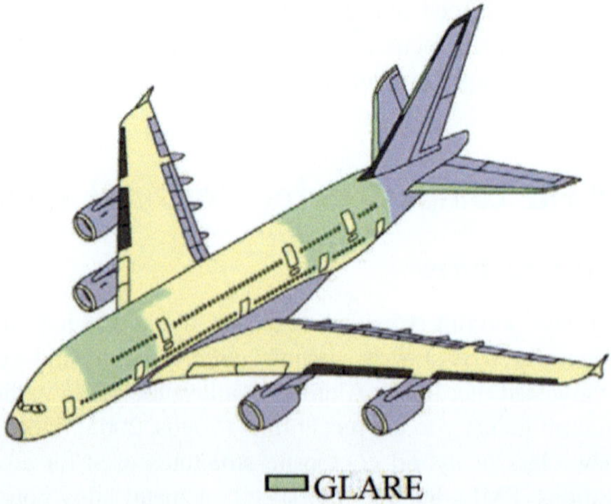

Figure 2.2 Location of GLARE laminate in fuselage sections of the Airbus A380.
Source: Sinmazçelik, T., Avcu, E., Bora M.Ö., and Çoban O. "A review: fibre metal laminates, background, bonding types and applied test methods." *Mater. Des.* 32, 7 (2011): 3671–3685.

Holbery and Houston, 2006; Badie et al., 2011). Glass together with carbon and aramid fiber-reinforced composites account for the majority of automotive components, but they are nondegradable and difficult to recycle (Marsh, 2003). Natural bast fibers such as jute, hemp, kenaf, and flax are good alternative to synthetic fibers because they are biodegradable and have the potential to reduce vehicle weight. Recycle thermoplastic resins such as polypropylene, polyolefin, polyethylene, polyurethane, and polyamide are commonly used in vehicles (Marsh, 2003). Mercedes-Benz was a pioneer in the use of jute fiber-reinforced plastic for the interior door panels of its E-class vehicles. Other major automotive manufacturers such as Audi, BMW, Daimler-Chrysler, Ford, Mitsubishi, Peugeot, Land Rover, Volkswagen, and Volvo have used biofibers for various components. Typical components are seat back, door panels, bumpers, boot liner, hat rack, spare tire lining, headliner panel, windshield, dashboard, business table, pillar cover panel, instrumental panels, boot lid finish panel, insulation, storage shelf/panel, sunroof sliders, and exterior underbody protection trim (Bledzki et al., 2006). Fig. 2.3 shows flax/polypropylene underbody components of the Mercedes Benz A-Class (Marsh, 2003).

2.2.3 Civil infrastructures

Deterioration of reinforced concrete and steel structures in the United States and worldwide has motivated the development of advanced and innovative materials and methods for structural rehabilitation, since replacement of these structures would be very costly and nearly prohibited. FRP composite materials appear to be

Figure 2.3 Flax/polypropylene underbody components of the Mercedes Benz A-Class.
Source: Marsh, G. "Next step for automotive materials." *Mater. Today* 6, 4 (2003): 36–43.

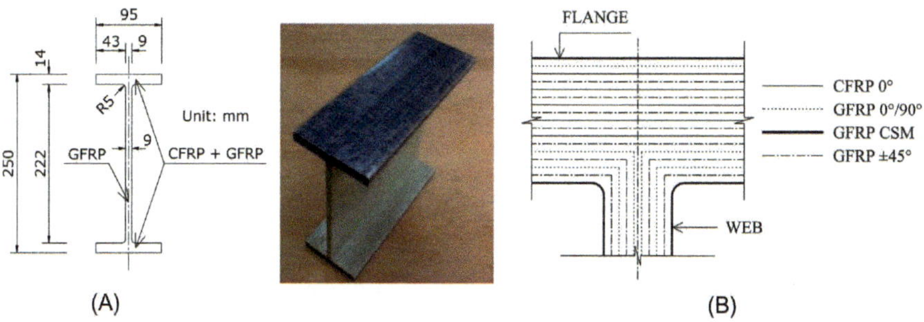

Figure 2.4 Details of I-shaped CFRP/GFRP girders: (A) cross-section and (B) fiber lay-up and stacking sequence.
Source: Nguyen, H., Mutsuyoshi, H., and Zatar, W. Push-out tests for shear connections between UHPFRC slabs and FRP girder. *J. Composite Struct.*, 118, 2014a, pp. 528–547.

good candidates to alternate conventional construction materials in various structural aspects (either strengthening of existing structures or new construction of infrastructures using entirely FRP). They possess many superior properties such as high-specific strength/stiffness, excellent corrosion resistance, and good durability. FRP have been widely used in many countries such as the United States, Canada, Switzerland, and Japan to mitigate salt-related deterioration resulting from airborne sea salt, winter snow, or the use of deicing salt and ice control. Recently, the combined use of carbon FRP (CFRP) and glass FRP (GFRP) in I-shaped girders (Fig. 2.4) have been investigated by the authors (Nguyen et al., 2010). The hybridization of the CFRP and GFRP exploits the lightweight and excellent mechanical properties of the CFRP/GFRP and the low cost of GFRP, which benefits structures in terms of strength and construction cost. An extensive

Figure 2.5 First HFRP pedestrian bridge using CFRP/GFRP I-girders in fishery harbor in Kure, Hiroshima, Japan, 2011.
Source: Nguyen, H., Zatar, W., and Mutsuyoshi, H. Hybrid FRP girders topped with segmental precast concrete slabs for accelerated bridge construction. *Trans. Res. Rec.: J. Trans. Res. Board*, 2, 2407, 2014b, pp. 83–93.

research program has been conducted to characterize the behavior of hybrid FRP (hereafter called "HFRP") composites (Nguyen et al., 2010, 2013, 2014a,b, 2015a,b). It was found that the developed HFRP I-girders are promising to replace short-span deteriorated pedestrian bridges, especially those exposed to chloride attacks by salt-laden marine environments. In addition, the HFRP girders can provide a competitive and sustainable option for accelerated bridge construction. These were proved by the successful application of the HFRP I-girders to an actual pedestrian bridge construction in Kure city, Hiroshima prefecture, Japan in 2011 (Fig. 2.5). The bridge was built to replace an existing deficient steel bridge. The FRP members were manufactured and transported to the site for assembly. This bridge consists of two HFRP I-girders topped with a GFRP gratings bridge deck. Each HFRP I-girder was assembled from three HFRP I-section units with two splice connections at quarter locations. The bridge was simply supported with a total length of 12 m and an effective width of 0.75 m. It was exposed to a high-corrosive environment by the ocean. The bridge's condition was periodically assessed by visual inspection for more than 3 years.

2.2.4 Energy sector components

According to US Department of Energy, wind power is sustainable and cost-effective because it is an abundant source of energy and a clean fuel source. Unlike coal (largest source of power in the United States), which produces greenhouse gas emissions, wind turbines do not produce atmospheric emissions that cause acid rain

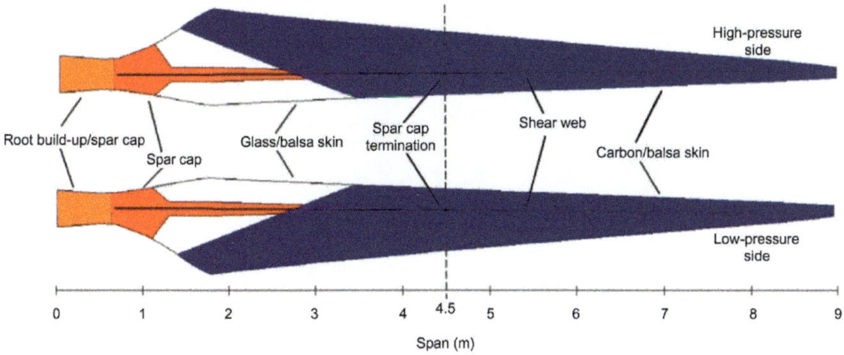

Figure 2.6 Structural elements of the TX-100 wind turbine blade.
Source: Rumsey, M.A., and Joshua A.P. "Structural health monitoring of wind turbine blades." In *The 15th International Symposium on: Smart Structures and Materials & Nondestructive Evaluation and Health Monitoring*, p. 69330E. International Society for Optics and Photonics, 2008.

or greenhouse gases. The majority of wind turbine blades are made of glass fibers embedded in either epoxy or polyester matrix, carbon/epoxy, or wood/epoxy composites. The use of CFRP to manufacture the turbine blade has also increased with increasing rotor size (Ciang et al., 2008). However, CFRP turbine blade has limited applications due to the high cost and manufacturing challenges. Fig. 2.6 shows 9-m-long wind turbine blade (TX-100) made of glass/epoxy and carbon/epoxy (Rumsey and Joshua, 2008). This blade was developed by Sandia National Laboratories and designed for a 100-kW stall-controlled turbine. The area of the blade skin contained carbon fiber (blue) and unidirectional fiberglass spar cap (red).

2.2.5 Marine transportation

FRP composites have been used for a wide range of naval structures including decks and superstructures of vessels, bulkheads, advanced mast systems, propellers, propulsion shafts, rudders, pipes, pumps, valves, machinery, and other equipment on large warships such as frigates, destroyers, and aircraft carriers (Mouritz et al., 2001). FRP composites are potential to other marine applications such as propulsors and control surfaces in submarines, patrol boats, hovercrafts, mine countermeasure ships, and corvettes. Researchers have found that the structural weight of a patrol boat made of all GFRP sandwich composite is about 10% and 36%, respectively, lighter than an aluminum boat and a steel boat of a similar size. In addition, the use of CFRP or utilizing latest fabrication techniques such as Seeman Composites Injection Molding Process may provide further savings in hull weight (Mouritz et al., 2001). Hybrid CFRP/GFRP composite laminates were used for face skins of the Visby Class corvette, which was designed by the Royal Swedish Navy (Fig. 2.7). The Visby class is a multipurpose vessel with capabilities of surveillance, combat, mine laying, mine countermeasures, and antisubmarine warfare operations.

Figure 2.7 Hybrid CFRP/GFRP composite laminates as face skins of Visby Class corvette. *Source*: From Teng, J.G., Chen, J.F., Smith, S.T. and Lam, L., FRP Strengthened RC Structures,Wiley, 2002.

The vessel is thus required to be lightweight, strong, resistant to underwater shock loads, and stealthy by having low radar and magnetic signatures. The Visby is the first naval vessel to make use of significant CFRP composite in the hull. The Royal Swedish Navy together with Royal Singapore Navy also uses Kevlar FRP in some structures of their ships (Mouritz et al., 2001).

2.2.6 Smart structures

A smart structure (a.k.a. intelligent structure, adaptive structure, and functional structure) is defined as a structure that is able to sense external stimuli such as pressure, velocity, density, or temperature change. It can process the information and respond in a controlled manner in real time (Thill et al., 2008). A smart structure consists of four key elements: actuators, sensors, control strategies, and power conditioning electronics. Many types of actuators and sensors such as piezoelectric materials, shape memory alloys (SMA) (alloys that can remember their original shapes), electrostrictive and magnetostrictive materials, and fiber optics are being considered for various applications (Chopra, 2002).

SMA reinforced composites have been studied for the past decades. NASA Langley research center (Turner et al., 2001) developed SMA hybrid composites consisting of a nickel-titanium alloy (a.k.a. nitinol or NiTi) embedded in a glass-epoxy composite matrix. Zhou and Lloyd (2009) investigated actuation characteristics of E-glass/epoxy and carbon/epoxy laminate beams embedded with nitinol wire actuators. The idea of embedding SMA actuators in a composite laminate for structural evaluation was first introduced by Rogers and Liang (1989). Since then, some researchers have embedded SMA actuators in laminates (e.g., glass/epoxy and Kevlar/epoxy prepreg) to obtain self-actuating structures (Thill et al., 2008; Rogers, 1990; Paine and Rogers, 1994; Turner et al., 2006; Liang et al., 1991). Shape memory polymers (SMPs) are another type of smart materials that work in a very similar way to SMAs. Compared with SMAs and shape memory ceramics, SMPs have unique characteristics such as light weight, low cost, low density, easy manufacture, high-shape deformability, biodegradability, and highly tailorable glass transition temperature (Liu et al., 2014). However, SMPs have relatively low-deformation stiffness and low-recovery stress. These drawbacks lead to studies on

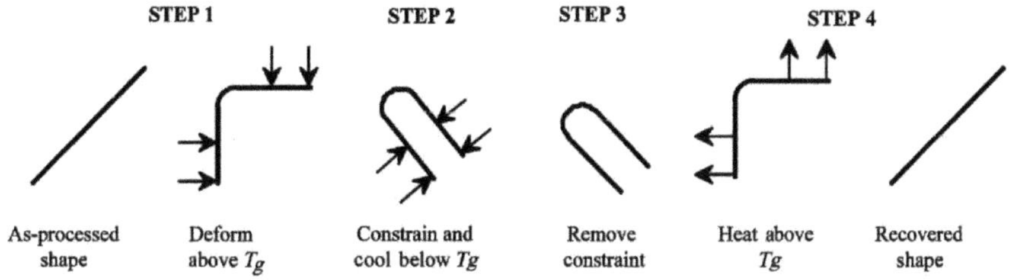

Figure 2.8 Idealized thermomechanical cycle for a shape memory polymer (SMP). *Source*: Gall, K., Martin, L.D., Liu, Y., Finch, D., Lake, M., and Munshi, N.A. "Shape memory polymer nanocomposites." *Acta Mater.* 50, 20 (2002): 5115–5126.

SMP composites (SMPCs) to enhance characteristics of SMPs. Extensive reviews on behavior and applications of SMAs, SMPs, and SMPCs have reported by few researchers (Liu et al., 2014; Mather et al., 2009; Gall et al., 2002). Fig. 2.8 schematically shows four steps of a shape recovery thermomechanical cycle for a SMP, where T_g is glass transition temperature.

2.2.7 Sporting goods

Composites have been increasingly used in sporting industry due to their high-specific strength, light weight, and excellent durability. Carbon/glass/Kevlar fiber-reinforced composites and carbon nanomaterials (e.g., carbon nanotubes, nanoclay, nanoparticles, and fullerenes) are of great interests for this market. Composite sporting equipment are usually manufactured by filament winding process. Typical sport products include golf shafts, tennis and squash racquets, canoes and kayaks, paddles, surfboards, hockey sticks, fishing rods, snowboards and skis, helmets, running shoes, bicycle frames, and others. Fig. 2.9 shows an example of a bicycle featuring carbon-based composite components, which are manufactured by 3T Design Ltd. 3T's VENTUS aerobar bicycles with the lightest, fastest carbon/epoxy racing aerobar (850 g) were ridden by some medal winners in 2008 Tour de France race (McConnell, 2008).

2.2.8 Telecommunication components

High growth of the telecommunication industry leads to increasing demands for higher speed data transmission and higher power capacity. Fiber optic technology such as fiber Bragg gratings have large number of applications in various fields. Fiber-optic communication is one of the most common methods used for high-speed communications. It is a method of transmitting light between two points through optical fibers (a.k.a. fiber optics). Optical fibers are flexible, long, thin strands of very pure glass fibers with a diameter of a human hair. Submarine optical fiber cables are known as a key component of transmission technology under the sea.

Figure 2.9 Bicycle featuring carbon-based composite components ridden by some cyclists in 2008 Tour de France race.
Source: McConnell, V.P. "Advanced performance elevates sporting spirit." *Reinf. Plast.* 52, 9 (2008): 24–29.

Li-ion batteries (LIBs) are among various energy and power technologies used for today's wireless and mobile applications. Liu et al. (2012) have reviewed the use of carbon nanotube (CNT)-based composites as an electrode material for LIBs. CNTs have one-dimensional tubular structure (Fig. 2.10) with high-electrical and thermal conductivities and extremely large surface area. CNTs are thus contributing to improve the electrochemical characteristics of both the anode and cathode of LIBs, resulting in their enhanced energy conversion and storage capacities.

2.3 Hybrid FRP–concrete composites structural members

FRP composites have wide acceptance in the construction industry due to their high-strength-to-weight ratio and good electrochemical corrosion resistance. Hybrid systems such as concrete-filled FRP circular tubes (CFFTs) are very effective in special types of applications such as piling, poles, highway overhead sign structures, and bridge components (Fam and Sami, 2002). Hybrid FRP–concrete systems benefit from the low cost and good compressive strength of concrete together with the high-tensile strength, low weight, and good durability of FRP. The following sections present some applications of the hybrid FRP–concrete systems for bridge components including deck slabs, beams/girders, and columns/piles.

2.3.1 Hybrid FRP–concrete girder/deck systems

Nguyen et al. (2013, 2014a,b, 2015a,b) have developed durable composite girder systems consisting of hybrid carbon/glass FRP (HFRP) I-girders (Fig. 2.4) and

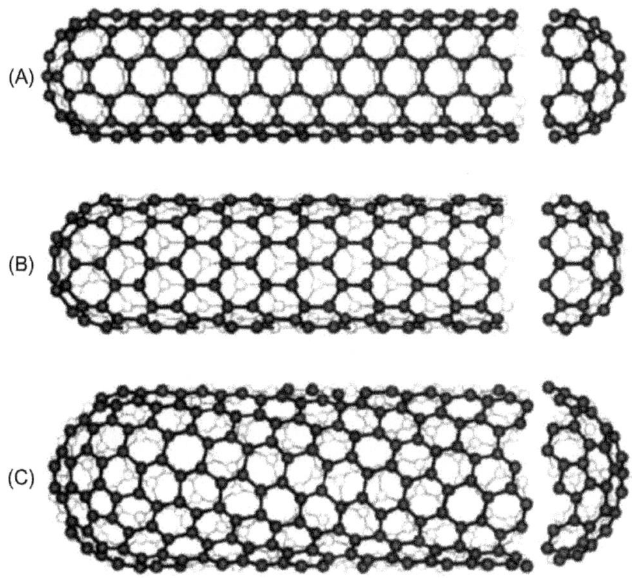

Figure 2.10 Structure of single-walled carbon nanotubes specified by a vector (m, n), which defines how the graphene sheet is rolled up: (A) armchair (m = n); (B) zigzag (m = 0, n # 0); (C) chiral (m # n # 0).
Source: Harris, P.J.F. "Carbon nanotube composites." *Int. Mater. Rev.* 49, 1 (2004): 31–43.

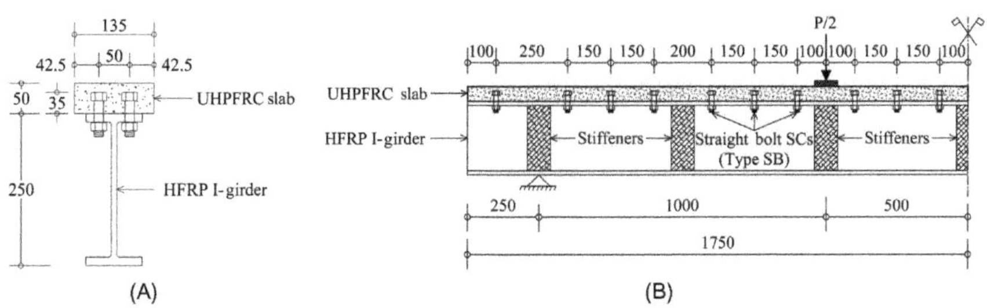

Figure 2.11 Hybrid FRP–UHPFRC composite girder: (A) cross-section and (B) elevation [unit: millimeter].
Source: Nguyen, H., Mutsuyoshi, H., and Zatar, W. Hybrid FRP–UHPFRC composite girders: Part 1—Experimental and numerical approach. *J. Compos. Struct.*, 125, 2015a, pp. 631–652.

precast ultra-high-performance fiber-reinforced concrete (UHPFRC) slabs (Fig. 2.11). Epoxy adhesive and/or bolt shear connectors were used to transfer horizontal shear force from the UHPFRC slabs to the HFRP I-girders. Comprehensive laboratory testing program and finite element analysis have been conducted to evaluate flexural behavior of the FRP–UHPFRC composite girders. It was found that although the stiffness of the FRP–UHPFRC composite girders is relatively low

Figure 2.12 Demonstrated GFRP–UHPFRC pedestrian bridge in Miyagi, Japan, 2012. *Source*: Nguyen, H., Mutsuyoshi, H., and Zatar, W. Hybrid FRP–UHPFRC composite girders: Part 1—Experimental and numerical approach. *J. Compos. Struct.*, 125, 2015a, pp. 631–652.

compared to concrete/steel girders of similar sizes, they can be applied to short-span bridges. The FRP and UHPFRC are very durable materials and the FRP–UHPFRC girders may therefore be good candidates for bridge structures, especially those are exposed to harsh environmental conditions. The FRP girders and the UHPFRC slabs can be prefabricated in a factory and transported to the construction site, resulting in fast construction of FRP–UHPFRC bridges. The developed FRP–UHPFRC composite girder system is thus competitive and attractive for accelerated bridge construction. In addition, the use of the FRP–UHPFRC girder system can reduce the overall weight of structures, resulting in smaller supporting concrete elements and a reduction of CO_2 emission (compared to traditional girder system). The UHPFRC slabs can effectively prevent premature delamination failure of the FRP girders and enhance overall stiffness of the FRP-UHPFRC girder system.

The applicability of the FRP–UHPFRC composite girder system is proven by an actual pedestrian bridge construction in Miyagi prefecture, Japan in 2012 (Fig. 2.12). The bridge was designed and erected by a research group from Saitama University in cooperation with FRP manufacturers, industries, and local governments in Japan. The effective width and the overall length of the bridge were 750 mm and 6000 mm, respectively. The bridge was built with two GFRP I-girders topped with an UHPFRC deck slab. The bridge was constructed in a fishery harbor, where chloride attack is an issue. Therefore, GFRP bolts were used in conjunction with epoxy adhesive to connect the GFRP I-girders and the UHPFRC slab. The use of the epoxy adhesive was necessary to obtain a full interaction for the FRP–UHPFRC composite girders (Nguyen et al., 2015b). The bridge was designed with a deflection limit of $L/500 = 12$ mm (where L is the bridge's span length). The

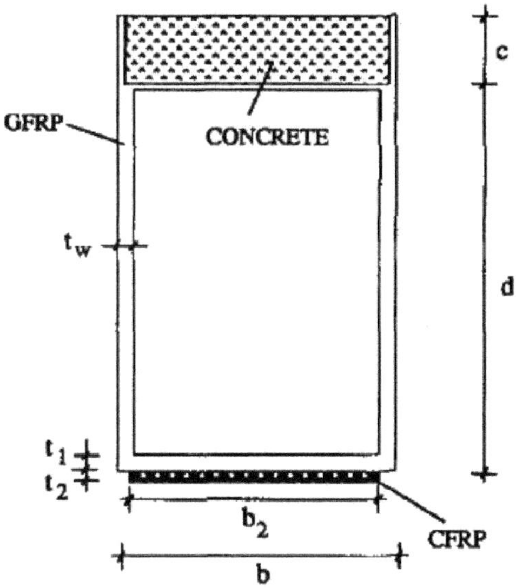

Figure 2.13 Cross-section of hybrid GFRP/CFRP–concrete beam.
Source: Deskovic, N., Triantafillou, T.C., and Meier, U. "Innovative design of FRP combined with concrete: short-term behavior." *J. Struct. Eng.* 121, 7 (1995): 1069–1078.

design live load of the bridge was 3.5 kN/m^2. Further details of this bridge can be found elsewhere (Nguyen et al., 2015a).

Deskovic et al. (1995) proposed hybrid beam sections consisting of: (1) a concrete layer substituted the GFRP compression flange of traditional pultruded box sections, thus reducing the material cost and increasing the beam's stiffness; (2) a GFRP box sections (with a very thin compression flange served as the formwork for casting the concrete layer); and (3) a thin layer of CFRP laminate externally bonded to the soffit of the GFRP box beam by epoxy adhesive to increase the section's rigidity. Fig. 2.13 shows the cross-section of the idealized hybrid GFRP/CFRP–concrete beam. It was found that the combination of FRP materials with concrete results in efficient and cost-effective hybrid members. These members were proven to have high strength and stiffness, low weight, and pseudoductility characteristics.

Keller et al. (2007) presented a new concept for a lightweight hybrid FRP bridge deck. The sandwich bridge deck consists of three layers including a FRP profile with T-upstands for the tensile skin, a lightweight concrete (LC) for the core, and a thin layer of UHPFRC as a compression skin (Fig. 2.14). The experimental investigation showed positive results regarding the feasibility of the proposed new hybrid bridge deck. The manufacturing of the deck was proved to be easy, rapid, and economical.

2.3.2 Hybrid FRP–concrete tubular columns/piles

Saafi et al. (1999) developed innovative types of structural columns for new construction. The columns were made of concrete-encased carbon and glass FRP tubes.

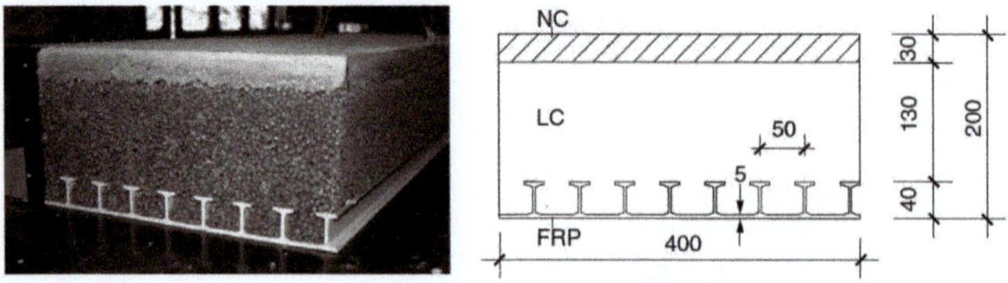

Figure 2.14 Cross section of a hybrid FRP and lightweight concrete sandwich bridge deck [unit: millimeter].
Source: Keller, T., Schaumann, E., and Vallée, T. "Flexural behavior of a hybrid FRP and lightweight concrete sandwich bridge deck." *Composites A: Appl. Sci. Manuf.* 38, 3 (2007): 879–889.

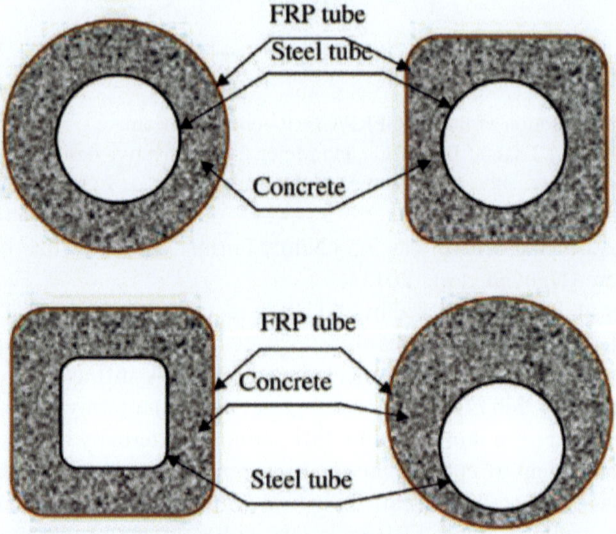

Figure 2.15 Typical sections of double-skin tubular members.
Source: From Teng, J.G., Yu, T., Wong, Y.L., and Dong, S.L. "Hybrid FRP–concrete–steel tubular columns: concept and behavior." Construct. Build. Mater. 21, 4 (2007): 846–854.

The concrete-filled FRP tubes were cast in place, where the tubes act as formwork, protective jacket, confinement, and flexural and shear reinforcements. Test results showed that external confinement of concrete by the FRP tubes significantly enhanced the strength, ductility, and energy absorption capacity of concrete.

Teng et al. (2007) proposed hybrid FRP–concrete–steel double-skin tubular columns consisting of concrete-filled between an outer FRP tube and an inner steel tube (Fig. 2.15). The three constituent materials in this new form of hybrid columns were optimally combined to obtain benefits that are not available with conventional reinforced concrete columns. The hybrid columns possess many advantages such as

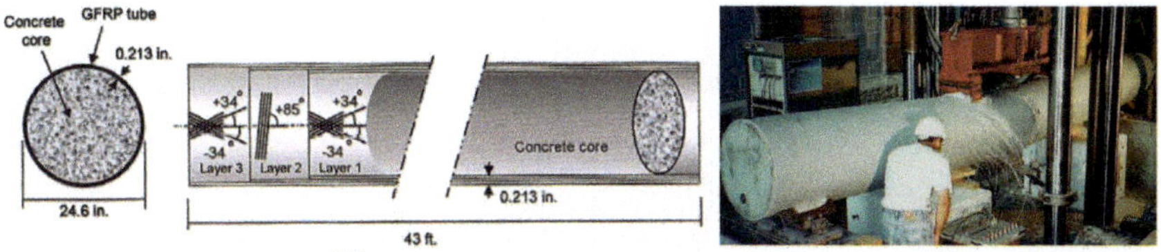

(A) (B)

Figure 2.16 Concrete-filled GFRP circular tube composite pile: (A) Details of piling and (B) Laboratory testing of the composite pile.
Source: Used with permission from PCI journal. From Fam, A., Pando, M., Filz, G., Rizkalla, S., 2003. Precast piles for Route 40 bridge in Virginia using concrete filled FRP tubes. PCI J. 48 (3), 32–45.

good ductility, corrosion resistance, and easy of fabrication and construction. The test results indicated that the concrete in the new hybrid column was effectively confined by the FRP and steel tubes. The local buckling of the inner steel tube was either delayed or suppressed by the surrounding concrete, resulting in a very ductile response of the hybrid column. Performance of hybrid column specimens with the inner steel tube shifting toward the tension side was examined to confirm the applicability of the proposed hybrid sections. These specimens were found to exhibit very ductile behavior. The outer FRP tube not only provides fire protection for the hybrid column but also provides both confinement to the concrete and additional shear resistance, resulting in an enhancement of the column behavior.

Fam et al. (2003) present a new generation of precast composite piles (Fig. 2.16) used for the Route 40 highway bridge over the Nottoway river in Virginia. The composite piles composed of 625 mm diameter concrete-filled GFRP circular tubes with a wall thickness of 5.3 mm. The composite piles were extended above the ground level and directly embedded into the reinforced concrete cap beam supporting the superstructure. Laboratory and field testing revealed that the use of concrete-filled FRP tubes as piling for bridge piers is feasible and practical. The FRP tubes eliminate the need for temporary formwork and internal reinforcements because they serve as both permanent formwork and reinforcements. The flexural strength of the composite piles was comparable to that of 508 mm square concrete piles prestressed with fourteen 12.7 mm diameter strands. The unit cost of the installed composite piles was 77% higher than that of the prestressed piles. However, the low maintenance and life-cycle costs and good durability of the precast FRP composite piles make them competitive and attractive for bridge structures subjected to harsh environments.

2.4 Conclusions

This chapter sheds light on key developments and structural applications of hybrid composites. Carbon-based hybrid composites such as carbon/Kevlar and glass/carbon have found many applications in high-performance structures such as aircrafts and naval ships. Thermosetting resins such as epoxy, polyester, phenolic, vinyl ester, polyurethane, polyamide, and polyamide-imide are most commonly used matrix materials. Glass, carbon, and Kevlar fiber-reinforced composites and their hybridizations have been widely used in various fields, especially construction industry. Current advancements and innovations in raw materials and manufacturing technologies have helped reducing the cost of hybrid composites and thus expanding their future applications.

References

Badie, M.A., Mahdi, E., Hamouda, A.M.S., 2011. An investigation into hybrid carbon/glass fiber reinforced epoxy composite automotive drive shaft. Mater. Des. 32 (3), 1485–1500.

Bledzki, A.K., Faruk, O., Sperber, V.E., 2006. Cars from bio-fibres. Macromol. Mater. Eng. 291 (5), 449–457.

Botelho, E.C., Silva, R.A., Pardini, L.C., Rezende, M.C., 2006. A review on the development and properties of continuous fiber/epoxy/aluminum hybrid composites for aircraft structures. Mater. Res. 9 (3), 247–256.

Chopra, I., 2002. Review of state of art of smart structures and integrated systems. AIAA J. 40 (11), 2145–2187.

Ciang, C.C., Lee, J.R., Bang, H.J., 2008. Structural health monitoring for a wind turbine system: a review of damage detection methods. Meas. Sci. Technol. 19 (12), 122001.

Deskovic, N., Triantafillou, T.C., Meier, U., 1995. Innovative design of FRP combined with concrete: short-term behavior. J. Struct. Eng. 121 (7), 1069–1078.

Fam, A., Pando, M., Filz, G., Rizkalla, S., 2003. Precast piles for Route 40 bridge in Virginia using concrete filled FRP tubes. PCI J. 48 (3), 32–45.

Fam, A.Z., Sami, H.R., 2002. Flexural behavior of concrete-filled fiber-reinforced polymer circular tubes. J. Compos. Constr. 6 (2), 123–132.

Gall, K., Martin, L.D., Liu, Y., Finch, D., Lake, M., Munshi, N.A., 2002. Shape memory polymer nanocomposites. Acta Mater. 50 (20), 5115–5126.

Harris, P.J.F., 2004. Carbon nanotube composites. Int. Mater. Rev. 49 (1), 31–43.

Holbery, J., Houston, D., 2006. Natural-fiber-reinforced polymer composites in automotive applications. JOM. 58 (11), 80–86.

Karbhari, V.M. Assessment of Durability of FRP Materials for Use in Civil Infrastructure. No. CA10-1130. 2011.

Keller, T., Schaumann, E., Vallée, T., 2007. Flexural behavior of a hybrid FRP and lightweight concrete sandwich bridge deck. Composites A: Appl. Sci. Manuf. 38 (3), 879–889.

Liang, C., Rogers, C.A., Fuller, C.R., 1991. Acoustic transmission and radiation analysis of adaptive shape-memory alloy reinforced laminated plates. J. Sound Vib. 145 (1), 23–41.

Liu, X.-M., dong Huang, Z., woon Oh, S., Zhang, B., Ma, P.C., Yuen, M.M.F., et al., 2012. Carbon nanotube (CNT)-based composites as electrode material for rechargeable Li-ion batteries: a review. Compos. Sci. Technol. 72 (2), 121–144.

Liu, Y., Du, H., Liu, L., Leng, J., 2014. Shape memory polymers and their composites in aerospace applications: a review. Smart Mater. Struct. 23 (2), 023001.

Marsh, G., 2003. Next step for automotive materials. Mater. Today. 6 (4), 36–43.

Mather, P.T., Luo, X., Rousseau, I.A., 2009. Shape memory polymer research. Annu. Rev. Mater. Res. 39, 445–471.
McConnell, V.P., 2008. Advanced performance elevates sporting spirit. Reinf. Plast. 52 (9), 24–29.
Mouritz, A.P., Gellert, E., Burchill, P., Challis, K., 2001. Review of advanced composite structures for naval ships and submarines. Compos. Struct. 53 (1), 21–42.
Nguyen, H., Mutsuyoshi, H., Zatar, W., 2013. Flexural behavior of hybrid composite beams. Trans. Res. Rec.: J. Transp. Res. Board. 2 (2332), 53–63.
Nguyen, H., Mutsuyoshi, H., Asamoto, S., Matsui, T., 2010. Structural behavior of hybrid FRP composite I-beam. J. Construct. Build. Mater. 24 (6), 956–969.
Nguyen, H., Mutsuyoshi, H., Zatar, W., 2014a. Push-out tests for shear connections between UHPFRC slabs and FRP girder. J. Composite Struct. 118, 528–547.
Nguyen, H., Mutsuyoshi, H., Zatar, W., 2015a. Hybrid FRP–UHPFRC composite girders: Part 1 – Experimental and numerical approach. J. Compos. Struct. 125, 631–652.
Nguyen, H., Zatar, W., Mutsuyoshi, H., 2014b. Hybrid FRP girders topped with segmental precast concrete slabs for accelerated bridge construction. Trans. Res. Rec.: J. Trans. Res. Board. 2 (2407), 83–93.
Nguyen, H., Zatar, W., Mutsuyoshi, H., 2015b. Hybrid FRP–UHPFRC composite girders: Part 2 – analytical approach. J. Compos. Struct. 125, 653–671.
Paine, J.S.N., Rogers, C.A., 1994. The response of SMA hybrid composite materials to low velocity impact. J. Intell. Mater. Syst. Struct. 5 (4), 530–535.
Rogers, C.A., 1990. Active vibration and structural acoustic control of shape memory alloy hybrid composites: experimental results. J. Acoust. Soc. Am. 88 (6), 2803–2811.
Rogers, C.A., Liang C., and Jia J. Behavior of shape memory alloy reinforced composite platesl Part 1: model formulation and control concepts. In Proceedings of the 30th Structures, Structural Dynamics and Materials Conference, pp. 2011–2017. 1989.
Rumsey, M.A., Joshua, A.P., 2008. Structural health monitoring of wind turbine blades. The 15th International Symposium on: Smart Structures and Materials & Nondestructive Evaluation and Health Monitoring. International Society for Optics and Photonics.
Saafi, M., Toutanji, H., Li, Z., 1999. Behavior of concrete columns confined with fiber reinforced polymer tubes. ACI Mater. J. 96 (4), 500–509.
Sinmazçelik, T., Avcu, E., Bora, M.Ö., Çoban, O., 2011. A review: fibre metal laminates, background, bonding types and applied test methods. Mater. Des. 32 (7), 3671–3685.
Soutis, C., 2005. Carbon fiber reinforced plastics in aircraft construction. Mater. Sci. Eng.: A. 412 (1), 171–176.
Stevens, C., 2010. In: Müssig, J. (Ed.), Industrial applications of natural fibres: structure, properties and technical applications, Vol. 10. John Wiley & Sons, Ltd, Chichester.
Teng, J.G., Yu, T., Wong, Y.L., Dong, S.L., 2007. Hybrid FRP–concrete–steel tubular columns: concept and behavior. Construct. Build. Mater. 21 (4), 846–854.
Thill, C., Etches, J., Bond, I., Potter, K., Weaver, P., 2008. Morphing skins. Aeronaut. J. 112 (1129), 117–139.
Turner, T.L., Lach, C.L., Cano, R.J., 2001. Fabrication and characterization of SMA hybrid composites. SPIE's 8th Annual International Symposium on Smart Structures and Materials. International Society for Optics and Photonics.
Turner, T.L., Buehrle, R.D., Cano, R.J., Fleming, G.A., 2006. Modeling, fabrication, and testing of a SMA hybrid composite jet engine chevron concept. J. Intell. Mater. Syst. Struct. 17 (6), 483–497.
Zhou, G., Lloyd, P., 2009. Design, manufacture and evaluation of bending behaviour of composite beams embedded with SMA wires. Compos. Sci. Technol. 69 (13), 2034–2041.

Hybrid polymer composites for electromagnetic absorption in electronic industry

3

Charalampos A. Stergiou[1], Marina Y. Koledintseva[2] and Konstantin N. Rozanov[3]

[1]Laboratory of Inorganic Materials, Centre for Research and Technology-Hellas, Thessaloniki, Greece, [2]Electromagnetic Compatibility Design Engineering, Oracle, Santa Clara, CA, United States, [3]Institute for Theoretical and Applied Electromagnetics, RAS, Moscow, Russia

Chapter Outline

3.1 Introduction 54

3.2 Shielding effectiveness definitions 60

3.3 Applications 64

3.3.1 Application of thin noise-suppressing sheet materials 64

3.3.2 Measurements of material parameters of noise-suppressing thin sheet materials 68

3.3.3 Comparison of noise-suppressing thin sheet materials 69

3.4 Specular absorber for far field operation 76

3.4.1 Introductory part 76

3.4.2 The bandwidth-to-thickness ratio 78

3.4.3 The fractional bandwidth 81

3.4.4 The angular performance 83

3.4.5 Conclusion 84

3.5 Material's design 84

3.5.1 Reflected wave absorbers 85

3.5.2 Transmitted wave absorbers 92

3.6 Summary 95

Acknowledgments 96

References 96

Hybrid Polymer Composite Materials: Applications. DOI: http://dx.doi.org/10.1016/B978-0-08-100785-3.00003-6

3.1 Introduction

Electromagnetic absorption in electronic industry is mainly associated with the concepts of shielding and filtering out electromagnetic interference (EMI) by using the proper absorbing electromagnetic materials. Electromagnetic shielding and filtering are intended to solve *electromagnetic compatibility* (EMC) problems. EMC can be defined as the ability of electronic units (digital or analog components, circuits, devices, blocks, and entire systems) or biological systems to function properly at the levels of the acceptable efficiency in the intended electromagnetic environment (Paul, 2006). *Electromagnetic shielding* is usually associated with any measure used to reduce electric and magnetic fields in the certain region of space (Ott, 2009).

Electromagnetic shielding plays two major roles. *The first role of shielding* is to protect susceptible electronic units, or "victims," from external undesirable electromagnetic disturbances, that is, EMI or noise. This is an *electromagnetic immunity* problem. Electromagnetic disturbances may penetrate an electronic unit due to various coupling paths either from the neighboring EMI sources inside the unit, or from external sources. This problem may be solved by designing a proper shielded case (a housing, a chassis, a screen, or other protecting enclosure, usually conducting, but not necessarily) to protect the electronic device from the external undesirable electromagnetic impacts. It is also important to properly shield all the input/output (I/O) ports, or connectors of the device, where cables or other units (peripheral devices, various computer components, or cards) are attached. This way, possible external electromagnetic disturbances would not be able to penetrate the device and affect the performance of the susceptible circuits. Inside an electronic device, the elimination of possible noise coupling paths may be achieved through the proper components placement. Apart from the optimal electronics layout, shielding of some noise-emitting components, for example, heatsinks on integrated circuits (ICs), may also result in reduction of coupling between a source and a victim.

The second role of shielding is to minimize undesirable *electromagnetic emissions* (EME) from an electronic unit into the external environment, so that this electronic unit does not disturb the operation of other susceptible devices, systems, or circuits. EME could be either conducted, or radiated, depending on the physics of the electric and magnetic fields formation and detection in the region of interest. Electromagnetic shielding should also provide the proper levels of communication, or information security.

As a rule, to prevent two circuits from electromagnetic coupling, it is recommended to place at least one of the circuits in a completely closed grounded enclosure made of a conducting magnetic material. However, such a shield is impractical, since real electronic devices always have slots, openings, perforation, hinges, screws, and other features needed from mechanical and thermal design points of view, which in many cases compromise electromagnetic shielding effectiveness (SE). Therefore, a proper design of an electronic device or a system is always a complex problem, where contradictory requirements from EMC, mechanical, and thermal points of view should be simultaneously satisfied in the most optimal and cost-effective way.

Moreover, many EMC problems are directly related to signal integrity (SI) and power integrity (PI) aspects. High-speed SI electronics designers are always concerned about possible unwanted electromagnetic fields and coupling paths affecting quality of the signals in their designs, especially as data rates of the digital designs have reached dozens gigabit-per-second (Gbps) and steadily increase as technology develops. Similarly, when PI is compromised, there could be emissions problems. Therefore, in electronics designs there are presently three correlated issues: EMC, SI, and PI.

The major *undesirable sources, or "culprits"* causing EMI and EMC problems are usually the following:

- radiated emissions;
- conducted emissions;
- fast transient burst, surge, electrostatic discharge;
- low-frequency magnetic fields emissions; and
- emissions of harmonic currents from power mains, voltage fluctuations, flicker, and inrush currents.

Correspondingly, the electronic units susceptible to electromagnetic disturbances should be immune to these sources.

The levels of acceptable EME and immunity are usually specified in various standards and regulations. Different requirements are applied to different groups of electronic products—commercial, industrial, or military. There are global and international requirements, but some countries have their own regional limits for EME and EMI. We are not going to discuss these regulations, but refer to Celozzi et al. (2008).

The most cost-effective approach is to provide the acceptable levels of EME and immunity of an electronics product at the stage of its design. However, in many cases, it is necessary to provide the proper shielding to an already designed and physically available unit. There is a number of known ways of EMC/EMI improvement:

- absorbing and reflecting electromagnetic (EM) waves by screens, or shields;
- diverting the EM coupling path;
- reducing the Q-factor of a cavity;
- shielding a source of EMI coupling;
- reducing surface current density; and
- attenuating a wave traveling through a long and wide slot.

For providing the proper shielding, one should identify all EMI sources, determine all susceptible circuits and critical coupling paths over the specific frequency ranges, where EMC problems could arise. It is important to remember that an electromagnetic shield (or a noise-suppressing structure) is always a combination of geometry and materials. For each particular application, the material type, its geometrical configuration and placement should be selected individually. This means that electromagnetic characteristics, shape, and quantity of the material should be determined based on the frequency and intensity of the unwanted electromagnetic disturbance to be eliminated. Since any electronic product operates over some frequency range, including a few higher harmonics of fundamental (critical) frequencies of its circuitry, the shield should be designed for the entire frequency range of the operation.

The main physical mechanisms of interaction between electromagnetic waves and material media to provide good shielding are *reflection* and *absorption*. However, the concept of reflection applies to electromagnetic waves rather than electromagnetic fields in the near region. In the near-field region, it is better to refer to *shielding using a screen*, which is often based on redirecting electric and/or magnetic fields. Let us consider these mechanisms starting from screening against quasi-static electric and magnetic fields.

For *low-frequency, or quasi-static magnetic and electric fields*, mechanisms of inducing opposing fields and diverting primary fields are used. For low-frequency magnetic fields, either magnetic flux shunting, or eddy current cancelation are used for shielding. High permeability of a screen provides higher concentration of magnetic flux inside the shield, usually a ferromagnetic metal. When a shield is exposed to a time-varying magnetic field, according to Faraday's law, a time-varying electric field appears. If the shield is conducting, it repulses the magnetic field due to the induced electric current producing a magnetic field opposing the incident one. If the conductor is imperfect, the current running on the surface produces magnetic induction and eddy currents within some skin layer.

A shield does not always need to completely enclose a product in order to be effective. For example, partial shields or screens are often employed to redirect near fields and isolate a source circuit from another circuit and to prevent coupling to cables or any unintentional antennas. However, for shielding against low-frequency magnetic fields, all the slots and joints in an enclosure should be eliminated to ensure good magnetic field containment within the enclosure.

There is a general rule for *time-varying fields in the near regions*. If electric field of the source is dominating, e.g., of electric dipole, then a shield or a screen should provide a barrier impedance which is significantly larger than the characteristic impedance of the surrounding area ($Z_{screen} \gg Z_{ambient}$). If the magnetic field is dominant, e.g., a loop with current, then the barrier admittance of the shield or screen should substantially exceed the characteristic admittance of the ambient medium ($Y_{screen} \gg Y_{ambient}$).

Absorption and reflection are the two mechanisms which should be employed for shielding from *electromagnetic waves of any nature*—plane, cylindrical, spherical, and with any polarization.

Reflection over a wide frequency range (from d.c. to tens of GHz and even up to THz frequencies) is usually provided by highly conductive materials—standard (nonmagnetic) and ferromagnetic. In majority of cases, these are metals. Overall, return loss is greatest at low frequencies and for metals with high conductivity, while magnetic materials degrade return loss (Paul, 2006). When an electromagnetic wave is incident upon a well-conducting metal screen, surface currents induced on this screen radiate the electromagnetic wave back. Only a small portion of the field penetrates inside the metal forming the skin layer. Skin effect is associated with electromagnetic dissipation, or *absorption*, inside an imperfect metal. The physics of the skin effect is related to the eddy currents. The eddy-current cancelation exists in any electrically conducting material, regardless of its magnetic properties. For an a.c. field penetrating a metal, the total magnetic induction decays exponentially in the direction perpendicular to the air–metal interface, and the characteristic delay length is the skin depth.

SE of metal screens is proportional to attenuation of electromagnetic waves incident upon a metallic medium with conductivity σ and relative permeability μ_r. The attenuation of an electromagnetic wave inside the metal is $\alpha = 1/\delta$, where $\delta = \sqrt{2/(\omega\sigma\mu_0\mu_r)}$ is skin depth, $\omega = 2\pi f$ is angular frequency associated with linear frequency f, and μ_0 is permeability of vacuum (Pozar, 1998, Chapter 1). Therefore, thickness of a metal screen should be chosen based on its conductivity σ and permeability μ_r, as well as the frequency range of electromagnetic waves. The higher the conductivity and permeability of a metal, the thinner screen should be used for the same frequency to achieve the same attenuation. Moreover, for the lower frequencies, one may need a thicker screen. Many ferromagnetic metals have very high permeability at the lower frequencies (in kHz or MHz ranges), and their application is preferable for reducing thickness of the protecting screens. Overall, absorption loss increases as $\sqrt{f\sigma\mu_r}$, while return loss behaves as a function of $\sigma/(f\mu_r)$ (Paul, 2006).

An electromagnetic shield or enclosure should minimize the penetration of electromagnetic waves inside the protected space region. It is thus important that the enclosure itself does not radiate inside the protected region. The geometry of the shield surface may be very complex, for example, corrugated, pyramidal, or irregular to increase SE through reflection. Since a closed enclosure forms a resonant cavity, possible slots or openings may compromise the SE of such an enclosure at the frequencies of the cavity resonant modes. The lower conductivity reduces surface currents and results in less reflection. Magnetic losses in a screen may additionally attenuate surface currents and therefore reduce radiation from any slots and openings on the screen.

Additional absorption, which occurs due to dissipation of electromagnetic energy inside the shield material, is associated with the dielectric and magnetic losses. They are described by the imaginary part of permittivity $\varepsilon_r''(\omega)$, or dielectric loss tangent $\tan\delta_\varepsilon(\omega) = \varepsilon_r''(\omega)/\varepsilon_r'(\omega)$, and similarly, the imaginary part of permeability $\mu_r''(\omega)$, or magnetic loss tangent $\tan\delta_\mu(\omega) = \mu_r''(\omega)/\mu_r'(\omega)$. Losses in the materials are associated with electromagnetic energy transformation into the other forms of energy, first of all, heat. In dielectrics, losses may be either dipolar (true dielectric loss associated with polarization of molecules and their dynamics in a.c. fields), or ohmic (due to the presence of impurities resulting in nonzero or inhomogeneous concentration of free charges). In magnetics, power is attenuated basically on account of the inability of dynamic magnetization mechanisms of domain wall movement and magnetization rotation to follow the excitation magnetic field at high frequencies. In composites, the skin effect on conducting inclusions may also be an important loss mechanism.

For specific applications, there may be a need for *materials absorbing electromagnetic energy* incident upon them rather than reflecting. In many applications, reflections from the protected surfaces are strongly undesirable. Some applications which require electromagnetic absorption are as follows:

- radar absorbing materials for radio frequency (RF) and microwave invisibility;
- materials to cover walls of anechoic chambers;
- materials for accumulating electromagnetic energy in industrial microwave furnaces;
- materials for medical applications where microwave heating is needed;
- flexible absorbing structures for optimal design of high-speed digital electronic equipment.

Overall, the following categories of materials are used for effective shielding and EMC/EMI improvement:

- highly conductive materials (e.g., for various types of gaskets); these could be metallic, or alloy-based conductors, nonmetallic conductors, conducting polymers, metal including fabrics, metal-infused paints;
- lossy dielectric materials (e.g., carbon-filled polymers, metal- or carbon-filled ceramics);
- lossy magnetic materials (e.g., ferrites, polymer-based ferrites, or other magnetic composites);
- band gap structures (e.g., 3D metamaterials or 2D metafilms, the latter are similar to frequency-selective surfaces with periodic prints);
- multilayered structures as combinations of materials listed above.

Engineering and synthesis of new shielding materials with advanced EM properties is a very active field of research and technology development, and the arsenal of available materials for practical applications is steadily increasing. Any material should satisfy requirements for specific applications. Choice of a material for a shield depends on many factors (Neelakanta, 1995):

- *EMC/EMI requirements*: what specific levels of SE the selected material should provide over the given frequency range; this also includes bandwidth of operation, that is, the effective frequency range with acceptable shielding parameters;
- *Suitability for specific shielding applications*: what types of fields should be eliminated or reduced (static or quasi-static electric or magnetic field; low-frequency electromagnetic fields; surface currents and near-region electromagnetic fields; radiations in RF, microwave, mm-wave, and higher frequency ranges). This factor also includes the preferable physics of operation (generation of opposing field, field diverting, reflection, absorption, energy transformation).
- *Geometrical considerations*: where shields are expected to be placed (around noise sources; on cables; on/around/near certain electronic components on printed-circuit boards (PCBs); around assemblies, devices, systems under test). This also includes the shape and size of shield/screen/enclosure and its possible perforation, slots, doors, and others.
- *Mechanical considerations*: rigidity, flexibility, compressibility, adhesive properties, ease of applying, weight, fastening and joints, capability to withstand vibration and mechanical stress, and others.;

- *Thermal stability*: performance under hostile thermal conditions;
- *Chemical considerations:* resistance to corrosion, oxidation, degradation of electrical contacts or dielectric/magnetic in aggressive environment; aging of material; nontoxicity;
- *Technological simplicity and repeatability*;
- *Cost-effectiveness.*

Shielding materials can be either *homogeneous (monolithic)*, or *heterogeneous (composite)* (Neelakanta, 1995). Composite shielding materials are very widespread nowadays. There are two groups of composite shielding materials depending on their morphology: the *host-inclusion* system and *the multilayered stackup* system. The first contains inclusions (particles of certain shape) in a host material, often called a matrix, or a base. The second group contains layers of homogeneous materials sandwiched together. Layers made of host-inclusion systems can also be used in the multilayered materials to achieve the desirable properties of a shielding structure.

Most absorbing host-inclusion composites contain metallic fillers (iron, silver, aluminum, nickel, copper, alloy particles) and nonmetallic fillers (iron-based magnetic oxides—ferrites, graphite, carbon fibers/sheets, indium/tin oxide powders, boron/boron tungstate) mixed in a host medium. These inclusion particles can be of various shapes. Some of them may possess one, two, or all three dimensions in the nanoscale. The first group includes fibers, or needles with nanosize diameter; the second group includes disks, or flakes with nanosize thickness, and the third are the nanospheres, spheroids, ellipsoids, or irregular-shaped particles with characteristic nanometer dimensions. The main requirements to the inclusions of absorbing EM shielding materials are the presence of significant electrical conductivity and/or permeability; significant electromagnetic loss over the given frequency ranges; availability in the desirable shape and with the certain aspect ratios; minimum chemical interaction with the host material; corrosion resistance; suitable size, weight, cost; shelf-life durability; no shedding-off from the composite; excellent abrasion strength; and ease of processing during preparation of the required mixture. With regard to the host materials, they are typically dielectrics, including various polymers, epoxy resins, or any other medium which provides good bonding with the inclusions. The important requirements to the host materials are their high strength and impact resistance, moldability, chemical stability, as well as coloration with pigments, for example, for esthetic needs.

Since all composites are essentially heterogeneous materials, their homogenized effective electromagnetic properties can be estimated using various effective medium theories and mixing rules (Sihvola, 1999; Lagarkov and Rozanov, 2009). This is possible if the constitutive electromagnetic parameters of phases (host and each type of inclusion material) are known, as well as information on the contents of each phase and the morphology of the composite is available. It is desirable that this data is available to the manufacturer of the material. The final user of the material usually deals with the effective conductivity, or permittivity, and/or permeability. These parameters are either measured at the user's end, or available from the manufacturer's datasheets, or could be retrieved from publications. To this effect, there are numerous measurement techniques to determine material parameters over specific frequency bands.

Herein, we are focusing on hybrid polymer EM absorbing composites, which comprise at least two different material phases as inclusions. The amount of the outlined parameters affecting their final electromagnetic performance indicates their advantage over monolithic shielding materials. To this end, in the present chapter, we will consider some practical applications and characterization techniques of absorbing materials for EMC/EMI purposes. We will also attempt an insight into the relation between the fundamental electromagnetic properties and shielding performance and propose some effective figures of merit. Finally, we will review some recent developments of hybrid composite shielding materials.

3.2 Shielding effectiveness definitions

The quantitative characteristic of any shielding structure is shielding effectiveness (*SE*). *SE* is actually a measure of comparison between magnitudes of fields on the entry and the exit sides of the barrier (shield) along some chosen direction. *SE* can express the comparison between magnitudes of

- electric field—incident $|E_{inc}|$ and transmitted through the barrier $|E_{tr}|$;
- magnetic field—incident $|H_{inc}|$ and transmitted through the barrier $|H_{tr}|$;
- electromagnetic field power P_{inc} before and after the barrier P_{tr}.

Assume, there is a slab of thickness t comprising a material with the constitutive parameters $\mu = \mu_0(\mu'_r - j\mu''_r)$, $\varepsilon = \varepsilon_0(\varepsilon'_r - j\varepsilon''_r)$, surrounded by free space. Dielectric loss includes true dielectric damping loss $\varepsilon''_{r,\mathrm{true}}$ and conductivity loss, so that $\varepsilon''_r = \varepsilon''_{r,\mathrm{true}} + \sigma/(\omega\varepsilon_0)$, where σ is the macroscopic conductivity of the material. The material intrinsic impedance is

$$\eta = \sqrt{\frac{\mu}{\varepsilon}}, \tag{3.1}$$

and the free space impedance is

$$\eta_0 = \sqrt{\frac{\mu_0}{\varepsilon_0}}. \tag{3.2}$$

The transmission and reflection coefficients for an interface of two media are defined by standard formulas (Pozar, 1998, Chapter 1). The transmission coefficient from air into the material is

$$T_m = \frac{2\eta}{(\eta + \eta_0)}, \tag{3.3}$$

and out of the material to the air is

$$T_\alpha = \frac{2\eta_0}{(\eta + \eta_0)}. \tag{3.4}$$

The reflection coefficient from the material into the air is

$$\Gamma_a = \frac{(\eta_0 - \eta)}{(\eta + \eta_0)}, \tag{3.5}$$

and from air back into the material is

$$\Gamma_m = \frac{(\eta - \eta_0)}{(\eta + \eta_0)}. \tag{3.6}$$

The propagation constant for a plane wave in the medium is

$$\gamma = \alpha + j\beta = j\omega\sqrt{\mu\varepsilon}. \tag{3.7}$$

SE can be defined in terms of electric field as

$$SE_E = 20\log_{10}|E_{\text{inc}}/E_t|, \tag{3.8}$$

and is always positive.

By analogy, in terms of magnetic field, *SE* is defined by

$$SE_H = 20\log_{10}|H_{\text{inc}}/H_t|, \tag{3.9}$$

For a uniform plane wave and the same media on both sides of the barrier, these two definitions are identical ($SE = SE_E = SE_H$), since electric and magnetic fields are related by the intrinsic impedance of the medium. However, for near fields and for different media, it is generally $SE_E \neq SE_H$, and these definitions cannot be taken as equivalent.

The transmission coefficient for a plane wave incident normally upon a 2D wide slab of thickness *t* is

$$T_{tr} = T_m T_a \frac{1}{1 + \Gamma_m \Gamma_a e^{-2\gamma t}} e^{-\gamma t}. \tag{3.10}$$

Then, *SE* due to this slab is calculated as

$$SE = -20\log_{10}|T_{\text{tr}}|. \tag{3.11}$$

An absorbing material layer is often backed by a highly conducting plane. Then the concept of return (reflection) loss (*RL*) is applicable. There is an analogy with a short-circuited transmission line, and the input impedance at the interface between absorber and air is

$$Z_{\text{in abs}} = j\eta \tan(\gamma t). \tag{3.12}$$

Then, the return loss (in dB) is

$$RL = 20\log_{10}\left|\frac{Z_{\text{in abs}} - \eta_0}{Z_{\text{in abs}} + \eta_0}\right|. \tag{3.13}$$

The most reasonable definition for both near-field and far field is in terms of power (P). There are the following equations to define the transmitted (P_{tr}), reflected (P_{refl}), and absorbed power (P_{refl}) normalized to the incident power (P_{inc}). Transmittance is given by

$$T_P = \frac{P_{\text{tr}}}{P_{\text{inc}}} = \left|T_m T_a \frac{1}{1-\Gamma_\alpha^2 e^{-2\gamma t}} e^{-\gamma t}\right|^2, \tag{3.14}$$

the reflectance is

$$R_P = \frac{P_{\text{refl}}}{P_{\text{inc}}} = \left|\Gamma_m + \frac{T_m T_a \Gamma_a}{1-\Gamma_\alpha^2 e^{-2\gamma t}} e^{-2\gamma t}\right|^2, \tag{3.15}$$

and the absorbance is determined from the power balance as

$$A_P = 1 - T_P - R_P. \tag{3.16}$$

An approach to calculate *SE* is Schelkunoff decomposition (Schelkunoff, 1943), which is also based on the power balance. The overall *SE* is composed of three terms: return loss—*RL*, multiple reflections loss—*MR*, and absorption loss—*AL*. The return loss is calculated as

$$RL = 20\log_{10}\left|\frac{1}{T_m T_a}\right|, \tag{3.17}$$

absorption loss is calculated as

$$AL = 20\log_{10}\left|e^{\alpha t}\right| \tag{3.18}$$

and multiple reflections loss is

$$MR = 20\log_{10}\left|1-\Gamma_m^2 e^{-2\gamma t}\right|. \tag{3.19}$$

The total *SE* is defined as

$$SE = -\log_{10}(T_P) = RL + MR + AL \tag{3.20}$$

Mismatch decomposition (McDowell and Hubing, 2014) can also be used to quantify shielding effectiveness in the plane-wave formulation. Based on this approach, the overall *SE* consists of the effective return (mismatch) loss and effective absorption (dissipation) loss,

$$SE = RL_{\text{eff}} + AL_{\text{eff}}. \tag{3.21}$$

In this case, the effective reflection embeds multiple reflections between the two interfaces of the barrier with free space. Then *SE* is calculated in terms of power with effective return loss

$$RL_{\text{eff}} = -10\log_{10}(1 - R_P) \tag{3.22}$$

and effective absorption loss

$$AL_{\text{eff}} = -10\log_{10}(1 - A_P) = -10\log_{10}\left[\frac{T_P}{(1 - R_P)}\right]. \tag{3.23}$$

For near fields, the plane wave formulas can be used as an approximation (Schulz et al., 1988). However, the wave impedances for elementary electric dipole and magnetic dipole should be used. Let's assume that both sources are directed along the same axis z. Considering a wave front with transverse components of E and H fields, one can get the following wave impedances for these elementary sources in free space:

$$Z_E = \frac{E_\Theta}{H_\Phi} = \eta_0 \Phi(\beta_0, r) \tag{3.24}$$

and

$$Z_H = \frac{E_\Phi}{H_\Theta} = -\frac{\eta_0^2}{Z_E} \tag{3.25}$$

with

$$\Phi(\beta_0, r) = \frac{(j/\beta_0 r) + (1/\beta_0 r)^2 + (j/\beta_0 r)^3}{(j/\beta_0 r) + (1/\beta_0 r)^2} \tag{3.26}$$

where η_0 is the free−space impedance, $\beta_0 = \omega\sqrt{\mu_0 \varepsilon_0}$ is the propagation constant in free space, and r is the distance from the dipole to the wave front. Then, the near-field *SE* approximations can be made by using these wave impedances in plane-wave *SE* formulas.

For the commonly used metal and good conductor barriers, material impedance is dominated by the conductivity, $\eta_C \approx \sqrt{j\omega\mu/\sigma}$. Assuming that the barrier thickness is much greater than the skin depth, the following return loss for uniform plane waves is obtained:

$$RL \approx 20\log_{10}\left|\frac{\eta_0}{4\eta_C}\right|. \tag{3.27}$$

The corresponding absorption loss depends then on the skin depth δ through

$$AL \approx 20\log_{10}\left|e^{t/\delta}\right|, \tag{3.28}$$

and the total *SE* is $SE = RL + AL$.

In general, absorption of electromagnetic energy relies on conductivity, dielectric loss, and/or magnetic loss of the absorber material. As follows from the Poynting theorem, the material absorption capability is determined by three time-averaged Joule power density components, as depicted in:

$$A\left[\frac{\mathrm{W}}{\mathrm{m}^3}\right] = \frac{\sigma|E|^2}{2} + \frac{\omega\varepsilon''\varepsilon_0|E|^2}{2} + \frac{\omega\mu''\mu_0|H|^2}{2}. \tag{3.29}$$

3.3 Applications

3.3.1 *Application of thin noise-suppressing sheet materials*

Modern electronic products, such as computer networking hardware, personal computers, laptops, tablets, cellular phones, smart watches, digital cameras, global-positioning systems, multimedia devices, data storage drives, and TV tuners, operate over ultra high frequency (UHF) and super high frequency (SHF) frequency bands. As high-speed digital electronics develops, clock frequencies increase, intensities of noise have a trend of steady growth, while components become downsized. Packing densities of digital and analog electronic devices increase, and it becomes impossible to assure sufficient space between EMI-emitting and other components. Even if noise is controlled within specifications, for example, US Federal Communications Commission (FCC) or Voluntary Control Council for Interference (VCCI) regulations, at the PCB level, secondary interference may cause functional faults when boards are connected through a bus or multiple PCBs mounted together. Moreover, product development times are shrinking: there is no time to redesign circuits. All these factors necessitate finding appropriate scientific and engineering solutions to numerous post-design EMC/EMI problems that arise at both levels—external (inter-system) and internal (intra-system). Thus, it is important to understand physics of

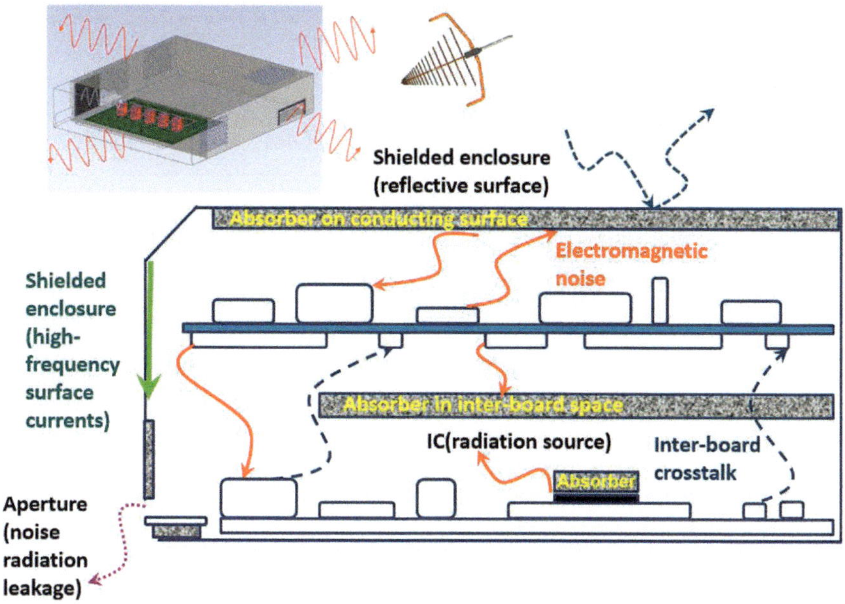

Figure 3.1 Some of the possible sources of unwanted EMI.

EMI coupling paths inside electronic devices, and how to eliminate these coupling paths and emissions. Some of the scenarios, where metal surfaces, slots, openings, and ICs can be sources of unintentional radiation or crosstalk, are shown in Fig. 3.1. Also, various connectors and cables attached to I/O ports on PCBs or on chassis are always vulnerable places where EMI problems arise.

Various RF and microwave electromagnetic wave absorbers (EMWA) are used to reduce unwanted radiation at some distance from the source. These materials are typically used to design walls of anechoic chambers, protecting screens, wallpaper, and coatings with specific filtering properties to protect susceptible devices, components, and circuits from external undesirable radiation. They can also be combined with conducting shielding enclosures and gaskets. EMWA mainly comprise lossy composite materials—both dielectric and magnetic—in single-layer or multilayer configuration. Such materials are the topic of the next section, but herein, we are going to consider absorbing materials for near-field applications.

In many cases, noise can be eliminated just at its source. In electronics industry, majority of hardware engineers know just two "remedies:" ferrites and copper tape. Indeed, various ferrite products are used to prevent internal circuit malfunction or external EME. Chips or chokes made of ferrites or magnetic alloys can be used as lumped noise suppressors placed in the direct proximity of noise sources. However, their effective operation is limited by comparatively low frequencies (up to a 1 GHz at most), where their permeability remains high. According to Snoek's law, the static permeability of a magnetic material is inversely proportional to the cutoff frequency, above which the material becomes inefficient (Snoek, 1948). Hence,

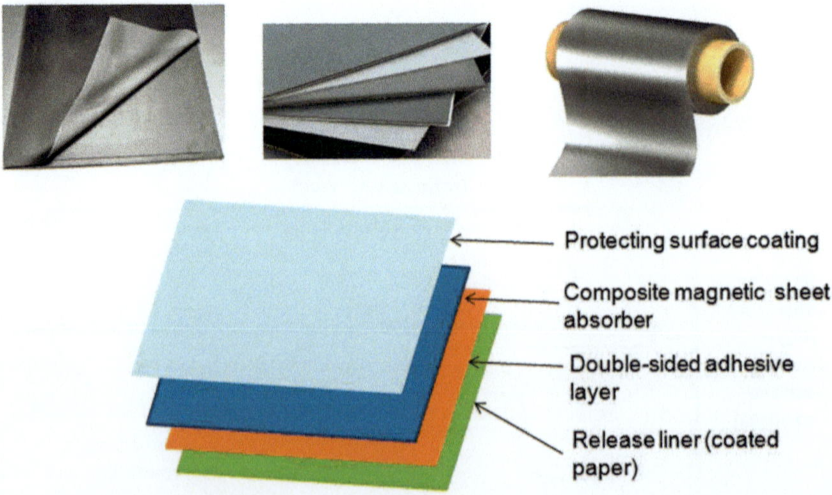

Figure 3.2 Typical structure of NSS materials.

ferrite components are unable to control all problems associated with noise at frequencies over a few gigahertz. Moreover, they cannot handle high-level noise intensities because of their magnetic saturation.

Another class of post-design EMI solutions uses flexible tape-like materials. Sometimes, a conventional copper tape or a conducting gasket may simply solve the problem by shielding slots, seams, or openings when the source is obvious. However, in many cases, the solution is not straightforward, and different shielding materials should be employed, including thin flexible noise suppressor sheet (NSS) absorbing materials (Suzuki et al., 2007; Yamaguchi et al., 2005; Maruta et al., 2006; Yoshida et al., 2001). These are 2D flat or sheet-shaped materials that could be affixed on a PCB or wrapped over its separate components, including ICs. An NSS material is usually multilayered as is shown in Fig. 3.2. The main difference between NSS and EMWA materials is that NSS suppresses noise at its source rather than absorbing noise at a distance. However, they both should be used for targeting existing EMI problems. For near-field applications, return loss from an NSS material is usually irrelevant. Absorption, or attenuation along the length of the absorbing patches, is of importance.

To apply on an enclosure, IC chip, heatsink, or other noise-emitting parts of an electronic device, special shapes of NSS may be needed. They could be die-cut. Water-jet cutting is also used, especially for fragile materials. Also, near-field absorbing materials could be custom-shaped using injection molding or 3D printing. Heat-shrink absorber tubes could be used on cables. Some custom-shaped absorbers are shown in Fig. 3.3.

The main mechanisms of NSS suppression are the inductive decoupling, filtering, common-mode current reduction, diverting fields from coupling paths, reduction of resonances, and attenuating waves traveling through long and wide slots.

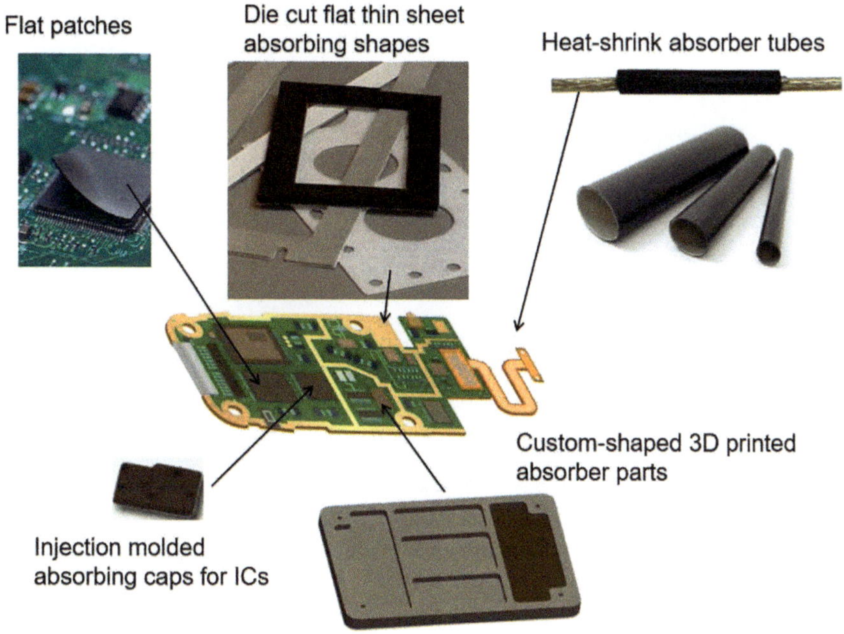

Figure 3.3 Custom-shaped noise-suppressing absorbers.

These effects act only upon noise, while normal signal operation remains intact (or almost intact). Some of these scenarios are illustrated in Fig. 3.4.

At present, there are various NSS materials on the market. Some manufacturers are NEC-Tokin, TDK, and MuRata Corporations of Japan; Intermark together with Kitagawa Industries, 3 M, ARC Technologies, Molex of USA, Laird Technologies with Emerson and Cuming Microwave Products in USA and Europe, Donghyun Electronics is Korea (NEC-Tokin; TDK; MuRata; Kitagawa Industries; ARC Technologies; 3M; Molex; Laird Technologies and Emerson Cuming Microwave Products; Donghyun Electronics). NSS are mainly high-loss magnetic insulating materials. Most of them are polymer-based materials to provide flexibility and better control over frequency characteristics during material manufacturing. An NSS material may contain ferrites as one of the possible ingredients, but not necessarily. To achieve good noise absorption in thin flexible sheets for near fields and surface currents, one needs high permeability, which is often due to conducting ferromagnetic inclusions. Magnetic or ferrite-containing materials have frequency-dependent losses due to domain-wall resonance, ferromagnetic resonance, and skin effect in conducting inclusions (Rozanov and Koledintseva, 2016). Flexible magnets with high permeability (static values $\mu_r > 100$, with record values reaching $\mu_r \rightarrow 200$) over as wide as possible frequency range are desirable for many noise-suppression situations. This can be achieved by high-density filling of magnetic powders into polymers, and by decreasing demagnetizing field of these powders, for example, using ferromagnetic flakes.

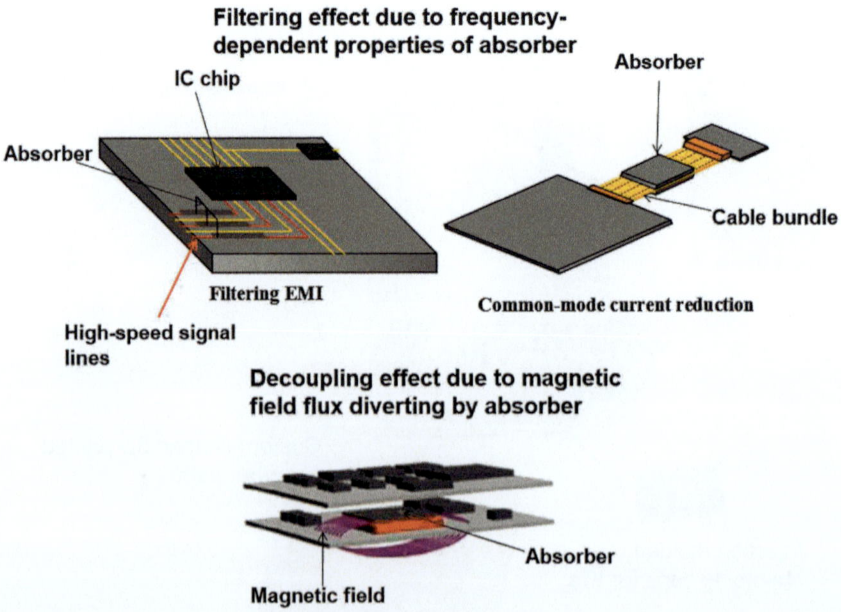

Figure 3.4 Scenarios of NSS absorber application.

Thin sheet flexible magnetic NSS materials are usually used on conducting surfaces to eliminate surface currents, which could be sources of unwanted radiation. Therefore, they can be a convenient and efficient alternative to ferrite chokes on cables. NSS could be placed on internal surfaces of enclosures of electronic devices to reduce Q-factors of resonances, as well as to interrupt noise coupling paths from a source to unintentional antennas. Main requirements to NSS are environmental consideration (halogen-free), thinness, flexibility to form any desired shape, and high performance over a desired frequency range. It is important to underline that practical choice of a type of a material depends on a particular application and specifics of a problem to be solved, and the following factors should be accounted for

- frequency range of interest;
- particular geometries of the sources, impedances, and proximity of conductors;
- particular materials with their frequency responses;
- noise and useful signal intensities; and
- required levels to which noise must be reduced.

3.3.2 Measurements of material parameters of noise-suppressing thin sheet materials

Knowledge of constitutive electromagnetic parameters is of great importance when choosing an appropriate absorbing material for EMC purposes. There are many techniques to measure permittivity and permeability of materials. A choice of a technique depends on the type of the material, frequency range of its application, and available measuring equipment. LCR meters and impedance analyzers are known to measure low-frequency permittivity and permeability (Agilent, 2014b). A coaxial shorted cavity completely filled with a magnetodielectric material (e.g., ferrite) is used to measure its parameters up to a few hundred megahertz (Xu et al., 2010). The high-frequency (over about 1 GHz) measurement techniques can be classified as resonator (cavity), "free-space," and transmission line methods (Baker-Jarvis, 2005). A good review of different electromagnetic material parameter measurement techniques is given in (Chen et al., 2004). Cavity methods are comparatively narrowband, and individual cavities should be designed for measurements at discrete frequency points over a wide frequency range. "Free-space" techniques are based on reflection and transmission of plane waves, incident upon a large, as compared to the longest wavelength, layer of a sample under test. As vector network analyzers (VNAs) for measuring *S*-parameters are widely used, the transmission/reflection measuring techniques have gained popularity (Baker-Jarvis, 1990; Agilent, 2014a), especially in conjunction with the Nicolson–Ross–Weir (NRW) method (Nicolson and Ross, 1970; Weir, 1974) using special coaxial or waveguide test fixtures. The measurement setup for the NRW technique with a coaxial air-filled line and washer-shaped test samples is shown in Fig. 3.5. However, transmission/reflection method has frequency limitations depending on the cross-sectional geometry of the test fixture. The lower cutoff frequency depends on the lowest mode propagating in the transmission line. The upper frequency limit is posed by the higher order modes excited and propagating in the lines, the increase of insertion losses with frequency, difficulties on the transmission line matching with the source and load, and possible volume resonances in the test samples. In the case of very thin sheet ferromagnetic material under test, the method described in Tosaka et al. (2005) can be used to retrieve permeability and conductivity. Extraction of material parameters for ferrites and flexible magnetodielectric materials using optimization techniques, for example, genetic algorithms, is also used (Jing et al., 2015).

3.3.3 Comparison of noise-suppressing thin sheet materials

Engineers encountering EMC/EMI problems in their designs may need to estimate how much of absorbing material to take to achieve sufficient mitigation of EMI noise. Obviously, far field values of *SE* and *RL* are not applicable for NSS materials placed on a surface. Engineers need to know how much electric and magnetic field attenuation, or electromagnetic power is produced due to the certain amount of an absorbing material per-unit-length (in cable applications), per-unit-area (in applications as patches), or even per-unit-volume (in cavity/enclosure applications as blocks).

When designing or choosing an appropriate absorbing material from a number of available candidates, a simple experiment-based technique to evaluate absorbing properties of materials is desirable. This may be an evaluation of attenuation of

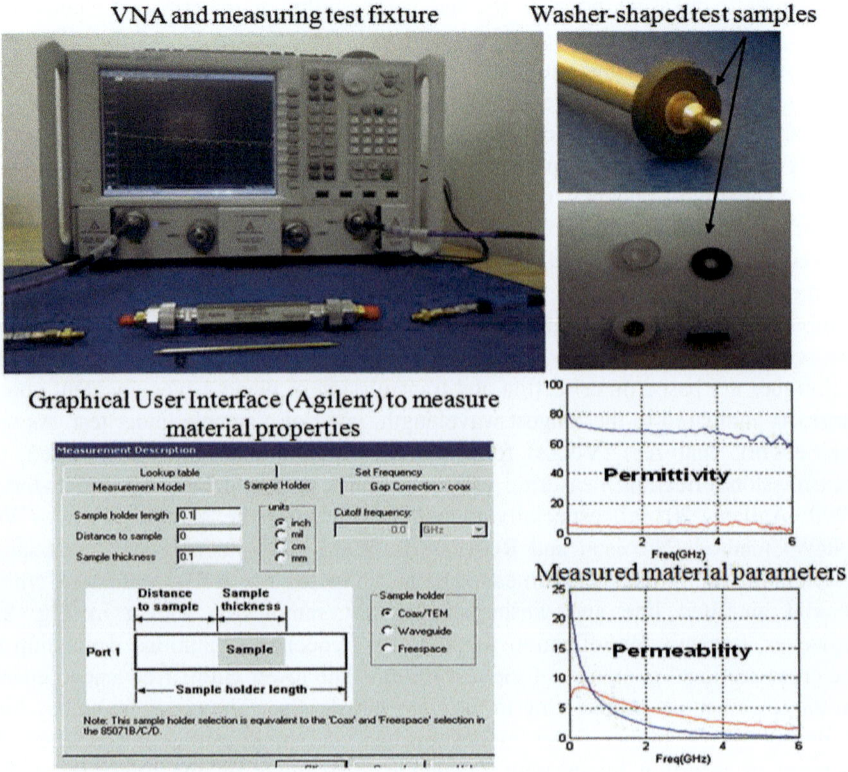

Figure 3.5 Measurement setup using VNA, coaxial air-filled test fixture, and washer-shaped samples.

propagating waves in a transmission line or waveguide structure with a cross-section completely or partially filled with the material under test. If a thin NSS material coats an extended metallic structure (a long metallic wire or a cable), the reduction of the radiation efficiency of this structure due to the suppression of the common-mode currents can be evaluated and correlated with the absorptive properties of this material per geometrical unit.

A simple approach to evaluate and compare absorptive properties of different thin NSS magnetic materials is to tightly wrap the center conductor of an air-filled coaxial line with a thin layer of the test material (its thickness should be $t < \lambda_{min}/2$, where λ_{min} is the minimum wavelength in the material), measure S-parameters on the line, and retrieve the attenuation due to the material (Koledintseva et al., 2012, 2011b). Fig. 3.6 illustrates this approach, and Fig. 3.7 shows the disassembled test fixture and a strip of a thin magnetic material to wrap the center conductor.

The mechanism of the wave attenuation in this technique is the appearance of the additional surface impedance due to the presence of a thin magnetodielectric layer on the conductor (Koledintseva et al., 2011b, 2012). In the other words, this

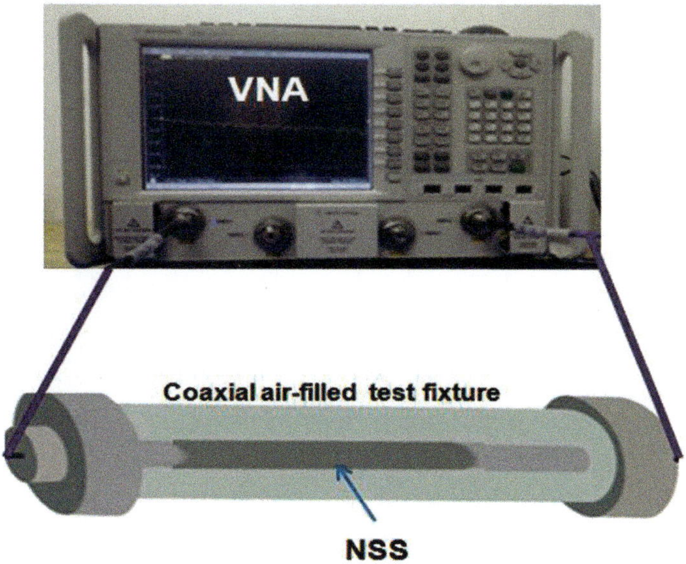

Figure 3.6 Test setup for evaluating absorbing properties of NSS absorbing materials.

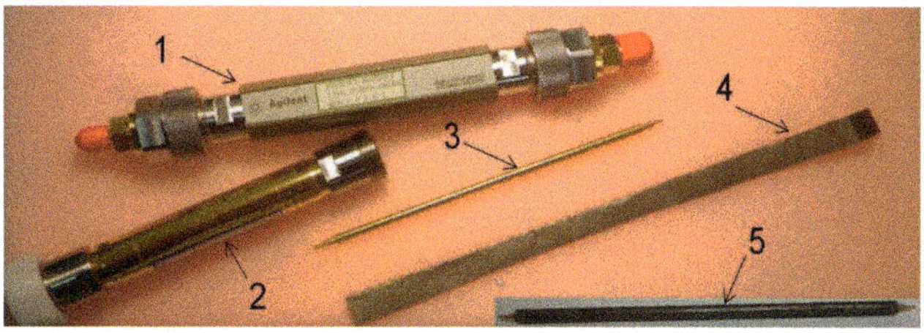

Figure 3.7 Coaxial air-filled test fixtures (1, 2), center conductor (3), a strip of a thin absorbing material to wrap the center conductor (4), and the wrapped center conductor (5).

layer contributes to the degradation of effective conductivity of the conductor, which is analogous to power loss on rough conductor interfaces (Sanderson, 1971).

The measured complex parameters μ_r and ε_r material parameters as functions of frequency for a few NSS materials are presented in Fig. 3.8. Fig. 3.9 shows the calculated attenuation in the coaxial structure with these absorbing materials. The computations were done using a transition from per-unit-length resistance, inductance, conductance, and capacitance (*RLGC*) parameters to *S*-parameters in empty line and with the coated center conductor, as described by Koledintseva et al. (2011b, 2012). This approach allows for comparing absorptive properties of different thin sheet magnetodielectric materials, as well as the materials with different thicknesses or lengths on the conductor. As a comparison, *SE* for far field of 0.3-mm-thick 2D panels made of the same materials as above is presented in Fig. 3.10.

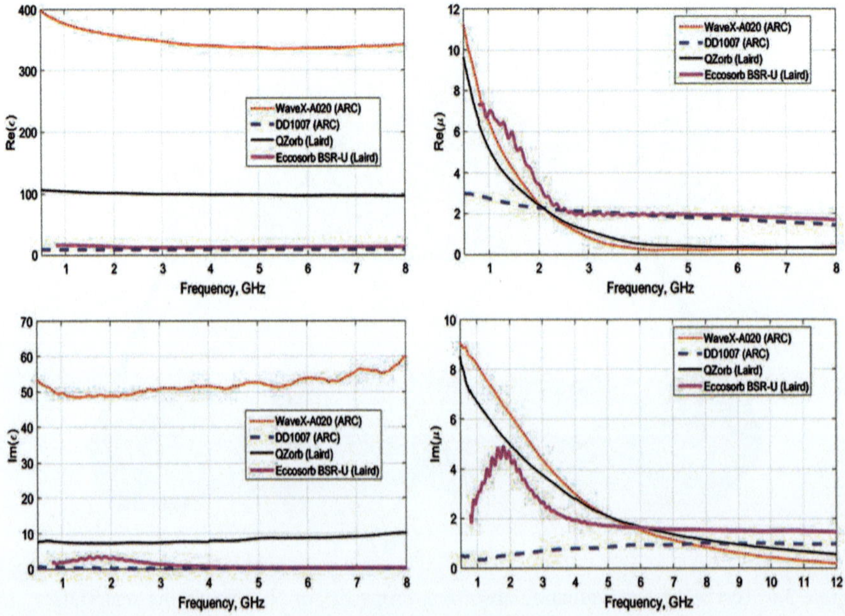

Figure 3.8 Material parameters of three NSS materials measured using the NRW technique in the 7/3-mm coaxial air-filled line with washer samples.

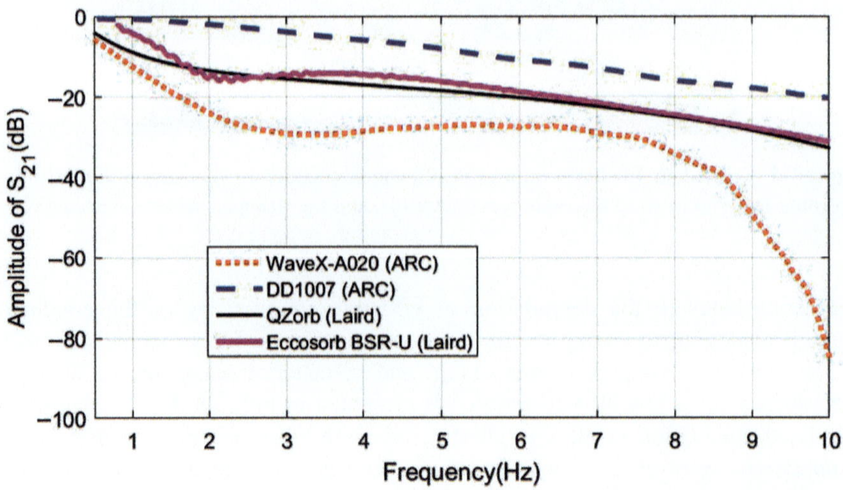

Figure 3.9 Transmittance of the propagating transverse electromagnetic wave (TEM) mode, expressed by the amplitude of S_{21} scattering parameter, in the air-filled 7/3-mm coaxial line with the coated center conductor; length of the coating is $L = 100$ mm; thickness of samples $t = 0.3$ mm.

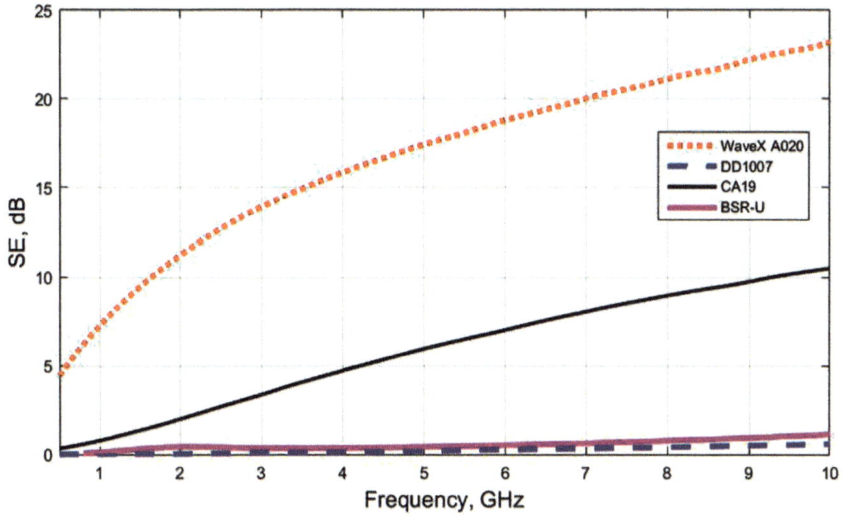

Figure 3.10 Calculated shielding effectiveness for 2D wide slabs of same thin sheet flexible materials as in the previous figure, $t = 0.3$ mm.

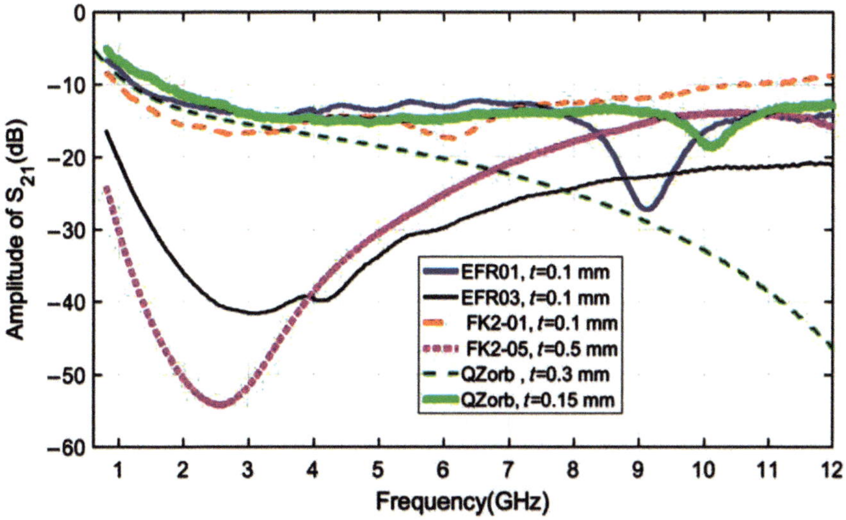

Figure 3.11 Transmittance of the propagating transverse electromagnetic wave (TEM) mode, expressed by the amplitude of S_{21} scattering parameter, in a coaxial line with coated center conductor.

Another example is the comparison of a number of NEC-Tokin materials and material QZorb (Laird). The attenuation on the air-filled coaxial line with wrapped center conductor is shown in Fig. 3.11. Thicknesses (t) of samples are different, and the coated length is the same, $L = 10$ cm.

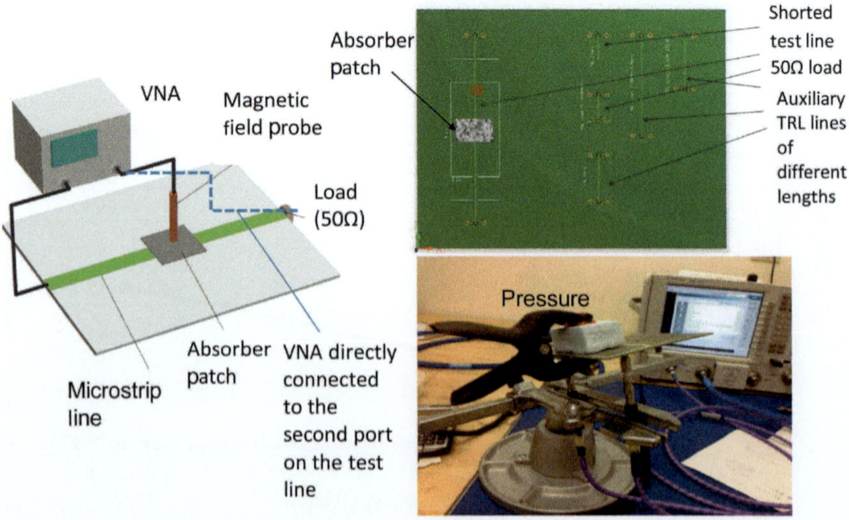

Figure 3.12 Microstrip line test fixture for evaluating absorptive properties of NSS materials.

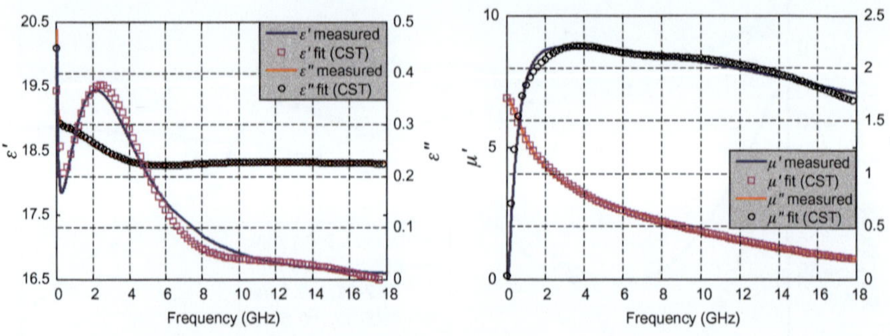

Figure 3.13 Material parameters of the tested material sample—measured and multiterm Debye fit (10 terms) in CST Studio (CST STUDIO SUITE®).

An alternative technique to evaluate absorptive properties of NSS materials uses a microstrip line. An NSS sample should be pressed tightly to the microstrip of the text fixture, as is shown in Fig. 3.12. One VNA port is connected to the microstrip test line, and the second port can be either attached to the magnetic field probe placed above the NSS patch, or directly connected to the output of the test line. The test PCB has auxiliary "Thru-reflect-line" pattern to calibrate connectors out (Pozar, 1998, Chapter 3). The material parameters of the absorber (Eccosorb BSR, Laird-Emerson&Cuming) are shown in Fig. 3.13. They were

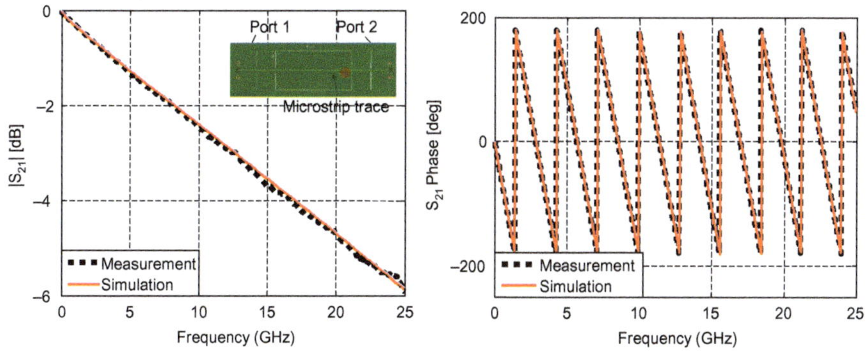

Figure 3.14 Measured and modeled S_{21} parameter on a bare microstrip structure.

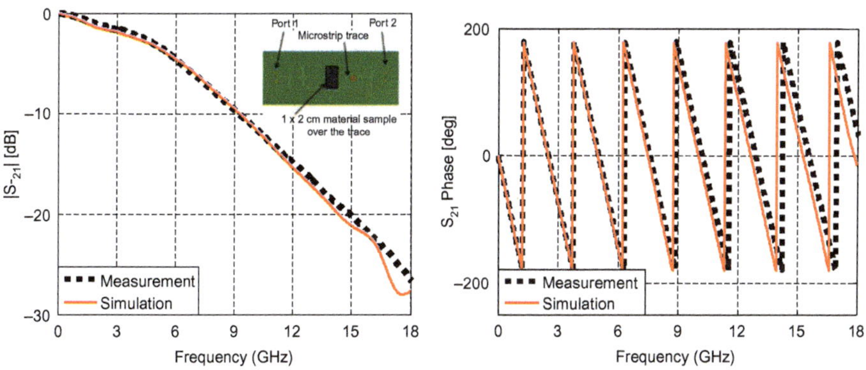

Figure 3.15 Measured and modeled S_{21} parameter on a microstrip structure with absorbing material.

measured using the NWR technique. Figs. 3.14 and 3.15 show measured and numerically modeled S_{21}-parameters of the bare microstrip line and with NSS patch (1 cm × 2 cm × 0.5 mm). Modeling was done in CST Studio (CST STUDIO SUITE®) using time-domain solver. This is a convenient way of comparing absorptive properties of NSS materials.

Numerical electromagnetic simulations could be useful for predicting how a specific absorber can reduce undesirable emissions. Accurate reproduction of material frequency characteristics is especially important for time-domain techniques, which are more efficient for wideband simulations that their frequency-domain counterparts (Radchenko et al., 2013). Some examples of modeled and measured EMI emission reduction due to NSS materials are given in Laird Technologies and Emerson Cuming Microwave Products, Koledintseva et al. (2011a), Radchenko et al. (2013), and Li et al. (2012a).

3.4 Specular absorber for far field operation

3.4.1 Introductory part

At present, microwave properties of various composites are under intensive studies. Most published reports refer to radar absorbers as possible applications, see Idris et al. (2016) for a recent review of the advances in the field. The conventional approach is to measure the effective relative permittivity, $\varepsilon_r = \varepsilon_r' - j\varepsilon_r''$, and permeability, $\mu_r = \mu_r' - j\mu_r''$, of a composite with certain concentrations of inclusions; calculate the frequency dependence of reflection coefficient of the composite slab backed by a metal substrate for the normal incidence of electromagnetic wave; and choose the slab thickness to minimize the reflection coefficient at a frequency of interest. The obtained deep minimum, typically of −30, −40 dB, or even −40 dB is considered as a proof for "a good absorbing ability" of the inclusions comprising the composite. Some recent examples of the approach are given in Choi et al. (2015), Gargama et al. (2016), and Wen et al. (2015).

The reflection coefficient R for an inhomogeneous slab of thickness t placed on a metal substrate is calculated as

$$R = \frac{j\sqrt{\mu_r/\varepsilon_r}\tan\left(2\pi t\sqrt{\varepsilon_r\mu_r}/\lambda\right) - 1}{j\sqrt{\mu_r/\varepsilon_r}\tan\left(2\pi t\sqrt{\varepsilon_r\mu_r}/\lambda\right) + 1}, \tag{3.30}$$

where λ is the free space wavelength. In Eq. (3.30), R is an amplitude ratio of the reflected and incident waves. As is mentioned in Section 3.2, return loss (RL) may be used instead of the reflection coefficient that is the squared module of R expressed in dB and having an opposite sign.

To facilitate the analysis of wavelength dependence of R defined by Eq. (3.30), representation of the dependence as a rational function that excludes the transcendental function of tangent from the equation is used. This may be done for a narrow-band case by expanding the tangent in series near the operating frequency. Following (Pottel, 1959), two cases may be considered, by expanding the tangent in the vicinity of either zero, or $\pi/2$.

In the first case, the radar absorber is referred to as a magnetic screen, and the reflection coefficient is

$$R = \frac{2\pi j\mu_r t/\lambda - 1}{2\pi j\mu_r t/\lambda + 1} \tag{3.31}$$

Permeability of materials is known to decrease rapidly with frequency over microwave band (Rozanov and Koledintseva, 2016). For this reason, efficient absorbers with the reflection coefficient governed by Eq. (3.31) are rare. Examples of these are sintered ferrites operating at megahertz frequencies (Naito and Suetake, 1971) and absorbers based on artificial microwave magnets (Lagarkov et al., 2003).

Most microwave radar absorbers comprising composite materials fall into the second type of absorbers, including quarter-wavelength ($\lambda/4$) absorbers.

For $\lambda/4$ absorbers, tangent in Eq. (3.30) with the argument close to $\pi/2$ may be represented as

$$\tan(x) \approx \frac{2x}{(\pi/2 - x)(\pi/2 + x)}. \tag{3.32}$$

Eq. (3.32) accounts for the quarter-wavelength reflection minimum at $x = \pi/2$, as well as for the null of the tangent at $x = 0$ and the symmetrical pole at $x = -\pi/2$. The inclusion of the additional null and pole allows for retaining odd/even properties of the reflection coefficient as a function of λ, which is of importance for estimating ultimate bandwidth of radar absorbers, as is explained below. The accuracy of this representation is reduced as compared to the case of retaining only $\pi/2$ pole because of the asymmetry of the retained singularities, but this is paid back by the simplicity of the resulting equation,

$$R = -\frac{16t^2\varepsilon_r\mu_r\pi + 16jt\mu_r\lambda - \pi\lambda^2}{16t^2\varepsilon_r\mu_r\pi - 16jt\mu_r\lambda - \pi\lambda^2}, \tag{3.33}$$

from which real and imaginary parts of the material parameters can be separated readily. Note that Eq. (3.33) does not describe magnetic screens. This is because for $x = \pi/2$, the agreement with the initial Eq. (3.30) is achieved for the numerator of Eq. (3.32) having the numerical factor of "2," which is a residue responsible for all the neglected nulls and poles. For $x = 0$, the residue is different.

The conditions, at which $R = 0$, follows from Eq. (3.33) as

$$\left(\frac{4t}{\lambda}\right)^2 (\varepsilon_r'\mu_r' - \varepsilon_r''\mu_r'') - \frac{16t\mu_r''}{\pi\lambda} = 1 \tag{3.34}$$

and

$$\frac{\pi t}{\lambda\mu_r'}(\varepsilon_r'\mu_r'' + \varepsilon_r''\mu_r') = 1 \tag{3.35}$$

In composites, where frequency dispersion of material parameters is typically negligible, Eq. (3.34) defines the wavelength, where the magnitude of the reflection coefficient has a minimum. Eq. (3.35) yields the condition when the reflection is zero in this minimum. Therefore, engineering of a radar absorber is a problem of matching rather than obtaining "good absorbing ability". To produce good absorption at a given frequency, real and imaginary parts of complex material parameters must have certain values. The more accurately Eqs. (3.34) and (3.35) are held, the lower value of the reflection coefficient is. Material loss may be either dielectric, or magnetic, or both. It is clearly seen from Eq. (3.35) that in composites filled with ferromagnetic particles, magnetic loss dominates, because the real permittivity is usually much larger as compared to the real permeability.

The last term in the left part of Eq. (3.34) is negligibly small near the frequency of the quarter-wavelength reflection minimum if the minimum is deep enough. If the reflection coefficient is not zero in its minimum, the location of the minimum is still given by Eq. (3.34), and the depth of the minimum $R_{\min}$ is approximately

$$R_{\min} = \left| \frac{(\pi t/\lambda \mu'_r)(\varepsilon'_r \mu''_r + \varepsilon''_r \mu'_r) - 1}{(\pi t/\lambda \mu'_r)(\varepsilon'_r \mu''_r + \varepsilon''_r \mu'_r) + 1} \right| \tag{3.36}$$

As follows from Eq. (3.36), the variation of $R_{\min}$ with slab thickness is an indication of the frequency dispersion of material parameters. Otherwise, the dependence on t in the right-hand part of Eq. (3.36) is completely equalized by the wavelength, where the $\lambda/4$ minimum appears, and which, according to Eq. (3.34), it is also proportional to t. Moreover, for any given $R_{\min}$, except for $R_{\min} = 0$, the two different sets of parameters can be found by turning $(\pi t/\lambda \mu'_r)(\varepsilon'_r \mu''_r + \varepsilon''_r \mu')$ to be either larger or smaller than unity.

Variation of effective material parameters over wide ranges is possible by varying inclusion concentrations, modifying inclusion form factors or morphology, for example, milling and deposition of a shell onto inclusion particles. The use of several types of inclusions in an absorbing composite provides more degrees of freedom in adjusting the material parameters. This may be employed for obtaining a good match of a composite radar absorber over a required frequency band. A classic example is one of the first absorbers patented in Holland in 1936 (Emerson, 1973). It was $\lambda/4$ composite slab operating at 2 GHz and containing carbon black inclusions to provide energy absorption and titanium oxide powder to adjust the operating frequency.

Therefore, the problem of obtaining deep reflection minimum at a given frequency is well studied. However, this problem is frequently of low importance. The priority is effectiveness of an absorber over a frequency range—the wider, the better. Another concern is that radar absorbers are of finite size, and edge diffraction is a limiting factor for obtaining low reflection coefficient (Perini and Cohen, 1993), see also Mosallaei and Rahmat-Samii (2000) for numerical results. Finally, because of unavoidable technological tolerances in material properties, the reflection coefficient may exceed the expected value. Because of these factors, characterization of quality of radar absorbers requires other criteria, some of which are discussed below.

3.4.2 The bandwidth-to-thickness ratio

Limitation on the thickness-to-bandwidth ratio of radar absorbers has been suggested in Rozanov (2000). It follows from the sum rule for the Kramers – Krönig relations governing the wavelength dependence of the complex reflection coefficient $R(\lambda)$ for normal incidence of a monochromatic electromagnetic wave onto a multilayer structure on a perfectly conducting substrate,

$$\left|\int_0^\infty \ln R(\lambda)d\lambda\right| \leq -2\pi^2 t\langle\mu_{r,s}\rangle, \tag{3.37}$$

where $\langle\mu_{r,s}\rangle$ is the static permeability value averaged over all the layers involved in the structure, and t is its total thickness. In contrast to the well-known Kramers − Krönig relations for material parameters, Eq. (3.37) includes the reflection coefficient as a function of the wavelength rather than of frequency, and Eq. (3.37) is an inequality rather than equality. The appearance of the inequality sigh in Eq. (3.37) is because the reflection coefficient may have nulls everywhere in the plane of complex wavelength, and ln R may therefore have poles in the upper semiplane of complex λ [provided that the monochromatic factor is accepted as exp $(-j\omega t)$], in contrast to the material parameters. The difference between the left- and right-hand parts of Eq. (3.37) is equal to the sum of all nulls of the reflection coefficient located in the upper semiplane of complex wavelength. If such nulls are not present, $R(\lambda)$ is referred to as a minimum phase-shift frequency dependence, and then equality in Eq. (3.37) is held.

It follows from Eq. (3.37) that, if a radar absorber has the reflection coefficient lower than R_0 within the wavelength range of λ_{min} to λ_{max}, then

$$\frac{\Delta\lambda}{t} < \frac{2\pi^2\langle\mu_{r,s}\rangle}{|\ln R_0|}, \tag{3.38}$$

where $\Delta\lambda = \lambda_{max} - \lambda_{min}$. As follows from Eq. (3.38), for $R_0 = 0.316$ (−10 dB), the bandwidth-to-thickness ratio of a radar absorber $\Delta\lambda/t$ cannot be larger than $17.2\langle\mu_{r,s}\rangle$. Notice that Eq. (3.38) contains the static permeability if the frequency dispersion of permeability is negligible over the waveband of interest and, therefore, most loss is concentrated within the operating band. If low-frequency loss is significant, the permeability value at the lowest frequency of the operating band should be used (Rozanov, 2000).

In contrast to Eq. (3.37), inequality Eq. (3.38) never turns to equality. For equating left- and right-hand parts of Eq. (3.38), the reflection coefficient must be equal to R_0 within the operating band and be equal to unity outside of it. This is impossible, because the reflection coefficient is a smooth function of the wavelength. However, the performance of practical absorbers may be sufficiently close to the limit defined by Eq. (3.38).

For a single-layer absorber with material parameters varying slowly with frequency and the reflection coefficient represented by Eq. (3.33), the bandwidth-to-thickness ratio is given by

$$\frac{\Delta\lambda}{t} < \frac{32}{\pi}\frac{R_0}{1-R_0^2}\mu_{r,s}, \tag{3.39}$$

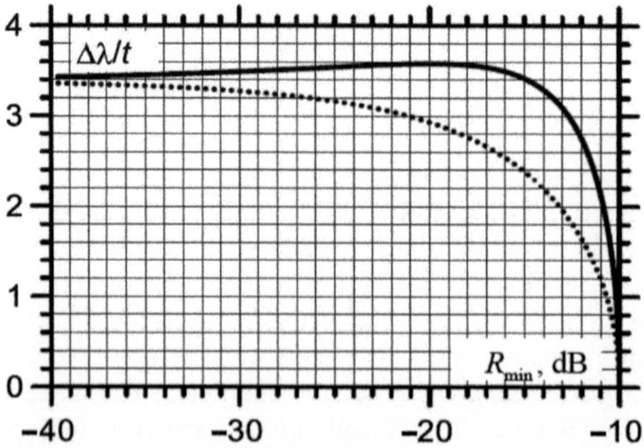

Figure 3.16 The dependence of bandwidth-to-thickness ratio on the value of $R_{\min}$ for a radar absorber with a single reflection minimum within the operating waveband for $R_0 = -10$ dB. Solid line is the case of a minimum phase-shift frequency dependence, dotted line is the case of the lower loss value of these defined by Eq. (3.36).

instead of Eq. (3.38). The values of $\Delta\lambda/t$ satisfying Eq. (3.39) are much lower than those for Eq. (3.38). For example, for $R_0 = 10$ dB, the coefficient of $\mu_{r,S}$ in the right part of Eq. (3.39) is 3.58, and the difference between these two limits is by a factor of 4.8. For lower values of R_0, the difference increases.

The equality in Eq. (3.39) is attained, if loss is the largest of the two values given by Eq. (3.36) (therefore, equality in Eq. (3.37) is also held) and $R_{\min} = R_0^2$. In other words, the least value of the reflection coefficient (in dB) must be twice as large as the reflection level kept over the operating band. However, the bandwidth-to-thickness ratio dependence on the depth of the reflection minimum is fairly smooth, at least as soon as total loss falls within two values corresponding to $R_{\min} = R_0^2$, see Fig. 3.16.

Operating waveband of a radar absorber can be extended as compared to Eq. (3.39) by employing frequency dispersion of material parameters, as well as using multilayer structures. For purely dielectric absorbers, a numerical study of these approaches has been made in Rozanov and Starostenko (1999), with the frequency dispersion of the permittivity assumed to be governed by the Lorentzian law. The optimal value of $\Delta\lambda/t$ appears to depend greatly on the reflectance level R_0 and the number of the minimums of the reflectivity n located within the operating waveband. The dependences on $\lambda_{\min}$, $\lambda_{\max}$, and $\Delta\lambda$ are weak, provided that no restrictions are imposed on the parameters of the Lorentzian curves. For a single-layer absorber, the dependence of the bandwidth-to-thickness ratio on the parameters of the problem is fitted closely by

$$\frac{\Delta\lambda}{t} = \frac{32}{\pi}\frac{3}{n+1}\frac{R_0^{1/n}}{1 - R_0^{2/n}}, \tag{3.40}$$

The number of minima n allows for all resonances of the reflection coefficient, independently of whether the resonance is related to the multiple reflections within the slab or to the intrinsic resonance of the permittivity of layers. Therefore, for a single-layer absorber, $n = k + 1$, where k is the number of the Lorentzian terms in the dispersion law of the material, and the unity is responsible for the $\lambda/4$ resonance. Eq. (3.40) is valid at $n \geq 2$; at $n = 1$, the numerical factor in the right part is different, see Eq. (3.39). For multilayer structures, the value of n may be increased due to $\lambda/4$ resonances on the layers of the structure, and the numerical factor in Eq. (3.40) is 4π instead of $32/\pi$, that is $\pi^2/8 \approx 1.23$ times greater than for a single-layer absorber. Eq. (3.40) allows for the bandwidth-to-thickness ratio increase to 8.4 at $n = 2$ and to 9.7 at $n = 3$.

Similar results are obtained in Rozanov and Starostenko (2003) for radar absorbers having frequency dispersion of permeability and, hence, magnetic loss. Note that frequency dispersion of material parameters is typically not very pronounced in composites filled with metal powders. The frequency dispersion of permittivity is observed only at the concentrations of inclusions close to the percolation threshold, and is usually due to imperfect electrical contacts comprising conducting clusters (Liu et al., 2007). Therefore, frequency dispersion in metal powders can hardly be employed for sufficient waveband broadening. Even though the permeability is frequency-dispersive, typically the permeability contrast is insufficient to obtain several distinct minima of the reflection coefficient within an operating waveband, especially if this waveband is comparatively narrow.

Multilayer composite radar absorbers have been suggested (Wang et al., 2012; Kim, 2011), but the gain in the thickness-to-bandwidth ratio appears to be lower than it may be predicted based on Eq. (3.40). The reason is that the broadband multilayer design implies a large difference in material parameters in neighboring layers. Therefore, a part of layers must contain lower concentration of the fillers, so the average permeability of the layer decreases, which, in turn, results in the decrease of $\Delta\lambda/t$ ratio. In fact, most multilayer designs employ separation of the absorbing slab into several layers with different parameters just to facilitate adjusting material parameters to obtain effective performance at the expense of broadbandness.

3.4.3 The fractional bandwidth

Effective and broadband performance of radar absorbers in many practical cases is of much more importance than its small thickness. Then the quality of the radar absorber may be characterized by the fractional bandwidth,

$$F = 2\frac{\lambda_{max} - \lambda_{min}}{\lambda_{max} + \lambda_{min}} = \frac{\Delta\lambda}{\lambda_0}, \tag{3.41}$$

where λ_0 is the middle of the operating waveband.

For radar absorbers employing composites with ferromagnetic powders, the frequency dispersion of material parameters is typically weak. In this case, the bandwidth is given by Eq. (3.38), and, for $\lambda/4$ absorber, the wavelength is $\lambda_0 = 4t\sqrt{\varepsilon_r\mu_r}$. Then, under the assumption of $R_0 \ll 1$, Eq. (3.38) yields

$$F = \frac{8}{\pi} R_0 \sqrt{\frac{\mu_r}{\varepsilon_r}}. \tag{3.42}$$

In composites with magnetic inclusions, the permeability and permittivity are governed by the same mixing rule. Therefore, if both these values are much larger than unity, then the ε_r/μ_r ratio is larger or equal to the relative permittivity of the host matrix of the composite. Standard polymer matrices have the permittivity of about 2, on the account of which the fractional bandwidth given by Eq. (3.42) is lower than 60%. In practice, this ratio is typically larger in composites with large filling concentrations, and the fractional bandwidth is lower.

Therefore, composites containing large volume fractions of tiny conducting particles do not allow for obtaining large bandwidth because of large permittivity. With decrease of the volume fraction, both the permittivity and permeability tend to unity, and the relative bandwidth increases. In this case, estimation Eq. (3.42) is no longer valid because the reflection in the $\lambda/2$ antiresonance is low, and the low-reflection band may incorporate not only $\pi/4$, but also $3\pi/4$ and higher resonances. In this case, $\lambda_{min} \to 0$ and $F \to 200\%$. Such absorbers comprising a porous matrix and a low concentration of conducting powders, typically carbon black, are also widely used, in particular for microwave anechoic chambers (Emerson, 1973). Though the permittivity is close to unity because of low concentration of inclusions, the bandwidth-to-thickness ratio as defined in the previous section is low. If the desired reflection level is not too low, for example, -10 dB, then composites filled with carbon fibers are good alternatives (Neo and Varadan, 2004).

Another perspective opportunity to obtain large fractional bandwidth is metamaterial absorbers. Several types of absorbers of this type are known in the literature: absorbers based on frequency-selective surfaces (FSS) (Kazantsev et al., 2010), high-impedance surfaces (Costa and Monorchio, 2012), artificial magnets (Lagarkov et al., 2003), circuit-analog absorbers (Choi et al., 2014), and others. All these structures are composed of metal inclusions of resonance size. Strictly speaking, most of these cannot be adequately described in terms of effective material parameters, because these parameters appear to be dependent on the thickness, as well as on electromagnetic environment of the material, when, for example, the response of a slab may be different when it is placed at a metal backing or in a free space (Vinogradov et al., 1999). Nevertheless, effective parameters for metamaterials are commonly used: metamaterials are defined as materials possessing negative values of the permittivity and permeability over some frequency bands. Due to the resonance scattering, metamaterials exhibit strong frequency dispersion of material parameters at microwaves. This, in contrast to conventional composites, allows for incorporating two or even three resonances into the operating waveband, and therefore the bandwidth-to-thickness ratio may be governed by Eq. (3.40). The value of the $\Delta\lambda/t$ ratio is not very high, because the static permeability of the metamaterials is unity, but the fractional bandwidth may be rather high, of about 80% to 120%, see Li et al. (2012b) for a review of fractional bandwidth of some recent designs. Radar absorbers incorporating FSS placed on magnetic composite substrates has also been suggested (Huang and Li, 2012; Zhang et al., 2013a), which results in both broadening the operating bandwidth according to Eq. (3.40) and increasing the bandwidth-to-thickness ratio due to nonunity effective permeability of the material.

3.4.4 The angular performance

One of important requirements to radar absorbers is good angular performance. It is impossible to obtain good absorbing efficiency at all incident angles φ, because of $R \to 0$ at $\varphi \to \pi/2$. Even for the case of $\varepsilon_r = \mu_r$ with sufficiently high losses, which is conventionally considered as a perfect radar absorber, the reflection coefficient equals to zero only for the normal incidence (Ruck et al., 1970), while it changes with φ as

$$R(\phi) = \frac{1 - \cos\phi}{1 + \cos\phi} \tag{3.43}$$

For analytical studies of the radar absorbing performance at oblique incidence, Eq. (3.37) is generalized as

$$\begin{aligned} &\left|\int_0^\infty \ln|R(\lambda)|d\lambda\right| \le 2\pi^2 \cos\phi \langle\mu_{r,s}\rangle t \\ &\left|\int_0^\infty \ln|R(\lambda)|d\lambda\right| \le \frac{2\pi^2}{\cos\phi}\langle\mu_{r,s}\rangle t, \end{aligned} \tag{3.44}$$

for transverse electric (TE) and transverse magnetic (TM) polarizations, respectively.

The second of Eq. (3.44) is not very helpful for obtaining estimates, because it predicts an unrestricted increase of the integral with the incident angle tending to $\pi/2$, though, as is well-known, $R = 1$, and therefore, the left part of equation must be zero at $\varphi = \pi/2$. It is readily obtained that when the incident angle approaches oblique incidence, the nulls of the reflection coefficient for the TM polarization move from the lower to the upper semiplane of the complex wavelength, which increase the difference between the left and right parts of the second of Eq. (3.44). For the first of Eq. (3.44) responsible for the performance at the TE polarization, equality is kept for all incident angles with a minimum phase-shift frequency dependence.

Alternatively, the analytical approach to the angular absorbing performance may be obtained based on the fractional representations of the reflection coefficient, analogous to Eqs. (3.31) and (3.33). However, systematic studies of this kind are not found in the literature. Available papers mostly provide qualitative analysis for obtaining slabs less sensitive to the incident angle, discussing, for example, the use of materials with high values of material parameters or employing of arrays of electrically small, resonant, and symmetric particles (Tretyakov, 2016). Note that the first of the concepts, based on the fact that the wave inside the layer propagates in the direction close to the normal to the interface even at oblique incidence angle, yields the input impedance of the layer which is nearly independent on the incident angle. It has been experimentally shown in Zhang et al. (2015b) that radar absorbers made of composites with ferromagnetic powders have slight angular dependence of the input impedance. However, the angular dependence of the reflection still retains due to the angular dependence of the characteristic impedance of the free space, which is also included in the calculation of the reflection coefficient.

For these reasons, specific results on the radar absorbers with good angular performance are typically obtained with the use of numerical optimization techniques, for example Yang et al. (2016), Micheli et al. (2011), and Zabri et al. (2014).

3.4.5 Conclusion

As is seen from the above consideration, the depth of reflection minimum is not a suitable measure for the characterization of radar absorbers. To describe the quality of an absorber, characterization of its bandwidth properties is required. The width of the operating frequency range cannot serve for bandwidth quantification. Indeed, if the bandwidth is 2 GHz, the operating range may be 0.1 to 2.1 GHz or 18 to 20 GHz, which is a huge difference from the standpoint of designing a radar absorber. A correct approach to define quality of absorbers should use one of the quantities. The first is the frequency bandwidth divided by the center frequency, that is, fractional bandwidth, in case when the absorber thickness is of not prime priority. The second is the frequency bandwidth divided by the squared center frequency, that is, the wavelength operating range, if requirement of thin absorber is of importance. As for angular performance quantification, no numerical criterion has been suggested in the literature so far.

3.5 Material's design

The preceding analysis leaves some room for a material's design strategy based on the combination of different types of filler materials to improve the performance of polymer-based microwave absorbers. This strategy aims at the optimization of EM and absorbing characteristics on the ground of morphological, interfacial, and chemical phenomena in the composite materials. Herein, the EM characteristics refer to the effective material properties of the composites, that is, their permeability and permittivity. In addition, enhanced absorber parameters can be attained by combining inclusion particles with intrinsically conductive, dielectric, and magnetic nature. The large amount of published research on the multiphase composites for microwave absorption applications reveals some similarities, despite the occasionally intuitive materials selection. Therefore, it is important to identify and record any present synergistic effects in the hybrid composites and extract the objectives for the design of effective microwave absorbers. Synergy in this case consists in the occurrence of improved return loss or absorption loss, which in certain occasions are higher than the sum of the individual performance of the components. Of course, there is an underlying physical mechanism, as those expressed in Eqs. (3.39), (3.34), and (3.35), which accounts for such an improvement. In the current review study, we have classified different composites into two major categories, with regard to the targeted applications, namely, the reflected and transmitted EM wave absorbers.

In order to have a solid reference level for the evaluation of hybrid composites, we focus on the single-layer configuration of planar absorbers realized by physical materials, excluding from our study the multilayer structures and artificial materials.

3.5.1 Reflected wave absorbers

Unintentional reflections of EM waves by metallic objects or structures in the far field has been identified in Section 3.1 as an impediment to the operation of telecommunication, identification, surveillance, and measurement systems functioning at the microwave frequency range. Therefore, materials with appropriate magnetic and dielectric properties are employed as microwave absorbing coatings to mitigate such implications.

In Section 3.4, we have deployed our approach for a more suitable characterization of metal-backed microwave absorbers, by using the bandwidth-to-thickness ratio or the fractional bandwidth. Since these figures-of-merit are not common in the literature so far, the review and evaluation of the published research will be based on the available bandwidth values, usually defined as $BW = f_{\max} - f_{\min}$, for a specified return loss (RL) level. Still, we need to mention that this measure cannot provide the reader with complete information on the absorbing quality.

We also wish to clearly state that we don't necessarily adopt the theoretical analyses or scientific explanations included in the cited articles, since this is a different issue from the presentation of measurement data.

3.5.1.1 Nonmagnetic filler materials

Among the nonmagnetic materials, the most significant grade of the microwave absorbing materials comprises polymer composites with carbon-based fillers. The advantage of such materials stems from the tailored dielectric properties, which originate in interfacial and dipolar polarization, rather than in macroscopic conductivity itself, as it has been commonly misinterpreted. In fact, the inherent conductive nature of these fillers sets an upper concentration limit for their use in low reflectivity applications, which occurs when a continuous conductive network is formed (the percolation threshold).

A typical member of this filler type is carbon black (CB), which appears as conductive C particles usually with size in the micro- or nanoscale. Different semiconductive or insulating compounds (SiC, $CaCO_3$, ZnO, nanoclay) may be added to CB-filled composites to obtain improved absorption characteristics in the

2–18 GHz range (Liu et al., 2011; Ling et al., 2011; Qin et al., 2013; Stergiou et al., 2015). In most of these cases, we notice the existence of an optimal composition with respect to the measured return loss. Liu et al. (2011) reported that *RL* of a composite with 5 wt% CB nanoparticles is increased by the addition of SiC nanoparticles, reaching the largest bandwidth of 6 GHz at the level of 10 dB (BW_{10dB}) for 50 wt% SiC. The positive impact of SiC is attributed to the improvement of CB dispersion in the polymer, which also reflects on the recorded lower percolation content. We estimate that for higher SiC or CB amount, permittivity is low or conductivity is high enough to attain impedance matching, respectively. Similarly, Ling et al. (2011) studied the CB/$CaCO_3$ bifiller system in resin medium and found the highest and broadest *RL* peaks ($RL_{max} = 25$ dB, $BW_{10dB} = 3$ GHz) for the composite filled with CB/$CaCO_3$ powder of 7:100 weight ratio. Composites loaded only with $CaCO_3$ microparticles show trivial *RL*, whereas composites with CB loading above the optimal exhibit inferior EM and mechanical properties, indicating possible agglomeration of the nanoparticles. Qin et al. (2013) reported that epoxy nanocomposite with 7 wt% CB and 10 wt% ZnO possess a stronger and broader *RL* peak ($RL_{max} = 19$ dB, $BW_{10dB} = 6$ GHz) compared to the composite with 15 wt% CB, both for $t = 3$ mm ($RL_{max} = 12$ dB, $BW_{10dB} = 2$ GHz). Therein, the tetrapod-like shape of the ZnO particles results in the increased porosity of the composite, which in combination with its lower ε_r values finally yields better impedance matching. The pursuit of low ε_r is not a global objective for the achievement of high *R*. In epoxy nanocomposites with low CB content (2 wt%) the addition of nanoclay up to 5 wt% reduces the effective ε_r and degrades the absorption performance (Stergiou et al., 2015). This occurs on account of the haloing effect, as CB nanoparticles are attracted to the surface of clay nanoplatelets and form nanoclusters.

A member of the carbonaceous materials group with enormous technical interest and wide utilization in absorption applications is the carbon nanotubes (CNT). Their fibrous shape with extremely large aspect ratio allows the fabrication of composite materials with improved mechanical properties, high conductivity and permittivity even at low concentrations. The suppression ability of EM reflections in the 2–18 GHz of CNT-loaded nanocomposites has been investigated in combination with other conductive, semiconductive or insulating compounds in the form of micro- and nanoparticles, nanowires or coating (Stergiou et al., 2015; Ting et al., 2013; Qing et al., 2014; Yao et al., 2009; Song et al., 2012; Zhu et al., 2011). For low concentrations of multiwalled CNT (MWCNT) up to 1 wt% the permittivity level is too low to foster high *RL* values from thin material layers. Thus, it was found that by introducing an optimal concentration of organomodified nanoclay (5 wt%) or MnO_2 (20 wt%) both dipolar polarization and ε_r increase sufficiently. Therefore, impedance matching is enhanced to achieve bandwidth $(BW)_{10dB} = 2.5$ GHz and $BW_{10dB} = 1.5$ GHz for layer thickness $t = 2$ mm, respectively (Stergiou et al., 2015; Ting et al., 2013). Qing et al. (2014) also found that in the MWCNT/Al_2O_3 bifiller system in epoxy matrix, the composite with 0.2 vol.% MWCNT and 40 vol.% Al_2O_3 possesses improved microwave absorption properties, with $RL > 10$ dB covering the whole 12–18 GHz band for $t = 1.84$ mm. For higher MWCNT content, ε_r continues to rise, shifting the *RL* peak to lower thickness, but beyond the monitored frequency window. In the high MWCNT content range, Yao et al. (2009) have studied a MWCNT/SiC paraffin nanocomposite containing up to 18 wt% MWCNT. Although ε_r gradually rises with the nanotube amount, the composite with 12 wt% shows the wider $BW_{10dB} = 4$ GHz for $t = 2$ mm. On the contrary, by coating CNT with insulating ZnO nanoparticles, the incorporation of higher MWCNT amount in wax was allowed (up to 40 wt%), which compromised ε_r and produced a broad *RL* peak of $BW_{10dB} > 6.5$ GHz for the same thickness (Song et al., 2012). Similarly, Zhu et al. (2011) explored the return loss of a composite containing 25 wt% MWCNT with attached SiC nanoparticles, synthesized by in-situ reaction of the two components. In this study, annealing of the MWCNT favors the formation of defects and SiC coating. Thus, polarization mechanisms grow with annealing temperature, finally yielding at 400 °C composites with improved absorption characteristics compared to the pristine MWCNT composite ($BW_{10dB} > 7.5$ GHz for $t = 2$ mm).

Since its discovery, graphene has attracted considerable attention due to its remarkable physical properties, comprising its huge surface area, extraordinary mechanical strength, and high electrical conductivity. Therefore, its ability to become structural block of new hybrid materials for absorption applications has been explored (Wang et al., 2013a, 2015d,g; Zhang et al., 2015a). In specific, the coexistence of graphene, in the form of reduced graphene oxide (rGO), along with NiO- or Ag-coated glass spheres raises dielectric polarization and permittivity to levels above the respective levels of the components (Wang et al., 2013a, 2015d). This synergy results in improved *RL* bandwidth values, which reach $BW_{10dB} = 5$ GHz for $t = 3$ mm and $BW_{10dB} = 4$ GHz for $t = 2$ mm, respectively. Furthermore, reported the advantages of rGO addition to MnO_2 and ZnO-containing composites (Wang et al., 2015g; Zhang et al., 2015a). By wrapping MnO_2 nanorods with rGO nanosheets or mixing them with ZnO tetrapod-like nanoparticles, the developed complex nanostructures, phase variations and high defects density on the rGO surface yield an increase of ε_r, which favors broadband impedance matching. To this end, the calculated reference bandwidths are accordingly $BW_{10dB} > 5$ GHz for $t = 2$ mm and $BW_{10dB} > 7.5$ GHz for $t = 3$ mm. Similarly to the previous conductive carbonaceous fillers, there is an optimal concentration above which ε_r' and ε_r'' are too high and satisfy only partly the matching conditions.

In order to fully exploit the potential of the reviewed C-based fillers and their advantageous impact on the microwave absorption, there have been attempts to use them in pair (Ren et al., 2016; Arooj and Yan, 2015). Claim that low concentration of inclusion mixture comprising graphene nanosheets and MWCNT (1.5 and 0.5 wt%) is enough to achieve high *RL* values exceeding 10 dB for $t = 3.5$ mm (Ren et al., 2016). For nanocomposites with higher MWCNT content (6 wt%), the addition of the insulating graphene oxide (GO) downgrades dielectric polarization, but by that means, it improves the impedance matching at quarter-wavelength thickness (Arooj and Yan, 2015). In this way, the hybrid epoxy nanocomposite with 0.5 wt% GO and 5.5 wt% MWCNT, possesses $RL > 10$ dB with $BW_{10dB} = 3$ GHz for $t = 2$ mm.

3.5.1.2 Magnetic-filler materials

Composite materials with ferrimagnetic inclusions form a large share of the implemented microwave absorbers, on account of the versatility of their EM properties. To this effect, ferrimagnetic oxides (ferrites) with cubic (spinel) or hexagonal structure and iron-based alloys and compounds are utilized, as their microwave magnetic behavior can be varied by control of the composition and processing parameters. By that means, the dynamic magnetization mechanisms of domain wall movement and ferromagnetic resonance are strongly affected and the complex permeability spectrum $\mu_r(f)$ is accordingly defined. Since permeability constitutes a decisive parameter in the fulfillment of the impedance matching conditions, this benefit can be reaped by combining different types of magnetic inclusions in the fabrication of polymer composites (Dosoudil et al., 2012; Lim et al., 2003; Wu et al., 2015; Han et al., 2012; Shimba et al., 2012; Yun et al., 2006; Yan et al., 2009; Li et al., 2014; Liu et al., 2006, 2013a; Drmota et al., 2012; Yang et al., 2014; Shen et al., 2012, 2015; Song et al., 2013; Jacobo et al., 2015; Zou et al., 2014; Guo et al., 2011; Zhang et al., 2013b; Cheng et al., 2013).

Spinel ferrites are a prevalent member of soft ferrimagnetic ceramic oxides, which have been also employed as a basic filler material for EM wave absorbers. Among them, composites containing Ni–Zn ferrite particles may exhibit their ferromagnetic resonance in the lower GHz range, depending on the particle composition, morphology and loading to the polymer matrix. To overcome this innate limitation and to extend the operational frequency band, Ni–Zn ferrite particles have been combined with carbonyl iron (CI), Fe–Si–B or Fe–B amorphous alloy, and Fe–Si or Fe–Si–Al alloy particles, to shift the effective ferromagnetic resonance to higher frequency into the microwave regime (Dosoudil et al., 2012; Lim et al., 2003; Wu et al., 2015; Han et al., 2012; Shimba et al., 2012; Yun et al., 2006). It is due to their higher magnetocrystalline or shape anisotropy values that the introduction of such metallic magnetic materials allows the preservation of high μ' and μ'' values in the frequency range of interest. Moreover, from the measured dielectric properties, we additionally notice the increase of $\varepsilon_r(f)$ by the decrease of the ferrite content. Thus, the enhanced constitutive EM parameters of these hybrid magnetic composites result in thinner microwave absorbers with higher matching frequency, on the basis of the quarter-wavelength principle. In terms of *RL* bandwidth, Dosoudil et al. (2012) have reported that BW_{10dB} of epoxy composites is increased in the 100 MHz–3 GHz range from 500 MHz to 2 GHz by the gradual substitution of the Ni–Zn ferrite nanofiller by Fe–Si alloy nanoparticles, with a constant total loading of 60 vol.%. On the contrary, Lim et al. (2003) have found that broad BW_{10dB}, fully covering the 2–4 GHz range, is favored by higher Ni–Zn ferrite content in epoxy composites filled with 50 vol.% of ferrite/Fe–Si–B alloy microparticles. In line with that improvement, Wu et al. (2015) reported that paraffin composites with 28 vol.% of Fe–Cu–Nb–Si–B flakes, coated with Ni–ferrite nanoparticles, exhibit better matching properties and higher *RL* than the uncoated alloy flakes. Despite the different approach of the optimal composition by these investigations, they all highlight the synergy between different magnetic compounds.

Ferrites with hexagonal crystal structure (hexaferrites) are well-known EM materials with intrinsic aptness for high frequency applications due to their stronger magnetocrystalline anisotropy, compared to their cubic counterparts. For their use as components for the preparation of hybrid composite materials, hexaferrites have been either mechanically mixed or grown together with other magnetic compounds, including Fe−Si−Al alloy, α−Fe, Fe oxide or spinel ferrite particles (Li et al., 2014; Liu et al., 2006; Drmota et al., 2012; Yang et al., 2014; Shen et al., 2012, 2015; Song et al., 2013; Jacobo et al., 2015) have studied the EM performance of hybrid polymer composites containing two soft magnetic filler compounds (Li et al., 2014). The filler was prepared by mixing $Ba_{1.56}Sr_{1.44}Co_2Fe_{24}O_{41}$ hexaferrite particles with Fe−Si−Al alloy flakes at different weight ratios. The addition of the alloy particles raises ε_r and shifts the $\mu_r(f)$ spectrum to lower frequency. In conclusion, all the hybrid composites show better absorption characteristics (RL_{max}, BW_{10dB}) for lower matching thickness, in comparison to the cases of separate fillers. The broadest absorbing performance is recorded for 90 wt% hexaferrite and 10 wt% Fe−Si−Al with $BW_{10dB} = 5$ GHz, for $t = 3$ mm. Similar conclusion was drawn by Liu et al. (2006), who studied the mixture of $Ba_3Co_{1.8}Fe_{23.6}Cr_{0.6}O_{41}$ hexaferrite/α-Fe nanoparticles. In this study, the composite with 38 vol.% α-Fe preserves a broad attenuation bandwidth ($BW_{10dB} = 6$ GHz) with the minimum thickness of 1.6 mm. Drmota et al. (2012) have investigated polymer composites filled at 80 wt% with materials of the $SrO-Fe_2O_3$ system. The filler nanoparticles of hard ferrite $SrFe_{12}O_{19}$ and various Fe oxides were synthesized with the coprecipitation technique and by varying the calcination temperature. It was found that, only at the temperature which enables the formation of the soft magnetic phase of Fe_3O_4, it was possible to obtain a $\mu_r(f)$ dispersion sufficient to produce $RL > 10$ dB.

The latter example introduces the case of the coexistence of hard and soft magnetic particles in close proximity, with interesting synergistic features. Yang et al. (2014) and Shen et al. (2015) have employed a thermal selective reduction process to produce fibers, which are formed by nanoparticles of the hard/soft magnetic system $Ba(Sr)Fe_{12}O_{19}$/α-Fe. For an optimal proportion of the two phases in the composite (70−77 wt% hexaferrite), exchange−coupling interaction takes place, which results in a broad $\mu''(f)$ loss peak. These optimal compositions of the $BaFe_{12}O_{19}$/α-Fe and $SrFe_{12}O_{19}$/α-Fe systems allow the production of hybrid composites with respective absorption bandwidths $BW_{10dB} = 7.5$ and 9.5 GHz for $t = 3$ mm, in the 2−18 GHz range. For higher α-Fe content, the EM properties are governed by the α-Fe contribution and the microwave absorption characteristics are degraded. In a contiguous way, Shen et al. (2012) and Song et al. (2013) have also explored the behavior of fibers comprising nanoparticles of the $BaFe_{12}O_{19}/Ni_{0.5}Zn_{0.5}Fe_2O_4$ and $Ba_{0.5}Sr_{0.5}Fe_{12}O_{19}/Ni_{0.5}Zn_{0.5}Fe_2O_4$ systems, by varying either the phases weight ratio or the annealing temperature. In both studies, the prepared fiber−wax composites also containing exchange-coupled inclusions possess improved microwave absorbing properties. Particularly in the former binary filler system, the composite loaded with 70 wt% of $BaFe_{12}O_{19}$ exhibits the maximum return loss ($RL = 35.5$ dB) with $BW_{10dB} > 10$ GHz for $t = 3$ mm. A similar interrelation between hard/soft phase exchange-coupling and microwave performance was reported by Jacobo et al. (2015). In this work, Ni-spinel ferrite nanoparticles were properly grown on $Sr_{0.5}Nd_{0.5}Co_{0.5}Fe_{10.5}O_{19}$ hexaferrite nanoparticles. The prepared epoxy-nanocomposite with equal hard/soft phase content, which shows the interphase exchange-coupling, has enhanced absorption performance with $RL > 10$ dB along the X-band (8–12 GHz).

In addition to the various ferrite materials, CI has been also widely used as an effective microwave absorbing material up to the 20 GHz range, due to its commercial availability and its low cost. Liu et al. (2013a) have prepared CI powder in the microscale with oxidized surface via a hydrothermal method, by varying the oxidation degree. The obtained CI particles, coated with a varying amount of Fe_3O_4, were mixed with wax (60 vol.%). By increasing the formed insulating Fe_3O_4 layer, both $\varepsilon_r(f)$ and $\mu_r(f)$ of the composites drop in the 500 MHz–18 GHz regime. The presence of the oxidized layer clearly improves the absorption performance of the composites, in terms of RL and its bandwidth, and leads to $BW_{10dB} = 7$ GHz for $t = 2$ mm. This research work was repeated with CI flakes, which were produced by ball-milling (Zou et al., 2014). In this case, a lower volume loading was applied (20 vol.%) and the optimal composite with thickness $t = 2$ mm possesses a more narrow reflection bandwidth ($BW_{10dB} = 3.5$ GHz). Synergy effect in the microwave absorption was also displayed by blending CI with Fe–Si alloy or Fe–Si–P–Sb alloy micropowders, as enhanced $RL(f)$ curves are drawn for the CI/alloy composites with 1:1 or 4:1 weight ratio (Guo et al., 2011; Zhang et al., 2013b). In both cases, the tempered EM parameters favor impedance matching, thus creating the optimal bandwidth $BW_{10dB} = 2$ GHz for $t = 3$ mm. Cheng et al. (2013) have characterized composites filled with different mixtures of CI micropowder and ferromagnetic $La(Sr)Mn_2O_3$ nanopowders. Despite the small thickness of the composites ($t = 1$ mm), for an optimal weight ratio of the two components (30:1), a maximum bandwidth of $BW_{10dB} = 2$ GHz is recorded in the X-band.

3.5.1.3 Mixed-type filler materials

From the above review analysis, we deduce that more flexibility adds up to materials design by the concurrent manipulation of both magnetic and dielectric effective properties. To this effect, it is useful to explore the potential synergies between the most prevalent types of magnetic particulates and the conductive carbonaceous species. By increasing the content of fillers of these two categories, higher EM polarization and losses normally occur. However, as it is explained in Section 3.4.1 and Stergiou and Litsardakis (2012), sufficient impedance matching over a wide frequency range theoretically necessitates moderate levels of polarization and losses. This conclusion is verified by experimental findings in numerous published studies.

In polymer composites loaded with spinel ferrites (including magnetite Fe_3O_4), hexagonal ferrites or CI powder, there appears an optimal concentration of added C-based fillers (CB, CNT, graphene and graphite) which is related with the maximization of the operational bandwidth (Liu et al., 2013b; Wang et al., 2013b, 2015b; Li et al., 2015; Zong et al., 2015; Huang et al., 2016a,b; Ghasemi, 2011;

Vinayasree et al., 2013, 2014; Gordani et al., 2015; He et al., 2015; Yang et al., 2015; Zhao et al., 2016; Alam et al., 2016; Xu et al., 2012). We thus notice that impedance matching occurs within a wider frequency range for those composites with higher permittivity values than the pure magnetic composite and with nonzero magnetic losses. He et al. (2015) investigated polyvinylidene fluoride-composites with rGO/$BaFe_{12}O_{19}$ hexaferrite inclusions and found that the rGO/ferrite composite with 5:100 weight ratio results in the maximum absorption bandwidth in the 0.5—18 GHz band ($BW_{10dB} = 9$ GHz for $t = 2$ mm). Moreover, in the single-walled CNT/$CoFe_2O_4$ filler system, Li et al. (2015) have found that with 10 wt% CNT, the maximum return loss and the broadest bandwidth over the 2–18 GHz range can be achieved ($BW_{10dB} = 7$ GHz for $t = 2$ mm). In these studies, further increase of the conductive-to-magnetic filler ratio raises ε_r to such levels which impedes the perfect fulfillment of the matching conditions. Wang et al. (2015e) have tested mixtures of $BaCo_{0.8}Zn_{1.2}Fe_{16}O_{27}$ hexaferrite with 5 and 10 wt% of MWCNT. It is displayed that, even with the lower MWCNT content, permittivity becomes high enough ($\varepsilon_r' > 15$), so that the best performance is attained with the pure magnetic composite ($BW_{10dB} = 7$ GHz for $t = 2$ mm). Another potential drawback of high CNT concentration is the tendency of nanotubes to agglomerate and form irregular nanostructures. This was observed by Gordani et al. (2015), in their study of single-walled CNT/$SrMgCoTi_2Fe_8O_{19}$ hexaferrite nanocomposites. Therein, loading with 4 vol.% of CNT results in $RL > 10$ dB in the whole monitored frequency band of 8–12 GHz, whereas for higher volume content, the observed agglomeration degrades the absorption characteristics.

By analogy to the previous approach, in polymer composites loaded with conductive forms of C, we may increase the absorption bandwidth by the addition of an optimal amount of magnetic fillers (Xiao et al., 2013; Ma et al., 2013; Liu et al., 2015, 2016; Gui et al., 2009). This improvement emanates from the enhanced complex permeability, as we may deduce from the characterization of rGO/Ni–Co–Zn ferrite nanocomposites in the 2–18 GHz region, which were studied by Liu et al. (2015, 2016). These nanocomposites were produced either by mixing separate fillers or via in-situ reaction of the precursors. Among the tested compositions with the two processes, the mass ratios of rGO/ferrite 1:15 and 1:10 exhibit the maximum attenuation bandwidth, $BW_{10dB} = 6$ GHz for $t = 2$ mm and $BW_{10dB} = 4.5$ GHz for $t = 2.5$ mm, respectively. In general, we notice that the optimal composition generally involves low carbon and high magnetic content. This is also derived by the characterization of CNT/$Ni_{0.5}Zn_{0.5}Fe_2O_4$ ferrite and graphene/CI composites in the 2–18 GHz region (Duan et al., 2014; Yuchang et al., 2015). The high aspect ratio of these carbonaceous materials introduces high dielectric polarization, so that more effective absorbers can be designed with the lowest tested filler content. Therefore, 10 wt% MWCNT loading to ferrite composite leads to $BW_{10dB} = 5.5$ GHz for $t = 2$ mm and 1 wt% graphene nanosheets loading to CI composite leads to $BW_{10dB} = 8.5$ GHz for $t = 1.1$ mm.

Another potentially advantageous materials design direction is the combination of magnetic with purely dielectric fillers (Mandal et al., 2013; Mandal and Das, 2014; Wang et al., 2015f; Shen et al., 2014; Qing et al., 2011; Sun et al.,

2015; Liu et al., 2004). The investigation of the Ni–Cu–Zn spinel/$PbZr_{0.52}Ti_{0.48}O_3$ and Ni–Co–Zn spinel/$BaTiO_3$ nanofiller systems has revealed that for 1:1 magnetic/dielectric mass ratio and 30 wt% loading to epoxy medium, high effective μ_r is preserved and ε_r is enhanced compared to the pure ferrite (Mandal et al., 2013; Mandal and Das, 2014). This synergy results in nanocomposites with $RL > 10$ dB in the whole X-band and maximum loss of $RL_{max} = 49$ dB ($t = 2.5$ mm) and $RL_{max} = 42.5$ dB ($t = 2$ mm), respectively. The same impact of $BaTiO_3$ nanoparticles on the EM properties of $Ba_{1.5}Sr_{1.5}Co_2Fe_{21.6}O_{41}$ hexaferrite composite was reported by Wang et al. (2015f). Hence, the optimal hybrid composite doped with 10 wt% $BaTiO_3$ nanoparticles has a broader absorption bandwidth for thinner material layer ($BW_{10dB} = 5$ GHz for $t = 2$ mm) than the optimal composite with pure ferrite. Without presenting the respective EM parameters, Shen et al. (2014) supported that coating $Ba_3Co_2Fe_{24}O_{41}$ hexaferrite with SiO_2 nanoparticles may strengthen and broaden $RL(f)$ peaks. As a consequence, the composite with hexaferrite platelet filler is optimally coated with triple SiO_2 layers and shows $BW_{10dB} = 1.5$ GHz for $t = 1$ mm. In addition, improved microwave absorbing materials with CI flake particulates were produced by the incorporation of dielectric oxide nanoparticles (Qing et al., 2011; Sun et al., 2015). In specific, by milling CI materials with 50 wt% $BaTiO_3$ or 25 wt% ZnO the effective EM parameters are diminished in the 2–18 GHz range, which enhances impedance matching and yields higher RL and attenuation bandwidth ($BW_{10dB} = 2.7$ GHz for $t = 2$ mm and $BW_{10dB} > 7$ GHz for $t = 2$ mm, respectively).

Apart from the filling degree of different types of materials, the impact of other parameters has been explored with regard to return loss. Particularly, Xu et al. (2012) studied MWCNT/CI silicone rubber composites, with different CI morphologies. Composites comprising spherical CI particles outperform their flaky counterparts in the 2–18 GHz band, since in the latter composites the enhanced interfaces generate strong dielectric polarization. Therefore, for 0.5 wt% of MWCNT and 45 wt% of spherical CI particles we achieve $BW_{10dB} = 6$ GHz for $t = 1$ mm. Asghari et al. (2013) have studied a MWCNT/$SrFe_{12-x}Cr_{x/2}Al_{x/2}O_{19}$ hexaferrite filler system at a fixed volume ratio, by varying the hexaferrite chemical composition ($x = 0$–2.5). The produced PVC nanocomposites were characterized in the 12–18 GHz and 50–74 GHz regimes, and the measurements indicate that loss higher than 10 dB in the whole frequency points can be achieved by the hexaferrite with $x = 2.5$ ($t = 1.8$ mm). The substitution of Fe by Cr and Al was reported to increase ferromagnetic anisotropy; however, the absence of the respective EM spectra does not allow the identification of the root cause of this improvement.

3.5.2 Transmitted wave absorbers

Apart from suppression of unwanted reflections, shielding of vulnerable electronic components or devices against traveling waves at microwave frequencies is of particular importance. To this end, the attenuation of the emitted EM power in a certain region at the far field is the primary objective for the design of shielding materials or structures. In actual shielding implementations, highly conductive materials may increase reflectivity either due to the impedance mismatch or through openings crossing the surface current paths. Therefore, effective shielding entails a compromise between improved impedance matching, weakened surface currents and strong material losses (electrical, dielectric, and magnetic). In this context, we will review the variation of this power balance in multicomponent hybrid composites.

The attenuation coefficient (α), defined as the real part of propagation constant (γ) in Eq. (3.7), is often selected to quantify the shielding performance of composites (Wang et al., 2015c,g; Huang et al., 2016a; Wen et al., 2011; Tong et al., 2013; Lv et al., 2015; Chuai et al., 2016; Feng et al., 2016; Zhang et al., 2016). However, the characterization of shielding materials in respect of the attenuation coefficient accounts only for the penetrated EM waves and excludes any impedance mismatches on the air–material interfaces. For a more complete description of these materials, the current review and evaluation of the developed composites will be based on the *SE*, the effective return loss (*RL* or SE_R) and absorption loss (*AL* or SE_A), as defined in Section 3.2.

3.5.2.1 Nonmagnetic filler materials

In shielding applications, we can exploit the vast variation of the electrical and dielectric properties of polymer composite materials, induced by the incorporation of conductive fillers. To this effect, C-based materials are the most commonly employed filler type. Various research works provide proof of the positive role of the MWCNT addition in increasing *SE* of composites, which also contain CB nanopowder (De Rosa et al., 2008; Al-Saleh and Saadeh, 2013; Dinesh et al., 2012), carbon micro- and nanofibers (De Rosa et al., 2008; Al-Saleh and Saadeh, 2013), or graphene nanosheets (Lin et al., 2016). The advantage of MWCNT stems from the enhancement of conductivity, due to the high aspect ratio of nanotubes. On this basis, Lin et al. (2016) reported that in the MWCNT/graphene bifiller system the optimal shielding performance in the 0.3–3 GHz range ($SE > 15$ dB for thickness $t = 0.5$ mm) is obtained by the composite containing 1.25 wt% MWCNT and 0.25 wt% graphene. Similarly, Dinesh et al. (2012) presented the increase of *SE* of a low-density polyethylene-composite with 20 wt% of CB ($SE > 8$ dB in the 8–12 GHz for $t = 0.35$ mm), caused by the addition of 1 wt% MWCNT.

The substitution of MWCNT by another conductive carbonaceous inclusion usually degrades the overall *SE*. Al-Saleh and Saadeh (2013) have found that MWCNT-doped styrene composites offer the maximum *SE* among the MWCNT/CB and MWCNT/C-nanofibers bifiller systems, with $SE > 30$ dB in the X-band for $t = 1.1$ mm. However, further improvement can be attained by the addition of conductive particles, while keeping a fixed MWCNT loading. By increasing the content of Ag flakes from 6 to 10 vol.% in a polyurethane (PU) composite with 0.2 vol.% MWCNT, *SE* is increased by 75 dB at 1 GHz for $t = 0.5$ mm (Kim et al., 2005). Chang et al. (2011) also improved *SE* of a 0.25-mm-thick MWCNT/PU composite by 8 dB up to 1.5 GHz by inserting 30 wt% of stainless steel microfibers. On the contrary, Al flakes are less effective than CB in increasing *SE* of rubber composites at a fixed mass content, despite their higher conductivity, due to their lower actual volume fraction (Tanrattanakul and Bunchuay, 2007).

As expected, the enhancement of total *SE* is accompanied by higher reflectivity, usually evident in composites with increased conductivity (Al-Hartomy et al., 2012; Singh et al., 2012; Maiti et al., 2013; Kong et al., 2014; Bera et al., 2015a,b; Petrova et al., 2016; Chen et al., 2016; Sharma et al., 2016). Even though, Sun et al. (2016) claim that microwave absorption plays a dominant role in determining the total *SE* of MWCNT/graphene composites, by simple calculations, we get that 90% of the impinging power is reflected. It is interesting to note that, in this study, conductivity apparently increases for higher porosity rate, as this facilitates the formation of conduction paths. Lower reflectance levels have been manifested by the incorporation of pure dielectrics (silica, clay and fly ash) in conductive composites (Petrova et al., 2016; Al-Ghamdi et al., 2016a; Gujral et al., 2016). Specifically, the addition of silica in CB-loaded rubber yields $SE > 15$ dB, while RL remains below 5 dB in the 2−12 GHz band for $t = 2.8$ mm (Al-Ghamdi et al., 2016a). Gujral et al. (2016) also reported that the insertion of 10 wt% fly ash powder in MWCNT-loaded PU matrix raised *SE* from 5 to 15 dB in the 12−18 GHz range, with $RL < 5$ dB for $t = 3$ mm.

Furthermore, composites homogeneity is another key parameter affecting the *SE* and effective conductivity. Lin et al. (2016) have highlighted the advantage of solution mixing in preparing composites with homogeneous filler dispersion and with higher *SE* by 12 dB, compared to those prepared with typical melt compounding. In this direction, higher *SE* values resulted from better dispersion degrees of conductive inclusions (rGO and MWCNT) by the addition of Fe_2O_3 nanoparticles (Singh et al., 2012), fly ash powder (Gujral et al., 2016), or by coating C-fibers with rGO nanoplatelets (Chen et al., 2016). The significance of homogeneous dispersion is also denoted by the drop of conductivity, and subsequently of *SE*, when agglomeration of platelet or fibrous conductive inclusions takes place (Chen et al., 2016; Sun et al., 2016; Li et al., 2016).

3.5.2.2 Mixed-type filler materials

On account of their diluted magnetic properties, pure ferrite-loaded composites can hardly produce sufficient attenuation of the travelling EM waves. To this end, for such applications, ferrite particles basically form hybrid composites in combination with conductive inclusions. In various conductive composites, the incorporation of particles with nonzero magnetic losses (Mn−Zn ferrite, Fe_3O_4, $Ba_3Co_2Fe_{23}O_{41}$ and $Ba_2Zn_2Fe_{11.5}O_{22}$ hexaferrites) has led to better impedance matching, which is subsequently displayed by the diminishing *RL* in the range up to 12 GHz (Phan et al., 2016; Al-Ghamdi et al., 2016b; Li et al., 2006). Therefore, at low ferrite content, *SE* is maximized following the maximization of total EM losses. Further, increase of the low-conductivity magnetic filler amount yields the corresponding decrease of absorbed power. On the same principle, polystyrene composites loaded with 2.24 vol.% of Fe_3O_4-decorated graphene nanosheets possess increased *SE* by 5 dB ($SE > 25$ dB in the X-band), while maintaining similar return loss with the pristine graphene composite (Chen et al., 2015b). By employing a polymer foam filled with

10 vol.% of Fe_3O_4-coated graphene, Shen et al. (2013) recorded an excellent combination of shielding properties ($SE > 15$ dB and $RL < 0.5$ dB) in the whole X-band for $t = 2.5$ mm.

Except for the ferromagnetic oxides, metallic magnetic particulates have been used in conjunction with different C-based fillers. Wang et al. (2013c) have shown that *SE* of a MWCNT epoxy composite in the 26.5–40 GHz region is increased by 10 dB by the inclusion of 1 wt% of Fe nanoparticles, in consequence of raised ε_r''. Huang et al. (2015) also investigated CoNi alloy-coated carbon fiber composites for different alloy compositions and recorded the maximum *SE* in the X-band ($SE > 25$ dB) for the composite with higher ε_r''. According to Labunov et al. (2012), in composites with these bifiller systems, reflection is the dominant shielding mechanism in the 8–12 GHz region, whereas absorption dominates in the 26.5–40 GHz. However, a low reflectance shield for the X-band ($RL < 2.5$ dB and $SE > 30$ dB for $t = 4$ mm) was prepared by Chen et al. (2015a) by blending 30 wt% of CI microparticles and 3 wt% of rGO nanosheets in epoxy medium. For even higher CI content (75 wt%) in MWCNT resin composite, inferior shielding characteristics are reported compared to the previous case ($RL < 5.5$ dB and $SE < 28$ dB for $t = 1.4$ mm) (Wang et al., 2015a).

3.6 Summary

Protection of electronic products and their susceptible circuits from external and internal sources of EMI plays an important role in consumer electronics, telecommunications, medical applications, space technology, as well as defense. Modern electronic circuit board design from EMC point of view always requires careful consideration of PCB layout, including placement of ICs, wiring patterns, ground pads, and noise source shielding to suppress noise emissions properly "in situ." However, post-design EMI-related problems may arise. Then flexible thin NSS and EMWA materials could be a good solution, with an ultimate reward of better product performance and shorter time-to-market.

Quantitative evaluation of specular microwave absorbers is also critical for their optimal design. It has been shown herein that the depth of reflection minimum is not a suitable measure for the characterization of radar absorbers. To describe the quality of an absorber, characterization of its bandwidth properties is required, whereas the width of the operating frequency range cannot serve for this purpose. A correct approach to define quality of absorbers should use one of the proposed quantities. The first is the frequency bandwidth divided by the center frequency, that is, fractional bandwidth, in case when the absorber thickness is of not prime priority. The second is the frequency bandwidth divided by the squared center frequency, that is, the wavelength operating range, if requirement of thin absorber is of importance.

From materials aspect, the primary purpose of this chapter is to introduce common multicomponent filler systems for shielding applications. This work in no way constitutes an exhaustive list of the studied species. Through a selective overview of the evolving field of composites, the potential advantages of hybrid composites were identified and discussed. Dielectric, magnetic, and absorbing properties of composites can be substantially modified by combining different inclusion types. The materials features yielding improved shielding performance are analyzed based on the electromagnetic theory of reflection and absorption outlined herein.

In conclusion, the direction toward more effective microwave absorbing and shielding composite materials is illuminated, by establishment of the application requirements, comprehension of the operation principles, and materials effects, and adoption of reliable quantitative evaluation criteria.

Acknowledgments

One of the authors, K. Rozanov, acknowledges financial support provided by Russian Scientific Foundation Project no. 16-19-10490. Moreover, one of the authors, C. Stergiou, wishes to thank Professor C. Delides for triggering his research on polymer nanocomposites.

References

3M, see NFC & RFID Materials at http://solutions.3m.com/wps/portal/3M/en_US/Electronics_NA/Electronics/Products/Electronics_Product_Catalog/ ~ /NFC-RFID-Materials?N = 8697920 + 8704979&rt = r3, access date January 31, 2017.

Agilent, 2014a, Basics of Measuring the Dielectric Properties of Materials, Application Note 5989-2589EN, available at cp.literature.agilent.com/litweb/pdf/5989-2589EN.pdf, Agilent, USA.

Agilent, 2014b, Solutions for Measuring Permittivity and Permeability with LCR Meters and Impedance Analyzers, Application Note 1369-1, 5980-2862 EN, available at cp.literature.agilent.com/litweb/pdf/5980-2862EN.pdf, Agilent, USA.

Alam, R.S., Moradi, M., Nikmanesh, H., 2016. Influence of multi-walled carbon nanotubes (MWCNTs) volume percentage on the magnetic and microwave absorbing properties of $BaMg_{0.5}Co_{0.5}TiFe_{10}O_{19}$/MWCNTs nanocomposites. Mater. Res. Bull. 73, 261–267.

Al-Ghamdi, A.A., Al-Hartomy, O.A., Al-Solamy, F.R., Dishovsky, N., Mihaylov, M., Atanasov, N., et al., 2016a. Microwave properties of natural rubber based composites comprising conductive carbon black/silica hybrid fillers. J. Polym. Res. 23, 180.

Al-Ghamdi, A.A., Al-Hartomy, O.A., Al-Solamy, F.R., Dishovsky, N., Malinova, P., Atanasova, G., et al., 2016b. Conductive carbon black/magnetite hybrid fillers in microwave absorbing composites based on natural rubber. Comp. B. 96, 231–241.

Al-Hartomy, O.A., Al-Ghamdi, A.A., Dishovsky, N., Shtarkova, R., Iliev, V., El-Tantawy, F., 2012. Comparative study of the influence of carbon nanotubes and graphene nanoplatelets on dielectric and microwave properties of conductive rubber composites. Res. Rev. Pol. 3, 107–116.

Al-Saleh, M.H., Saadeh, W.H., 2013. Hybrids of conductive polymer nanocomposites. Mater. Des. 52, 1071–1076.

ARC Technologies, see Products/Wave-X at http://arc-tech.com/wave-x/, access date January 31, 2017.

Arooj Y., Z. Yan, Electromagnetic wave absorbing characteristics of graphene-oxide dispersed carbon nanotubes/epoxy Composites, IEEE Proc. 12th Intern. Bhurban Conf. Appl. Sci. Tech., Islamabad, Pakistan, Jan. 2015.

Asghari, M., Ghasemi, A., Paimozd, E., Morisako, A., 2013. Evaluation of microwave and magnetic properties of substituted $SrFe_{12}O_{19}$ and substituted $SrFe_{12}O_{19}$/multi-walled carbon nanotubes nanocomposites. Mater. Chem. Phys. 143, 161–166.

Baker-Jarvis J., "Transmission/reflection and short-circuit line permittivity measurements", Technical Note 1341, NIST, U.S. Department of Commerce, Boulder, CO, USA, July 1990.

Baker-Jarvis J., Measuring the permittivity and permeability of lossy materials: solids, liquids, metals, building materials, and negative-index materials, Tech. Report 1536, NIST, U.S. Department of Commerce, Boulder, CO, USA, 2005.

Bera, R., Suin, S., Maiti, S., Shrivastava, N.K., Khatua, B.B., 2015a. Carbon nanohorn and graphene nanoplate based polystyrene nanocomposites for superior electromagnetic interference shielding applications. J. Appl. Pol. Sci. 132, 42803.

Bera, R., Karan, S.K., Das, A.K., Paria, S., Khatua, B.B., 2015b. Single wall carbon nanohorn (SWCNH)/graphene nanoplate/poly(methyl methacrylate) nanocomposites: a promising material for electromagnetic interference shielding applications. RCS Adv. 5, 70482–70493.

Celozzi, S., Araneo, R., Lovat, G., 2008. Electromagnetic Shielding. IEEE Press, Wiley, Hoboken, NJ, USA, pp. 317–352., ISBN 978-0-470-05536-6, Appendix C "Standards and Measurement Methods".

Chang, H., Kao, M.-J., Huang, K.-D., Kuo, C.-G., Huang, S.-Y., 2011. Electromagnetic shielding effectiveness of thin film with composite carbon nanotubes and stainless steel fibers. J. Nanosci. Nanotech. 11, 1754–1757.

Chen, J., Wu, J., Ge, H., Zhao, D., Liu, C., Hong, X., 2016. Reduced graphene oxide deposited carbon fiber reinforced polymer composites for electromagnetic interference shielding. Composites, A. 82, 141–150.

Chen, L.F., Ong, C.K., Neo, C.P., Varadan, V.V., Varadan, V.K., 2004. Microwave Electronics: Measurement and Materials Characterisation. Wiley, England, ISBN 0-470-84492-2.

Chen, Y., Zhang, H.-B., Huang, Y., Jiang, Y., Zheng, W.-C., Yu, Z.-Z., 2015a. Magnetic and electrically conductive epoxy/graphene/carbonyl iron nanocomposites for efficient electromagnetic interference shielding. Comp. Sci. Tech. 118, 178–185.

Chen, Y., Wang, Y., Zhang, H.-B., Li, X., Gui, C.-X., Yu, Z.-Z., 2015b. Enhanced electromagnetic interference shielding efficiency of polystyrene/graphene composites with magnetic Fe_3O_4 nanoparticles. Carbon. 82, 67–76.

Cheng Y., K. Wei, P. Xia, Q. Bai, Preparation and microwave absorption properties of carbonyl iron/$La_{1-x}Sr_xMnO_3$ ($x = 0.3$, 0.4 and 0.5) composites, IEEE Proc. 4th Intern. Conf. on Dig. Manuf. Autom., Shandong, China, June 2013.

Choi, I., Lee, D.Y., Lee, D.G., 2015. Radar absorbing composite structures dispersed with nano-conductive particles. Compos. Struct. 122, 23–30.

Choi, W.H., Shin, J.H., Song, T.H., Kim, J.B., Cho, C.M., Lee, W.J., et al., 2014. Design of circuit-analog (CA) absorber and application to the leading edge of a wing-shaped structure. IEEE Trans. Electromagn. Compat. 56, 599–607.

Chuai, D., Liu, X., Yu, R., Ye, J., Shi, Y., 2016. Enhanced microwave absorption properties of flake-shaped FePCB metallic glass/graphene composites. Composites, A. 89, 33–39.

Costa, F., Monorchio, A., 2012. Electromagnetic absorbers based on high-impedance surfaces: from ultra-narrowband to ultra-wideband absorption. Adv. Electromagnetics. 1, 7–12.

CST STUDIO SUITE®, CST and Dassault Systèmes, www.cst.com, access date January 31, 2017.

De Rosa, I.M., Sarasini, F., Sarto, M.S., Tamburrano, A., 2008. EMC impact of advanced carbon fiber/carbon nanotube reinforced composites for next-generation aerospace applications. IEEE Trans. EMC. 50, 556–563.

Dinesh, P., Renukappa, N.M., Siddaramaiah, Rajan, J.S., 2012. Electrical properties and EMI shielding characteristics of multiwalled carbon nanotubes filled carbon black-high density polyethylene nanocomposites. Comp. Interf. 19, 121–133.

Donghyun Electronics, see EMI Suppressor at http://changsung.com/_new/wp-content/themes/changsung/images/product_PDF/EMCSolution.pdf, access date January 31, 2017.

Dosoudil, R., Franek, J., Slama, J., Usakova, M., Gruskova, A., 2012. Electromagnetic wave absorption performances of metal alloy/spinel ferrite/polymer composites. IEEE Trans. Magn. 48, 1524–1527.

Drmota, A., Koselj, J., Drofenik, M., Znidarsic, A., 2012. Electromagnetic wave absorption of polymeric nanocomposites based on ferrite with a spinel and hexagonal crystal structure. J. Magn. Magn. Mater. 324, 1225–1229.

Duan, M.C., Yu, L.M., Sheng, L.M., An, K., Ren, W., Zhao, X.L., 2014. Electromagnetic and microwave absorbing properties of SmCo coated single-wall carbon nanotubes/NiZn-ferrite nanocrystalline composite. J. Appl. Phys. 115, 174101.

Emerson, W.H., 1973. Electromagnetic wave absorbers and anechoic chambers through the years. IEEE Trans. Antennas Propagat., v. 21, 484–490.

Feng, J., Pu, F., Li, Z., Li, X., Hu, X., Bai, J., 2016. Interfacial interactions and synergistic effect of CoNi nanocrystals and nitrogen-doped graphene in a composite microwave absorber. Carbon. 104, 214–225.

Gargama, H., Thakur, A.K., Chaturvedi, S.K., 2016. Polyvinylidene fluoride/nanocrystalline iron composite materials for EMI shielding and absorption applications. J. Alloys Compd. 654, 209–215.

Ghasemi, A., 2011. Remarkable influence of carbon nanotubes on microwave absorption characteristics of strontium ferrite/CNT nanocomposites. J. Magn. Magn. Mater. 323, 3133–3137.

Gordani, G.R., Ghasemi, A., Saidi, A., 2015. High frequency electromagnetic reflection loss performance of substituted Sr-hexaferrite nanoparticles/SWCNTs/epoxy nanocomposite. J. Magn. Magn. Mater. 391, 184–190.

Gui, X., Ye, W., Wei, J., Wang, K., Lv, R., Zhu, H., et al., 2009. Optimization of electromagnetic matching of Fe-filled carbon nanotubes/ferrite composites for microwave absorption. J. Phys. D: Appl. Phys. 075002.

Gujral, P., Varshney, S., Dhawan, S.K., 2016. Designing of multiphase fly ash/MWCNT/PU composite sheet against electromagnetic environmental pollution. J. Electron. Mater. 45, 3142–3148.

Guo, J., Duan, Y., Liu, L., Chen, L., Liu, S., 2011. Electromagnetic and microwave absorption properties of carbonyl-iron/$Fe_{91}Si_9$ composites in gigahertz range. J. Electromagn. Anal. Appl. 3, 140–146.

Han, R., Gong, L.-Q., Wang, T., Qiao, L., Li, F.-S., 2012. Complex permeability and microwave absorbing properties of planar anisotropy carbonyl–iron/$Ni_{0.5}Zn_{0.5}Fe_2O_4$ composite in quasimicrowave band. Mater. Chem. Phys. 131, 555–560.

He, H., Luo, F., Qian, N., Wang, N., 2015. Improved microwave absorption and electromagnetic properties of $BaFe_{12}O_{19}$-poly(vinylidene fluoride) composites by incorporating reduced graphene oxides. J. Appl. Phys. 117, 085502.

Huang, R.F., Li, Z.W., 2012. Broadband and ultrathin screen with magnetic substrate for microwave reflectivity reduction. Appl. Phys. Lett. 101, 154101.

Huang, X., Dai, B., Ren, Y., Xu, J., Zhao, C., 2015. Controllable synthesis and electromagnetic interference shielding properties of magnetic CoNi alloy nanoparticles coated on biocarbon nanofibers. J. Mater. Sci.: Mater. Electron. 26, 2584–2588.

Huang, X., Zhang, J., Rao, W., Sang, T., Song, B., Wong, C., 2016a. Tunable electromagnetic properties and enhanced microwave absorption ability of flaky graphite/cobalt zinc ferrite composites. J. Alloys Compd. 662, 409–414.

Huang, Y., Ding, X., Li, S., Zhang, N., Wang, J., 2016b. Magnetic reduced graphene oxide nanocomposite as an effective electromagnetic wave absorber and its absorbing mechanism. Ceram. Int. Available from: http://dx.doi.org/10.1016/j.ceramint.2016.07.223.

Idris, F.M., Hashim, M., Abbas, Z., Ismail, I., Nazlan, R., Ibrahim, I.R., 2016. Recent developments of smart electromagnetic absorbers based polymer-composites at gigahertz frequencies. J. Magn. Magn. Mater. 405, 197–208.

Jacobo, S.E., Bercoff, P.G., Herme, C.A., Vives, L.A., 2015. Sr hexaferrite/Ni ferrite nanocomposites: magnetic behavior and microwave absorbing properties in the X-band. Mater. Chem. Phys. 157, 124–129.

Jing, S., Zhang, Y.-J., Li, J., Liu, D., Koledintseva, M.Y., Pommerenke, D.J., et al., 2015. Extraction of permittivity and permeability for ferrites and flexible magneto-dielectric materials using a genetic algorithm. IEEE Trans. Electromag. Compat. 57, 349–356.

Kazantsev, Y.N., Lopatin, A.V., Kazantseva, N.E., Shatrov, A.D., Mal'tsev, V.P., Vicáková, J., et al., 2010. Broadening of operating frequency band of magnetic-type radio absorbers by FSS incorporation. IEEE Trans. Antennas Propag. 58, 1227–1235.

Kim, S.S., 2011. Microwave absorbing properties of magnetic composite sheets for oblique incidence angles. IEEE Trans. Magn. 47, 4314–4317.

Kim, Y.J., An, K.J., Suh, K.S., Choi, H.-D., Kwon, J.H., Chung, Y.-C., et al., 2005. Hybridization of oxidized MWNT and silver powder in polyurethane matrix for electromagnetic interference shielding application. IEEE Trans. EMC. 47, 872–879.

Kitagawa Industries, see EMI Absorbers at http://kgs-ind.com/products/emc/, access date January 31, 2017.

Koledintseva, M., Rozanov, K.N., Drewniak, J., 2011a. Engineering, modeling and testing of composite absorbing materials for EMC applications. In: Attaf, B. (Ed.), Advances in Composite Materials-Ecodesign and Analysis. InTech, Rijeka, Croatia, ISBN 978-953-307-150-3, Chapter 13, pp. 291–316.

Koledintseva, M.Y., Razmadze, A.G., Gafarov, A.Y., Khilkevich, V.V., Drewniak, J.L., Tsutaoka, T., 2011b. Attenuation in extended structures coated with thin magneto-dielectric absorber layer. Prog. Electromagn. Res. 118, 441–459.

Koledintseva, M., Khilkevich, V.V., Razmadze, A.G., Gafarov, A.Y., De, S., Drewniak, J.L., 2012. Evaluation of absorptive properties and permeability of thin sheet magneto-dielectric materials. J. Magn. Magn. Mater. 324, 3389–3392.

Kong, L., Yin, X., Yuan, X., Zhang, Y., Liu, X., Cheng, L., et al., 2014. Electromagnetic wave absorption properties of graphene modified with carbon nanotube/poly(dimethyl siloxane) composites. Carbon. 73, 185–193.

Labunov, V.A., Danilyuk, A.L., Prudnikava, A.L., Komissarov, I., Shulitski, B.G., Speisser, C., et al., 2012. Microwave absorption in nanocomposite material of magnetically functionalized carbon nanotubes. J. Appl. Phys. 112, 024302.

Lagarkov, A.N., Rozanov, K.N., 2009. High-frequency behavior of magnetic composites. J. Magn. Magn. Mater. 321, 2082–2092.

Lagarkov, A.N., Semenenko, V.N., Kisel, V.N., Chistyaev, V.A., 2003. Development and simulation of microwave artificial magnetic composites utilizing nonmagnetic inclusions. J. Magn. Magn. Mater. 258–259, 161–166.

Laird Technologies and Emerson Cuming Microwave Products, see RF/Microwave Absorbers & Dielectrics at http://www.lairdtech.com/products and http://www.eccosorb.com/, access date January 31, 2017.

Li, B.-W., Shen, Y., Yue, Z.-X., Nan, C.-W., 2006. Enhanced microwave absorption in nickel/hexagonal-ferrite/polymer composites. Appl. Phys. Lett. 89, 132504.

Li, G., Sheng, L., Yu, L., An, K., Ren, W., Zhao, X., 2015. Electromagnetic and microwave absorption properties of single-walled carbon nanotubes and $CoFe_2O_4$ nanocomposites. Mater. Sci. Eng. B. 193, 153–159.

Li, J., Zhang, Y.-J., Gafarov, A.Y., De, S., Koledintseva, M.Y., Marchand, J., et al., 2012a. EMI reduction evaluation with flexible absorbing materials and ferrite cores applied on cables. IEEE Symp. Electromag. Compat. 646–651, Pittsburg, PA.

Li, M., Xiao, S.Q., Bai, Y.Y., Wang, B.Z., 2012b. An ultrathin and broadband radar absorber using resistive FSS. IEEE Antennas Wireless Propagat. Lett. 11, 748–751.

Li, Q., Feng, Z., Yan, S., Nie, Y., Wang, X., 2014. Electromagnetic properties and impedance matching effect of flaky Fe–Si–Al/Co_2Z ferrite composite. J. Electr. Mater. 43, 3688–3694.

Li, Y., Zhao, Y., Sun, J., Hao, Y., Zhang, J., Han, X., 2016. Mechanical and electromagnetic interference shielding properties of carbon fiber/graphene nanosheets/epoxy composite. Polym. Comp. 37, 2494–2502.

Lim, K.M., Kim, M.C., Lee, K.A., Park, C.G., 2003. Electromagnetic wave absorption properties of amorphous alloy–ferrite–epoxy composites in quasi-microwave band. IEEE Trans. Magn. 39, 1836–1841.

Lin, J.-H., Lin, Z.-I., Pan, Y.-J., Chen, C.-K., Huang, C.-L., Huang, C.-H., et al., 2016. Improvement in mechanical properties and electromagnetic interference shielding effectiveness of PVA-based composites: synergistic effect between graphene nano-sheets and multi-walled carbon nanotubes. Macromol. Mater. Eng. 301, 199–211.

Ling, Q., Sun, J., Zhao, Q., Zhou, Q., 2011. Effects of carbon black content on microwave absorbing and mechanical properties of linear low density polyethylene/ethylene-octene copolymer/calcium carbonate composites. Pol. Plast. Tech. Eng. 50, 89–94.

Liu, J.R., Itoh, M., Horikawa, T., Itakura, M., Kuwano, N., Machida, K., 2004. Complex permittivity, permeability and electromagnetic wave absorption of α-Fe/C(amorphous) and Fe_2B/C(amorphous) nanocomposites.. J. Phys. D: Appl. Phys. 37, 2737–2741.

Liu, J.R., Itoh, M., Machida, K., 2006. Magnetic and electromagnetic wave absorption properties of α-Fe/Z-type Ba-ferrite nanocomposites. Appl. Phys. Lett. 88, 062503.

Liu, L., Matitsine, S.M., Gan, Y.B., Chen, L.F., Kong, L.B., Rozanov, K.N., 2007. Frequency dependence of effective permittivity of carbon nanotube composites. J. Appl. Phys. 101, 094106.

Liu, P., Yao, Z., Zhou, J., 2015. Preparation of reduced graphene oxide/$Ni_{0.4}Zn_{0.4}Co_{0.2}Fe_2O_4$ nanocomposites and their excellent microwave absorption properties. Ceram. Int. 41, 13409–13416.

Liu, P., Yao, Z., Zhou, J., 2016. Fabrication and microwave absorption of reduced graphene oxide/ $Ni_{0.4}Zn_{0.4}Co_{0.2}Fe_2O_4$ nanocomposites. Ceram. Int. 42, 9241–9249.

Liu, Q., Zi, Z., Zhang, M., Pang, A., Dai, J., Sun, Y., 2013a. Enhanced microwave absorption properties of carbonyl iron/Fe_3O_4 composites synthesized by a simple hydrothermal method. J. Alloys Compd. 561, 65–70.

Liu, Q.C., Dai, J.M., Zi, Z.F., Pang, A.B., Liu, Q.Z., Wu, D.J., et al., 2013b. Low temperature solution synthesis and microwave absorption properties of multiwalled carbon nanotubes/Fe_3O_4 composites. J. Low Temp. Phys. 170, 261–267.

Liu, X., Zhang, Z., Wu, Y., 2011. Absorption properties of carbon black/silicon carbide microwave absorbers. Comp.: B. 42, 326–329.

Lv, H., Ji, G., Liang, X., Zhang, H., Du, Y., 2015. A novel rod-like MnO_2@Fe loading on graphene giving excellent electromagnetic absorption properties. J. Mater. Chem. C. 3, 5056–5064.

Ma, E., Li, J., Zhao, N., Liu, E., He, C., Shi, C., 2013. Preparation of reduced graphene oxide/Fe_3O_4 nanocomposite and its microwave electromagnetic properties. Mater. Lett. 91, 209–212.

Maiti, S., Shrivastava, N.K., Suin, S., Khatua, B.B., 2013. Polystyrene/MWCNT/graphite nanoplate nanocomposites: efficient electromagnetic interference shielding material through graphite nanoplate – MWCNT – graphite nanoplate networking. Appl. Mater. Interfaces. 5, 4712–4724.

Mandal, A., Das, C.K., 2014. Effect of $BaTiO_3$ on the microwave absorbing properties of co-doped Ni-Zn ferrite nanocomposites. J. Appl. Pol. Sci. 131, 39926.

Mandal, A., Ghosh, D., Malas, A., Pal, P., Das, C.K., 2013. Synthesis and microwave absorbing properties of Cu-doped nickel zinc ferrite/$Pb(Zr_{0.52}Ti_{0.48})O_3$ nanocomposites. J. Eng. 2013, 391083.

Maruta, K., Sugawara, M., Shimada, Y., Yamaguchi, M., 2006. Analysis of optimum sheet for integrated electromagnetic noise suppressors. IEEE Trans. Magn. 42, 3377–3379.

McDowell, A.J., Hubing, T.H., 2014. Analysis and comparison of plane wave shielding effectiveness decompositions. IEEE Trans. Electromag. Compat. 56, 1711–1714.

Micheli, D., Pastore, R., Apollo, C., Marchetti, M., Gradoni, G., Primiani, V.M., et al., 2011. Broadband electromagnetic absorbers using carbon nanostructure-based composites. IEEE Trans. Microwave Theory Tech. 59, 2633–2646.

Molex, see EMI/Noise Suppression Sheets at http://www.molex.com/molex/product-search/products.action, access date January 31, 2017.

Mosallaei, H., Rahmat-Samii, Y., 2000. RCS reduction of canonical targets using genetic algorithm synthesized RAM. IEEE Trans. Antennas Propag. 48, 1594–1606.

MuRata, see Microwave Absorber at http://www.murata.com/en-us/products/emc/absorber, access date January 31, 2017.

Naito, Y., Suetake, K., 1971. Application of ferrite to electromagnetic wave absorber and its characteristics. IEEE Trans. Microw. Theory Technol. 19, 65–72.

NEC-Tokin, see EMC & Noise Countermeasure/Inductors/Transformers at https://www.nec-tokin.com/english/product/dl_emc.html, access date January 31, 2017.

Neelakanta, P.S., 1995. Handbook of Electromagnetic Materials: Monolithic and Composite Versions and Their Applications. CRC Press, Boca Raton, USA, pp. 491–508. ISBN 0-8493-2500-5, Chapter 21 "Electromagnetic Shielding Materials", pp. 447-490; Chapter 22 "Electromagnetic Wave Absorbing Materials".

Neo, C.P., Varadan, V.K., 2004. Optimization of carbon fiber composite for microwave absorber. IEEE Trans. Electromagn. Compat. 46, 102–106.

Nicolson, A.M., Ross, G., 1970. Measurement of the intrinsic properties of materials by time domain techniques. IEEE Trans. Instrum. Meas. 19, 377–382.

Ott, H.W., 2009. Electromagnetic Compatibility Engineering. Wiley, Hoboken, NJ, USA, ISBN 978-0-470-18930-6, Ch. 6 "Shielding".

Paul, C.R., 2006. Introduction to Electromagnetic Compatibility. 2nd ed Wiley, Hoboken, NJ, USA, ISBN-13: 978-0-471-75500-5, ISBN-10: 0-471-75500-1, Ch. 10 "Shielding".

Perini, J., Cohen, L.S., 1993. Design of broad-band radar-absorbing materials for large angles of incidence. IEEE Trans. EMC. 35, 223–230.

Petrova, I., Kotsilkova, R., Ivanov, E., Kuzhir, P., Bychanok, D., Kouravelou, K., et al., 2016. Nanoscale reinforcement of polypropylene composites with carbon nanotubes and clay: dispersion state, electromagnetic and nanomechanical properties. Pol. Eng. Sci. 56, 269–277.

Phan, C.H., Mariatti, M., Koh, Y.H., 2016. Electromagnetic interference shielding performance of epoxy composites filled with multiwalled carbon nanotubes/manganese zinc ferrite hybrid fillers. J. Magn. Magn. Mater. 401, 472–478.

Pottel, P., 1959. Uber die erhohung der frequenzbandbreite danner "λ/4-schicht" absorber fur electromagnetische zentimmeterwellen. Z. Angew. Phys. 11, 46–51.

Pozar, D.M., 1998. Microwave Engineering. second ed. Wiley, New York, ISBN 0-471-17096-8.

Qin, H., Liao, Q., Zhang, G., Huang, Y., Zhang, Y., 2013. Microwave absorption properties of carbon black and tetrapod-like ZnO whiskers composites. Appl. Surf. Sci. 286, 7–11.

Qing, Y., Zhou, W., Luo, F., Zhu, D., 2011. Optimization of electromagnetic matching of carbonyl iron/$BaTiO_3$ composites for microwave absorption. J. Magn. Magn. Mater. 323, 600–606.

Qing, Y., Wang, X., Zhou, Y., Huang, Z., Luo, F., Zhou, W., 2014. Enhanced microwave absorption of multi-walled carbon nanotubes/epoxy composites incorporated with ceramic particles. Compos. Sci. Technol. 102, 161–168.

Radchenko A., J. Bishop, R. Johnson, P. Dixon, M. Koledintseva, R. Jobava, et al., Sheet absorbing material modeling and application for enclosures, IEEE Symp. Electromag. Compat., Denver, CO, Aug. 5–9, 2013, pp. 645–650.

Ren, F., Zhu, G., Wu, G., Wang, K., Cui, X., 2016. Effects of surfactant treatment on mechanical and microwave absorbing properties of graphene nanosheets/multiwalled carbon nanotubes/cyanate ester composites. Pol. Compos. Available from: http://dx.doi.org/10.1002/pc.23908.

Rozanov, K.N., Koledintseva, M.Y., 2016. Application of generalized Snoek's law over a finite frequency range: a case study. J. Appl. Phys. 119, 073901.

Rozanov, K.N., Starostenko, S.N., 1999. Numerical study of bandwidth of radar absorbers. Eur. Phys. J.: Appl. Phys. 8, 147–151.

Rozanov, K.N., Starostenko, S.N., 2003. Influence of permeability dispersion on the bandwidth of magnetic radio absorbers. J. Commun. Technol. Electron. 48, 652–659.

Rozanov, K.N., 2000. Ultimate thickness to bandwidth ratio of radar absorbers. IEEE Trans. Antennas Propag. 48, 1230–1234.

Ruck, G., Barrick, D.E., Stuart, W.D., Krichbaum, C.K., 1970. Radar Cross Section Handbook, vol. 2. Plenum, New York, ch. 8.

Sanderson, A.E., 1971. Effect of surface roughness on propagation of the TEM mode. Adv. Microwave. 7, 2–57.

Schelkunoff, S.A., 1943. Electromagnetic Waves. D. Van Nostrand, Princeton, NJ.

Schulz, R.B., Plantz, V.C., Brush, D.R., 1988. Shielding theory and practice. IEEE Trans. Electromagn. Compat. 30, 187–201.

Sharma, S.K., Gupta, V., Tandon, R.P., Sachdev, V.K., 2016. Synergic effect of graphene and MWCNT fillers on electromagnetic shielding properties of graphene—MWCNT/ABS nanocomposites. RCS Adv. 6, 18257–18265.

Shen, B., Zhai, W., Tao, M., Ling, J., Zheng, W., 2013. Lightweight, multifunctional polyetherimide/graphene@Fe_3O_4 composite foams for shielding of electromagnetic pollution. Appl. Mater. Interfaces. 5, 11383–11391.

Shen, J., Chen, K., Li, L., Duan, Y., Li, J., Kong, W., 2014. Fabrication of Z-type barium ferrite/silica composites with enhanced microwave absorption. Sci. China Technol. Sci. 57, 1858–1864.

Shen, X., Song, F., Xiang, J., Liu, M., Zhu, Y., Wang, Y., 2012. Shape anisotropy, exchange-coupling interaction and microwave absorption of hard/soft nanocomposite ferrite microfibers. J. Am. Ceram. Soc. 95, 3863–3870.

Shen, X., Song, F., Yang, X., Wang, Z., Jing, M., Wang, Y., 2015. Hexaferrite/a-iron composite nanowires: Microstructure, exchange-coupling interaction and microwave absorption. J. Alloys Compd. 621, 146–153.

Shimba, K., Tezuka, N., Sugimoto, S., 2012. Magnetic and microwave absorption properties of polymer composites with amorphous Fe-B/Ni–Zn ferrite nanoparticles. Mater. Sci. Eng., B. 177, 251–256.

Sihvola A., 1999, Electromagnetic Mixing Formulas and Applications, IEE Electromagnetic Waves Series 47, The Institution of Electrical Engineers, ISBN 0-85296-772-1, London, UK.

Singh, A.P., Garg, P., Alam, F., Singh, K., Mathur, R.B., Tandon, R.P., et al., 2012. Phenolic resin-based composite sheets filled with mixtures of reduced graphene oxide, γ-Fe_2O_3 and carbon fibers for excellent electromagnetic interference shielding in the X-band. Carbon. 50, 3868–3875.

Snoek, J.L., 1948. Dispersion and absorption in magnetic ferrites at frequencies above one Mc/s. Physica. 14, 207–217.

Song, F., Shen, X., Yang, X., Meng, X., Xiang, J., Liu, R., et al., 2013. Bandwidth enhancement in microwave absorption of binary nanocomposite ferrites hollow microfibers. J. Nanosci. Nanotechnol. 13, 3115–3120.

Song, W.-L., Cao, M.-S., Wen, B., Hou, Z.-L., Cheng, J., Yuan, J., 2012. Synthesis of zinc oxide particles coated multiwalled carbon nanotubes: dielectric properties, electromagnetic interference shielding and microwave absorption. Mater. Res. Bull. 47, 1747–1754.

Stergiou, C., Litsardakis, G., 2012. Design of microwave absorbing coatings with new Ni and La doped $SrCo_2$-W hexaferrites. IEEE Trans. Magn. 48, 1516–1519.

Stergiou, C.A., Stimoniaris, A.Z., Delides, C.G., 2015. Hybrid nanocomposites with organoclay and carbon-based fillers for EMI suppression. IEEE Trans. EMC. 57, 470–476.

Sun X., H. Wang, Z. Chao, L. Zhang, Y. Wang, Y. He, et al., Preparation and electromagnetic characteristics of flake shaped carbonyl iron–zinc oxide nanocomposites, IEEE Proc., UIC-ATC-ScalCom-CBDC-IoP, Beijing, China, 2015.

Sun, X., Liu, X., Shen, X., Wu, Y., Wang, Z., Kim, J.-K., 2016. Graphene foam/carbon nanotube/poly(dimethyl siloxane) composites for exceptional microwave shielding. Composites A. 85, 199–206.

Suzuki H., T. Hotchi, M. Inoue, S. Takeda, Radiation suppression ratio measurement of noise suppression sheet by compact 3m method, Proc. 18th Int. Zurich Symp. on Electromagnetic Compatibility, Munich, pp. 281–283, 2007.

Tanrattanakul, V., Bunchuay, A., 2007. Microwave absorbing rubber composites containing carbon black and aluminum powder. J. Appl. Pol. Sci. 1015, 2036–2045.

TDK, see Noise Suppressing/ Magnetic Sheet at https://product.tdk.com/info/en/products/noise_magnet-sheet/index.html, access date January 31, 2017.

Ting, T.-H., Chiang, C.-C., Lin, P.-C., Lin, C.-H., 2013. Optimisation of the electromagnetic matching of manganese dioxide/multi-wall carbon nanotube composites as dielectric microwave-absorbing materials. J. Magn. Magn. Mater. 339, 100–105.

Tong, S.-Y., Tung, M.-J., Ko, W.-S., Huang, Y.-T., Wang, Y.-P., Wang, L.-C., et al., 2013. Effect of Ni fillers on microwave absorption and effective permeability of NiCuZn ferrite/Ni/polymer functional composites. J. Alloys Compd. 550, 39–45.

Tosaka, T., Nagano, I., Yagitani, S., Yoshimura, Y., 2005. Determining the relative permeability and conductivity of thin materials. IEEE Trans. Electromagn. Compat. 47, 352–360.

Tretyakov, S., 2016. Thin absorbers: operational principles and various realizations. IEEE Electromagn. Compat. Mag. 5, 61–66, Quarter 2.

Vinayasree, S., Soloman, M.A., Sunny, V., Mohanan, P., Kurian, P., Anantharaman, M.R., 2013. A microwave absorber based on strontium ferrite–carbon black–nitrile rubber for S and X-band applications. Compos. Sci. Technol. 82, 69–75.

Vinayasree, S., Soloman, M.A., Sunny, V., Mohanan, P., Kurian, P., Joy, P.A., et al., 2014. Flexible microwave absorbers based on barium hexaferrite, carbon black, and nitrile rubber for 2–12 GHz applications. J. Appl. Phys. 116, 024902.

Vinogradov, A.P., Machnovskii, D.P., Rozanov, K.N., 1999. Effective boundary layer in composite materials. J. Commun. Technol. Electron. 44, 317–322.

Wang, H., Zhu, D., Zhou, W., Luo, F., 2015a. Effect of multiwalled carbon nanotubes on the electromagnetic interference shielding properties of polyimide/carbonyl Iron composites. Ind. Eng. Chem. Res. 54, 6589–6595.

Wang, H., Zhu, D., Zhou, W., Luo, F., 2015b. Synthesis and microwave absorbing properties of Ni–Cu ferrite/MWCNTs composites. J. Mater. Sci.: Mater. Electron. 26, 7698–7704.

Wang, J., Wang, J., Xu, R., Sun, Y., Zhang, B., Chen, W., et al., 2015c. Enhanced microwave absorption properties of epoxy composites reinforced with $Fe_{50}Ni_{50}$-functionalized graphene. J. Alloys Compd. 653, 14–21.

Wang, J., Sun, Y., Chen, W., Wang, T., Xu, R., Wang, J., 2015d. Enhanced microwave absorption performance of lightweight absorber based on reduced graphene oxide and Ag-coated hollow glass spheres/epoxy composite. J. Appl. Phys. 117, 154903.

Wang, L., Zhang, J., Zhang, Q., 2015e. The effect of MWCNTs on the microwave electromagnetic properties of ferrite–MWCNTs composites. J. Mater. Sci.: Mater. Electron. 26, 1895–1899.

Wang, L., Huang, Y., Ding, X., Liu, P., Zong, M., 2013a. Ternary nanocomposites of graphene@SiO_2@NiO nanoflowers: synthesis and their microwave electromagnetic properties. Micro Nano Lett. 8, 391–394.

Wang, T., Qiao, L., Han, R., Zhang, Z.Q., 2012. The origin of reflection loss peaks in the double-layer electromagnetic wave absorber. J. Magn. Magn. Mater. 324, 3209–3212.

Wang, X., Li, Q., Su, Z., Gong, W., Gong, R., Chen, Y., et al., 2015f. Enhanced microwave absorption of multiferroic Co_2Z hexaferrite–$BaTiO_3$ composites with tunable impedance matching. J. Alloys Compd. 243, 111–115.

Wang, Y., Guan, H., Du, S., Wang, Y., 2015g. A facile hydrothermal synthesis of MnO_2 nanorod–reduced graphene oxide nanocomposites possessing excellent microwave absorption properties. RCS Adv. 5, 88979–88988.

Wang, Y., Huang, Y., Wang, Q., Zong, M., 2013b. Preparation and electromagnetic properties of graphene-supported $Ni_{0.8}Zn_{0.2}Ce_{0.06}Fe_{1.94}O_4$ nanocomposite. Powder Technol. 249, 304–308.

Wang, Z., Wei, G., Zhao, G.-L., 2013c. Enhanced electromagnetic wave shielding effectiveness of Fe doped carbon nanotubes/epoxy composites. Appl. Phys. Lett. 103, 183109.

Weir, W.B., 1974. Automatic measurement of complex dielectric constant and permeability at microwave frequencies. Proc. IEEE. 62, 33–36.

Wen, F., Zhang, F., Liu, Z., 2011. Investigation on microwave absorption properties for multiwalled carbon nanotubes/Fe/Co/Ni nanopowders as lightweight absorbers, J. Phys. Chem. C, 115, 14025–14030.

Wen, S.L., Liu, Y., Zhao, X.C., 2015. Facile chemical synthesis, electromagnetic response, and enhanced microwave absorption of cobalt powders with controllable morphologies. J. Chem. Phys. 143, 084707.

Wu, Y., Han, M., Deng, L., 2015. Enhancing the microwave absorption properties of Fe–Cu–Nb–Si–B nanocomposite flakes by coating with spinel ferrite $NiFe_2O_4$. IEEE Trans. Magn. 51, 2802204.

Xiao, Q., Hao, B., Li, L., Mao, M., Zhou, Y., Xu, F., 2013. Preparation and microwave absorbing performance of Co–Cu–Ni ferrite/multi-walled carbon nanotubes composites. Adv. Mater. Res. 652–654, 327–330.

Xu, J., Koledintseva, M., Zhang, Y., He, Y., DuBroff, R.E., Drewniak, J.L., et al., 2010. Complex permittivity and permeability measurements and finite-difference time domain simulation of ferrite materials. IEEE Trans. Electromagn. Compat. 52, 878–887.

Xu, Y., Zhang, D., Cai, J., Yuan, L., Zhang, W., 2012. Effects of multi-walled carbon nanotubes on the electromagnetic absorbing characteristics of composites filled with carbonyl iron particles. J. Mater. Sci. Technol. 28, 34–40.

Yamaguchi, M., Maruta, K., Ono, H., 2005. Operating mechanism for RF electromagnetic noise suppression sheets. IEEE Trans. Magn. 41, 3565–3567.

Yan, L., Wang, J., Ye, Y., Hao, Z., Liu, Q., Li, F., 2009. Broadband and thin microwave absorber of nickel–zinc ferrite/carbonyl iron composite. J. Alloys Compd. 487, 708–711.

Yang, H., Ye, T., Lin, Y., Liu, M., 2015. Preparation and microwave absorption property of graphene/$BaFe_{12}O_{19}$/$CoFe_2O_4$ nanocomposite. Appl. Surf. Sci. 357, 1289–1293.

Yang, R.B., Liang, W.F., Wu, C.H., Chen, C.C., 2016. Synthesis and microwave absorbing characteristics of functionally graded carbonyl iron/polyurethane composites. AIP Adv. 6, 055910.

Yang, X., Wang, Z., Jing, M., Liu, R., Song, F., Shen, X., 2014. Magnetic nanocomposite Ba-ferrite/α-iron hollow microfiber: A multifunctional 1D space platform for dyes removal and microwave absorption. Ceram. Int. 40, 15585–15594.

Yao, J., Wang, L., Liu, G., Hua, S., 2009. Electromagnetic properties of nanometer SiC/CNTs composite. Adv. Mater. Res. 79–82, 349–352.

Yoshida, S., Ono, H., Ando, S., Tsuda, F., Ito, T., Shimada, Y., et al., 2001. High-frequency noise suppression in downsized circuits using magnetic granular films. IEEE Trans. Magn. 37, 2401–2403.

Yuchang, Q., Dandan, M., Yingying, Z., Fa, L., Wancheng, Z., 2015. Graphene nanosheet- and flake carbonyl iron particle-filled epoxy–silicone composites as thin–thickness and wide-bandwidth microwave absorber. Carbon. 86, 98–107.

Yun, Y.W., Kim, S.W., Kim, G.Y., Kim, Y.B., Yun, Y.C., Lee, K.S., 2006. Electromagnetic shielding properties of soft magnetic metal and ferrite composites for application to suppress noise in a radio frequency range. J. Electroceram. 17, 467–469.

Zabri, N., Cahill, R., Schuchinsky, A., 2014. Polarisation independent resistively loaded frequency selective surface absorber with optimum oblique incidence performance. IET Microwave Antennas Propag. 8, 1198–1203.

Zhang, B., Wang, J., Wang, J., Huo, S., Zhang, B., Tang, Y., 2016. Microwave absorption properties of lightweight absorber based on $Fe_{50}Ni_{50}$-coated poly(acrylonitrile) microspheres and reduced graphene oxide composites. J. Magn. Magn. Mater. 413, 81–88.

Zhang, H.B., Deng, L.W., Zhou, P.H., Zhang, L., Cheng, D.M., Chen, H.Y., et al., 2013a. Low frequency needlepoint-shape metamaterial absorber based on magnetic medium. J. Appl. Phys. 113, 013903.
Zhang, L., Zhang, X., Zhang, G., Zhang, Z., Liu, S., Li, P., et al., 2015a. Investigation on the optimization, design and microwave absorption properties of reduced graphene oxide/ tetrapod-like ZnO composites. RCS Adv. 5, 10197−10203.
Zhang, L.B., Lu, H.P., Zhou, P.H., Xie, J.L., Deng, L.J., 2015b. Oblique incidence performance of microwave absorbers based on magnetic polymer composites. IEEE Trans. Magn. 51, 7100604.
Zhang, Z., Ji, Z., Duan, Y., Gu, S., Guo, J., 2013b. The superior electromagnetic properties of carbonyl-iron/$Fe_{91.2}Si_{3.1}P_{2.9}$Sb2.8 composites powder and impedance match mechanism. J. Mater. Sci.: Mater. Electron. 24, 968−973.
Zhao C., M. Shen, Z. Li, R. Sun, A. Xia, X. Liu, 2016, Green synthesis and enhanced microwave absorption property of reduced graphene oxide-SrFe12O19 nanocomposites, J. Alloys Compd., 689, 1037−1043.
Zhu, H.-L., Bai, Y.-J., Liu, R., Lun, N., Qi, Y.-X., Han, F.-D., et al., 2011. In situ synthesis of one-dimensional MWCNT/SiC porous nanocomposites with excellent microwave absorption properties. J. Mater. Chem. 21, 13581−13587.
Zong, M., Huang, Y., Zhang, N., Wu, H., 2015. Influence of (RGO)/(ferrite) ratios and graphene reduction degree on microwave absorption properties of graphene composites. J. Alloys Compd. 664, 491−501.
Zou, J., Liu, Q., Zi, Z., Dai, J., 2014. Enhanced electromagnetic wave absorption properties of planar anisotropy carbonyl-iron/Fe_3O_4 composites in gigahertz range. Mater. Res. Innovations. 18, 304−309.

Hybrid polysaccharide-based systems for biomedical applications

4

Paula I.P. Soares[1], Coro Echeverria[1], Ana C. Baptista[1], Carlos F.C. João[1], Susete N. Fernandes[1], Ana P.C. Almeida[1], Jorge C. Silva[2], Maria H. Godinho[1] and João P. Borges[1]
[1]i3N/CENIMAT, Department of Materials Science, Faculty of Science and Technology, Universidade NOVA de Lisboa, Caparica, Portugal, [2]i3N/CENIMAT, Physics Department, Faculty of Science and Technology, Universidade NOVA de Lisboa, Caparica, Portugal

Chapter Outline

4.1 Introduction 107
4.2 Biosensors and actuators 110
- 4.2.1 Biosensors 110
- 4.2.2 Actuators 114

4.3 Theranostic systems 117
- 4.3.1 Polysaccharides in theranostic applications 118
- 4.3.2 Hybrid polysaccharide systems for theranostic applications 120

4.4 Tissue engineering 123
- 4.4.1 Tissue-engineering applications of native cellulose and chitin/chitosan 123
- 4.4.2 Tissue-engineering applications of nanocellulose and nanochitin 127
- 4.4.3 Biomimetic scaffolds for tissue engineering 129

4.5 Conclusions and future perspectives 134
Acknowledgments 135
References 135

4.1 Introduction

Polysaccharides, among other polymers, have been widely used for biomedical applications including tissue engineering, drug delivery, wound healing, blood plasma expansion, and injectable drug release (Peng et al., 2015). The presence of diverse functional groups facilitates the modification of polysaccharides to obtain systems for diverse biomedical applications. Moreover, this class of polymers is

Hybrid Polymer Composite Materials: Applications. DOI: http://dx.doi.org/10.1016/B978-0-08-100785-3.00004-8

known to be nontoxic, biocompatible, biodegradable, and renewable. These macromolecules are composed of repeating units of mono- or disaccharides linked by glycosidic bonds that form either linear or branched structures (Dias et al., 2011).

Surveying the literature, one finds several references reviewing the different types of polysaccharides and their applications (Zong et al., 2012; Ferreira et al., 2015; Liu et al., 2015; Shi, 2016). These different types can be divided into two main groups according to their source: natural polysaccharides that are obtained from algae, plants, microorganisms, and animals, and semisynthetic polysaccharides that are produced by chemical or enzymatic modification of the parent macromolecules (Zong et al., 2012).

Among the natural polysaccharides, cellulose and chitin have a major highlight as these are the most abundant natural origin biomaterials (João et al., 2015). Cellulose is found primarily in plants and is a linear chain of ringed glucose molecules [poly-β-(1 → 4)-D-glucosamine units] (Fig. 4.1) with a flat ribbon-like conformation (Moon et al., 2011). During biosynthesis, the hydrogen bond between hydroxyl groups and oxygen of adjacent molecules induces the

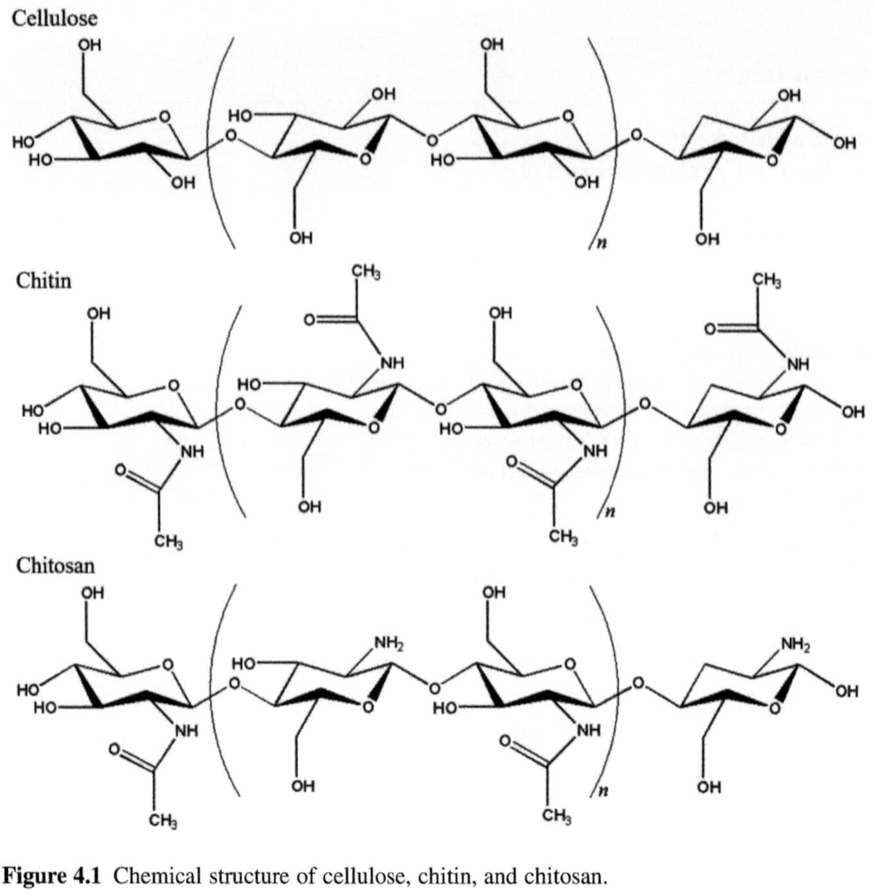

Figure 4.1 Chemical structure of cellulose, chitin, and chitosan.

formation of a parallel stacking of multiple cellulose chains. The continuous packing of the long chains gives rise to the formation of hierarchic structures starting from elementary fibrils and going to macrofibers. Cellulosic fibers have been extracted from different sources (wood, plants, tunicates, algae, and bacteria) and considered with different purposes. Its main use as reinforcement is due to easy renewability, biodegradability, availability, high toughness, low specific gravity, acceptable specific strength, and enhanced energy recovery (Bledzki and Gassan, 1999; Faruk et al., 2012; Thakur et al., 2013).

Chitin is a polysaccharide structurally similar to cellulose with vast biological and chemical attributes. Composed of poly-β-(1 → 4)-*N*-acetyl-D-glucosamine units (Fig. 4.1), it is found mainly in the exoskeleton of crustaceans and insects but also in mollusks and fungi (Muzzarelli, 1977). Because of its linear structure with two hydroxyl groups and an acetamide group, chitin is predominantly crystalline with strong hydrogen bonding. Chitosan (CS) is the deacetylated form of chitin (Fig. 4.1), being a polycationic polymer with one amino group and two hydroxyl groups in the repeating glycosidic residue. The carbohydrate backbone contributes to the rigid crystalline structure obtained after CS refinement. This structure is dependent on inter- and intramolecular hydrogen bounding (Dash et al., 2011). CS maintains the majority of chitin's properties but displays an utmost hydrophilicity and increased water absorption ability. CS can be dissolved in acidic solvents improving its processability and facilitating chemical modification when compared with pristine chitin (Desbrieres and Babak, 2010; Kim, 2013; Payne and Raghavan, 2007). CS is nontoxic, hydrophilic, biocompatible, biodegradable, and antibacterial and has been widely used as a biomaterial, prepared as films, three-dimensional (3D) structures and microspheres, among others, and as a pharmaceutical excipient in drug formulations (Liu et al., 2007; Dash et al., 2011).

Polysaccharides have also been used in the last decades as effective colloidal stabilizers, agents for increased biocompatibility, providing chemical functionality to nanostructures, namely inorganic nanomaterials such as carbon nanotubes (CNTs) (Bandyopadhyaya et al., 2002), colloidal silver (Balantrapu and Goia, 2009), gold nanoparticles (AuNPs) (Guo et al., 2010), quantum dots (QDs) (Shen et al., 2012), and iron oxide (Fe_3O_4) nanoparticles (NPs) (Soares et al., 2016b). These combinations led to the development of hybrid materials, composed of at least two components, an organic and an inorganic that has had a huge impact in several research fields, including in biomedical applications.

In this chapter, a highlight is made of the most recent publications concerning the use of hybrid materials composed of polysaccharides for the following biomedical applications: biosensors, actuators, theranostics, and tissue engineering. We will focus on composite materials based on cellulose and chitin/CS, although some other polysaccharides are also mentioned throughout the chapter. One cannot finish this chapter without giving the reader a glimpse on the use of liquid crystals-based materials, most notably liquid crystalline nanocellulose, within the field of scaffolding for tissue engineering.

4.2 Biosensors and actuators

4.2.1 Biosensors

Biosensors have become an emerging area of interdisciplinary research with particularly interest in the field of healthcare for the detection of various kinds of targeted analytes in biological systems. Fig. 4.2A shows the evolution of articles published with *Biosensors* keyword from 2006 to 2016. A biosensor is defined as an analytical device that integrates the specificity of biomolecules with a physicochemical transducer for conversion of a biochemical signal into a measurable electrical signal (Turner, 2013). The selectivity and specificity of a biosensor highly depend on biological recognition systems connected to a suitable transducer. Fig. 4.2B is representative of the typical architecture of biosensors.

As an ideal biosensor has to be stable for long-term application, several biomolecule immobilization strategies have been investigated such as adsorption, covalent linking, entrapment, cross-linking, or affinity (Sassolas et al., 2012). Chemical properties of a desired support decide the method of immobilization and the operational stability of a biosensor.

Naturally occurring macromolecular materials are a valuable support for biomolecules immobilization giving their biocompatibility (Suginta et al., 2013; Baptista et al., 2013b). For instance, polysaccharides such as cellulose and CS are often used as enzyme immobilization materials, due to their biocompatibility, chemical stability, mechanical and physical properties, low cost and availability as renewable natural resources. In this section, we will explore the recent advances made in the development of biosensors for glucose, cholesterol, and other biological molecules detection using cellulose and CS as matrices for enzyme immobilization.

4.2.1.1 Polysaccharide-based biosensors for glucose detection

Blood glucose monitoring has been established as a valuable clinical indicator of diabetes. Thus, significant research efforts are focused on developing improved

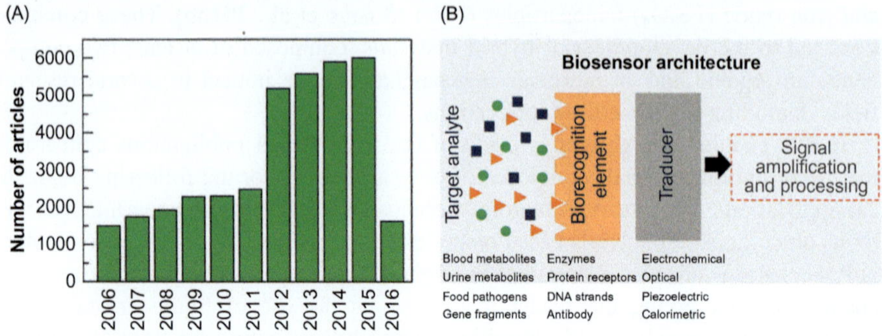

Figure 4.2 (A) Number of scientific articles found on ISI Web of Science using *Biosensors* keyword and (B) general scheme of biosensor architecture.

methods to measure glucose. Enzymatic amperometric glucose biosensors are the most common devices commercially available and have been widely studied over the last few decades. A glucose sensor based on Fe_3O_4 NPs−CS composite films has been reported by Kaushik et al. (2008). Glucose oxidase (GOx) was immobilized via physical adsorption onto a composite film made of Fe_3O_4 NPs dispersed in CS, and the nanocomposite was then deposited on an indium−tin oxide (ITO) glass plate. The reported GOx/CS−Fe_3O_4/ITO bioelectrode showed a fast response time (5 s) attributed to a faster electron communication feature of CS−Fe_3O_4 nanocomposite. This electrode can be used to detect glucose from 0.5 to 22 mM with a sensitivity of 9.3 $\mu A\ mg^{-1}\ dL^{-1}\ cm^{-2}$. Ren et al. (2009) studied the feasibility of an amperometric glucose biosensor based on the immobilization of GOx in gold nanorod−cellulose acetate (CA) composite. The biosensor exhibited a sensitivity of 8.4 $\mu A\ mM^{-1}\ cm^{-2}$, a low glucose detection limit of 0.2 μM and a glucose detection range between 0.2 μM and 2.2 mM. Wu et al. (2009) investigated the immobilization of GOx in a cellulose−multiwalled CNT (MWCNT) matrix reconstituted with a room temperature ionic liquid (IL) (1-ethyl-3-methylimidazolium acetate, [EMIM][CH_3COO]). The porous cellulose−MWCNT matrix allows a large amount of enzyme to be immobilized close to the electrode surface, where direct electron communication between active site of enzyme and the electrode is enabled. The encapsulated GOx showed good bioelectrochemical activity, enhanced biological affinity as well as good stability and repeatability. Later, the immobilization of GOx into cellulose−tin oxide (SnO_2) hybrid nanocomposite was reported by Mahadeva and Kim (2011) using a physical absorption method. The glucose biosensor reported has displayed a linear response in the range of 0.5−12 mM, which covers the clinical region of glucose concentration.

Matsuhisa et al. (2013) developed an amperometric glucose biosensor by covalently immobilizing GOx within a protein−polysaccharide hybrid sol−gel silicate-film. This hybrid sol−gel film consists of three organo-silanes (3-aminopropyltriethoxysilane, tetraethoxysilane, and triethoxy-1*H*,1*H*,2*H*,2*H*-tridecafluoro-*n*-octylsilane) and two biomacromolecules (bovine serum albumin and CS). The reported organic−inorganic-hybrid biosensor exhibited a sensitivity of 1.84 $\mu A\ mM^{-1}$, a glucose detection limit of 0.032 mM and highly selective determination of glucose. Recently, Esmaeili et al. (2015) explored the use of a polypyrrole−cellulose nanocrystal (PPy−CNC)-based composite in GOx adsorption for glucose sensing. The porous structure formed in the presence of nanocellulose provided strong adsorption ability for the immobilization of GOx, and the crystalline structure of nanocellulose also facilitates electron transfer between the electrode and GOx. The biosensor showed good linearity in the detection of glucose within the concentration range of 1−20 mM with a sensitivity of 0.73 $\mu A\ mM^{-1}$.

4.2.1.2 Polysaccharide-based biosensors for cholesterol detection

Cholesterol is a fundamental parameter in the diagnosis of coronary heart disease, atherosclerosis, and other clinical disorders, and in the risk assessment of thrombosis and myocardial infarction (Arya et al., 2008). Tsai et al. (2008)

developed a cholesterol biosensor by the combination of MWCNTs, CS, platinum (Pt) NPs, and cholesterol oxidase (ChOx). The Pt NPs were electrochemically deposited in MWCNT–CS matrix followed by ChOx loading. The proposed cholesterol biosensor retained 60% of its initial activity after 7 days when stored at 4°C and exhibited a sensitivity of 0.044 A M^{-1} cm^{-2} and a fast response time of about 8 s.

An organic–inorganic CS–SnO_2 nanobiocomposite film was developed by Ansari et al. (2009). CS and SnO_2 NPs were deposited onto an ITO glass plate to immobilize ChOx for cholesterol detection. This ChOx/CS–SnO_2/ITO cholesterol sensor was found to retain 95% of the enzyme activity for 4–6 weeks and exhibited a sensitivity of 34.7 mA mg^{-1} dL^{-1} cm^{-2}, a linear response to cholesterol in the range of 0.26–10.36 mM and a low detection limit of 0.13 mM. Moreover, the authors suggested that the nanobiocomposite can be used in order to estimate lipoproteins, triglycerides and urea, among other analytes.

Later, a polymer–polymer bioelectrode was reported by Barik et al. (2010). ChOx was covalently immobilized onto polyaniline–carboxymethyl cellulose (PANI–CMC) nanocomposite film deposited onto ITO-coated glass plate. The ChOx/PANI–CMC/ITO bioelectrode displayed sensitivity to cholesterol of 0.14 mA mM^{-1} cm^{-2}, linear response between 0.5 mM and 22 mM and a response time of 10 s. Gomathi et al. (2011) have described the development of an amperometric cholesterol biosensor by the immobilization of ChOx onto a CS nanofibers/AuNPs (CSNFs/AuNPs) composite network. The CSNF–AuNPs/ChOx biosensor exhibited excellent long-term stability, a linear response to cholesterol in the range of 1–45 μM, a sensitivity of 1.02 μA μM^{-1}, and a short response time of, approximately 5 s.

Recently, cellulose can be found not only in the bioelectrode composition but also as the substrate of biosensors. Cholesterol paper-based biosensors have emerged as a new, cheap, and fast way of diagnosis. Paper-based biosensors offer several advantages when compared to traditional substrates such as low cost, high abundance, biocompatibility, and disposability. Furthermore, paper-based analysis only requires a small amount of samples and reagents, which make it suitable for biosensor applications. A graphene (G), polyvinylpyrrolidone (PVP), and PANI composite has been successfully prepared by Ruecha et al. (2014) and used for the modification of paper-based biosensors via electrospraying. ChOx is attached to the G/PVP/PANI-modified electrode for the amperometric determination of cholesterol. Under optimum conditions, a linear range of 50 μM to 10 mM was achieved and the detection limit was found to be 1 μM. It was also concluded that this paper-based biosensor retained 89% of its initial response after a storage period of 2 weeks indicating a good stability. Also, a paper-based analytical device coupled with a silver NP-modified boron-doped diamond (AgNP/BDD) electrode was developed by Nantaphol et al. (2015) as a cholesterol sensor. An electrodeposition method was used to deposit AgNP onto the BDD electrode surface followed by drop casting of ChOx. The fabricated device demonstrated a good linearity from 0.01 to 7 mM, a cholesterol detection limit of 0.006 mM and a sensitivity of 49.6 μA mM^{-1} cm^{-2}.

4.2.1.3 Polysaccharide-based biosensors for the detection of other molecules

Monitoring the level of other biological molecules, such as urea, creatinine, and dopamine (DA), in the human body is also a vital issue in biochemical and medical diagnostics. Ali et al. (2013) studied the immobilization of urease onto a CS–Fe_3O_4 nanobiocomposite. The proposed urea biosensor demonstrated a good sensitivity with a potentiometric response of ~42 mV per decade at room temperature over a wide range of urea concentration (from 0.1 to 80 mM); a fast time response (<12 s) for the selected urea concentrations; and a good stability since the electrode retained 90% of its initial activity. Lately, a highly sensitive nonenzymatic urea sensor using nickel oxide deposited on a cellulose/CNT composite was reported by Nguyen and Yoon (2016). The electrode showed high stability and exhibited a sensitivity of 371 $\mu A\ mM^{-1}\ cm^{-2}$ with a fast response time of 4 s.

A three-enzyme creatinine biosensor was investigated by Yadav et al. (2011, 2012). Creatinine is an important clinical analyte for the determination of renal and muscular dysfunction (Killard and Smyth, 2000). Three enzymes, creatinine amidohydrolase (CA), creatine amidinohydrolase (CI), and sarcosine oxidase (SO) were immobilized on a zinc oxide (ZnO) NPs/CS/carboxylated MWCNT/PANI composite film electrochemically deposited on a Pt electrode (Yadav et al., 2011). The fabricated biosensor showed a linear response to creatinine in the range of 10–650 M with a sensitivity of 0.030 $\mu A\ \mu M^{-1}\ cm^{-2}$. When stored at 4°C after a period of 120 days, the biosensor showed a loss of 15% of its initial response. Later, the same authors reported the immobilization of CA, CI, and SO onto a composite film of Fe_3O_4 NPs and CS-*graft*-PANI (Fe_3O_4-NPs/CHIT-*g*-PANI) electrodeposited on the surface of a Pt electrode through glutaraldehyde coupling (Yadav et al., 2012). The biosensor revealed a better performance than the previous one exhibiting a creatinine sensitivity of 3.9 $\mu A\ \mu M^{-1}\ cm^{-2}$ with a rapid response of 2 s, approximately. Long-term stability was observed since only 10% of its initial response is lost after 120 uses over 200 days, when stored at 4°C.

Advances in the development of electrochemical DA sensors have been made in recent years. DA is one of the most important neurotransmitters and a key indicator for schizophrenia and Parkinson's disease (Jackowska and Krysinski, 2013). A tyrosinase biosensor based on Fe_3O_4–CS nanocomposite was developed by Wang et al. (2010) for the amperometric detection of DA. The fabricated biosensor exhibited a linear response between 2.0×10^{-8} M and 7.5×10^{-5} M and a sensitivity of 46 $\mu A\ mM^{-1}$. In 2016, palladium NPs-bacterial cellulose (PdBC) hybrid nanofibers were combined with laccase and Nafion to construct a novel biosensing platform to detect DA. This biosensor developed by Dawei et al. (Dawei et al., 2016) displayed an excellent electrocatalysis toward DA with a sensitivity of 38.4 $\mu A\ mM^{-1}$, a detection limit estimated to be 1.26 μM and wide linear range between 5 μM and 167 μM. The storage stability of the modified electrode was studied through one month of storage at pH 7 and 4°C, and the response current value retained 93.2% of its original value, suggesting satisfactory storage stability.

4.2.2 Actuators

Actuators, materials, or systems able to react to the external environment can transduce an input energy (electrical, thermal, light, and chemical gradients) into external mechanical response. These kind of systems are applicable in a wide range of areas such as robotics, automation, fluidics, optics, acoustics, biotechnology, and, of course, biomedical engineering that is the focus of the present chapter (Carpi and Smela, 2009).

A good inspiration for the development of actuators is found in Nature. Indeed, Nature has many examples, such as jellyfish, sea cucumber (see Section 4.3), sea anemones, and those found in the human body such as stomach, tongue, heart, and, of course, the typical soft actuator that is musculature.

The breakthrough that polymer science has accomplished in this subject is paramount, being possible to mimic the morphology and functionality of muscle-tendon-ligament structure by designing a wearable robotic device made of artificial muscles (Park et al., 2014b). The interest in the field is growing exponentially as deduced from Fig. 4.3, where the evolution of citation to research involving the words "polymeric actuators" is analyzed.

For instance, electroactive polymers (EAPs) actuators also known as "artificial muscles" have been developed and improved fulfilling the needs of practical industrial applications. Furthermore, their low actuation voltage made possible biomedical applications such as heart compression devices, surgical tools, and ocular muscles, among others (Carpi and Smela, 2009).

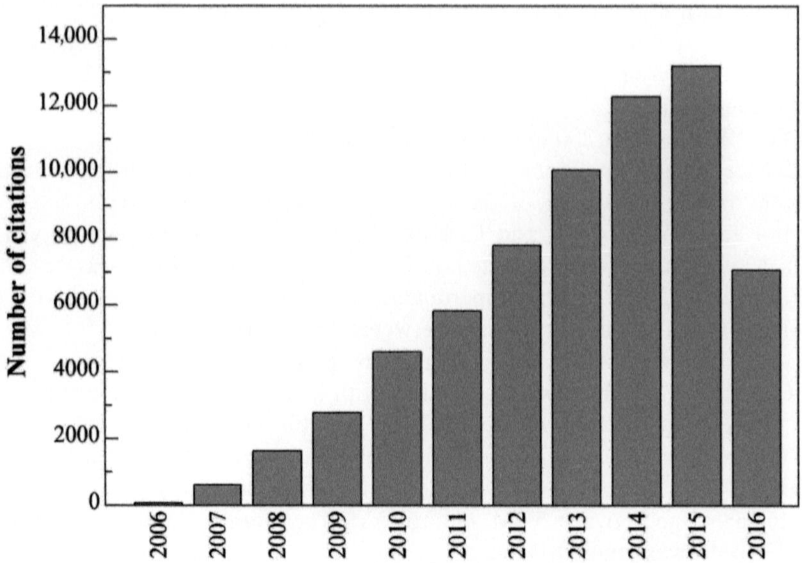

Figure 4.3 Evolution of citations in the last 10 years for "polymer actuators" keyword obtained from ISI Web of Science.

In this regard, biological polymers could impulse a step forward offering degrees of functionalities not available in most synthetic polymers, in addition to biocompatibility. Besides, most of the biopolymers exhibit shear piezoelectricity due to their natural orientation. Piezoelectricity, which is the coupling between electrical and mechanical properties, is exhibited by crystal structures that do not have a center of symmetry.

In this section, we will explore the recent development on biopolymer-based actuators with potential biomedical application, highlighting the contribution of polysaccharides such as cellulose and CS.

4.2.2.1 Biopolymer–polymer hybrid actuators

Kim et al. (2006) reported for the first time structural changes in cellulose by applying an electric field and presented cellulose as a smart material. In particular, the authors demonstrated the applicability of cellophane (cellulose xanthate) as an EAP and described it as electroactive paper (EAPap) by building a bending actuator using cellulose paper with gold electrodes deposited on both sides of the paper. The authors concluded that the responsiveness of the system lays on two factors: piezoelectricity of cellulose and ion transport. Regenerated cellulose possesses ordered and disordered regions, being the ordered domains crystalline. In such organization, disordered regions are able to retain water molecules but also contributed to the dipolar orientation stabilizing dipoles giving rise to a permanent polarization that ends up in a piezoelectric behavior. The ions (Na^+) in the system are mobile and migrate when exposed to an electric field. Ion and water transport across the polymer induced the volumetric change and thus the bending (Kim et al., 2006). Although large bending of the EAPap was achieved with low actuation, the system was extremely sensitive to humidity, factor that facilitated its degradation. In order to overcome this problem, they hybridized the cellulose EAPap with CS, by coating cellulose film with a CS layer (Wang et al., 2007). The bending displacement of the EAPap actuators was evaluated with respect to voltage, frequency, humidity, and acetic acid quantity. The authors found that the actuation of the hybrid cellulose/CS film was based on acetic acid anion migration and dependent on the voltage and even thickness of the film. However, it is indeed true that no noticeable improvements were achieved.

Seeking for a radical improvement in both low power consumption and larger electrochemical displacement, Li et al. (2011) developed an electrospun fullerenol–CA hybrid biocompatible actuator to be used in biomedical applications. They combine the properties of fullerenol, namely water-soluble and biodegradable with the biocompatibility of CA. The successful hybrid system presents improved actuation derived from the combination of three factors: an increased crystallization, piezoelectric effect (due to CA), and electrostrictivity of fullerenol. In addition, the morphological analysis of the obtained membranes revealed a porous structure mimicking the extracellular matrix of natural muscles.

For the development of artificial muscles, conductive polymers have received special attention due to their large active strain and stress, high power/weight ratio, moderate response, and excellent life cycle (Otero and Sansiñena, 1995; Lu et al., 2004; Qi et al., 2004). One of the strategies followed to improve their lack of mechanical stability and flexibility consists in hybridization of conductive polymers with hydrogels (Ismail et al., 2008; Sekine et al., 2010; Dai et al., 2008). In this line, Ismail et al. (2011) hybridized the conductive polymer PPy with CS; the authors combined the electroactivity of PPy that provide conductivity to the system with large swelling strain, mechanical stability, and biocompatibility of the CS hydrogel. The authors fabricated a 3D microfiber CS structure by means of wet spinning followed by the polymerization of pyrrole. Such architecture favored the electrical activity and improved the response time (Gowda et al., 2011). The hybrid system developed ended up in an artificial muscle able to sense applied current, electrolyte concentration, and temperature while actuating. The device obtained overcame a linear actuation strain of 0.54% and presented high enough conductivity to allow PPy oxidation and reduction reactions causing sensing and actuation.

BC is a nanomaterial widely studied, cost-efficient, and synthetized extracellularly by the bacterium *Acetobacter xylinum*. BC presents distinctive properties such as high purity, a fine fiber network, high crystallinity, high tensile strength, and tissue biocompatibility (Esa et al., 2014; Fu et al., 2013). BC has also been explored as an electroactive hybrid actuator material. Kim (2013) using freeze-dried BC, ILs, and poly(3,4-ethylenedioxythiophene)−poly(styrenesulfonate) (PEDOT:PSS) obtained a sponge-type porous system with durable actuation, without delamination and showing good electrochemical properties. The authors were able to demonstrate that 1-ethyl-3-methylimidazolium tetrafluoroborate (EMI-BF4) could be used to induce significant bending deformations (Kim, 2013). More recently, Kim et al. (2015) present a bioelectronic muscular actuator based on chemically modified graphene and TEMPO-oxidized BC, IL, and dimethyl sulfoxide (DMSO)-doped PEDOT:PSS with enhanced electrochemical properties and remarkably improved mechanical stiffness. Lately, Wang et al. (2016b) exploited the potential of using modified BC membrane with conductive polymers to build soft biomolecules with exceptionally high-performance actuation. The carboxylated BC combined with IL and PEDOT:PSS electrodes allowed the authors to prepare an actuator with improved ionic conductivity and ionic exchange capacity up to 22.8 and 1.5 times, respectively, when compared with pure BC, and with a larger bending deformation. If poly(2-acrylamido-2-methyl-1-propanesulfonic acid) is added to the same system, a cross-linked composite membrane could be obtained (Wang et al., 2016a). The authors showed that this eco-friendly ionic artificial muscle presents higher ionic conductivity, tensile strength, and specific capacitance (32.5%, 44.9%, and 160%, respectively) and bending deformation (4.5 times larger) than that of the CBC-IL actuator.

4.2.2.2 Biopolymer-carbonaceous particles hybrid actuators

Baughman et al. (1999) described the use of single-wall CNTs sheets as actuators (electrodes) where no dopant intercalation was required. By changing the applied voltage, electric charges were injected to the CNT electrode that was compensated by electrolyte ions giving rise to the so-called double layer. In the past decade, various CNT actuators were developed (Madden et al., 2006; Ebron et al., 2006), but nowadays the strategy followed toward optimized systems also comprises the hybridization of CNT with biopolymer composite actuators (Sukrut et al., 2008). CNT contributed to the hybrid system enhancing electrical, thermal, and mechanical properties. In this line, Hu et al. report a simple approach toward the fabrication of CS/CNT actuators by blending and casting CNT−CS films. In addition to its biocompatibility, low cost, and responsiveness, CS was chosen because of its dispersant ability for CNT. The actuator obtained presented motion controlled by low alternating voltage under atmospheric conditions, offering implementation opportunities in artificial muscles and biomimetic applications.

Lu and Chen (2010) took advantage of the compatibility of ILs with CS, the dispersant role of CS with CNT and the excellent electrochemical and electromechanical performance of CNT in ILs to fabricate an actuator from all biocompatible components. The electrode membrane was prepared from an aqueous solution of MWCNTs covered by CS and the ionic diffusion layer was obtained from IL−CS−glycerol solution. Glycerol was used to impart some flexibility to the membranes to facilitate further bending. The final dehydrated composite actuator was obtained by combining them (bimorph structure). The good compatibility of ILs in the system allowed the effective transportation between anode and cathode demonstrating actuation at very low applied voltage.

In the framework of actuators formed of hybrid biopolymer/CNT system, in the past year our group developed and optimized a humidity-driven cellulose-based liquid crystalline soft-motor (Geng et al., 2013; Echeverria et al., 2015). The principles of actuation laid in the cholesteric structure and anisotropy of liquid crystalline cellulose derivative, hydroxypropylcellulose (HPC) as this polysaccharide is able to from lyotropic liquid crystalline phases in water solutions (42 to 70 wt.%). Films of HPC prepared from liquid crystalline solutions were proven to produce anisotropic liquid crystal networks able to actuate when exposed to humidity gradients (Geng et al., 2013). The mechanism behind the motion is based on order/disorder transitions induced by water molecules. In a recent study, our group demonstrated that when hybridizing LC-HPC with highly anisotropic NPs such as CNT, the responsiveness to humidity was improved as a consequence of the increased anisotropy and orientational order of the solid films.

4.3 Theranostic systems

Recently the concept of precision medicine has emerged as a promising way to address unmet medical needs. This concept is based on prevention and treatment strategies to take individual genetic and epigenetic variability into account, which allows the production of better therapeutic outcomes and simultaneously reducing patient discomfort and undesirable side-effects (Jain et al., 2015; Collins and Varmus 2015). Based on the methods developed for precision medicine, the word "theranostic" was first introduced in 2002 by Funkhouser (2002) by coupling therapy and diagnostic imaging during breast cancer treatment. This method was designed to either predict individual drug response or select a patient subpopulation for clinical trials in order to determine a specific therapy such as treatment selection or dose optimization. In this regard, the development of theranostic systems may revolutionize the entire healthcare scene by developing personalized medicine for each patient. The use of nanoscale systems is a potential useful tool to achieve the ultimate aim of theranostics: to combine in a single agent both diagnostic and therapeutic capabilities (Wang et al., 2013c; Jain et al., 2015; Prabhu et al., 2015).

The development of nanotechnology in the last decades led to the discovery or improvement of several techniques in the biomedical field. In the diagnostics area, the development of new contrast agents based on NPs, particularly of superparamagnetic Fe_3O_4 NPs (SPIONs) has shown great potential to design theranostic agents where the detection modality can be used not only before or after but also during the treatment regimen (Wang et al., 2012). The use of NPs as theranostic agents has several advantages ranging from the size of the NPs to their composition. On the one hand, NPs with sizes above 10 nm are able to avoid first pass renal clearance thus leading to extended circulation times and also avoid extravasation from intact blood vessels into healthy tissues decreasing the side-effects. On the other hand, NPs with sizes below 100 nm are able to extravasate into tumor tissue through the irregularly dilated and leaky tumor blood vessels, which in combination with the poor lymphatic drainage is called the enhanced permeability and retention effect (Svenson, 2013). Regarding their composition, the development of theranostic systems is generally based on the combination of different materials to achieve a multifunctional system. In this context, polysaccharide-based systems are widely used. First, polysaccharides are natural biomaterials, which can incorporate inorganic materials such as metals to minimize the concern related to toxicity, biodegradability, and physiological stability. Furthermore, polysaccharide-based NPs have shown to decrease uptake by the mononuclear phagocytic system, thus increasing the NPs' circulating time and consequently increasing the possibility of disease site accumulation. Finally, polysaccharides are not only a cost-effective type of biomaterial due to its abundance in nature but also possess derivable groups that can be used for further functionalization. With the later property, it is possible to conjugate targeting, therapeutic, and imaging agents therefore producing multifunctional systems (Peng et al., 2015; Xiao, 2016; Swierczewska et al., 2016).

In this section, we will focus on the theranostic applications of hybrid polysaccharide-based composites, with emphasis on the different combinations of polysaccharides/inorganic materials.

4.3.1 Polysaccharides in theranostic applications

For theranostic applications, both therapeutic and imaging agents are combined through covalent bonding to the backbone of polysaccharides or physical encapsulation into polysaccharide-based NPs (Swierczewska et al., 2016).

CS and its derivatives have attracted much attention in the last decades due to the presence of amino groups in its backbone, which makes this biopolymer able to respond to pH. Moreover, CS-based NPs were shown to protect drugs from acidic denaturation and enzyme degradation, exhibiting mucoadhesive properties and enhancing absorption of molecular drugs (Xiao, 2016; Chen et al., 2013). A variety of chemical modifications have been made into CS's backbone to produce its derivatives, such as carboxymethyl CS, trimethyl CS, PEGylated CS, or glycol CS (Sahu et al., 2010; Kim et al., 2010; Guan et al., 2012; Mao et al., 2001). The later CS's derivative has the advantage of being water-soluble, which in combination with hydrophobic 5β-cholanic acid conjugates self-assembled to form tumor-homing CS-based NPs (Park et al., 2007; Kwon et al., 2003; Yhee et al., 2014). For example, Kim et al. (2010) used glycol CS NPs labeled with a near-infrared (NIR) fluorescence (NIRF) dye, Cy5.5 for imaging, and paclitaxel, an anticancer drug, aiming to simultaneously execute cancer diagnostics and therapy, that is, cancer theranostic. These NPs exhibited significantly increased tumor-homing ability with low nonspecific uptake by other tissues in in vivo studies. In fact, CS-based NPs have been widely used as drug delivery systems for paclitaxel (Koo et al., 2013; Wei et al., 2013), doxorubicin (DOX) (Yoon et al., 2014; Soares et al., 2016c), and siRNA (Yoon et al., 2014; Yhee et al., 2015; Wei et al., 2013; Lee et al., 2012) among others, as well as for photosensitizers for photodynamic therapy for cancer treatment (Oh et al., 2013; Lee et al., 2011), tissue engineering (Depan and Misra, 2012; Depan et al., 2014; Qin et al., 2009; Matsuda et al., 2005), and magnetic hyperthermia (Zamora-Mora et al., 2015; Zamora-Mora et al., 2014; Soares et al., 2016b).

Besides CS, other polysaccharides are also used for theranostic applications. Dextran, a natural glucose-containing polysaccharide, has an antithrombotic effect by attracting platelets, erythrocytes, and vascular endothelium to coagulate around the polysaccharide, and consequently interfering with the normal blood clotting cascade, being used extensively in anticoagulation therapy (Dias et al., 2011). Dextran sulfate can be used in combination with the amphiphilic copolymer CS-*g*-polylactic acid to form NPs via polyelectrolyte complex technique to produce a drug delivery system for both DOX and temozolomide (Di Martino and Sedlarik, 2014). Among many applications, this biopolymer is often used as a coating for SPIONs, which will be further discussed.

Alginate, a polysaccharide mainly derived from brown seaweed, possesses anionic groups that can be introduced onto the backbone of polysaccharides through chemical modification. For example, alginate is used in combination with CS to form NPs through ionic interactions that act as plasmid vectors. The interaction between the anionic phosphate backbone of DNA with the primary amino groups of CS protects DNA from degradation, whereas the interaction between CS and alginate reduces the strength of the CS–DNA ionic interaction thus allowing transfection (Xiao, 2016; Rafiee et al., 2014).

Another example of a polysaccharide used in theranostic systems is hyaluronic acid or hyaluronate (HA), a glycosaminoglycan that is the major component of the extracellular matrix. It has a specific receptor, CD44 that is found to be overexpressed on many cancer cells, being an excellent polymeric material for anticancer agents delivery (Lee et al., 2008; Homma et al., 2010) or imaging of cancerous cells (Wang et al., 2013c; Surace et al., 2009; Zoller, 1995).

CNCs are another example of a polysaccharide that due to their physicochemical properties, relatively low cost and the ability to penetrate cancer cells are gaining attention in the research community for its use as theranostic agent (Colombo et al., 2015).

4.3.2 Hybrid polysaccharide systems for theranostic applications

Although polysaccharides can be used as theranostic systems without other agents, these macromolecules are often used in hybrid systems in combination with inorganic materials. Furthermore, this combination extends the range of application and facilitates the production of multifunctional systems for theranostic applications. The most common inorganic materials that are used in combination with polysaccharides are gold, QDs, and SPIONs (Vivero-Escoto and Huang, 2011; Krasia-Christoforou and Georgiou, 2013), among others.

4.3.2.1 Polysaccharides and gold hybrid systems

The use of AuNPs dates back to 1857 when Faraday discovered the properties of colloidal gold that differs from bulk gold (Edwards and Thomas, 2007). Nowadays, AuNPs are used in analytical chemistry (Haes et al., 2004), electronics (Murray, 2008), as well as biology and nanomedicine due to their characteristic properties such as tunable optical properties, high surface area and surface modification (Vivero-Escoto and Huang, 2011).

Choi et al. (2010) prepared thiolated dextran-coated gold nanorods for targeted delivery to inflammatory macrophages and their photothermal ablation under NIR light irradiation. The in vitro results showed that dextran has superior ability to bond the gold nanorods surface against macrophages compared to PEGylated gold nanorods. In addition, in vitro photothermal irradiation experiments showed significant cell-killing efficacy, even with a low concentration of Au and a low-power light source. Cai and Yao (2013) also used dextran to prepare AuNPs-loaded lysozyme-dextran nanogels for both optical cell imaging and DOX release.

Wang et al. (2014) also developed a gold-based nanosystem for photothermal therapy. These authors prepared multistimuli responsive platforms based on drug loaded gold nanocages with hyaluronic acid. These gold-based platforms are able to release DOX only in intracellular environments, such as endosomal and lysosomal vesicles, by the degradation of HA by intracellular lysosomal enzyme hyaluronidase. In addition, NIR light irradiation improved the therapeutic efficacy.

Another research group developed multifunctional nanocarriers based on CS/gold nanorods hybrid nanospheres. The in vitro experiments revealed that these nanocarriers can be used for real-time cell imaging as well as a NIR thermotherapy nanodevice and a drug delivery device for cisplatin (Guo et al., 2010).

Hu et al. recently produced in situ CNC-based gene vectors with AuNPs, from bifunctional CNC conjugated with heterogeneous polymer brushes, such as poly (poly(ethylene glycol)ethyl ether methacrylate)—PPEGEEMA- and/or poly(2-(dimethylamino)ethyl methacrylate)—PDMAEMA as a potential multifunction therapy system. From the results obtained, the authors could see that genes were complexed effectively by the cationic PDMAEMA chains, whereas the presence of uncharged PPEGEEMA brushes significantly reduces the CNC-based material cytotoxicity and the AuNPs imparted imaging function (Hu et al., 2016a).

4.3.2.2 Polysaccharides and quantum dots hybrid systems

Quantum dots are fluorescent semiconductor nanocrystals with a size range of 2–100 nm with unique tunable optical and targeting properties that can be composed by inorganic entities from groups III–V (InP and InAs) or II–VI (CdSe and CdTe) (Baptista et al., 2013a; Jain et al., 2015). Tan et al. (2007) developed QDs encapsulated into CS NPs to deliver HER2/neu siRNA to two human breast cancer cell lines. Using this composite NPs, the authors were able to monitor the delivery and transfection of the siRNA. Another group (Yuan et al., 2010) combined blue-light emitting ZnO QDs with CS to prepare a targeted drug release system. The results showed that CS increases the stability of the nontoxic water dispersible QDs, while acting as a drug delivery system for an anticancer agent.

Shen et al. (2012) used a combination of CdTe quantum dots, SPIONs, and CS gelled into ternary hybrid nanogels and conjugated with tetrapeptides and folate to obtain multifunctional hybrid NPs. These nanometric particles proved to be useful for magnetic guidance, thus accumulating into the tumor tissue, and for camptothecin controlled release, while having low cytotoxicity and favorable cell biocompatibility.

4.3.2.3 Polysaccharides and SPIONs hybrid systems

The use of magnetic NPs as theranostic agents is very advantageous due to their inherent biocompatibility and cost-effectiveness, together with the unique magnetic properties that enables their use as contrast agents for magnetic resonance imaging (MRI). Among the different types of magnetic NPs, SPIONs are the most commonly used for biomedical applications (Baptista et al., 2013a; Wang et al., 2012).

Högemann-Savellano et al. (2003) produced monocrystalline magnetic NPs coated with dextran and conjugated to holo-transferrin to detect overexpression of engineered transferrin receptors, therefore facilitating the visualization of tumors selectively in vivo at very high spatial resolution. On a different study, You et al. (2014) prepared SPIONs coated with dextran sulfate-*b*-poly(glycerol methacrylate) copolymer as a potential contrast agent for atherosclerosis MRI. These composite NPs are highly stable and are specifically taken up by the activated macrophages via receptor-mediated endocytosis and produced distinct contrast enhancement in the T_2-weighted Mr cellular imaging of activated macrophages. DeNardo et al. (2007) conjugated a bioprobe (^{111}In-DOTA-ChL6) to polyethylene glycol-Fe_3O_4-impregnated dextran 20 nm particles (^{111}In-probes) and tested its effectiveness in human breast cancer xenografts in mice. The results showed tumor response with heat dose dependence without toxicity. In fact, a member of the same work group developed magnetic NPs coated with dextran and embedded with a specific ligand for use in targeted thermotherapy (Ivkov, 2006).

SPIONs are also easily functionalized with hyaluronic acid to produce theranostic systems mainly for controlled drug delivery and MRI. El-Dakdouki et al. (2012) used this combination for targeted imaging and drug delivery for cancer applications. In this case, the authors concluded that HA enhances the internalization of SPIONs, while having high magnetic relaxivity, thus enabling MRI of cancer cells. In a different study, Park et al. (2014a) used nanohybrid liposomes coated with amphiphilic hyaluronic acid-ceramide containing DOX and Magnevist (an MRI contrast agent) for the same applications. In vivo studies demonstrated that the nanohybrid liposome can be used as a tumor-targeting MR imaging probe for cancer diagnosis, and it is able to increase the circulation of the drug in the blood stream, thus improving the therapeutic efficacy of the nanohybrid liposome. Using the same materials, it is also possible to produce HA polymeric micelles encapsulating oleic acid-coated SPIONs. These micelles are selectively cytotoxic against some human cancer cell lines and accumulate preferentially in tumor tissue after intravenous administration (Šmejkalová et al., 2014).

Agyare et al. (2014) developed a theranostic magnetic nanovehicle composed of Magnevist conjugated with CS and loaded with cyclophosphamide and functionalized with antiamyloid antibody IgG4.2 [pF(ab′)24.1]. This theranostic nanovehicle was designed to target cerebrovascular amyloid and reduce inflammation in Alzheimer's disease. The in vivo studies in mice demonstrated contrast imaging of cerebrovascular amyloid with MRI and single photon emission computed tomography (SPECT), and a reduction in the production of proinflammatory cytokine. In a different study, Zarrin et al. (2015) developed a new trimodally targeted nanomagnetic oncotheranostic system for simultaneous early diagnosis and efficient treatment of cancer. This system is composed of SPIONs coated with folic acid-conjugated CS loaded with DOX. The results showed that drug release from the produced system is pH-sensitive, while simultaneously showing potential as T_2 MRI contrast agents and treatment of folate receptor-positive cancers.

CS can also be conjugated with polyethylene glycol (PEG) to form a copolymer for SPIONs coating. Zhang's group (Veiseh et al., 2010) used this type of NPs for efficient siRNA delivery and noninvasive monitoring through MRI. These complexes also include chlorotoxin, a tumor-targeting peptide to improve tumor specificity and potency. MRI results showed that this nanovector is able to generate specific contrast enhancement of glioblastoma cells. More recently, the same group used these copolymers modified through covalent attachment of O(6)-benzylguanine and chlorotoxin to obtain a redox-responsive, cross-linked, biocompatible surface coating for SPIONs. With these NPs, the authors were able to perform tumor cell specific O(6)-benzylguanine delivery and improve in vivo efficacy to achieve inhibition of the upregulated DNA repair protein responsible for chemotherapy resistance in glioblastoma multiform (Stephen et al., 2014).

Wang et al. (2013a) used an interesting combination of SPIONs and graphene coated with CS to create a platform for simultaneous gene/drug and SPION delivery to tumor. The MRI results suggest that the nanoplatform has a strong T_2

contrast-enhancing effect and preferentially accumulates in the tumors. The incorporation of CS improves the stability, solubility, and biocompatibility of graphene sheets.

In our group, we have studied the thermal and magnetic properties of CS-coated SPIONs, previously designed for magnetic hyperthermia and as contrast agents for MRI (Soares et al., 2016a). The developed SPIONs are highly stable due to the presence of surfactants and are able to generate heat under the application of an external alternating current magnetic field (Soares et al., 2014, 2015). Furthermore, the presence of CS does not compromise the physicochemical and magnetic properties of the SPIONs, maintaining the ability of these NPs to significantly rise the surrounding temperature (Soares et al., 2016b). These CS-based NPs are also efficient drug vehicles for DOX, allowing a controlled and pH-sensitive drug release (Soares et al., 2016c). With these studies, we are developing a multifunctional theranostic system for cancer based on a combination of SPIONs as MRI contrast and magnetic hyperthermia agents, and a biocompatible and biodegradable CS-based coating.

Table 4.1 summarizes some examples of hybrid polysaccharide systems for theranostic applications described in this chapter.

4.4 Tissue engineering

The main focus of current studies on Tissue Engineering and Regenerative Medicine is to develop new therapeutic solutions based on materials that enhance tissue repair and regeneration. Most recent studies in these fields used polymers due to their ease of fabrication, structural control, low cost, and availability (Thomas et al., 2012; Thakur and Thakur, 2014; Thakur et al., 2014a,b). Among these, polysaccharides such as cellulose and chitin/CS have been widely used as bio-based polymers from renewable sources for tissue-engineering applications. Some examples are given in the following section.

4.4.1 Tissue-engineering applications of native cellulose and chitin/chitosan

4.4.1.1 Bone and cartilage

BC fibers in nanometer scale are microscopically similar to collagen nanofibers in natural tissue, and this makes BC a promising collagen mimicking component in tissue engineering (Wang et al., 2013b). BC has been used in several applications namely as scaffold component for bone tissue engineering. Fang et al. (2009) produced a scaffold with potential use for bone tissue engineering using hydroxyapatite/BC (HAp/BC) nanocomposite. The proliferation and differentiation of stromal cells derived from human bone marrow (hBMSC) was studied, showing higher adherence and activity of the hBMSC when compared to pure BC due to the larger pore sizes and improved inorganic component. The study reveals that the attachment, proliferation, and differentiation of in vitro cultured hBMSC can be

Table 4.1 Examples of hybrid polysaccharide systems for theranostic applications

Polysaccharide	Inorganic component	Biomedical applications	Ref.
Hyaluronic acid or hyaluronate (HA)	–	Drug delivery	Lee et al. (2008), Homma et al. (2010)
		Imaging	Wang et al. (2013c), Surace et al. (2009), Zoller (1995)
	Gold	Drug delivery, photothermal therapy, and chemotherapy	Wang et al. (2014)
	SPIONs	MRI, drug delivery	El-Dakdouki et al. (2012), Park et al. (2014a), Šmejkalová et al. (2014)
Dextran	–	Drug delivery	Di Martino and Sedlarik (2014)
	Gold	Targeted drug delivery, photothermal ablation	Choi et al. (2010)
		Optical cell imaging, drug delivery	Cai and Yao (2013)
	SPIONs	MRI	Högemann-Savellano et al. (2003), You et al. (2014)
		Magnetic hyperthermia	DeNardo et al. (2007), Ivkov (2006)
		Drug delivery, MRI	Saboktakin et al. (2010)
Chitosan (CS)	–	Drug delivery	Kim et al. (2010), Koo et al. (2013), Wei et al. (2013), Yoon et al. (2014), Soares et al. (2016c)
		siRNA delivery	Wei et al. (2013), Yoon et al. (2014), Yhee et al. (2015), Lee et al. (2012)
	Gold	Photothermal therapy	Oh et al. (2013), Lee et al. (2011)
		Photothermal therapy, optical cell imaging	Guo et al. (2010)
	Quantum dots (QD)	siRNA delivery, cell imaging	Tan et al. (2007)
		Targeted drug release, cell imaging	Yuan et al. (2010)
	QDs/ SPIONs	Magnetic guidance, drug release	(Shen et al., 2012)
	SPIONs	MRI, Alzheimer' disease treatment	(Agyare et al., 2014)

(*Continued*)

Table 4.1 (Continued)

Polysaccharide	Inorganic component	Biomedical applications	Ref.
		MRI, drug delivery	Zarrin et al. (2015), Saboktakin et al. (2010), Tagami et al. (2011)
		MRI, siRNA delivery	Veiseh et al. (2010), Stephen et al. (2014)
		Magnetic hyperthermia	Zamora-Mora et al. (2014, 2015), Soares et al. (2016b)
		Magnetic hyperthermia, drug release	Anirudhan and Sandeep (2012)
	SPIONs/ graphene	MRI, gene/drug delivery	Wang et al. (2013a)

modulated by the HAp/BC nanocomposite properties (Fang et al., 2009). More recently, HAp bioactivated by BC membrane was reported to promote osteoblast growth and formation of bone nodules. Osteoblast adhesion and growth were significantly increased on BC/HAp membranes compared with BC alone (Tazi et al., 2012). Other strategies for bone tissue engineering using BC include the combination of BC/HAp and gelatin (Gel) (Wang et al., 2013b), BC/HAp and calcium phosphate (Hu et al., 2016b), BC/Hap and graphene oxide composite (Ramani and Sastry, 2014), otoliths/collagen/BC networks (OCBC) (Olyveira et al., 2011), BC/Hap and CMC (Zimmermann et al., 2011), BC, and agarose (Yang et al., 2011).

Cellulose and cellulose-based materials are also reported for bone tissue engineering, such as hydroxypropylmethylcellulose (HPMC) (Gauthier et al., 1999, 2003; Zhang et al., 2014; Trojani et al., 2006; Liu et al., 2014), CA (Gouma et al., 2012), CMC (Sa et al., 2014) and hydroxyethyl cellulose (HEC) (Chahal et al., 2015, 2016). Liuyun et al. (2009) produced nano-HAp, CS, and CMC scaffolds by freeze–drying method, which presented a structure with interconnected pores (100–500 μm), high compressive strength (up to 3.5 MPa), good structural stability, and degradation. The results of cell culture experiment in vitro and a scaffold implantation after 4 weeks in vivo indicated that the scaffold had good tissue biocompatibility, biodegradation, and osteoconduction (Jiang et al., 2009).

Cartilage has low capacity for spontaneous repair and for the development of appropriate vehicles for chondrocytes implantation is of great interest. Injectable cellulose-based hydrogel (silanized hydroxypropyl methylcellulose—Si-HPMC) was reported as a good candidate for tissue engineering for the repair of articular cartilage (Vinatier et al., 2007; Vinatier et al., 2009). More recently, the same authors combined the Si-HPMC hydrogel with a marine exopolysaccharide, GY785, which stimulates the in vitro chondrogenesis of adipose stromal cells (Rederstorff et al., 2015). Poly(*N*-isopropylacrylamide)-*g*-methylcellulose, a thermoreversible hydrogel, was reported as good candidate as 3D support for cell encapsulation toward the regeneration of articular cartilage using a minimal invasive strategy (Sa-Lima et al., 2011).

4.4.1.2 Skin

Vatankhah et al. (2014) developed a nanofibrous electrospun CA/Gel scaffolds that mimic both morphological and structural features of normal skin for wound dressing or tissue engineering depending on the composition of CA/Gel. Cellulose/Gel sponges constructed directly from a cellulose solution effectively promote wound healing as a result of the presence of macro- and microporous architecture that could fit the requirements of oxygen permeability, controlled water vapor evaporation, and wound exudate absorption (Pei et al., 2015). Panichpakdee et al. (2014) developed CA fiber mats containing inclusion complexes of asiaticoside in 2-hydroxypropyl-β-cyclodextrin for potential usage as wound dressing that were effective in upregulating the production of collagen of cultured human dermal fibroblasts. Electrospun nanofibrous mats prepared from poly(ε-caprolactone), CA, and dextran blend solution with tetracycline hydrochloride incorporated in a small amount improved cell proliferation, enhanced blood clotting, and cell attachment as well as antimicrobial activity of the composite mat (Liao et al., 2015).

Biocompatible nanofibrous mats using HEC with PVA showed improved cellular adhesion profiles and stability to be used as scaffolds for skin tissue engineering. The scaffolds showed good cell proliferation and attachment after culturing (Zulkifli et al., 2014, 2015).

BC–hyaluronan nanocomposite with 3D network structure were obtained through a solution impregnation method. These composite films facilitated the growth of primary human fibroblast cells, and in vivo experiments indicated that the composites with 0.1% hyaluronan had the shortest wound healing time while the composites with 0.05% hyaluronan yielded best tissue repair results (Li et al., 2015).

4.4.1.3 Other tissues

Cellulose–heparin fibers offer promise in the preparation of woven fabrics for use in the construction of artificial vessels with excellent blood compatibility. Cellulose and cellulose–heparin composite fibers prepared from nonvolatile room temperature IL solvents by electrospinning showed anticoagulant activity, with heparin' bioactivity unaffected even under exposure to the high voltage involved in electrospinning (Viswanathan et al., 2006). More recently, in 2011, a heparin–BC (Hep–BC) hybrid nanofiber has been developed via the cosynthesis technique revealing the presence of anticoagulant sulfate groups in Hep–BC hybrid nanofiber (Wan et al., 2011).

Polyelectrolyte complex CS/CMC scaffolds with great potential as pulp tissue scaffolds were prepared using a freeze–drying process. The morphology and distribution of pulp cells on these 3D scaffolds was investigated thought primary pulp cells culture and expression of osteonectin and dentin sialophosphoprotein genes in vitro showing that after implanted in vivo the scaffold recruited surrounding tissue to grow on it (Chen and Fan, 2007). Osteogenic and odontogenic differentiation of dental pulp stem cells (DPSCs) cultured on a CMC–HAp hybrid hydrogel showed good adhesion and viability in cultured cells, and the in vitro data confirm the ability of DPSCs to differentiate toward osteogenic and odontogenic lineages in presence of CMC–HAp hybrid hydrogel (Teti et al., 2015).

Bilayered nanofibrous material for the production of hollow organ tissue engineering was already produced using potato starch (PS) to interrupt BC assembly. Muscle cells were cultured on the surface of the BC/PS scaffolds and mechanical characteristics similar to traditional BC were obtained. In vivo and in vitro studies revealed that BC/PS membranes permit muscle cells infiltration with enhancement of wound healing. This study provides preclinical evidence that muscle cell-seeded tubularized scaffolds can be used in the reconstruction of urethral defects. This biomaterial could also be suitable for numerous other types of hollow organ tissue engineering grafts, including vascular, bladder, ureter, esophagus, and intestine (Lv et al., 2016).

4.4.2 Tissue-engineering applications of nanocellulose and nanochitin

Notwithstanding their substantial applicability, the native forms of both cellulose and CS have important limitations regarding their properties, functionality, durability, and uniformity, which have led to the search for the highly crystalline regions that compose these natural fibers–nanofibrils. Cellulose and chitin nanofibrils are anisotropic particles with high degree of biodegradability that exhibit extremely high modulus and strength along their axes, allowing also facile chemical modifications of their surfaces (Moon et al., 2011).

CNCs, cellulose nanofibrilated (CNF), or BC in its unmodified or modified forms are another class of nanomaterials that due to their physicochemical properties, high surface area, surface chemistry richness, biocompatibility, high strength, liquid-crystalline behavior, biodegradability, relatively low cost, and nontoxic carbohydrate-based nature have gained pronounced interest in the scientific community toward biomedical applications (Shah et al., 2013; Domingues et al., 2014; Lin and Dufresne, 2014). Most recently, one can find good reviews in the literature (Kucińska-Lipka et al., 2015; Guise and Fangueiro, 2016; Ul-Islam et al., 2015) where the authors present the most recent state-of-the-art regarding the use of nanocellulose in applications as diversified as wound dressing, tissue engineering, or enzymatic immobilization, either on its own or in a hybrid and composite form.

Nanochitin can be obtained from the dissolution of the amorphous region of chitin by acid hydrolysis in often called chitin whiskers, chitin nanocrystals, or chitin nanofibrils. Chitin nanofibrils are usually used as matrix reinforcements or bioactive components in composite materials. These composite materials found several applications in the several fields like tissue engineering, drug delivery (Kim, 2013), cosmetic (Morganti, 2008), adsorbents in industry, and water purification or even protein biosensors (Mincea et al., 2012). Some interesting reviews can be found in the literature describing some of the possible applications of nanochitin (João et al., 2015; Muzzarelli et al., 2014).

4.4.2.1 Nanocellulose-based scaffolds

3D bioprint of human chondrocytes using bioink consisting of CNFs and alginate has been reported: several structures, including a human ear, representing cartilage tissue were printed with high fidelity and stability. However, different compositions of bioink may need to be developed according to the different usage and mechanical properties requirements. Toxicity tests and cell-viability analysis of the 3D CNFs alginate printed ink with human nasoseptal chondrocytes showed that the bioink is biocompatible and a suitable material for cell culture (Markstedt et al., 2015). These combination of nanocellulose and alginate has been reported already as good candidate for cell encapsulation engineering (Park et al., 2015; Krontiras et al., 2015). A three-component porous nanocomposite scaffold based on cellulose nanofibers and cross-linked matrix of Gel and CS was reported by Naseri et al. (2016). The scaffold produced has interconnected pores and nanoscaled pore wall roughness. These favorable features for cell interactions aiming at cartilage repair combined with good cytocompatibility toward chondrocytes make the 3D nanocomposite scaffold a good candidate for cell attachment and extracellular matrix production with mechanical properties closer to those of natural cartilage after implantation (Naseri et al., 2016).

In some studies, the authors use a combination of nanocellulose with different matrices to produce biocomposites for skin tissue engineering, including collagen (Zhijiang and Guang, 2011), Gel (Nakayama et al., 2004), alginate (Lin et al., 2012), CS (Kim et al., 2011; Lin et al., 2013), poly(ethylene glycol) (Cai and Kim, 2010), and poly(vinyl alcohol) (Gonzalez et al., 2014).

In a recent review, the authors exploit the most recent advances in nanocellulose for biomedical applications. In the tissue-engineering field, the most commonly used nanocellulose is BC in combination with HAp (Jorfi and Foster, 2015). In addition, CNCs with high aspect-ratio deposited onto glass substrates have also been used to grow myoblasts oriented according to the bulk direction of the CNC, demonstrating the potential of CNCs for muscle tissue engineering (Dugan et al., 2010, 2013).

4.4.2.2 Nanochitin/chitosan-based scaffolds

Recently, Liu et al. (Liu et al., 2016) dispersed chitin nanocrystals with 300 and 20 nm of length and width, respectively, in a CS solution, followed by a dispersion-based freeze–drying approach. These composite structures were tested as MC3T3-E1 osteoblast cells scaffolds, demonstrating excellent biocompatibility and low toxicity, while promoting cell adhesion and proliferation.

Muzzarelli et al. (2014) published a review related to the biomedical applications of nanochitin and nano-CS, including the methods for obtaining such products. In this review, one can find examples of bone tissue engineering using CSNFs and nanochitin hydrogels, demonstrating the potential of these materials in this application. Moreover, the authors review the use of nano-CS in epithelial tissue regeneration, where nano-CS is used in combination with Gel, collagen, or fibroin. In one example, hydrogels composed of hemicelluloses, chitin nanofibrils, and poly(vinyl alcohol) (PVA) with an average length of 200 nm and width of 40 nm showed remarkable swelling, good mechanical strength, and thermal stability (Guan et al., 2014a, 2014b). Some other examples are related to bone and dental tissue regeneration by using nanofibrous CS with genipin (Norowski et al., 2012) and HAp NPs (Gentile et al., 2012; Frohbergh et al., 2012).

In another review, the authors focused on chitin and CSNFs together with other materials. In one example, fibrous mats of chitin/poly-glycolic acid and chitin/silk fibroin were tested as tissue-engineering scaffolds, demonstrating high spreading rates of normal human epidermal fibroblasts and normal human epidermal keratinocytes (Park et al., 2006a, 2006b). Nanofibrous scaffolds of CS/polyethylene oxide also promoted the attachment of human osteoblasts and chondrocytes (Bhattarai et al., 2005; Subramanian et al., 2005). Some other recent studies related to the application of nanochitin and nano-CS can be found in this review (Azuma et al., 2014).

4.4.3 Biomimetic scaffolds for tissue engineering

Natural systems have always attracted high interest and considered as source of inspiration in solving of human problems. The ability to exploit intricate multiscale design giving rise to the formation of complex structures with adaptable functionalities has shown that natural structures are far more developed than those constructed by men. Inspired by natural species, the science of emulating and mimicking nature has been an emerging field that researchers have been exploring through the combination of bioinspired engineering and nanomaterials (Aziz and El sherif, 2016). This type of approach has found great acceptance in the biomedical and tissue-engineering fields (Egan et al., 2015). From all the biologically mineralized composites that we are able to find in Nature, bone has been highly studied taking into account not only the structural architecture but also mechanical performance and functionality.

Bone is a hybrid tissue of organic (collagen) and mineral (HAp) matter and a highly hierarchical structure (Fig. 4.4) constructed by a complex process involving cells, proteins, growth factors, and hormones. Collagen molecules are created by specific cells, the osteoblasts, segregated into the extracellular matrix space in a dense state and with the continuous growth of the polymeric chains they escape from cellular control to self-assembly in local cholesteric domains (Belamie et al., 2006). The liquid crystalline environment allows the production of an oriented organic network that is further mineralized by osteoblasts cells with the deposit of HAp crystals (Giraud-Guille et al., 2003, 2004). The natural design, also found in

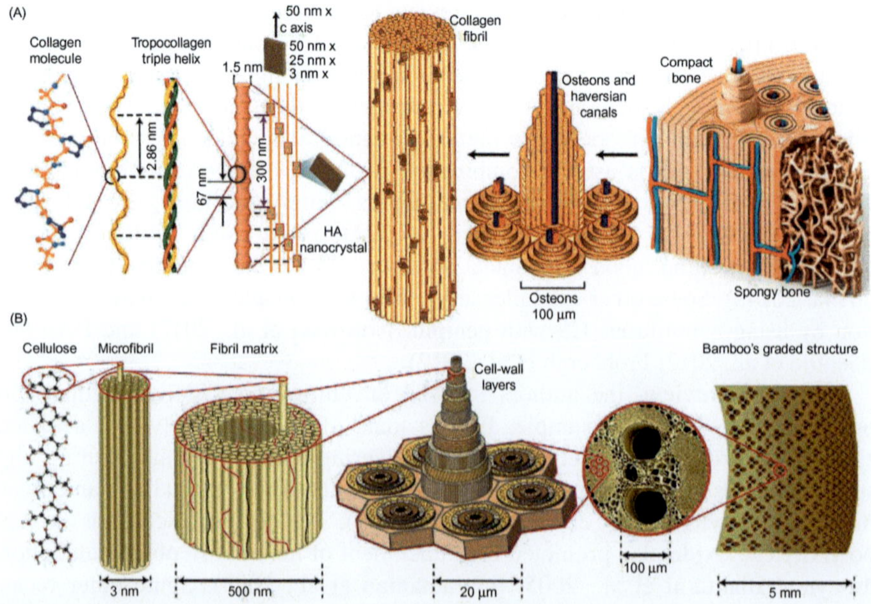

Figure 4.4 Similarities between the hierarchical structure of bone (A) and bamboo (B). Reprinted with permission from Macmillan Publishers Ltd © (2014): Wegst, U.G.K., Bai, H., Saiz, E., Tomsia, A.P. & Ritchie, R.O., 2015. Bioinspired structural materials. Nat. Mater., 14, 23–36 ⟨http://www.nature.com/nmat/journal/v14/n1/full/nmat4089.html⟩.

other Nature materials like bamboo (Fig. 4.4) and engineering path toward the formation of bone structure, allows the unique combination of desirable properties like strength, toughness, and lightweight that are not available from a simple sum of the constituent materials properties but yet achievable with a multiscale formation mechanism (Wegst et al., 2015).

Based on bone's structure features, several authors have focused on developing materials that could somehow mimic the natural structures behavior. Taking the organic phase of the insect cuticle and the exoskeleton of crustaceans as case study, Jin et al. (2013) produced a biomimetic composite from the coassembled solution of self-assembled ultrafine (3 nm) chitin nanofibers in a silk matrix that replicates the structures found in living tissues. The biocomposite thus structured revealed strong hydrogen bonding between the chitin nanofibers and the surrounding silk matrix, resulting in an increased elastic modulus. In a different set of studies (Capadona et al., 2007, 2008, 2009, 2012; Shanmuganathan et al., 2010a, 2010b; Potter et al., 2014; Jorfi et al., 2013, 2014; Weder et al., 2014; Breuer-Thal et al., 2014), the authors developed a new class of biologically inspired, mechanically adaptive cellulose nanocomposites that could be selectively tuned between soft and stiff states, imitating the sea cucumber dermis' architecture (Fig. 4.5). These invertebrates exhibit the ability to change their skin stiffness from 5 to 50 MPa under threat conditions. This extraordinary capability is achieved by a nanocomposite that

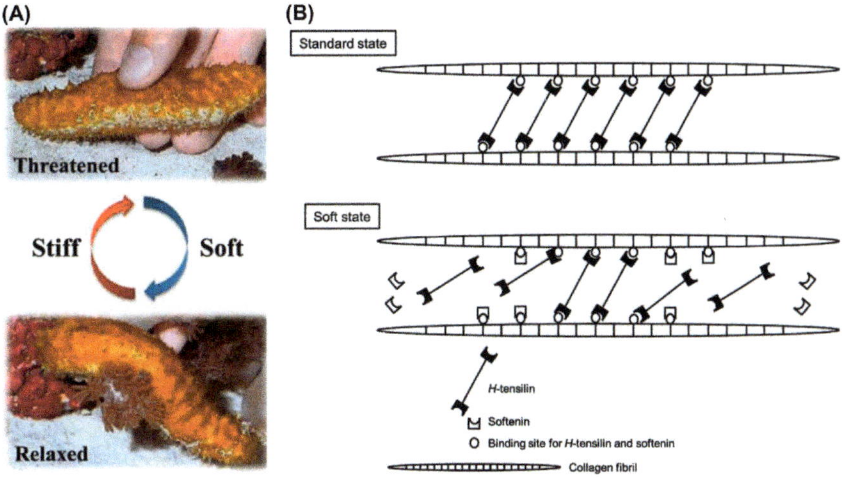

Figure 4.5 (A) Photographs of a sea cucumber under threat (stiff) and relaxed (soft); (B) hypothetical model of the stiffness change mechanism in the sea cucumber dermis.
(A) Reprinted with permission from John Wiley and Sons © (2015): Jorfi, M. & Foster, E.J., 2015. Recent advances in nanocellulose for biomedical applications. J. Appl. Polym. Sci., 132, n/a—n/a ⟨http://onlinelibrary.wiley.com/doi/10.1002/app.41719/abstract⟩; (B) Reprinted with permission from Takehana, Y., Yamada, A., Tamori, M. & Motokawa, T., 2014. Softenin, a novel protein that softens the connective tissue of sea cucumbers through inhibiting interaction between collagen fibrils. PLoS One, 9, e85644.

relies on changeable interactions of stiff collagen fibers dispersed throughout a soft fibrillin matrix. In the developed composite, similar behavior was achieved by exploring the CNCs' surface hydroxyl groups that strongly interact with each other by hydrogen bonding and/or van der Walls forces. However, in the presence of hydrogen-bond-forming liquids such as water, the CNC—CNC interactions are strongly reduce due to competitive hydrogen bonding or interfacial interactions with intermolecular van der Waals' forces (Trotter et al., 2000; Motokawa, 1994; Thurmond and Trotter, 1996; Wilkie, 2002; Takehana et al., 2014). Further on, Harris et al. (2011) explored the in vivo application of these mechanically switchable materials as substrates for penetrating brain implants. The developed microprobes composed of 15% v/v PVAc/CNC nanocomposite that were inserted into the cerebral cortex of a rat through the pia mater without any assistive devices. In this procedure, the authors confirmed the adaptability of the nanocomposite stiffness that was able to rapidly decrease once implanted into the rodent brain, closely matching the brain tissue mechanical properties.

Although the idea of mimicking natural materials is of great interest and will most likely be the future direction in materials development for tissue-engineering applications, a lot of work is still need to accomplish the complex capabilities of natural tissues. Nevertheless, the first steps have been taken in characterization, modeling, and manufacturing. The reader can find further information related to the design approaches and the difficulties associated with the development of biomimetic structures in this recent review (Wegst et al., 2015).

4.4.3.1 Liquid-crystalline-based scaffolds for tissue engineering

The rod-like anisotropic biomaterials obtained from collagen, chitin or cellulose can, in aqueous suspension, originate chiral nematic liquid crystalline phases (N*) and upon solvent evaporation the N* phase can be imprinted in 2D or 3D materials (Lagerwall et al., 2014; Salas et al., 2014; Klemm et al., 2011; Fernandes et al., 2013; Giese et al., 2015; Giraud-Guille et al., 2004; Borges et al., 2014). The N* structure observed in these hierarchical self-assembled structures produced by chitin or cellulose nanorods is reminiscent of the twisted plywood structure (also referred to as Bouligand structure) and can be observed in the organic matrix of crustacean cuticles, insects' exoskeleton, and less commonly in plants but nevertheless appearing for example in some iridescent berries. The observation of these structures, by polarizing optical microscopy (POM), shows characteristic cholesteric textures revealing the nanorods assemble in regular *lamine* a gradually rotating around the normal direction. This long-ranged assembly mechanism leads to the formation of patterned materials with strong anisotropy, which is at the origin of the observed crustaceans' mechanical strength and plants' or fruits' structural colors (Bouligand, 1972; Giraud-Guille et al., 2004; Vignolini et al., 2013). The textures observed by POM are characteristic of chiral nematic liquid crystals, and these findings gave the researchers some insight regarding the formation of anisotropic structures observed in biological materials, as for instance in living tissue, and the possibility of the participation of the liquid crystalline state of matter in its formation (Belamie et al., 2006; Price et al., 2016). The fact that using the nanofibrils' self-assembly allows recreating the natural arrangement of living tissues opens a new and promising approach to the use of biomimetic materials in tissue regeneration (Price et al., 2016; Woltman et al., 2007; Sanchez et al., 2005; Egan et al., 2015; Giraud Guille et al., 2005). Despite the simplicity of obtaining such biomimetic organized structures, a survey in the literature results in scarce examples of these systems regarding these applications (Matsumura et al., 2015; Nguyen et al., 2016; Yamamoto et al., 2010).

Nguyen et al. (2016) recently described the preparation of a mussel-inspired membrane system that comprises carboxylated-CNCs, catechol moiety of DA and AgNP. The use of catechol moiety DA allows in situ Ag^+ reduction and consequently AgNPs formation onto CNCs surface without additional chemical treatments. The authors obtained an anisotropic aligned membrane, derived from the self-assembly of the modified cellulose, with improved mechanical activity in the alignment direction, antimicrobial activity, and enhanced electric conductivity when compared with the pristine modified CNC.

In 2010, inspired by the biomineralization process of crustacean cuticles, Yamamoto et al. (2010) were able to prepare hybrid $CaCO_3$/chitin nanorods gels (Fig. 4.6) with hierarchical structures derived from the chitin liquid crystalline aqueous suspension. The authors showed that the amorphous calcium carbonate

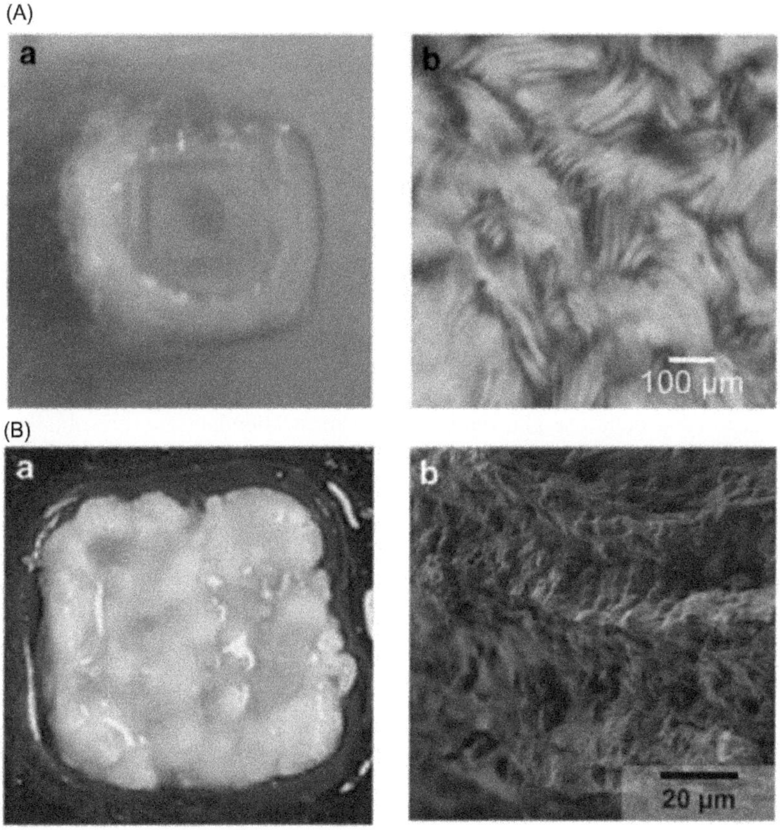

Figure 4.6 (A) Images of the cholesteric chitin gel: (a) photograph, (b) polarized optical micrograph image; (B) images of the mineralized chitin gel: (a) photograph, (b) scanning electron microscope image.
Reprinted with permission from Macmillan Publishers Ltd © (2010): Yamamoto, Y., Nishimura, T., Saito, T. & Kato, T., 2010. $CaCO_3$/chitin-whisker hybrids: formation of $CaCO_3$ crystals in chitin-based liquid-crystalline suspension. Polym. J., 42, 583−586 ⟨http://www.nature.com/pj/journal/v42/n7/abs/pj201032a.html⟩.

(ACC) was fully crystallized into $CaCO_3$ within the 3D organic matrix, whereas in a randomly organized isotropic matrix of the same material this crystallization process was not fully accomplished. However, the mechanical strength observed in the crustacean cuticles was absent. The authors attributed this drawback to the small size of the chitin nanorods (Yamamoto et al., 2010). Recently, within the group guided by Professor T. Kato, using the same concept and materials cross-linked with polymer network of poly(acrylic acid) (PAA), it was possible to obtain a self-standing gel with a helically ordered structure (Matsumura et al., 2015). In this research study, the authors greatly improved the crystallization process time, and the results suggest that after 7 days ACC NPs where homogeneously distributed inside the chitin/PAA matrix and crystalized to calcite nanocrystals.

Despite the small amount of liquid-crystalline structures presented here, the observed enhancement in the materials' properties or processing conditions is a driving force that certainly will lead to further development of these biomimetic materials.

4.5 Conclusions and future perspectives

Polysaccharides are a class of biopolymers from renewable sources with several advantages like nontoxicity, biocompatibility, and biodegradability. From all the polysaccharides, cellulose and chitin are the most abundant and are obtained from plants and animals, respectively. When combined with inorganic materials, these biopolymers originate composite hybrid materials that found applications in a wide range of biomedical applications.

In this chapter, the authors presented a glimpse of the possible combinations of hybrid materials containing polysaccharides. Although the major focus was on cellulose and chitin/CS, in some sections other polysaccharides were also described. In the three subsections of this chapter, the reader can understand the different development stages of those materials. For example, in the biosensors field, some studies of almost ready-to-use biosensors were given. For glucose or cholesterol detection, cellulose and chitosan can be conjugated with Fe_3O_4 NPs, gold nanorods, and CNTs, among others. In all cases, the use of a hybrid composite not only improves the immobilization of GOx but also increases the system sensitivity. The research in the actuators' field is still in an early stage and although some examples such as the EAPap, an actuator based on cellulose paper and gold, have demonstrated large bending, it still demonstrates high sensitivity to humidity. This limitation is enough for stopping such material from entering the market. Regarding theranostics, even so it is a field with huge potential and several research is being made, we are still far from clinical application. Nevertheless, designing prodrugs and polymers carrying conjugated diagnostic and therapeutic agents is already possible. The huge amount of research studies related to theranostic, along with the "modern" idea of developing means to achieve a state of precision medicine, will probably bring new developments in the near future.

In the third field reviewed, tissue engineering, hybrid systems can be roughly divided into two groups: composite structures composed of macromolecules such as native cellulose, chitin or CS, and nanocomposites where the polysaccharide is in its "nano" state, that is, as nanofibers, nanorods, or nanostructures. One can even consider a third group in which composite scaffolds are produced from both native polysaccharides and their nanoderivatives. The advantage of using the nanoderivatives of polysaccharides relies on the possibility of developing biomimetic structures, which resemble biological structures and as a consequence promoting a better and enhanced tissue regeneration in a more natural way. Moreover, the use of liquid crystalline phases allows the formation of hierarchical self-assembled structures that are similar to those observed for example in the organic matrix of crustacean cuticles, insects' exoskeleton, enhancing the development of biomimetic structures for tissue engineering.

As a conclusion, there is still a long road ahead toward the market entry and clinical application of hybrid polysaccharide-based materials. Regardless of their limitations, the use of biodegradable and biocompatible renewable materials like hybrid polysaccharide-based materials persist as focus of current research studies, being transversal to all fields in biomedical applications.

Acknowledgments

This work is funded by FEDER funds through the COMPETE 2020 Program and National Funds through FCT-Portuguese Foundation for Science and Technology under the project number POCI-01-0145-FEDER-007688, Reference UID/CTM/50025. The authors also acknowledge the Minister of Science, Technology and Higher Education for National Funds, European Social Funds and Portuguese Science Foundation—FCT for the following grants: PhD grant SFRH/BD/80860/2011 (C.F.C. João), Post-Doc grants SFRH/BPD/104407/2014 (A. Baptista), SFRH/BPD/88779/2012 (C. Echeverria), and SFRH/BDP/78430/2011 (S. Fernandes).

References

Agyare, E.K., Jaruszewski, K.M., Curran, G.L., Rosenberg, J.T., Grant, S.C., Lowe, V.J., et al., 2014. Engineering theranostic nanovehicles capable of targeting cerebrovascular amyloid deposits. J. Control. Release. 185, 121–129.

Ali, A., Alsalhi, M.S., Atif, M., Anees, A.A., Israr, M.Q., Sadaf, J.R., et al., 2013. Potentiometric urea biosensor utilizing nanobiocomposite of chitosan-iron oxide magnetic nanoparticles. J. Phys. Conf. Ser. 414, 012024.

Anirudhan, T.S., Sandeep, S., 2012. Synthesis, characterization, cellular uptake and cytotoxicity of a multi-functional magnetic nanocomposite for the targeted delivery and controlled release of doxorubicin to cancer cells. J. Mater. Chem. 22, 12888–12899.

Ansari, A.A., Kaushik, A., Solanki, P.R., Malhotra, B.D., 2009. Electrochemical cholesterol sensor based on tin oxide-chitosan nanobiocomposite film. Electroanalysis. 21, 965–972.

Arya, S.K., Datta, M., Malhotra, B.D., 2008. Recent advances in cholesterol biosensor. Biosens. Bioelectron. 23, 1083–1100.

Aziz, M.S., EL Sherif, A.Y., 2016. Biomimicry as an approach for bio-inspired structure with the aid of computation. Alexandria Eng. J. 55, 707–714.

Azuma, K., Ifuku, S., Osaki, T., Okamoto, Y., Minami, S., 2014. Preparation and biomedical applications of chitin and chitosan nanofibers. J. Biomed. Nanotechnol. 10, 2891–2920.

Balantrapu, K., Goia, D.V., 2009. Silver nanoparticles for printable electronics and biological applications. J. Mater. Res. 24, 2828–2836.

Bandyopadhyaya, R., Nativ-Roth, E., Regev, O., Yerushalmi-Rozen, R., 2002. Stabilization of individual carbon nanotubes in aqueous solutions. Nano Lett. 2, 25–28.

Baptista, A., Soares, P., Ferreira, I., Borges, J.P., 2013a. Nanofibers and nanoparticles in biomedical applications. In: TIWARI, A., TIWARI, A. (Eds.), Bioengineered Nanomaterials. CRC Press, New York.

Baptista, A.C., Ferreira, I., Borges, J.P., 2013b. Cellulose-based bioelectronic devices. In: VAN DE VEN, T., GODBOUT, L. (Eds.), *Cellulose—Medical, Pharmaceutical and Electronic Applications*. InTech, Hoboken, NJ, USA.

Barik, A., Solanki, P.R., Kaushik, A., Ali, A., Pandey, M.K., Kim, C.G., et al., 2010. Polyaniline–carboxymethyl cellulose nanocomposite for cholesterol detection. J. Nanosci. Nanotechnol. 10, 6479–6488.

Bledzki, A.K., Gassan, J., 1999. Composites reinforced with cellulose based fibres. Prog. Polym. Sci. 24, 221–274.

Belamie, E., Mosser, G., Gobeaux, F., Giraud-Guille, M.M., 2006. Possible transient liquid crystal phase during the laying out of connective tissues: α-chitin and collagen as models. J. Phys.: Condens. Matter. 18, S115.

Bhattarai, N., Edmondson, D., Veiseh, O., Matsen, F.A., Zhang, M., 2005. Electrospun chitosan-based nanofibers and their cellular compatibility. Biomaterials. 26, 6176–6184.

Borges, J.P., Canejo, J.P., Fernandes, S.N., Brogueira, P., Godinho, M.H., 2014. Cellulose-based liquid crystalline composite systems. Nanocellulose Polymer Nanocomposites. John Wiley & Sons, Inc.

Bouligand, Y., 1972. Twisted fibrous arrangements in biological materials and cholesteric mesophases. Tissue Cell. 4, 189–217.

Breuer-Thal, B., Witt, R., Weder, C., Foster, J., Jorfi, M., Roberts, M.N. 2014. Medical injection device. WO2014040886 A1.

Cai, H., Yao, P., 2013. In situ preparation of gold nanoparticle-loaded lysozyme-dextran nanogels and applications for cell imaging and drug delivery. Nanoscale. 5, 2892–2900.

Cai, Z., Kim, J., 2010. Bacterial cellulose/poly(ethylene glycol) composite: characterization and first evaluation of biocompatibility. Cellulose. 17, 83–91.

Capadona, J.R., Van Den Berg, O., Capadona, L.A., Schroeter, M., Rowan, S.J., Tyler, D.J., et al., 2007. A versatile approach for the processing of polymer nanocomposites with self-assembled nanofibre templates. Nat. Nanotechnol. 2, 765–769.

Capadona, J.R., Shanmuganathan, K., Tyler, D.J., Rowan, S.J., Weder, C., 2008. Stimuli-responsive polymer nanocomposites inspired by the sea cucumber dermis. Science. 319, 1370–1374.

Capadona, J.R., Shanmuganathan, K., Trittschuh, S., Seidel, S., Rowan, S.J., Weder, C., 2009. Polymer nanocomposites with nanowhiskers isolated from microcrystalline cellulose. Biomacromolecules. 10, 712–716.

Capadona, J.R., Tyler, D.J., Zorman, C.A., Rowan, S.J., Weder, C., 2012. Mechanically adaptive nanocomposites for neural interfacing. MRS Bull. 37, 581–589.

Carpi, F., Smela, E., 2009. Front Matter. *Biomedical Applications of Electroactive Polymer Actuators*. John Wiley & Sons, Ltd, Chippenham, Wiltshire.

Chahal, S., Hussain, F.S.J., Kumar, A., Yusoff, M.M., Rasad, M.S.B.A., 2015. Electrospun hydroxyethyl cellulose nanofibers functionalized with calcium phosphate coating for bone tissue engineering. RSC Adv. 5, 29497–29504.

Chahal, S., Hussain, F.S.J., Kumar, A., Rasad, M.S.B.A., Yusoff, M.M., 2016. Fabrication, characterization and in vitro biocompatibility of electrospun hydroxyethyl cellulose poly (vinyl) alcohol nanofibrous composite biomaterial for bone tissue engineering. Chem. Eng. Sci. 144, 17–29.

Chen, H.Q., Fan, M.W., 2007. Chitosan/carboxymethyl cellulose polyelectrolyte complex scaffolds for pulp cells regeneration. J. Bioact. Compat. Polym. 22, 475–491.

Chen, M.-C., Mi, F.-L., Liao, Z.-X., Hsiao, C.-W., Sonaje, K., Chung, M.-F., et al., 2013. Recent advances in chitosan-based nanoparticles for oral delivery of macromolecules. Adv. Drug Delivery Rev. 65, 865–879.

Choi, R., Yang, J., Choi, J., Lim, E.-K., Kim, E., Suh, J.-S., et al., 2010. Thiolated dextran-coated gold nanorods for photothermal ablation of inflammatory macrophages. Langmuir. 26, 17520–17527.

Collins, F.S., Varmus, H., 2015. A new initiative on precision medicine. N. Engl. J. Med. 372, 793–795.

Colombo, L., Zoia, L., Violatto, M.B., Previdi, S., Talamini, L., Sitia, L., et al., 2015. Organ distribution and bone tropism of cellulose nanocrystals in living mice. Biomacromolecules. 16, 2862–2871.

Dai, T., Jiang, X., Hua, S., Wang, X., Lu, Y., 2008. Facile fabrication of conducting polymer hydrogels via supramolecular self-assembly. Chem. Commun. 4279–4281.

Dash, M., Chiellini, F., Ottenbrite, R.M., Chiellini, E., 2011. Chitosan—a versatile semi-synthetic polymer in biomedical applications. Prog. Polym. Sci. 36, 981–1014.

Dawei, L., Kelong, A., Qingqing, W., Pengfei, L., Qufu, W., 2016. Preparation of Pd/bacterial cellulose hybrid nanofibers for dopamine detection. Molecules. 21, 618–628.

Denardo, S.J., Denardo, G.L., Natarajan, A., Miers, L.A., Foreman, A.R., Gruettner, C., et al., 2007. Thermal Dosimetry Predictive of Efficacy of ^{111}In-ChL6 nanoparticle AMF-induced thermoablative therapy for human breast cancer in mice. J. Nucl. Med. 48, 437–444.

Depan, D., Misra, R.D., 2012. Processing–structure–functional property relationship in organic–inorganic nanostructured scaffolds for bone-tissue engineering: the response of preosteoblasts. J. Biomed. Mater. Res. A. 100, 3080–3091.

Depan, D., Pratheep Kumar, A., Singh, R.P., Misra, R.D.K., 2014. Stability of chitosan/montmorillonite nanohybrid towards enzymatic degradation on grafting with poly(lactic acid). Mater. Sci. Technol. 30, 587–592.

Desbrieres, J., Babak, V., 2010. Interfacial properties of chitin and chitosan based systems. Soft Matter. 6, 2358–2363.

Dias, A.M.G.C., Hussain, A., Marcos, A.S., Roque, A.C.A., 2011. A biotechnological perspective on the application of iron oxide magnetic colloids modified with polyssacharides. Biotechnol. Adv. 29, 142–155.

Di Martino, A., Sedlarik, V., 2014. Amphiphilic chitosan-grafted-functionalized polylactic acid based nanoparticles as a delivery system for doxorubicin and temozolomide co-therapy. Int. J. Pharm. 474, 134–145.

Domingues, R.M.A., Gomes, M.E., Reis, R.L., 2014. The potential of cellulose nanocrystals in tissue engineering strategies. Biomacromolecules. 15, 2327–2346.

Dugan, J.M., Gough, J.E., Eichhorn, S.J., 2010. Directing the morphology and differentiation of skeletal muscle cells using oriented cellulose nanowhiskers. Biomacromolecules. 11, 2498–2504.

Dugan, J.M., Collins, R.F., Gough, J.E., Eichhorn, S.J., 2013. Oriented surfaces of adsorbed cellulose nanowhiskers promote skeletal muscle myogenesis. Acta Biomater. 9, 4707–4715.

Ebron, V.H., Yang, Z., Seyer, D.J., Kozlov, M.E., Oh, J., Xie, H., et al., 2006. Fuel-powered artificial muscles. Science. 311, 1580–1583.

Echeverria, C., Aguirre, L.E., Merino, E.G., Almeida, P.L., Godinho, M.H., 2015. Carbon nanotubes as reinforcement of cellulose liquid crystalline responsive networks. ACS Appl. Mater. Interfaces. 7, 21005–21009.

Edwards, P.P., Thomas, J.M., 2007. Gold in a metallic divided state—from faraday to present-day nanoscience. Angew. Chem. Int. Ed. 46, 5480−5486.

Egan, P., Sinko, R., Leduc, P.R., Keten, S., 2015. Therole of mechanics in biological and bio-inspired systems. Nat. Commun. 6, 138.

El-Dakdouki, M.H., Zhu, D.C., El-Boubbou, K., Kamat, M., Chen, J., Li, W., et al., 2012. Development of multifunctional hyaluronan-coated nanoparticles for imaging and drug delivery to cancer cells. Biomacromolecules. 13, 1144−1151.

Esa, F., Tasirin, S.M., Rahman, N.A. 2014. Overview of bacterial cellulose production and application. 2nd International Conference on Agricultural and Food Engineering (Cafe 2014)—New Trends Forward, 2, 113−119.

Esmaeili, C., Abdi, M.M., Mathew, P.A., Jonoobi, M., Oksman, K., Rezayi, M., 2015. Synergy effect of nanocrystalline cellulose for the biosensing detection of glucose. Sensors 15.

Fang, B., Wan, Y.Z., Tang, T.T., Gao, C., Dai, K.R., 2009. Proliferation and osteoblastic differentiation of human bone marrow stromal cells on hydroxyapatite/bacterial cellulose nanocomposite scaffolds. Tissue Eng. Part A. 15, 1091−1098.

Faruk, O., Bledzki, A.K., Fink, H.-P., Sain, M., 2012. Biocomposites reinforced with natural fibers: 2000−2010. Prog. Polym. Sci. 37 (11), 1552−1596.

Fernandes, S.N., Geng, Y., Vignolini, S., Glover, B.J., Trindade, A.C., Canejo, J.P., et al., 2013. Structural color and iridescence in transparent sheared cellulosic films. Macromol. Chem. Phys. 214, 25−32.

Ferreira, S.S., Passos, C.P., Madureira, P., Vilanova, M., Coimbra, M.A., 2015. Structure−function relationships of immunostimulatory polysaccharides: a review. Carbohydr. Polym. 132, 378−396.

Frohbergh, M.E., Katsman, A., Botta, G.P., Lazarovici, P., Schauer, C.L., Wegst, U.G., et al., 2012. Electrospun hydroxyapatite-containing chitosan nanofibers crosslinked with genipin for bone tissue engineering. Biomaterials. 33, 9167−9178.

Fu, L.N., Zhang, J., Yang, G., 2013. Present status and applications of bacterial cellulose-based materials for skin tissue repair. Carbohydr. Polym. 92, 1432−1442.

Funkhouser, J., 2002. Reinventing Pharma: the theranostic revolution. Curr. Drug Discovery. 2, 17−19.

Gauthier, O., Bouler, J.M., Weiss, P., Bosco, J., Aguado, E., Daculsi, G., 1999. Short-term effects of mineral particle sizes on cellular degradation activity after implantation of injectable calcium phosphate biomaterials and the consequences for bone substitution. Bone. 25, 71s−74s.

Gauthier, O., Khairoun, I., Bosco, J., Obadia, L., Bourges, X., Rau, C., et al., 2003. Noninvasive bone replacement with a new injectable calcium phosphate biomaterial. J. Biomed. Mater. Res., Part A. 66A, 47−54.

Geng, Y., Almeida, P.L., Fernandes, S.N., Cheng, C., Palffy-Muhoray, P., Godinho, M.H., 2013. A cellulose liquid crystal motor: a steam engine of the second kind. Sci. Rep. 3, 1028.

Gentile, P., Mattioli-Belmonte, M., Chiono, V., Ferretti, C., Baino, F., Tonda-Turo, C., et al., 2012. Bioactive glass/polymer composite scaffolds mimicking bone tissue. J. Biomed. Mater. Res. A. 100, 2654−2667.

Giese, M., Blusch, L.K., Khan, M.K., Maclachlan, M.J., 2015. Functional materials from cellulose-derived liquid-crystal templates. Angew. Chem. Int. Ed. 54, 2888−2910.

Giraud-Guille, M.M., Besseau, L., Martin, R., 2003. Liquid crystalline assemblies of collagen in bone and in vitro systems. J. Biomech. 36, 1571−1579.

Giraud-Guille, M.-M., Belamie, E., Mosser, G., 2004. Organic and mineral networks in carapaces, bones and biomimetic materials. C.R. Palevol. 3, 503–513.

Giraud Guille, M.M., Mosser, G., Helary, C., Eglin, D., 2005. Bone matrix like assemblies of collagen: From liquid crystals to gels and biomimetic materials. Micron. 36, 602–608.

Gomathi, P., Ragupathy, D., Choi, J.H., Yeum, J.H., Lee, S.C., Kim, J.C., et al., 2011. Fabrication of novel chitosan nanofiber/gold nanoparticles composite towards improved performance for a cholesterol sensor. Sens. Actuators, B: Chem. 153, 44–49.

Gonzalez, J.S., Ludueña, L.N., Ponce, A., Alvarez, V.A., 2014. Poly(vinyl alcohol)/cellulose nanowhiskers nanocomposite hydrogels for potential wound dressings. Mater. Sci. Eng., C. 34, 54–61.

Gouma, P., Xue, R., Goldbeck, C.P., Perrotta, P., Balazsi, C., 2012. Nano-hydroxyapatite—cellulose acetate composites for growing of bone cells. Mater. Sci. Eng., C. 32, 607–612.

Gowda, S.R., Reddy, A.L.M., Shaijumon, M.M., Zhan, X., Ci, L., Ajayan, P.M., 2011. Conformal coating of thin polymer electrolyte layer on nanostructured electrode materials for three-dimensional battery applications. Nano Lett. 11, 101–106.

Guan, M., Zhou, Y., Zhu, Q.L., Liu, Y., Bei, Y.Y., Zhang, X.N., et al., 2012. *N*-trimethyl chitosan nanoparticle-encapsulated lactosyl-norcantharidin for liver cancer therapy with high targeting efficacy. Nanomedicine. 8, 1172–1181.

Guan, Y., Bian, J., Peng, F., Zhang, X.-M., Sun, R.-C., 2014a. High strength of hemicelluloses based hydrogels by freeze/thaw technique. Carbohydr. Polym. 101, 272–280.

Guan, Y., Zhang, B., Bian, J., Peng, F., Sun, R.-C., 2014b. Nanoreinforced hemicellulose-based hydrogels prepared by freeze–thaw treatment. Cellulose. 21, 1709–1721.

Guise, C., Fangueiro, R., 2016. Biomedical applications of nanocellulose. In: Fangueiro, R., Rana, S. (Eds.), Natural Fibres: Advances in Science and Technology Towards Industrial Applications: From Science to Market. Springer Netherlands, Dordrecht.

Guo, R., Zhang, L., Qian, H., Li, R., Jiang, X., Liu, B., 2010. Multifunctional nanocarriers for cell imaging, drug delivery, and near-IR photothermal therapy. Langmuir. 26, 5428–5434.

Haes, A.J., Stuart, D.A., Nie, S., Van Duyne, R.P., 2004. Using solution-phase nanoparticles, surface-confined nanoparticle arrays and single nanoparticles as biological sensing platforms. J. Fluoresc. 14, 355–367.

Harris, J.P., Capadona, J.R., Miller, R.H., Healy, B.C., Shanmuganathan, K., Rowan, S.J., et al., 2011. Mechanically adaptive intracortical implants improve the proximity of neuronal cell bodies. J. Neural Eng. 8, 066011–066011.

Högemann-Savellano, D., Bos, E., Blondet, C., Sato, F., Abe, T., Josephson, L., et al., 2003. The transferrin receptor: a potential molecular imaging marker for human cancer. Neoplasia (New York, N.Y.). 5, 495–506.

Homma, A., Sato, H., Tamura, T., Okamachi, A., Emura, T., Ishizawa, T., et al., 2010. Synthesis and optimization of hyaluronic acid-methotrexate conjugates to maximize benefit in the treatment of osteoarthritis. Bioorg. Med. Chem. 18, 1062–1075.

Hu, H., Hou, X.-J., Wang, X.-C., Nie, J.-J., Cai, Q., Xu, F.-J., 2016a. Gold nanoparticle-conjugated heterogeneous polymer brush-wrapped cellulose nanocrystals prepared by combining different controllable polymerization techniques for theranostic applications. Polym. Chem. 7, 3107–3116.

Hu, Y., Zhu, Y.J., Zhou, X., Ruan, C.S., Pan, H.B., Catchmark, J.M., 2016b. Bioabsorbable cellulose composites prepared by an improved mineral-binding process for bone defect repair. J. Mater. Chem. B. 4, 1235–1246.

Ismail, Y.A., Shin, S.R., Shin, K.M., Yoon, S.G., Shon, K., Kim, S.I., et al., 2008. Electrochemical actuation in chitosan/polyaniline microfibers for artificial muscles fabricated using an in situ polymerization. Sens. Actuators, B: Chem. 129, 834–840.

Ismail, Y.A., Martínez, J.G., Al Harrasi, A.S., Kim, S.J., Otero, T.F., 2011. Sensing characteristics of a conducting polymer/hydrogel hybrid microfiber artificial muscle. Sens. Actuators, B: Chem. 160, 1180–1190.

Ivkov, R. 2006. Magnetic nanoscale particle compositions, and therapeutic methods related thereto. US patent application.

Jackowska, K., Krysinski, P., 2013. New trends in the electrochemical sensing of dopamine. Anal. Bioanal. Chem. 405, 3753–3771.

Jain, T., Kumar, S., Dutta, P.K., 2015. Theranostics: a way of modern medical diagnostics and the role of chitosan. J. Mol. Genet. Med. 9, 151.

Jin, J., Hassanzadeh, P., Perotto, G., Sun, W., Brenckle, M.A., Kaplan, D., et al., 2013. A biomimetic composite from solution self-assembly of chitin nanofibers in a silk fibroin matrix. Adv. Mater. 25, 4482–4487.

João, C.F.C., Silva, J.C., Borges, J.P., 2015. Chitin-based nanocomposites: biomedical applications. In: Thakur, K.V., Thakur, K.M. (Eds.), Eco-friendly Polymer Nanocomposites: Chemistry and Applications. Springer India, New Delhi.

Jorfi, M., Foster, E.J., 2015. Recent advances in nanocellulose for biomedical applications. J. Appl. Polym. Sci. 132, n/a–n/a.

Jorfi, M., Roberts, M.N., Foster, E.J., Weder, C., 2013. Physiologically responsive, mechanically adaptive bio-nanocomposites for biomedical applications. ACS Appl. Mater. Interfaces. 5, 1517–1526.

Jorfi, M., Voirin, G., Foster, E.J., Weder, C., 2014. Physiologically responsive, mechanically adaptive polymer optical fibers for optogenetics. Opt. Lett. 39, 2872–2875.

Kaushik, A., Khan, R., Solanki, P.R., Pandey, P., Alam, J., Ahmad, S., et al., 2008. Iron oxide nanoparticles–chitosan composite based glucose biosensor. Biosens. Bioelectron. 24, 676–683.

Killard, A.J., Smyth, M.R., 2000. Creatinine biosensors: principles and designs. Trends Biotechnol. 18, 433–437.

Kim, J., Yun, S., Ounaies, Z., 2006. Discovery of cellulose as a smart material. Macromolecules. 39, 4202–4206.

Kim, J., Cai, Z., Lee, H.S., Choi, G.S., Lee, D.H., Jo, C., 2011. Preparation and characterization of a bacterial cellulose/chitosan composite for potential biomedical application. J. Polym. Res. 18, 739–744.

Kim, K., Kim, J.H., Park, H., Kim, Y.-S., Park, K., Nam, H., et al., 2010. Tumor-homing multifunctional nanoparticles for cancer theragnosis: Simultaneous diagnosis, drug delivery, and therapeutic monitoring. J. Controlled Release. 146, 219–227.

Kim, S.-K., 2013. Chitin and Chitosan Derivatives. CRC Press.

Kim, S.-S., Jeon, J.-H., Kim, H.-I., Kee, C.D., Oh, I.-K., 2015. High-fidelity bioelectronic muscular actuator based on graphene-mediated and TEMPO-oxidized bacterial cellulose. Adv. Funct. Mater. 25, 3560–3570.

Klemm, D., Kramer, F., Moritz, S., Lindstrom, T., Ankerfors, M., Gray, D., et al., 2011. Nanocelluloses: a new family of nature-based materials. Angew. Chem. Int. Ed. Engl. 50, 5438–5466.

Koo, H., Min, K.H., Lee, S.C., Park, J.H., Park, K., Jeong, S.Y., et al., 2013. Enhanced drug-loading and therapeutic efficacy of hydrotropic oligomer-conjugated glycol chitosan nanoparticles for tumor-targeted paclitaxel delivery. J. Controlled Release. 172, 823–831.

Krasia-Christoforou, T., Georgiou, T.K., 2013. Polymeric theranostics: using polymer-based systems for simultaneous imaging and therapy. J. Mater. Chem. B. 1, 3002–3025.
Krontiras, P., Gatenholm, P., Hagg, D.A., 2015. Adipogenic differentiation of stem cells in three-dimensional porous bacterial nanocellulose scaffolds. J. Biomed. Mater. Res., Part B: Appl. Biomater. 103, 195–203.
Kucińska-Lipka, J., Gubanska, I., Janik, H., 2015. Bacterial cellulose in the field of wound healing and regenerative medicine of skin: recent trends and future prospectives. Polym. Bull. 72, 2399–2419.
Kwon, S., Park, J.H., Chung, H., Kwon, I.C., Jeong, S.Y., Kim, I.-S., 2003. Physicochemical characteristics of self-assembled nanoparticles based on glycol chitosan bearing 5β-cholanic acid. Langmuir. 19, 10188–10193.
Lagerwall, J.P.F., Schütz, C., Salajkova, M., Noh, J., Hyun Park, J., Scalia, G., et al., 2014. Cellulose nanocrystal-based materials: from liquid crystal self-assembly and glass formation to multifunctional thin films. NPG Asia Mater. 6, e80.
Lee, H., Lee, K., Park, T.G., 2008. Hyaluronic acid-paclitaxel conjugate micelles: synthesis, characterization, and antitumor activity. Bioconjugate Chem. 19, 1319–1325.
Lee, S.J., Koo, H., Lee, D.-E., Min, S., Lee, S., Chen, X., et al., 2011. Tumor-homing photosensitizer-conjugated glycol chitosan nanoparticles for synchronous photodynamic imaging and therapy based on cellular on/off system. Biomaterials. 32, 4021–4029.
Lee, S.J., Huh, M.S., Lee, S.Y., Min, S., Lee, S., Koo, H., et al., 2012. Tumor-homing poly-siRNA/glycol chitosan self-cross-linked nanoparticles for systemic siRNA delivery in cancer treatment. Angew. Chem. Int. Ed. 51, 7203–7207.
Li, J., Vadahanambi, S., Kee, C.-D., Oh, I.-K., 2011. Electrospun fullerenol-cellulose biocompatible actuators. Biomacromolecules. 12, 2048–2054.
Li, Y., Jiang, H., Zheng, W., Gong, N., Chen, L., Jiang, X., et al., 2015. Bacterial cellulose–hyaluronan nanocomposite biomaterials as wound dressings for severe skin injury repair. J. Mater. Chem. B. 3, 3498–3507.
Liao, N.N., Unnithan, A.R., Joshi, M.K., Tiwari, A.P., Hong, S.T., Park, C.H., et al., 2015. Electrospun bioactive poly(epsilon-caprolactone)-cellulose acetate-dextran antibacterial composite mats for wound dressing applications. Colloids Surf., A: Physicochem. Eng. Aspects. 469, 194–201.
Lin, N., Bruzzese, C., Dufresne, A., 2012. TEMPO-oxidized nanocellulose participating as crosslinking aid for alginate-based sponges. ACS Appl. Mater. Interfaces. 4, 4948–4959.
Lin, N., Dufresne, A. 2014. Nanocellulose in biomedicine: current status and future prospect. Eur. Polym. J., 59, 302–325.
Lin, W.-C., Lien, C.-C., Yeh, H.-J., Yu, C.-M., Hsu, S.-H., 2013. Bacterial cellulose and bacterial cellulose–chitosan membranes for wound dressing applications. Carbohydr. Polym. 94, 603–611.
Liu, C., Tan, Y., Liu, C., Chen, X., Yu, L., 2007. Preparations, characterizations and applications of chitosan-based nanoparticles. J. Ocean Univ. Chin. 6, 237–243.
Liu, J., Willför, S., Xu, C., 2015. A review of bioactive plant polysaccharides: Biological activities, functionalization, and biomedical applications. Bioact. Carbohydr. Diet. Fibre. 5, 31–61.
Liu, M., Zheng, H., Chen, J., Li, S., Huang, J., Zho, C., 2016. Chitosan-chitin nanocrystal composite scaffolds for tissue engineering. Carbohydr. Polym. 152, 832–840.
Liu, W.Z., Zhang, J.T., Rethore, G., Khairoun, K., Pilet, P., Tancret, F., et al., 2014. A novel injectable, cohesive and toughened Si-HPMC (silanized-hydroxypropyl methylcellulose)

composite calcium phosphate cement for bone substitution. Acta Biomater. 10, 3335–3345.

Liuyun, J., Yubao, L., Chengdong, X., 2009. Preparation and biological properties of a novel composite scaffold of nano-hydroxyapatite/chitosan/carboxymethyl cellulose for bone tissue engineering. J. Biomed. Sci. 16, 65.

Lu, L., Chen, W., 2010. Biocompatible composite actuator: a supramolecular structure consisting of the biopolymer chitosan, carbon nanotubes, and an ionic liquid. Adv. Mater. 22, 3745–3748.

Lu, W., Smela, E., Adams, P., Zuccarello, G., Mattes, B.R., 2004. Development of solid-in-hollow electrochemical linear actuators using highly conductive polyaniline. Chem. Mater. 16, 1615–1621.

Lv, X.G., Yang, J.X., Feng, C., Li, Z., Chen, S.Y., Xie, M.K., et al., 2016. Bacterial cellulose-based biomimetic nanofibrous scaffold with muscle cells for hollow organ tissue engineering. ACS Biomater. Sci. Eng. 2, 19–29.

Madden, J.D.W., Barisci, J.N., Anquetil, P.A., Spinks, G.M., Wallace, G.G., Baughman, R. H., et al., 2006. Fast carbon nanotube charging and actuation. Adv. Mater. 18, 870–873.

Mahadeva, S.K., Kim, J., 2011. Conductometric glucose biosensor made with cellulose and tin oxide hybrid nanocomposite. Sens. Actuators, B: Chem. 157, 177–182.

Mao, H.Q., Roy, K., Troung-Le, V.L., Janes, K.A., Lin, K.Y., Wang, Y., et al., 2001. Chitosan-DNA nanoparticles as gene carriers: synthesis, characterization and transfection efficiency. J. Controlled Release. 70, 399–421.

Markstedt, K., Mantas, A., Tournier, I., Avila, H.M., Hagg, D., Gatenholm, P., 2015. 3D bioprinting human chondrocytes with nanocellulose-alginate bioink for cartilage tissue engineering applications. Biomacromolecules. 16, 1489–1496.

Matsuda, A., Kobayashi, H., Itoh, S., Kataoka, K., Tanaka, J., 2005. Immobilization of laminin peptide in molecularly aligned chitosan by covalent bonding. Biomaterials. 26, 2273–2279.

Matsuhisa, H., Tsuchiya, M., Hasebe, Y., 2013. Protein and polysaccharide-composite sol–gel silicate film for an interference-free amperometric glucose biosensor. Colloids Surf., B: Biointerfaces. 111, 523–529.

Matsumura, S., Kajiyama, S., Nishimura, T., Kato, T., 2015. Formation of helically structured chitin/$CaCO_3$ hybrids through an approach inspired by the biomineralization processes of crustacean cuticles. Small. 11, 5127–5133.

Mincea, M., Negrulescu, A., Ostafe, V., 2012. Preparation, modification, and applications of chitin nanowhiskers: a review. Rev. Adv. Mater. Sci. 30, 225–242.

Moon, R.J., Martini, A., Nairn, J., Simonsen, J., Youngblood, J., 2011. Cellulose nanomaterials review: structure, properties and nanocomposites. Chem. Soc. Rev. 40, 3941–3994.

Morganti, P., Morganti, G., 2008. Chitin nanofibrils for advanced cosmeceuticals. Clin. Dermatol. 26 (4), 334–340.

Motokawa, T., 1994. Effects of ionic environment on viscosity of Triton-extracted catch connective tissue of a sea cucumber body wall. Comp. Biochem. Physiol., B: Biochem. Mol. Biol. 109, 613–622.

Murray, R.W., 2008. Nanoelectrochemistry: metal nanoparticles, nanoelectrodes, and nanopores. Chem. Rev. 108, 2688–2720.

Muzzarelli, R.A.A., 1977. Chitin. first ed Pergamon, Great Britain.

Muzzarelli, R.A., El Mehtedi, M., Mattioli-Belmonte, M., 2014. Emerging biomedical applications of nano-chitins and nano-chitosans obtained via advanced eco-friendly technologies from marine resources. Mar. Drugs. 12, 5468–5502.

Nakayama, A., Kakugo, A., Gong, J.P., Osada, Y., Takai, M., Erata, T., et al., 2004. High mechanical strength double-network hydrogel with bacterial cellulose. Adv. Funct. Mater. 14, 1124–1128.

Nantaphol, S., Chailapakul, O., Siangproh, W., 2015. A novel paper-based device coupled with a silver nanoparticle-modified boron-doped diamond electrode for cholesterol detection. Anal. Chim. Acta. 891, 136–143.

Naseri, N., Poirier, J.M., Girandon, L., Frohlich, M., Oksman, K., Mathew, A.P., 2016. 3-Dimensional porous nanocomposite scaffolds based on cellulose nanofibers for cartilage tissue engineering: tailoring of porosity and mechanical performance. RSC Adv. 6, 5999–6007.

Nguyen, H.-L., Jo, Y., Cha, M., Cha, Y., Yoon, D., Sanandiya, N., et al., 2016. Mussel-inspired anisotropic nanocellulose and silver nanoparticle composite with improved mechanical properties, electrical conductivity and antibacterial activity. Polymers. 8, 102.

Nguyen, N.S., Yoon, H.H., 2016. Nickel oxide-deposited cellulose/CNT composite electrode for non-enzymatic urea detection. Sens. Actuators, B: Chem. 236, 304–310.

Norowski, P.A., Mishra, S., Adatrow, P.C., Haggard, W.O., Bumgardner, J.D., 2012. Suture pullout strength and in vitro fibroblast and RAW 264.7 monocyte biocompatibility of genipin crosslinked nanofibrous chitosan mats for guided tissue regeneration. J. Biomed. Mater. Res. A. 100, 2890–2896.

Oh, I.-H., Min, H.S., Li, L., Tran, T.H., Lee, Y.-K., Kwon, I.C., et al., 2013. Cancer cell-specific photoactivity of pheophorbide a—glycol chitosan nanoparticles for photodynamic therapy in tumor-bearing mice. Biomaterials. 34, 6454–6463.

Olyveira, G.M., Valido, D.P., Costa, L.M.M., Gois, P.B.P., Filho, L.X., Basmaji, P., 2011. First otoliths/collagen/bacterial cellulose nanocomposites as a potential scaffold for bone tissue regeneration. J. Biomater. Nanobiotechnol. 2, 239–243.

Otero, T.F., Sansiñena, J.M., 1995. Artificial muscles based on conducting polymers. Bioelectrochem. Bioenerg. 38, 411–414.

Panichpakdee, J., Pavasant, P., Supaphol, P., 2014. Electrospinning of asiaticoside/2-hydroxypropyl-β-cyclodextrin inclusion complex-loaded cellulose acetate fiber mats: release characteristics and potential for use as wound dressing. Polym. Korea. 38, 338–350.

Park, J.H., Cho, H.J., Yoon, H.Y., Yoon, I.S., Ko, S.H., Shim, J.S., et al., 2014a. Hyaluronic acid derivative-coated nanohybrid liposomes for cancer imaging and drug delivery. J. Controlled Release. 174, 98–108.

Park, K., Kim, J.H., Nam, Y.S., Lee, S., Nam, H.Y., Kim, K., et al., 2007. Effect of polymer molecular weight on the tumor targeting characteristics of self-assembled glycol chitosan nanoparticles. J. Controlled Release. 122, 305–314.

Park, K.E., Jung, S.Y., Lee, S.J., Min, B.-M., Park, W.H., 2006a. Biomimetic nanofibrous scaffolds: preparation and characterization of chitin/silk fibroin blend nanofibers. Int. J. Biol. Macromol. 38, 165–173.

Park, K.E., Kang, H.K., Lee, S.J., Min, B.M., Park, W.H., 2006b. Biomimetic nanofibrous scaffolds: preparation and characterization of PGA/chitin blend nanofibers. Biomacromolecules. 7, 635–643.

Park, M., Lee, D., Hyun, J., 2015. Nanocellulose-alginate hydrogel for cell encapsulation. Carbohydr. Polym. 116, 223–228.

Park, Y.-L., Chen, B.-R., Young, D., Stirling, L., Wood, R.J., Goldfield, E.C., et al., 2014b. Design and control of a bio-inspired soft wearable robotic device for ankle? foot rehabilitation. Bioinspiration Biomimetics. 9, 016007.

Payne, G.F., Raghavan, S.R., 2007. Chitosan: a soft interconnect for hierarchical assembly of nano-scale components. Soft Matter. 3, 521–527.

Pei, Y., Ye, D.D., Zhao, Q., Wang, X.Y., Zhang, C., et al., 2015. Effectively promoting wound healing with cellulose/gelatin sponges constructed directly from a cellulose solution. J. Mater. Chem. B. 3, 7518–7528.

Peng, H., Liu, X., Wang, G., Li, M., Bratlie, K.M., Cochran, E., et al., 2015. Polymeric multifunctional nanomaterials for theranostics. J. Mater. Chem. B. 3, 6856–6870.

Potter, K.A., Jorfi, M., Householder, K.T., Foster, E.J., Weder, C., Capadona, J.R., 2014. Curcumin-releasing mechanically adaptive intracortical implants improve the proximal neuronal density and blood-brain barrier stability. Acta Biomater. 10, 2209–2222.

Prabhu, R.H., Patravale, V.B., Joshi, M.D., 2015. Polymeric nanoparticles for targeted treatment in oncology: current insights. Int. J. Nanomed. 10, 1001–1018.

Price, J.C., Roach, P., El Haj, A.J., 2016. Liquid crystalline ordered collagen substrates for applications in tissue engineering. ACS Biomater. Sci. Eng. 2, 625–633.

Qi, B., Lu, W., Mattes, B.R., 2004. Strain and energy efficiency of polyaniline fiber electrochemical actuators in aqueous electrolytes. J. Phys. Chem. B. 108, 6222–6227.

Qin, L., Dichen, L., Zhongmin, J., Jue, W., Aimin, L., Zhen, W., 2009. Fabrication and in vitro evaluation of calcium phosphate combined with chitosan fibers for scaffold structures. J. Bioact. Compat. Polym. 24, 113–124.

Rafiee, A., Alimohammadian, M.H., Gazori, T., Riazi-Rad, F., Fatemi, S.M.R., Parizadeh, A., et al., 2014. Comparison of chitosan, alginate and chitosan/alginate nanoparticles with respect to their size, stability, toxicity and transfection. Asian Pac. J. Trop. Dis. 4, 372–377.

Ramani, D., Sastry, T.P., 2014. Bacterial cellulose-reinforced hydroxyapatite functionalized graphene oxide: a potential osteoinductive composite. Cellulose. 21, 3585–3595.

Rederstorff, E., Rethore, G., Weiss, P., Sourice, S., Beck-Cormier, S., Mathieu, E., et al., 2015. Enriching a cellulose hydrogel with a biologically active marine exopolysaccharide for cell-based cartilage engineering. J. Tissue Eng. Regener. Med http://dx.doi:10.1002/term.2018.

Ren, X., Chen, D., Meng, X., Tang, F., Du, A., Zhang, L., 2009. Amperometric glucose biosensor based on a gold nanorods/cellulose acetate composite film as immobilization matrix. Colloids Surf., B: Biointerfaces. 72, 188–192.

Ruecha, N., Rangkupan, R., Rodthongkum, N., Chailapakul, O., 2014. Novel paper-based cholesterol biosensor using graphene/polyvinylpyrrolidone/polyaniline nanocomposite. Biosens. Bioelectron. 52, 13–19.

Sa-Lima, H., Tuzlakoglu, K., Mano, J.F., Reis, R.L., 2011. Thermoresponsive poly(*N*-isopropylacrylamide)-*g*-methylcellulose hydrogel as a three-dimensional extracellular matrix for cartilage-engineered applications. J. Biomed. Mater. Res., Part A. 98, 596–603.

Sa, M.W., Kim, S.E., Yun, Y.P., Song, H.R., Kim, J.Y., 2014. Fabrication of hybrid scaffolds by polymer deposition system and its in-vivo evaluation with a rat tibial defect model. Tissue Eng. Regener. Med. 11, 439–445.

Saboktakin, M.R., Tabatabaie, R., Maharramov, A., Ramazanov, M.A., 2010. Synthesis and characterization of superparamagnetic chitosan–dextran sulfate hydrogels as nano carriers for colon-specific drug delivery. Carbohydr. Polym. 81, 372–376.

Sahu, S.K., Mallick, S.K., Santra, S., Maiti, T.K., Ghosh, S.K., Pramanik, P., 2010. In vitro evaluation of folic acid modified carboxymethyl chitosan nanoparticles loaded with doxorubicin for targeted delivery. J. Mater. Sci. Mater. Med. 21, 1587–1597.

Salas, C., Nypelö, T., Rodriguez-Abreu, C., Carrillo, C., Rojas, O.J., 2014. Nanocellulose properties and applications in colloids and interfaces. Curr. Opin. Colloid Interface Sci. 19, 383–396.

Sanchez, C., Arribart, H., Giraud Guille, M.M., 2005. Biomimetism and bioinspiration as tools for the design of innovative materials and systems. Nat. Mater. 4, 277–288.

Sassolas, A., Blum, L.J., Leca-Bouvier, B.D., 2012. Immobilization strategies to develop enzymatic biosensors. Biotechnol. Adv. 30, 489–511.

Sekine, S., Ido, Y., Miyake, T., Nagamine, K., Nishizawa, M., 2010. Conducting polymer electrodes printed on hydrogel. J. Am. Chem. Soc. 132, 13174–13175.

Shah, N., Ul-Islam, M., Khattak, W.A., Park, J.K., 2013. Overview of bacterial cellulose composites: a multipurpose advanced material. Carbohydr. Polym. 98, 1585–1598.

Shanmuganathan, K., Capadona, J.R., Rowan, S.J., Weder, C., 2010a. Bio-inspired mechanically-adaptive nanocomposites derived from cotton cellulose whiskers. J. Mater. Chem. 20, 180–186.

Shanmuganathan, K., Capadona, J.R., Rowan, S.J., Weder, C., 2010b. Biomimetic mechanically adaptive nanocomposites. Prog. Polym. Sci. 35, 212–222.

Shen, J.-M., Guan, X.-M., Liu, X.-Y., Lan, J.-F., Cheng, T., Zhang, H.-X., 2012. Luminescent/magnetic hybrid nanoparticles with folate-conjugated peptide composites for tumor-targeted drug delivery. Bioconjugate Chem. 23, 1010–1021.

Shi, L., 2016. Bioactivities, isolation and purification methods of polysaccharides from natural products: a review. Int. J. Biol. Macromol. 92, 37–48.

Šmejkalová, D., Nešporová, K., Huerta-Angeles, G., Syrovátka, J., Jirák, D., Gálisová, A., et al., 2014. Selective in vitro anticancer effect of superparamagnetic iron oxide nanoparticles loaded in hyaluronan polymeric micelles. Biomacromolecules. 15, 4012–4020.

Soares, P.I., Alves, A.M., Pereira, L.C., Coutinho, J.T., Ferreira, I.M., Novo, C.M., et al., 2014. Effects of surfactants on the magnetic properties of iron oxide colloids. J. Colloid Interface Sci. 419, 46–51.

Soares, P.I., Lochte, F., Echeverria, C., Pereira, L.C., Coutinho, J.T., Ferreira, I.M., et al., 2015. Thermal and magnetic properties of iron oxide colloids: influence of surfactants. Nanotechnology. 26, 425704.

Soares, P.I.P., Laia, C.A.T., Carvalho, A., Pereira, L.C.J., Coutinho, J.T., Ferreira, I.M.M., et al., 2016a. Iron oxide nanoparticles stabilized with a bilayer of oleic acid for magnetic hyperthermia and MRI applications. Appl. Surface Sci. 383, 240–247.

Soares, P.I.P., Machado, D., Laia, C., Pereira, L.C.J., Coutinho, J.T., Ferreira, I.M.M., et al., 2016b. Thermal and magnetic properties of chitosan-iron oxide nanoparticles. Carbohydr. Polym. 149, 382–390.

Soares, P.I.P., Sousa, A.I., Silva, J.C., Ferreira, I.M.M., Novo, C.M.M., Borges, J.P., 2016c. Chitosan-based nanoparticles as drug delivery systems for doxorubicin: optimization and modelling. Carbohydr. Polym. 147, 304–312.

Stephen, Z.R., Kievit, F.M., Veiseh, O., Chiarelli, P.A., Fang, C., Wang, K., et al., 2014. Redox-responsive magnetic nanoparticle for targeted convection-enhanced delivery of O6-benzylguanine to brain tumors. ACS Nano. 8, 10383–10395.

Subramanian, A., Vu, D., Larsen, G.F., Lin, H.Y., 2005. Preparation and evaluation of the electrospun chitosan/PEO fibers for potential applications in cartilage tissue engineering. J. Biomater. Sci. Polym. Ed. 16, 861–873.

Suginta, W., Khunkaewla, P., Schulte, A., 2013. Electrochemical biosensor applications of polysaccharides chitin and chitosan. Chem. Rev. 113, 5458–5479.

Sukrut, O., Manjeet, J., Ashwini, K.A., 2008. pH and electrical actuation of single walled carbon nanotube/chitosan composite fibers. Smart Mater. Struct. 17, 055016.

Surace, C., Arpicco, S., Dufay-Wojcicki, A., Marsaud, V., Bouclier, C., Clay, D., et al., 2009. Lipoplexes targeting the CD44 hyaluronic acid receptor for efficient transfection of breast cancer cells. Mol. Pharm. 6, 1062–1073.

Svenson, S., 2013. Theranostics: are we there yet? Mol. Pharmaceutics. 10, 848–856.

Swierczewska, M., Han, H.S., Kim, K., Park, J.H., Lee, S., 2016. Polysaccharide-based nanoparticles for theranostic nanomedicine. Adv. Drug Delivery Rev. 99, 70–84, Part A.

Tagami, T., Ernsting, M.J., Li, S.-D., 2011. Efficient tumor regression by a single and low dose treatment with a novel and enhanced formulation of thermosensitive liposomal doxorubicin. J. Controlled Release. 152, 303–309.

Takehana, Y., Yamada, A., Tamori, M., Motokawa, T., 2014. Softenin, a novel protein that softens the connective tissue of sea cucumbers through inhibiting interaction between collagen fibrils. PLoS One. 9, e85644.

Tan, W.B., Jiang, S., Zhang, Y., 2007. Quantum-dot based nanoparticles for targeted silencing of HER2/neu gene via RNA interference. Biomaterials. 28, 1565–1571.

Tazi, N., Zhang, Z., Messaddeq, Y., Almeida-Lopes, L., Zanardi, L.M., Levinson, D., et al., 2012. Hydroxyapatite bioactivated bacterial cellulose promotes osteoblast growth and the formation of bone nodules. AMB Express. 2, 61.

Teti, G., Salvatore, V., Focaroli, S., Durante, S., Mazzotti, A., Dicarlo, M., et al., 2015. In vitro osteogenic and odontogenic differentiation of human dental pulp stem cells seeded on carboxymethyl cellulose–hydroxyapatite hybrid hydrogel. Front. Physiol. 6, 297.

Thakur, V.K., Singha, A.S., Thakur, M.K., 2013. Ecofriendly biocomposites from natural fibers: mechanical and weathering study. Int. J. Polym. Anal. Charact. 18 (1), 64–72.

Thakur, V.K., Thakur, M.K., 2014. Journal of Cleaner Production. J. Cleaner Prod. 82 (C), 1–15.

Thakur, V.K., Thakur, M.K., Gupta, R.K., 2014a. Review: raw natural fiber–based polymer composites. Int. J. Polym. Anal. Charact. 19 (3), 256–271.

Thakur, V.K., Thakur, M.K., Raghavan, P., Kessler, M.R., 2014b. Progress in green polymer composites from lignin for multifunctional applications: a review. ACS Sustainable Chem. Eng. 2 (5), 1072–1092. Available from: http://dx.doi.org/10.1021/sc500087z.

Thurmond, F., Trotter, J., 1996. Morphology and biomechanics of the microfibrillar network of sea cucumber dermis. J. Exp. Biol. 199, 1817–1828.

Thomas, S., Visakh, P.M., Mathew, A.P., 2012. Advances in Natural Polymers. Springer, http://books4.elsevierproofcentral.com/authorproofs/c7da238d7dd9481571ae0ade59991-ca5/MC.

Trojani, C., Boukhechba, F., Scimeca, J.C., Vandenbos, F., Michiels, J.F., Daculsi, G., et al., 2006. Ectopic bone formation using an injectable biphasic calcium phosphate/Si-HPMC hydrogel composite loaded with undifferentiated bone marrow stromal cells. Biomaterials. 27, 3256–3264.

Trotter, J.A., Tipper, J., Lyons-Levy, G., Chino, K., Heuer, A.H., Liu, Z., et al., 2000. Towards a fibrous composite with dynamically controlled stiffness: lessons from echinoderms. Biochem. Soc. Trans. 28, 357–362.

Tsai, Y.-C., Chen, S.-Y., Lee, C.-A., 2008. Amperometric cholesterol biosensors based on carbon nanotube–chitosan–platinum–cholesterol oxidase nanobiocomposite. Sens. Actuators, B: Chem. 135, 96–101.

Turner, A.P.F., 2013. Biosensors: sense and sensibility. Chem. Soc. Rev. 42, 3184–3196.

Ul-Islam, M., Khan, S., Ullah, M.W., Park, J.K., 2015. Bacterial cellulose composites: Synthetic strategies and multiple applications in bio-medical and electro-conductive fields. Biotechnol. J. 10, 1847–1861.

Vatankhah, E., Prabhakaran, M.P., Jin, G., Mobarakeh, L.G., Ramakrishna, S., 2014. Development of nanofibrous cellulose acetate/gelatin skin substitutes for variety wound treatment applications. J. Biomater. Appl. 28, 909–921.

Veiseh, O., Kievit, F.M., Fang, C., Mu, N., Jana, S., Leung, M.C., et al., 2010. Chlorotoxin bound magnetic nanovector tailored for cancer cell targeting, imaging, and siRNA delivery. Biomaterials. 31, 8032–8042.

Vignolini, S., Moyroud, E., Glover, B.J., Steiner, U., 2013. Analysing photonic structures in plants. J. R. Soc. Interface. 10, 20130394.

Vinatier, C., Magne, D., Moreau, A., Gauthier, O., Malard, O., Vignes-Colombeix, C., et al., 2007. Engineering cartilage with human nasal chondrocytes and a silanized hydroxypropyl methylcellulose hydrogel. J. Biomed. Mater. Res., Part A. 80, 66–74.

Vinatier, C., Gauthier, O., Fatimi, A., Merceron, C., Masson, M., Moreau, A., et al., 2009. An injectable cellulose-based hydrogel for the transfer of autologous nasal chondrocytes in articular cartilage defects. Biotechnol. Bioeng. 102, 1259–1267.

Viswanathan, G., Murugesan, S., Pushparaj, V., Nalamasu, O., Ajayan, P.M., Linhardt, R.J., 2006. Preparation of biopolymer fibers by electrospinning from room temperature ionic liquids. Biomacromolecules. 7, 415–418.

Vivero-Escoto, J.L., Huang, Y.-T., 2011. Inorganic-organic hybrid nanomaterials for therapeutic and diagnostic imaging applications. Int. J. Mol. Sci. 12, 3888–3927.

Wan, Y.Z., Gao, C., Han, M., Liang, H., Ren, K.J., Wang, Y.L., et al., 2011. Preparation and characterization of bacterial cellulose/heparin hybrid nanofiber for potential vascular tissue engineering scaffolds. Polym. Adv. Technol. 22, 2643–2648.

Wang, C., Ravi, S., Garapati, U.S., Das, M., Howell, M., Mallelamallela, J., et al., 2013a. Multifunctional chitosan magnetic-graphene (CMG) nanoparticles: a theranostic platform for tumor-targeted co-delivery of drugs, genes and MRI contrast agents. J. Mater. Chem. B: Mater. Biol. Med. 1, 4396–4405.

Wang, F., Jeon, J.-H., Kim, S.-J., Park, J.-O., Park, S., 2016a. An eco-friendly ultra-high performance ionic artificial muscle based on poly(2-acrylamido-2-methyl-1-propanesulfonic acid) and carboxylated bacterial cellulose. J. Mater. Chem. B. 4, 5015–5024.

Wang, F., Jeon, J.-H., Park, S., Kee, C.-D., Kim, S.-J., Oh, I.-K., 2016b. A soft biomolecule actuator based on a highly functionalized bacterial cellulose nano-fiber network with carboxylic acid groups. Soft Matter. 12, 246–254.

Wang, J., Yang, C.X., Wan, Y.Z., Luo, H.L., He, F., Dai, K.R., et al., 2013b. Laser patterning of bacterial cellulose hydrogel and its modification with gelatin and hydroxyapatite for bone tissue engineering. Soft Mater. 11, 173–180.

Wang, L.S., Chuang, M.C., Ho, J.A., 2012. Nanotheranostics—a review of recent publications. Int. J. Nanomed. 7, 4679–4695.

Wang, N., Chen, Y., Kim, J., 2007. Electroactive paper actuator made with chitosan-cellulose films: effect of acetic acid. Macromol. Mater. Eng. 292, 748–753.

Wang, Y., Zhang, X., Chen, Y., Xu, H., Tan, Y., Wang, S., 2010. Detection of dopamine based on tyrosinase-Fe_3O_4 nanoparticles-chitosan nanocomposite biosensor. Am. J. Biomed. Sci. 209–216.

Wang, Z., Niu, G., Chen, X., 2013c. Polymeric materials for theranostic applications. Pharm. Res. 31, 1358–1376.

Wang, Z., Chen, Z., Liu, Z., Shi, P., Dong, K., Ju, E., et al., 2014. A multi-stimuli responsive gold nanocage–hyaluronic platform for targeted photothermal and chemotherapy. Biomaterials. 35, 9678–9688.

Weder, C., Foster, J., Jorfi, M., Roberts, M.N. 2014. *Polymer nanocomposite having switchable mechanical properties*. WO2014040885 A3.

Wegst, U.G.K., Bai, H., Saiz, E., Tomsia, A.P., Ritchie, R.O., 2015. Bioinspired structural materials. Nat. Mater. 14, 23–36.

Wei, W., Lv, P.-P., Chen, X.-M., Yue, Z.-G., Fu, Q., Liu, S.-Y., et al., 2013. Codelivery of mTERT siRNA and paclitaxel by chitosan-based nanoparticles promoted synergistic tumor suppression. Biomaterials. 34, 3912–3923.

Wilkie, I.C., 2002. Is muscle involved in the mechanical adaptability of echinoderm mutable collagenous tissue? J. Exp. Biol. 205, 159–165.

Woltman, S.J., Jay, G.D., Crawford, G.P., 2007. Liquid-crystal materials find a new order in biomedical applications. Nat. Mater. 6, 929–938.

Wu, X., Zhao, F., Varcoe, J.R., Thumser, A.E., Avignone-Rossa, C., Slade, R.C.T., 2009. Direct electron transfer of glucose oxidase immobilized in an ionic liquid reconstituted cellulose–carbon nanotube matrix. Bioelectrochemistry. 77, 64–68.

Xiao, C., 2016. Development of stimuli-responsive polysaccharides-based nanotheranostics. Curr. Nanosci. 12, 33–37.

Yadav, S., Devi, R., Kumar, A., Pundir, C.S., 2011. Tri-enzyme functionalized ZnO-NPs/CHIT/c-MWCNT/PANI composite film for amperometric determination of creatinine. Biosens. Bioelectron. 28, 64–70.

Yadav, S., Devi, R., Bhar, P., Singhla, S., Pundir, C.S., 2012. Immobilization of creatininase, creatinase and sarcosine oxidase on iron oxide nanoparticles/chitosan-g-polyaniline modified Pt electrode for detection of creatinine. Enzyme Microb. Technol. 50, 247–254.

Yamamoto, Y., Nishimura, T., Saito, T., Kato, T., 2010. $CaCO_3$/chitin-whisker hybrids: formation of $CaCO_3$ crystals in chitin-based liquid-crystalline suspension. Polym. J. 42, 583–586.

Yang, C.X., Gao, C., Wan, Y.Z., Tang, T.T., Zhang, S.H., Dai, K.R., 2011. Preparation and characterization of three-dimensional nanostructured macroporous bacterial cellulose/agarose scaffold for tissue engineering. J. Porous Mater. 18, 545–552.

Yhee, J.Y., Son, S., Kim, N., Choi, K., Kwon, I.C., 2014. Theranostic applications of organic nanoparticles for cancer treatment. MRS Bull. 39, 239–249.

Yhee, J.Y., Song, S., Lee, S.J., Park, S.-G., Kim, K.-S., Kim, M.G., et al., 2015. Cancer-targeted MDR-1 siRNA delivery using self-cross-linked glycol chitosan nanoparticles to overcome drug resistance. J. Controlled Release. 198, 1–9.

Yoon, H.Y., Son, S., Lee, S.J., You, D.G., Yhee, J.Y., Park, J.H., et al., 2014. Glycol chitosan nanoparticles as specialized cancer therapeutic vehicles: sequential delivery of doxorubicin and Bcl-2 siRNA. Sci. Rep. 4, 6878.

You, D.G., Saravanakumar, G., Son, S., Han, H.S., Heo, R., Kim, K., et al., 2014. Dextran sulfate-coated superparamagnetic iron oxide nanoparticles as a contrast agent for atherosclerosis imaging. Carbohydr. Polym. 101, 1225–1233.

Yuan, Q., Hein, S., Misra, R.D.K., 2010. New generation of chitosan-encapsulated ZnO quantum dots loaded with drug: Synthesis, characterization and in vitro drug delivery response. Acta Biomater. 6, 2732–2739.

Zamora-Mora, V., Fernández-Gutiérrez, M., Román, J., Goya, G., Hernández, R., Mijangos, C., 2014. Magnetic core–shell chitosan nanoparticles: rheological characterization and hyperthermia application. Carbohydr. Polym. 102, 691698.

Zamora-Mora, V., Soares, P., Echeverria, C., Hernández, R., Mijangos, C., 2015. Composite chitosan/agarose ferrogels for potential applications in magnetic hyperthermia. Gels. 1, 69.

Zarrin, A., Sadighian, S., Rostamizadeh, K., Firuzi, O., Hamidi, M., Mohammadi-Samani, S., et al., 2015. Design, preparation, and in vitro characterization of a trimodally-targeted nanomagnetic onco-theranostic system for cancer diagnosis and therapy. Int. J. Pharm.

Zhang, J.T., Liu, W.Z., Schnitzler, V., Tancret, F., Bouler, J.M., 2014. Calcium phosphate cements for bone substitution: chemistry, handling and mechanical properties. Acta Biomater. 10, 1035–1049.

Zhijiang, C., Guang, Y., 2011. Bacterial cellulose/collagen composite: Characterization and first evaluation of cytocompatibility. J. Appl. Polym. Sci. 120, 2938–2944.

Zimmermann, K.A., Leblanc, J.M., Sheets, K.T., Fox, R.W., Gatenholm, P., 2011. Biomimetic design of a bacterial cellulose/hydroxyapatite nanocomposite for bone healing applications. Mater. Sci. Eng., C. 31, 43–49.

Zoller, M., 1995. CD44: physiological expression of distinct isoforms as evidence for organ-specific metastasis formation. J. Mol. Med. (Berl.). 73, 425–438.

Zong, A., Cao, H., Wang, F., 2012. Anticancer polysaccharides from natural resources: A review of recent research. Carbohydr. Polym. 90, 1395–1410.

Zulkifli, F.H., Hussain, F.S., Rasad, M.S., Mohd Yusoff, M., 2014. Nanostructured materials from hydroxyethyl cellulose for skin tissue engineering. Carbohydr. Polym. 114, 238–245.

Zulkifli, F.H., Jahir Hussain, F.S., Abdull Rasad, M.S., Mohd Yusoff, M., 2015. Improved cellular response of chemically crosslinked collagen incorporated hydroxyethyl cellulose/poly(vinyl) alcohol nanofibers scaffold. J. Biomater. Appl. 29, 1014–1027.

Hybrid multifunctional composites—recent applications

5

Naheed Saba[1], *Mohammad Jawaid*[1,2,3], *Mohamed Thariq Hameed Sultan*[2,4] *and Othman Alothman*[2]
[1]Laboratory of Biocomposite Technology, Institute of Tropical Forestry and Forest Products (INTROP), Universiti Putra Malaysia, Serdang, Selangor, Malaysia, [2]Aerospace Manufacturing Research Centre, Level 7, Tower Block, Faculty of Engineering, Universiti Putra Malaysia, Serdang, Selangor Darul Ehsan, Malaysia, [3]Department of Chemical Engineering, College of Engineering, King Saud University, Riyadh, Saudi Arabia, [4]Department of Aerospace Engineering, Faculty of Engineering, Universiti Putra Malaysia, Serdang, Selangor Darul Ehsan, Malaysia

Chapter Outline

5.1 Introduction 151
5.1.1 Polymer composites 151
5.1.2 Nanocomposites 153
5.2 Multifunctional materials 153
5.3 Multifunctional composites 154
5.3.1 Multifunctional natural fibers composites 155
5.3.2 Multifunctional synthetic fibers composites 155
5.3.3 Multifunctional nanocomposites 156
5.3.4 Multifunctional hybrid composites 157
5.3.5 Multifunctional hybrid nanocomposites 158
5.4 Applications of multifunctional materials and multifunctional composites 161
5.5 Conclusion 163
Acknowledgment 164
References 164

5.1 Introduction

5.1.1 Polymer composites

The development of polymer composites materials displays keen interest as they are lightweight, flexible, and tougher and is usually fabricated on large size into intricately shaped components for industrial applications by combining large varieties of

Hybrid Polymer Composite Materials: Applications. DOI: http://dx.doi.org/10.1016/B978-0-08-100785-3.00005-X

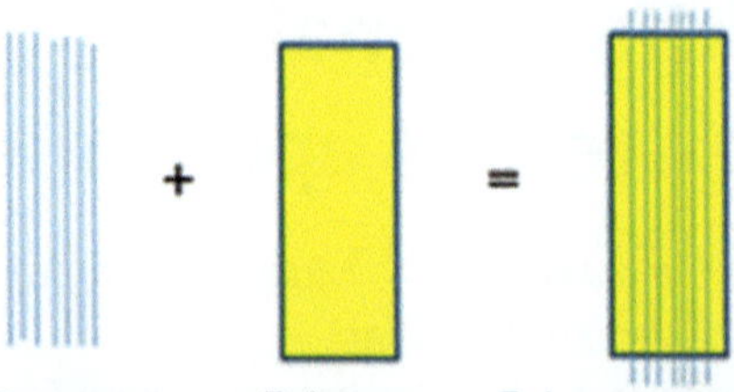

Figure 5.1 Components of fibers reinforced polymer composites.

reinforcements and polymers (Saba et al., 2015, 2014). Polymer and the reinforcements such as fibers (natural or synthetic) are the major components of polymer composites as displayed in Fig. 5.1. Polymer composite offers several advantages as compared to metallic structures, but besides this, they also have severe shortcomings with a major concerned on their wave propagation response, damage modeling and detection, joining, fatigue life prediction, and prestress issues (Benjeddou et al., 2016). Currently, polymer composite structures are widely used for manufacturing aeronautic components and are being extended to other industrial applications including railway, ship, automotive, civil engineering, and constructions. However, in most cases, they do not satisfy the growing demand of advanced and active applications, thus directing the emergence and innovation of "hybrid polymer composites" (Lopes et al., 2014). Hybrid polymer composite materials provide combination of properties such as tensile strength, modulus, impact strength, compressive strength, and favorable thermal stability which cannot be realized in polymer composite materials (Saba et al., 2014, 2016). The development of hybrid polymer composites by the combinations of polymer and reinforcements is being explored from past decades for smart, high performance, and functional material applications. Furthermore, the continuous requirement for the development of materials and structures that can perform simultaneously, drive the interest in development of multifunctional materials and structures. The multifunctional materials or structures possess (1) multiple structural functions with high stiffness, high strength, high damping, and high fracture toughness, (2) mutual structural and nonstructural functions including load-bearing structure with the promising tendency of providing noise and vibration control, thermal insulation, self-repair, and energy storage/harvesting or (3) includes (both 1 and 2) (Gibson, 2010). Multifunctional materials, multifunctional composites, and multifunctional structures all belong to the group of multifunctional material systems as the major components, as shown in Fig. 5.2 (Ferreira et al., 2016).

Currently, there is a growing demand for multifunctional composites and multifunctional nanocomposites material in the advanced engineering and biomedical applications. The growing applications of composites direct the requirements of multifunctional design materials that are referred to as multifunctional composite materials. Multifunctionality arises both from the material and the structure. Multifunctional materials best exemplified by carbon nanotubes (CNTs) having inherently high mechanical strength and electrical conductivity providing both actuation and sensing. Multifunctional composites are exemplified by the incorporation

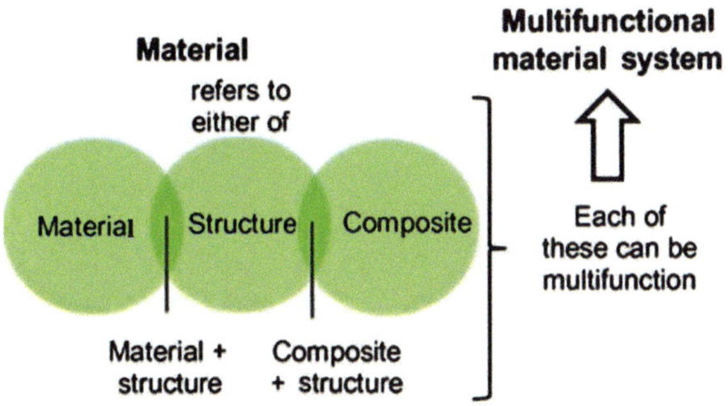

Figure 5.2 Components of multifunctional material systems (Ferreira et al., 2016).

of the CNTs into the polymers like epoxy, polyester, and others together with iron powder having mechanical strength and electrical conductivity. Remarkably, multifunctionality is grouped into two groups, structural and nonstructural based on their functions. Materials with multiple structural functions such as high toughness, stiffness, strength, ductility, fatigue, resistance, and lowered damping properties are grouped as structural multifunctional materials (Gibson, 2010; Ferreira et al., 2016), whereas others are simply grouped as nonstructural functions including thermal conductivity and electrical resistivity (Matic, 2003).

5.1.2 Nanocomposites

Nanocomposite materials are the potential substitutes to overcome the drawbacks of conventional composites (Saba et al., 2014), and currently, receiving higher attention in fundamental and applied materials science field (Nicolay et al., 2015). A nanocomposite obtained when a polymeric matrix (thermoplastics or thermosets) reinforced with nanoparticles/nanosized fillers in order to improve a particular property of the material (Saba et al., 2015). Considerable changes in the physical, mechanical, chemical, and electrical properties of nanocomposites are realized because of the combination of nanoparticles with the polymer matrix (Nicolay et al., 2015).

5.2 Multifunctional materials

A multifunctional material system integrates the functions of more than two different components and/or composites/materials/structures increasing the total system's efficiency.

Multifunctional materials also referred to as smart materials and are regarded as the materials having the function to withstand particular loads besides high mechanical strength, good fatigue resistance, and high thermal stability with relatively poor chemical and environmental interactions. Their composites are characterized by the distinctive feature of being able to detect environmental changes or external stimuli at the most optimum conditions and can actively respond according to these modifications (Benjeddou et al., 2016). The multifunctional materials include carbonaceous nanoparticles such as CNTs, polyhedral oligomeric silsesquioxane (POSS), graphite, inorganic nanoparticles silica, nanoclay, and magnetic-noble metal based nanoparticles having high aspect ratio and unique properties (Saba et al., 2016). The incorporation of these nanoparticles on micro-gel or imprinted polymers results in the fabrication of multifunctional nanocomposites (Sierra-Martin and Fernandez-Barbero, 2015). Recently, researchers developed multifunctional materials performing the dual functions of structural and nonstructural groups. The nanofillers were developed from oil palm empty fruit bunch fibers (OPEFB) (Saba et al., 2015) having the tendency to improve both the flame retardancy (nonstructural functions), tensile, and impact strength (structural functions) of the epoxy composites. Furthermore, a new inorganic nanofiller called boron nitride (BN) also delivers extensive applications due to high thermal conductivity ($280\ W\ m^{-1}\ K^{-1}$) and high electrical resistivity of short-circuiting in certain systems and for TIMs (Raza et al., 2015).

5.3 Multifunctional composites

In the most traditional way, the development of composite structures involves the load-carrying function and other functional requirements separately. However, the requirement of performing the two functions simultaneously directs the innovation of multifunctional composites. Multifunctional composites represent the advanced, hi-tech, and next generation of engineered materials. Multifunctional composite materials perform two or more functions simultaneously, such as carrying mechanical loads along with delivering or storing electrical energy (Asp and Greenhalgh, 2014). It can be ascribed by the fact that the constituents of multifunctional composites can simultaneously and synergistically provide structural and electrochemical energy storage functions to the final system. A multifunctional composite materials can also be realized and characterized in terms of electromagnetic shielding effectiveness and ballistic properties (Micheli et al., 2016). Multifunctionality of composite materials also vary from electrical to thermal and mechanical functions (Gibson, 2010). Such composites are categorized into three different types based on their functions,

1. Multifunctional composites performing structural applications having unique tendency of balancing mechanical properties under different loading conditions by activating different energy dissipation mechanisms or load-bearing features. Nano-reinforced composites and multiscaled systems are the best example of this group.

   Recently, multifunctional carbon fibers reinforced polymer composites have been reported by the researchers that possess tendency of carrying mechanical loads and can store electrical energy simultaneously. Fabricated multifunctional composites can act as a supercapacitor with sufficient good mechanical properties including Young's modulus, shear stiffness, compression strength, and peel/shear toughness. The schematic of multifunctional structural power storage material is represented in Fig. 5.3. http://www3.imperial.ac.uk/portal/page/portallive/polymersandcompositesengineering/aboutpace/multifunctionalmaterials.

Figure 5.3 Multifunctional structural power storage material.

2. Multifunctional composites performing nonstructural applications, such as self-healing, power harvesting, and structural health monitoring.
3. Composites performing a combination of structural and nonstructural functions to enhance mechanical properties and enable specific features to extend product life and feasibility.

The advantage of using multifunctional composite materials is that they can simultaneously carry out two or more functions as they possess energy storage, mechanical, and electrical energy properties. Thus, it provides alternative materials for conventional energy storage systems to reduce weight and improve redundancy or enhance performance of the final product (Shirshova et al., 2009). Moreover, these composites provide unique opportunities to create structural materials with added functionality for sensing or self-repair and new material architectures by incorporating novel fibers and nanomaterials.

5.3.1 Multifunctional natural fibers composites

Multifunctional composite materials also include natural fibers reinforced conductive polymer composites having wider technological applications. These composites are being used as highly potential materials in chemical/electrochemical catalysts, chemical detecting sensors, antennas, photothermal optical recording, electronic noses, and insulation materials (Al-Oqla et al., 2015). They also show extensive used as components in connectors, terminals, fuel cells, switches, printed circuit boards, insulators, self-regulating heaters, panels, industrial, and household plugs. Besides this, they are also involved in antibacterial packaging, disposable accessories, tissue engineering, drug delivery, neural probes, bioactuators, biosensors, furniture, building components, and automotive industries.

5.3.2 Multifunctional synthetic fibers composites

Multifunctional glass fibers reinforced poly(methyl methacrylate) PMMA-$BaTiO_3$ structural/dielectric composites have been reported by the researchers for potential use as structural dielectrics in multifunctional capacitors having better energy storage and excellent mechanical properties (Stefanescu et al., 2011). The scheme for the fabrication of PMMA-based composites reinforced glass fibers and particulate filler $BaTiO_3$ are displayed in Fig. 5.4. The fabricated PMMA-based multifunctional composites possess high dielectric constant and higher stiffness showing promising applications in airplanes, space shuttles, and in hybrid ground vehicles (Stefanescu et al., 2011).

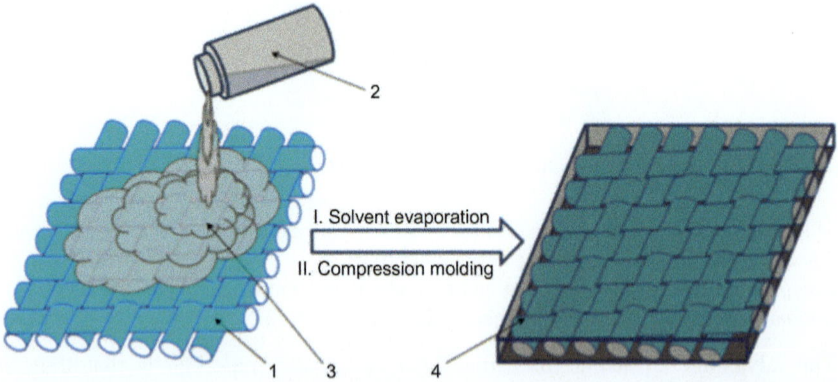

Figure 5.4 Fabrication of PMMA/fiberglass/$BaTiO_3$ hybrid composite (Stefanescu et al., 2011).

5.3.3 Multifunctional nanocomposites

Multifunctional or multistructural nanocomposites developed by adding characteristics additional phase or functional nanoparticles. The addition of nanoparticles completely modified the mechanical, thermal, barrier, and electrical behavior of a traditional laminate without altering the total weight of the structure. These composites can be used in various fields ranging from electronics to aerospace industries (Jia et al., 2011; Chrissafis and Bikiaris, 2011; Riquelme et al., 2016).

Multifunctional polymer nanocomposites fabrication by utilizing carbon fillers have attracted widespread attention (Spitalsky et al., 2010; Dang et al., 2012; Sengupta et al., 2011; Bhowmick et al., 2016). However, among many carbon fillers, CNTs, carbon nanofibers (CNFs), POSS, graphite, graphene, and nanoclay are the most potential and promising nanofillers for developing hybrid nanocomposites with multifunctional applications (Ghamsari et al., 2014). Graphene and CNTs are the most widely used carbon nanomaterials (CNMs) due to their unique morphology, structure, and physicochemical properties (Chau et al., 2015). All of them possess different spatial configurations of covalently bonded carbon atoms and are used extensively as multifunctional materials besides CNFs and cellulose nanofibers are illustrated in Fig. 5.5.

Graphene nanoplatelets (GNPs) with two-dimensional structure and high aspect ratio consisting of graphite nanocrystals layers are ideal reinforcing and conducting fillers (Li et al., 2013). Addition of CNTs improves the mechanical and electrical properties of the polymer composites due to their excellent conductivity and high elastic modulus. CNTs show extensive applications in automotive and aerospace applications as polymer composite materials reinforcement, where constraints as lightweight, mechanical resistance, and electromagnetic shielding efficiency are satisfied simultaneously (Micheli et al., 2016).

Huge varieties of metal oxide fillers such as TiO_2, ZrO_2, SiO_2, besides the expensive carbon-based fillers have been successively added into polymers to attain the required properties improvement. Currently, nano-ZnO also received higher attention as multifunctional inorganic nanofiller on account of its unique physical

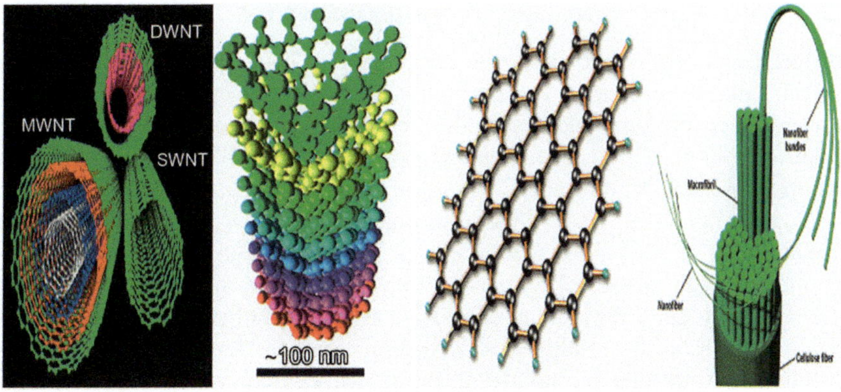

Figure 5.5 (A) CNTs, (B) CNFs, (C) graphene, and (D) cellulose nanofiber.
Modified from Ferreira, A.D.B., Nóvoa, P.R., Marques, A.T., Multifunctional material systems: a state-of-the-art review, Compos. Struct. (2016), 151, 3–35.

and chemical properties, such as wide band gap of 3.37 eV at room temperature, a low dielectric constant, and a high exciton binding energy of approximately 60 meV in the ultraviolet (UV) region. Nano-ZnO can be used in a variety of applications including piezoelectric transducers, gas sensors, sunscreens, and cosmetics, as it is an effective UV absorber with a good transparency in the visible range (Shafei and Abou-Okeil, 2011). The nonstructural functions including photocatalytic and antibacterial activities of ZnO nanoparticles are also of great interest for composite industries (Nicolay et al., 2015; Amornpitoksuk et al., 2012; Zhang et al., 2007).

5.3.4 Multifunctional hybrid composites

Several substantial efforts have been made from past decades to find new and effective alternative for existing metallic materials, in terms of better performance, weight, and cost effectiveness, by fabricating promising hybrid polymer composites. Hybrid composite materials have extensive engineering and broad spectrum application with a combination of properties such as compressive strength, tensile modulus, impact strength, and thermal stability that cannot be realized in polymers or polymer composite materials. The multifunctional hybrid composites shows widespread and diverse functional applications ranging from environmental, catalytic, electrical, biological, automotive, space shuttles, and energy conversion to energy storage sectors. An effective research study has been reported in the literature indicating, considerable improvements of multiple structural functions in the case of hybrid polymer composites by incorporating nanoscale reinforcements along with conventional micron fibers or particles reinforcements (Gibson, 2010).

5.3.5 Multifunctional hybrid nanocomposites

Advancements in polymer composite science and technology allow the introduction of nanomaterials to design the exciting multifunctional materials and structures with enhanced physical and mechanical properties for wider end use application (Quaresimin et al., 2015). The ideal nanocomposite materials confer multifunctionality such as considerable and effective mechanical and antimicrobial properties, biocompatibility, nontoxicity, and surface functionality to favor cell proliferation (Bhowmick et al., 2016). The development of hybrid multifunctional nanocomposites having both nanolevel reinforcements along with conventional micron level reinforcements, generate great possibility to achieve simultaneous improvements in multiple structural and multiple nonstructural functions properties. The nonstructural functions include electrical and/or thermal conductivity, energy harvesting/storage, sensing and actuation, electromagnetic interference shielding, self-healing capability, biodegradability, and recyclability (Gibson, 2010).

Recently, researchers also developed multifunctional hybrid nanocomposites by incorporating nanooil palm filler into the kenaf reinforced epoxy polymer composites, displaying high mechanical strength, toughness, impact strength along with improved thermal stability (Saba et al., 2016). The resulting hybrid nanocomposite materials consume comparatively lesser thermal and electrical energy. Developed material also shows structural function by replacing conventional structural building materials such as steel, cement, aluminum, woods, and iron rods/plates/beams.

Researchers also reviewed that multifunctional nanocomposites also been developed by combining carbon nanofillers/fibers (CNFs), with suitable inorganic fillers or dispersing different types of CNFs in a polymer matrix (Raza et al., 2015). Recently, developed CNTs/ionic liquid (ILs) hybrids also represent a promising group of multifunctional materials to fabricate multifunctional Bucky syntactic foam composites. Bucky syntactic foam shows specific applications having featured mechanical and electrical properties due to synergistic properties in hybrid offered by both CNTs and ILs (Ghamsari et al., 2014). In other research study, the reaction of vanadium acetyl−acetonate and a suspension of preformed titanium dioxide or cerium dioxide nanoparticles via chemical depositions result in the development of vanadium dioxide nanocomposite thin films on glass substrates. The thin films are thermochromics with considerable photocatalytic activity under irradiation with 254-nm light (Warwick et al., 2011). In other interesting work, nanocomposites with carbon-based hybrid nanostructures synthesized from ferrocene possess magnetic properties and displayed broader applications in transducers, sensors, magnetic storage, electronic devices, electromagnetic, microwave absorption, magnetic actuators, treatment of diseases, and environmental remediation (Riquelme et al., 2016). Incorporation of BN on vapor grown CNF/rubbery epoxy composites also revealed thermal interface and multifunctional applications owing to their low thermal resistance, improved thermal conductivity, and better mechanical properties (Raza et al., 2015). In other interesting study, addition of GNPs and carbon black, shows synergistic effects in ethylene−propylene−diene terpolymer rubber (EPDM)-based nanocomposites (Valentini et al., 2016).

Recent advances in the synthesis and characterization of engineered nanomaterials highlighted the concepts of designing multifunctional membranes with increased selectivity, permeability and resistance to fouling (Crock et al., 2013). Nanomaterials such as CNTs, zeolites, metal oxide nanoparticles, stacked graphene films, and C_{60} fullerenes are being widely used in manufacturing multifunctional membranes. Developed membranes show diverse enhanced environmental catalytic reactions function such as dehalogenation, denitrification, and perchlorate removal (Crock et al., 2013). Moreover, recent and ongoing technological aspects of structural materials for developing multifunctional high performance and advanced self-sensing cementitious materials convey effective structural and nonstructural functionality as it acts as sensors to monitor the health of the structures along with construction materials simultaneously (Danoglidis et al., 2016).

Currently, there is a growing demand for multifunctional biodegradable materials in biomedical applications as imitator of tissues properties in the human body for efficient repair and regeneration. The multifunctional biodegradable materials are developed with varieties of natural and synthetic polymers for load-bearing orthopedic applications having enhanced mechanical properties (Kumar, et al., 2016). Silver (Ag) and reduced graphene oxide in the hybrid particles reinforced in the poly(ε-caprolactone) matrix illustrates synergistic effect and form multifunctional biodegradable material for potential use in fracture fixation devices and tissue engineering (Kumar, et al., 2016).

Research study revealed that polymers derived from petroleum-based products or naturally derived can be used to produce hydrogel, shown in Fig. 5.6

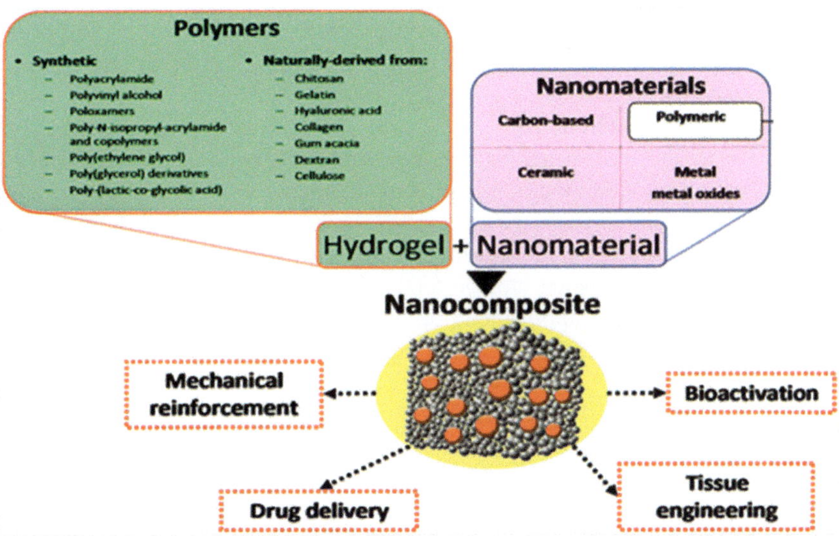

Figure 5.6 Processing of multifunctional nanocomposite from hydrogel and nanomaterials (Biondi et al., 2015).

(Biondi et al., 2015). The hydrogel is then used to fabricate multifunctional nanocomposite materials by embedding variety of nanomaterials for extensive biomedical applications for efficient repair and regeneration (Biondi et al., 2015). They can also show extensive and attractive applications in electronic components and as effective packaging materials with ecofriendly recycling and sustainable feature.

Recently, a multifunctional hybrid material in the form of a sandwich epoxy composites panel faces, reinforced with the glass fibers has been reported with the core consisting of CNTs reinforced polymer foam filling metallic honeycomb (Bollen et al., 2016). The developed sandwiched material shows combined mechanical and electromagnetic absorption performances.

Hybrid polymer composites based on aluminosilicates particles, including clays, zeolites, and mesoporous aluminosilicates, show wider applications in sensors, adsorbents, actuators, water softeners, filtration membranes, catalysis, automotive (Kurauchi et al., 1991; Wang and Xiao, 2008), textile (Solarski et al., 2008), construction (Rodriguez et al., 2011), aerospace (Chen and Curliss, 2001), and food packaging (Ray et al., 2006) besides energy storage applications. These properties are reflected because of their high shapes electivity, excellent thermal/hydrothermal stability, high surface area, and superior ion-exchange ability (Lopes et al., 2014). In other findings, multifunctional composite films are fabricated by compounding multifunctional fillers linked by covalent bonding in hindered amine light stabilizer (HAS)-*f*-CNTs with ultra-high molecular weight polyethylene (UHMWPE) polymer (Dintcheva et al., 2015).

Researchers in the other study developed nanohybrid (phPOSS-*f*-CNTs) multifunctional filler, solely based on CNTs and phenyl polyhedral olygomenric silsesquioxane (phPOSS), for the fabrication of nanocomposites (Dintcheva et al., 2016). The hybrid filler phPOSS-*f*-CNTs are incorporated into UHMWPE by hot compaction to develop multifunctional polymer nanocomposites.

Interestingly, the growing needs to produce advance materials for energy storage lead to the emergence of technologies to use multifunctional composites with electrospun fibers bring revolution in nanoscience and nanotechnology. Electrospun carbon nanofibers (ECNFs) produced by electrospinning technique and their associated hybrid nanocomposites are the advanced materials electrodes for high efficiency energy conversion with better storage tendency (Peng et al., 2016). Fig. 5.7 illustrates the energy storage applications of ECNFs- and ECNFs-based hybrid nanocomposites (Peng et al., 2016). ECNFs hybrid nanocomposites can also be used to develop electrodes and electrochemical capacitors (ECs), due to its advantages of higher power densities, longer cycle lives, and faster recharge tendency as compared with batteries (Peng et al., 2016).

Presently, ECNFs and its hybrid nanocomposites show intensive applications in energy storage devices, such as supercapacitors, batteries, and hydrogen storage, due to their characteristics chemical and physical properties, including chemical stability, large surface area, and high electrical conductivity (Peng et al., 2016).

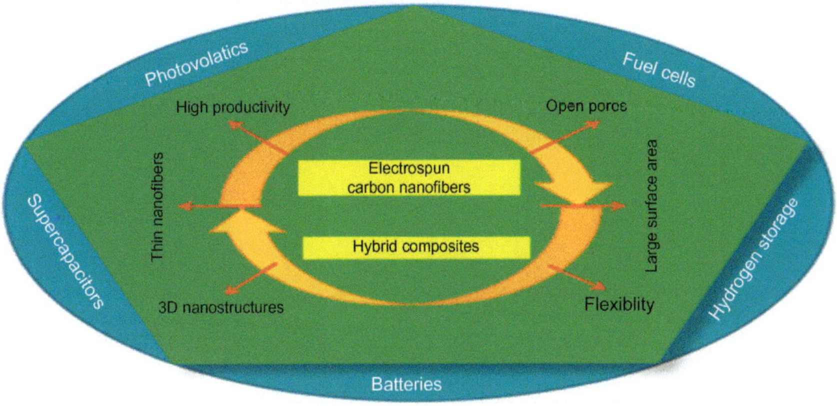

Figure 5.7 Energy conversion and storage applications of ECNFs- and ECNFs-based hybrid nanocomposites (Peng et al., 2016).

5.4 Applications of multifunctional materials and multifunctional composites

Multifunctional material system possesses improved system's efficiency as it incorporates a wide range of material or components combinations having different properties and functions (Ferreira et al., 2016). The formations of multifunctional system or composites (Ferreira et al., 2016) are illustrated in Fig. 5.8. The multifunctional material system consents in savings number of parts and reduces the joining operations thus impeccably increases the efficiency of the system. Effective integration of multifunctional material system perfectly eliminates the consumption of conventional connectors, sockets, boards, and bulky cables that provides major weight and volume.

The diverse applications of multifunctional composites are displayed in Fig. 5.9. Multifunctional composites also offer their extension in aerospace engineering, secondary structures applications, aircraft interior, solar cells, and in wind energy generation. Multifunctional composites applied in wind energy generation application involves shape-changing aerodynamic panels for flow control, reconfigurable aircraft wings, turbine blade, wind turbine configuration, variable geometry engine exhausts, mechanical memory cells, microelectromechanical systems (microswitches), valves, micropumps, flexible direction panel position in solar cells, energy storage, and its conversion. Besides this, they are also used in flexible and foldable electronic devices and optics (shape-changing mirrors for active focusing in adaptive optical systems) and in designing innovative architecture (adaptive shape panels for roofs and windows) (Gibson, 2010). Multifunctional composite materials are also associated with aerospace, portable electronics, and electric/hybrid ground transport in which battery life and power management are the restrictive factors (Asp and Greenhalgh, 2014). In addition, its applications are also been

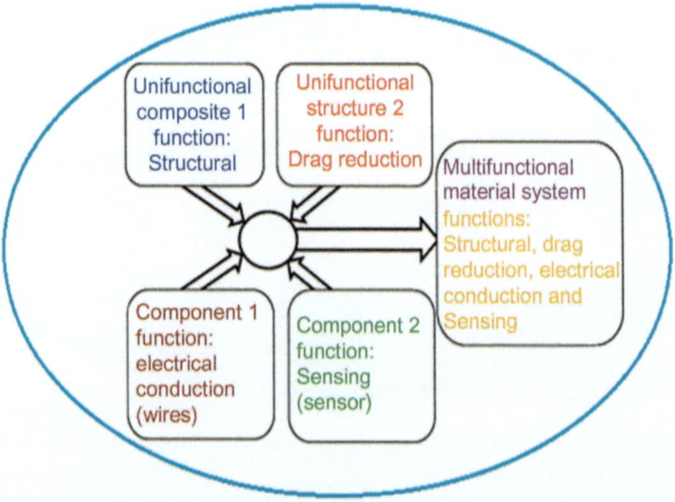

Figure 5.8 Components of multifunctional material system.
Modified from Ferreira, A.D.B., Nóvoa, P.R., Marques, A.T., Multifunctional material systems: a state-of-the-art review, Compos. Struct. (2016), 151, 3–35.

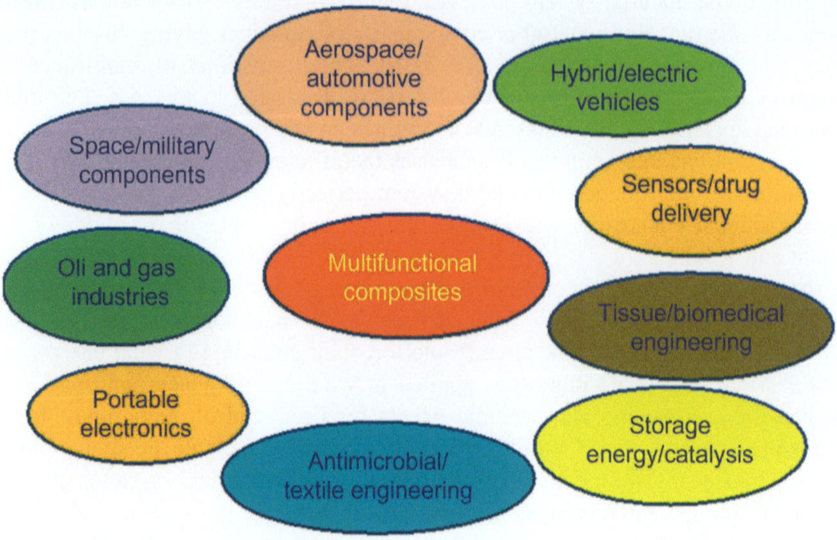

Figure 5.9 Wider applications of multifunctional composites.

stretched toward constructional/structural aspects, water/air pollution control, health monitoring/vector, treatment/remediation, disease diagnosis/screening, food processing/storage factors, pest detection/control, agricultural productivity, and its related issues (Ferreira et al., 2016).

Some of the recent and exclusive multifunctional composites and their specific multifunctional applications are summarized as follows:

1. Researchers currently reported multifunctional composite scaffolds for tissue engineering applications by using poly(3-hydroxybutyrate) foams having highly interconnected porosity (85% porosity) (Misra et al., 2010). Poly(3-hydroxybutyrate) foams were prepared using a unique combination of solvent casting and particulate leaching techniques using commercially available sugar cubes as porogen.
2. Variety of polymers both synthetic and naturally derived can be used to produce the hydrogel for fabricating nanocomposites by incorporating nanofillers/particles (Biondi et al., 2015). Developed multifunctional nanocomposites show a number of biomedical application such as actuators, tissue engineering scaffolds, tissue regeneration, biosensors, electrically stimulated drug delivery, and drug delivery (Biondi et al., 2015).
3. Multifunctional composites having conducting polymers filled with different metals shows numerous nonstructural and structural functions such as cold solders, electromagnetic shielding of electronic equipment, switching devices, conducting adhesives in electronics, static charge dissipating materials, surge protection devices, under fill for flip chips, and food packaging (Al-Oqla et al., 2015).
4. Carbon fibers reinforced polymer composites show multifunctional and exclusive applications in aerospace industries. The produced composites that offered multifunctional properties such as improved mechanical strength, better load-carrying tendency besides poor conductivity of extreme electrical currents generated by a lightning strike.
5. Grafted silane and graphene oxide poly(*p*-phenylene benzobisoxazole) (PBO) fibers reinforced resin composites exhibit excellent performance combined with high strength, high modulus, and lightweight (Chen et al., 2015). Developed multifunctional composites exhibited great potential in space applications, owing to their improved interfacial, high-energy UV resistant, and excellent atomic oxygen resistant properties (Hu et al., 2012; Leal et al., 2007).
6. Multifunctional Bucky syntactic foam based composites consisting of CNTs and ILs hybrid possess can effectively improve conductivity and are widely utilized in transportation to protect electro static discharge sensitive (ESD sensitive) components, electronic assembly, aerospace/marine structures/devices, deep see exploration vehicles, food-packaging containers, sports equipment, furniture, and structure-health monitoring (Ghamsari et al., 2014).

5.5 Conclusion

The continuous growing demand for a material that display and performs simultaneous different functions such as load-bearing, high strength, stiffness, energy storage/conversion, and electrical conductivity consequences the advent of multifunctional materials based polymer composites. Multifunctional materials such as nanosized particles including CNTs, graphene, POSS, inorganic nanoparticles, silica, ZnO, and even cellulose nanofiber are widely used as reinforcing filler for fabricating polymer composites. Multifunctional nanofiller possesses high aspect ratio, electrooptical behavior, high electrical/thermal conductivity, flame retardancy, significant chemical reactivity, and strong mechanical stiffness properties.

Incorporation of multifunctional materials in the polymer composites leads the way toward the development of innovative and highly advanced multifunctional polymer nanocomposites and hybrid nanocomposites together with natural and synthetic fibers, having structural and nonstructural functions. Multifunctional hybrid nanocomposites have gained a great interest among academicians and industries from the past decades as promising alternatives for metallic component in terms of performance, cost, weight, and volume improvements. Multifunctional composites shows wider, exceptional, and advanced emerging applications in ultra-filtration, batteries, circuit boards, supercapacitors, structural battery, structural separators, electrodes, photovoltaic cells, optical electronic components, and in device architectures. They also displayed applications in aerospace shuttles, hybrid ground vehicles, biomedical engineering, military, optical sensing, and imaging as they exhibit superior mechanical properties, physical, antimicrobial, and high dielectric constants.

Future prospective would be the development of multifunctional nanomaterials from renewable and biodegradable sources, and their incorporation in natural fibers reinforced bio-based polymer to developed green multifunctional hybrid composites and nanocomposites for biomedical and green constructional applications.

Acknowledgment

The authors extend their appreciation to the International Scientific Partnership Program ISPP at King Saud University for funding this research work through ISPP# 0011.

References

Al-Oqla, F.M., Sapuan, S.M., Anwer, T., Jawaid, M., Hoque, M., 2015. Natural fiber reinforced conductive polymer composites as functional materials: a review. Synth. Metals. 206, 42–54.

Amornpitoksuk, P., Suwanboon, S., Sangkanu, S., Sukhoom, A., Muensit, N., 2012. Morphology, photocatalytic and antibacterial activities of radial spherical ZnO nanorods controlled with a diblock copolymer. Superlattices Microstruct. 51, 103–113.

Asp, L.E., Greenhalgh, E.S., 2014. Structural power composites. Compos. Sci. Technol. 101, 41–61.

Benjeddou, A., Araújo, A.L., Carrera, E., Reddy, J., Marques, A.T., Soares, C.M.M., 2016. Smart composites and composite structures In honour of the 70th anniversary of Professor Carlos Alberto Mota Soares. Compos. Struct. 151, 1–2.

Bhowmick, A., Jana, P., Pramanik, N., Mitra, T., Banerjee, S.L., Gnanamani, A., et al., 2016. Multifunctional zirconium oxide doped chitosan based hybrid nanocomposites as bone tissue engineering materials. Carbohydr. Polym. 151, 879–888.

Biondi, M., Borzacchiello, A., Mayol, L., Ambrosio, L., 2015. Nanoparticle-integrated hydrogels as multifunctional composite materials for biomedical applications. Gels. 1, 162–178.

Bollen, P., Quiévy, N., Detrembleur, C., Thomassin, J.-M., Monnereau, L., Bailly, C., et al., 2016. Processing of a new class of multifunctional hybrid for electromagnetic absorption based on a foam filled honeycomb. Mater. Des. 89, 323–334.

Chau, N.D.Q., Ménard-Moyon, C., Kostarelos, K., Bianco, A., 2015. Multifunctional carbon nanomaterial hybrids for magnetic manipulation and targeting. Biochem. Biophys. Res. Commun. 468, 454–462.

Chen, C., Curliss, D., 2001. Resin matrix composites: organoclay-aerospace epoxy nanocomposites. Part II. SAMPE J. (USA). 37, 11–18.

Chen, L., Wei, F., Liu, L., Cheng, W., Hu, Z., Wu, G., et al., 2015. Grafting of silane and graphene oxide onto PBO fibers: multifunctional interphase for fiber/polymer matrix composites with simultaneously improved interfacial and atomic oxygen resistant properties. Compos. Sci. Technol. 106, 32–38.

Chrissafis, K., Bikiaris, D., 2011. Can nanoparticles really enhance thermal stability of polymers? Part I: an overview on thermal decomposition of addition polymers. Thermochim. Acta. 523, 1–24.

Crock, C.A., Rogensues, A.R., Shan, W., Tarabara, V.V., 2013. Polymer nanocomposites with graphene-based hierarchical fillers as materials for multifunctional water treatment membranes. Water Res. 47, 3984–3996.

Dang, Z.M., Yuan, J.K., Zha, J.W., Zhou, T., Li, S.T., Hu, G.H., 2012. Fundamentals, processes and applications of high-permittivity polymer–matrix composites. Progr. Mater. Sci. 57, 660–723.

Danoglidis, P.A., Konsta-Gdoutos, M.S., Gdoutos, E.E., Shah, S.P., 2016. Strength, energy absorption capability and self-sensing properties of multifunctional carbon nanotube reinforced mortars. Constr. Build. Mater. 120, 265–274.

Dintcheva, N.T., Arrigo, R., Morici, E., Gambarotti, C., Carroccio, S., Cicogna, F., et al., 2015. Multi-functional hindered amine light stabilizers-functionalized carbon nanotubes for advanced ultra-high molecular weight polyethylene-based nanocomposites. Compos. B: Eng. 82, 196–204.

Dintcheva, N.T., Arrigo, R., Carroccio, S., Curcuruto, G., Guenzi, M., Gambarotti, C., et al., 2016. Multi-functional polyhedral oligomeric silsesquioxane-functionalized carbon nanotubes for photo-oxidative stable ultra-high molecular weight polyethylene-based nanocomposites. Eur. Polym. J. 75, 525–537.

Ferreira, A.D.B., Nóvoa, P.R., Marques, A.T., 2016. Multifunctional material systems: a state-of-the-art review. Compos. Struct. 151, 3–35.

Ghamsari, A.K., Wicker, S., Woldesenbet, E., 2014. Bucky syntactic foam; multi-functional composite utilizing carbon nanotubes-ionic liquid hybrid. Compos. B: Eng. 67, 1–8.

Gibson, R.F., 2010. A review of recent research on mechanics of multifunctional composite materials and structures. Compos. Struct. 92, 2793–2810.

Hu, Z., Li, J., Tang, P., Li, D., Song, Y., Li, Y., et al., 2012. One-pot preparation and continuous spinning of carbon nanotube/poly(p-phenylene benzobisoxazole) copolymer fibers. J. Mater. Chem. 22, 19863–19871.

Jia, Y., Peng, K., Gong, X.-L., Zhang, Z., 2011. Creep and recovery of polypropylene/carbon nanotube composites. Int. J. Plast. 27, 1239–1251.

Kumar, S., Raj, S., Jain, S., Chatterjee, K., 2016. Multifunctional biodegradable polymer nanocomposite incorporating graphene-silver hybrid for biomedical applications. Mater. Des. 108, 319–332.

Kurauchi, T., Okada, A., Nomura, T., Nishio, T., Saegusa, S., Deguchi, R., 1991. Nylon 6-clay hybrid-synthesis, properties and application to automotive timing belt cover. SAE Tech. Paper. Available from: http://dx.doi:10.4271/910584.

Leal, A.A., Deitzel, J.M., Gillespie, J.W., 2007. Assessment of compressive properties of high performance organic fibers. Compos. Sci. Technol. 67, 2786–2794.

Li, W., Dichiara, A., Bai, J., 2013. Carbon nanotube–graphene nanoplatelet hybrids as high-performance multifunctional reinforcements in epoxy composites. Compos. Sci. Technol. 74, 221–227.

Lopes, A., Martins, P., Lanceros-Mendez, S., 2014. Aluminosilicate and aluminosilicate based polymer composites: present status, applications and future trends. Progr. Surf. Sci. 89, 239–277.

Matic, P., 2003. Conference Volume 5053. Proc. SPIE 5053, Smart Structures and Materials 2003: Active Materials: Behavior and Mechanics, 61 (August 12, 2003); http://dx.doi.org/10.1117/12.498546

Micheli, D., Vricella, A., Pastore, R., Delfini, A., Giusti, A., Albano, M., et al., 2016. Ballistic and electromagnetic shielding behaviour of multifunctional Kevlar fiber reinforced epoxy composites modified by carbon nanotubes. Carbon. 104, 141–156.

Misra, S.K., Ansari, T.I., Valappil, S.P., Mohn, D., Philip, S.E., Stark, W.J., et al., 2010. Poly (3-hydroxybutyrate) multifunctional composite scaffolds for tissue engineering applications. Biomaterials. 31, 2806–2815.

Nicolay, A., Lanzutti, A., Poelman, M., Ruelle, B., Fedrizzi, L., Dubois, P., et al., 2015. Elaboration and characterization of a multifunctional silane/ZnO hybrid nanocomposite coating. Appl. Surf. Sci. 327, 379–388.

Peng, S., Li, L., Lee, J.K.Y., Tian, L., Srinivasan, M., Adams, S., et al., 2016. Electrospun carbon nanofibers and their hybrid composites as advanced materials for energy conversion and storage. Nano Energy. 22, 361–395.

Quaresimin, M., Bertani, R., Zappalorto, M., Pontefisso, A., Simionato, F., Bartolozzi, A., 2015. Multifunctional polymer nanocomposites with enhanced mechanical and anti-microbial properties. Compos. B: Eng. 80, 108–115.

Ray, S., Quek, S.Y., Easteal, A., Chen, X.D., 2006. The potential use of polymer-clay nanocomposites in food packaging. Int. J. Food Eng. 2 (4),

Raza, M., Westwood, A., Stirling, C., Ahmad, R., 2015. Effect of boron nitride addition on properties of vapour grown carbon nanofiber/rubbery epoxy composites for thermal interface applications. Compos. Sci. Technol. 120, 9–16.

Riquelme, J., Garzón, C., Bergmann, C., Geshev, J., Quijada, R., 2016. Development of multifunctional polymer nanocomposites with carbon-based hybrid nanostructures synthesized from ferrocene. Eur. Polym. J. 75, 200–209.

Rodriguez, L., Silvestre, A., Sifuentes-Gallardo, P., Sorto Castañon, C., Díaz Flores, L.L., Hernández Rivera, M.A., et al., 2011. Crosslinking and reinforcement of PET/TiO_2/clay composites for pavement applications. Adv. Mater. Res. Trans. Tech. Publ. 168, 2340–2343.

Saba, N., Jawaid, M., Alothman, O.Y., Paridah, M.T., 2016. A review on dynamic mechanical properties of natural fibre reinforced polymer composites. Constr. Build. Mater. 106, 149–159.

Saba, N., Paridah, M.T., Abdan, K., Ibrahim, N.A., 2015. Preparation and characterization of fire retardant nano-filler from oil palm empty fruit bunch fibers. BioResources. 10, 4530–4543.

Saba, N., Paridah, M.T., Abdan, K., Ibrahim, N.A., 2016. Effect of oil palm nano filler on mechanical and morphological properties of kenaf reinforced epoxy composites. Constr. Build. Mater. 123, 15–26.

Saba, N., Paridah, M.T., Jawaid, M., 2014. A review on potentiality of nano filler/natural fiber filled polymer hybrid composites. Polymers. 6, 2247–2273.

Saba, N., Paridah, M.T., Jawaid, M., 2015. Mechanical properties of kenaf fibre reinforced polymer composite: a review. Constr. Build. Mater. 76, 87–96.

Sengupta, R., Bhattacharya, M., Bandyopadhyay, S., Bhowmick, A.K., 2011. A review on the mechanical and electrical properties of graphite and modified graphite reinforced polymer composites. Progr. Polym. Sci. 36, 638–670.

Shafei, A.E., Abou-Okeil, A., 2011. ZnO/carboxymethyl chitosan bionano-composite to impart antibacterial and UV protection for cotton fabric. Carbohydr. Polym. 83, 920–925.

Shirshova, N., Greenhalgh, E., Shaffer, M., Steinke, J., Curtis, P., Bismarck, A., Structured multifunctional composites for power storage devices, Proc. 17th Int. Conf. on 'Composite Materials Proceedings', IOM Communication Ltd, 2009.

Sierra-Martin, B., Fernandez-Barbero, A., 2015. Inorganic/polymer hybrid nanoparticles for sensing applications. Adv. Colloid Interface Sci. 233, 25–37.

Solarski, S., Ferreira, M., Devaux, E., Fontaine, G., Bachelet, P., Bourbigot, S., et al., 2008. Designing polylactide/clay nanocomposites for textile applications: effect of processing conditions, spinning, and characterization. J. Appl. Polym. Sci. 109, 841–851.

Spitalsky, Z., Tasis, D., Papagelis, K., Galiotis, C., 2010. Carbon nanotube–polymer composites: chemistry, processing, mechanical and electrical properties. Progr. Polym. Sci. 35, 357–401.

Stefanescu, E.A., Tan, X., Lin, Z., Bowler, N., Kessler, M.R., 2011. Multifunctional fiberglass-reinforced PMMA-$BaTiO_3$ structural/dielectric composites. Polymer. 52, 2016–2024.

Valentini, L., Bon, S.B., Lopez-Manchado, M., Verdejo, R., Pappalardo, L., Bolognini, A., et al., 2016. Synergistic effect of graphene nanoplatelets and carbon black in multifunctional EPDM nanocomposites. Compos. Sci. Technol. 128, 123–130.

Wang, Z., Xiao, H., 2008. Nanocomposites: recent development and potential automotive applications. SAE Int. J. Mater. Manuf. 1, 631–640.

Warwick, M.E., Dunnill, C.W., Goodall, J., Darr, J.A., Binions, R., 2011. Hybrid chemical vapour and nanoceramic aerosol assisted deposition for multifunctional nanocomposite thin films. Thin Solid Films. 519, 5942–5948.

Zhang, L., Jiang, Y., Ding, Y., Povey, M., York, D., 2007. Investigation into the antibacterial behaviour of suspensions of ZnO nanoparticles (ZnO nanofluids). J. Nanopart. Res. 9, 479–489.

Conducting polymer-based thermoelectric composites: principles, processing, and applications

6

Temesgen A. Yemata[1,2], *Qun Ye*[1], *Hui Zhou*[1], *Aung K.K. Kyaw*[1], *Wee S. Chin*[2] *and Jianwei Xu*[1,2]

[1]Institute of Materials Research and Engineering, Agency for Science, Technology and Research (A*STAR), Singapore, Singapore, [2]Department of Chemistry, National University of Singapore, Singapore

Chapter Outline

6.1 Introduction 169
6.2 Preparation and processing of TE composites 172
6.3 Conductive polymers for TE materials 175
6.4 Polymer–inorganic hybrid TE materials 175
- 6.4.1 PANi–inorganic TE nanocomposites 175
- 6.4.2 PTH-Inorganic TE Nano composites 176
- 6.4.3 PEDOT:PSS–Inorganic TE nanocomposites 177

6.5 Carbon-based nanostructures/polymer nanocomposites TE materials 179
- 6.5.1 CNTs and polymer composites TE materials 179
- 6.5.2 Conducting polymer/graphene composites 183
- 6.5.3 Conducting composites of polymer/graphite 186

6.6 Conclusions and outlook 186
Acknowledgment 191
References 191

6.1 Introduction

The demand for sustainable renewable energy sources is apparent as the energy demand around the globe rises and a replaceable fossil fuels supply diminishes (Snyder and Toberer, 2008; Sootsman et al., 2009). As a result, to meet the emerging energy challenges, significant contributions can be made by a wide spectrum of revolutionary technologies that are able to effectively harvest waste heat (Snyder and Toberer, 2008; Sootsman et al., 2009; Bell, 2008). Thermoelectric (TE) materials

Hybrid Polymer Composite Materials: Applications. DOI: http://dx.doi.org/10.1016/B978-0-08-100785-3.00006-1

could be fabricated in a solid-state device that utilizes electrons as a working fluid, different from an air conditioner which utilizes two-phase fluids like the regular refrigerant. Attractive features including absence of physical mobile phase, high reliability, and environmental friendliness can be potentially utilized for a large-scale waste heat recovery (Snyder and Toberer, 2008; Bell, 2008; Zhao et al., 2014; Zhao and Tan, 2014; Rowe, 1999). TE devices can be operated either in a power generator or refrigeration mode (Chen et al., 2012; Pei et al., 2011). TE power generators have been established to generate electricity from the waste heat in car exhaust gas (Pei et al., 2011; Zhao et al., 2014), drive a wristwatch by changing body heat into the electrical power, and harvest heat for other low power applications, that is, wireless sensor networks, mobile apparatus, and medical devices. Although TE devices that operate in the refrigeration mode (Zhao et al., 2014; Tritt, 1999) can be found in dormitory, small refrigerators and coolers can be used for laser systems, car seat cooler/heaters, and chip-scale coolers for photonics and microelectronics at the present time (Bell, 2008; Zhao and Tan, 2014).

Historically in 1821, T. J. Seebeck found that a circuit formed from two different metals with junctions at different temperatures would cause a compass magnet to deflect. At the beginning, Seebeck thought this was because of magnetism induced by the temperature change. But, he understood that it was an induced electric current, which led to deflection of the magnet. More precisely, the temperature gradient yields an electric potential (voltage) which can push an electric current to flow in a closed circuit system. Today, we called this effect as the Peltier–Seebeck effect.

The Seebeck voltage originates from the perturbation of the equilibrium distribution of charge carriers, designated by the Fermi–Dirac distribution. When a temperature gradient is established, charge carriers diffuse from a high-temperature end to a low-temperature region to reach a new equilibrium distribution. When two chemically different materials in a thermal gradient are connected, a circuit is formed in which the induced voltage drives the migration of charge carriers (i.e., a current) that can be used to perform useful work (Fig. 6.1A). On the other hand, the thermoelectricity could be used for solid-state heating or cooling. As shown in Fig. 6.1B, an electric current applied to a thermocouple results in heat to be given from or absorbed at the junction between the two different electrically conducting materials (Fig. 6.1B). If the direction of the applied electric current is reversed, the operation of the junction can be switched from heating to cooling. This phenomenon of Peltier effect in TE refrigeration is accomplished through one or more combinations of p- and n-type semiconductor materials when a direct current is passed, as shown in Fig. 6.1B (DiSalvo, 1999). Moreover, in 1838, the accurate nature of the Peltier effect was described by a German physicist H. Lentz, in which heat is either absorbed or generated as a current flows through a conducting circuit (Alam and Ramakrishna, 2013; Tritt and Subramanian, 2006).

The complete description on the thermodynamic association between the Seebeck and Peltier effects was announced in 1851 by W. Thomson (later known as L. Kelvin). The Peltier coefficient that managed the Peltier effect was described as the product of the absolute temperature and Seebeck coefficient. This thermodynamic source led Thomson to forecast a third TE phenomenon called the Thomson effect (Alam and Ramakrishna, 2013; Rowe, 2005; MacDonald, 1962).

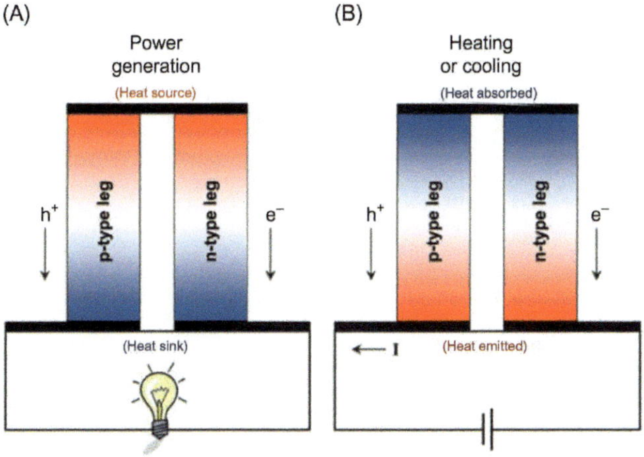

Figure 6.1 Diagrammatic illustrations of thermoelectric devices used for (A) power generation and (B) heating and cooling applications (Szczech et al., 2011).
Reproduced from Szczech, J.R., Higgins, J.M., and Jin, S. Enhancement of the thermoelectric properties in nanoscale and nanostructured materials. J. Mater. Chem., 2011. 21(12): p. 4037–4055, with permission of The Royal Society of Chemistry.

The performance of TE materials is governed with the dimensionless TE figure of merit (ZT) which is defined as the following equation:

$$\mathrm{ZT} = \frac{S^2 \sigma T}{\kappa}$$

where κ, S, T, and σ are the thermal conductivity (W m^{-1} K^{-1}), the Seebeck coefficient (V K^{-1}), the absolute temperature (K), and the electrical conductivity (S cm^{-1}), respectively. Moreover, σS^2 is known as the power factor (PF) (Zhao et al., 2014; Ohta, 2007; Biswas et al., 2012; Heremans et al., 2013). In order to achieve a TE module with a large ZT value, it is imperative that TE materials have a high Seebeck coefficient that supplies significant voltage, and a high electrical conductivity that permits effective electron transport, and at the same time, thermal conductivity should be kept as small as possible to reduce heat losses (Snyder and Toberer, 2008; Sootsman et al., 2009; Kanatzidis, 2009). Thermal conductivity κ comprises electronic (κ_{el}) and lattice (κ_{latt}) components, and total κ is the sum of both parts (i.e., $\kappa = \kappa_{el} + \kappa_{latt}$). Most of the electronic term (κ_{el}) is directly related to the electrical conductivity through the Wiedemann–Franz law. Plenty of investigation has been conducted to reduce κ_{latt} so as to diminish the κ and hence enhance the overall ZT (Alam and Ramakrishna, 2013; Tritt and Subramanian, 2006; Martín-González et al., 2013).

TE generators are wonderful alternatives to transform waste heat into useful energy, and earlier investigations on TE materials as well as related devices have been mainly focused on inorganic materials, for instance, Bi–Te alloys, but the high processing cost and low manufacturability restrict the wide use of conventional TE

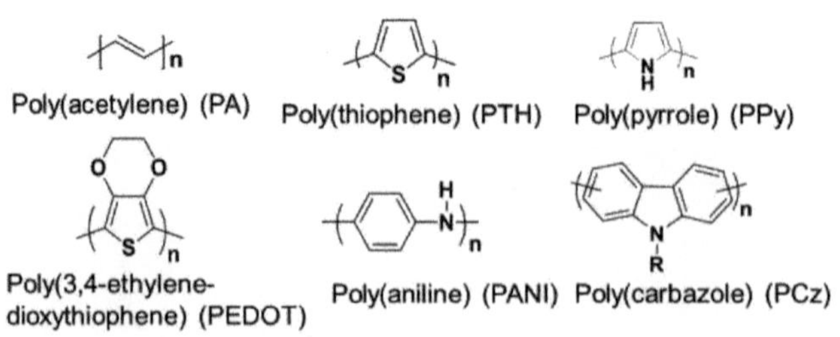

Chart 6.1 The chemical structures of commonly used electrically conducting polymers.

inorganic TE materials (Snyder and Toberer, 2008; Zebarjadi et al., 2012). On the contrary, polymer-based materials are possible alternatives for TE uses due to their relatively low material cost, good processability, and reasonable malleability. The polymer-based TE devices also could simply be modified with diverse functions (Wei et al., 2015; Du et al., 2012b; Yao et al., 2010). Numerous organic semiconductor polymers including, polypyrrole (Maddison et al., 1988, 1989), polyparaphenylene (Moreau et al., 1997; Moliton et al., 1999), polycarbazole (Aïch et al., 2009), polythiophene (Shinohara et al., 2007), poly(3,4-ethylenedioxythiophene) (PEDOT) (Wei et al., 2015), polyaniline (PANi) (Bhadra et al., 2009; Holland and Monkman, 1995), and polyacetylene (Choi et al., 1997; Li et al., 2010) have been explored for TE applications. The chemical structures of commonly used conducting polymers together with their chemical names and abbreviations are given in Chart 6.1. Furthermore, polymer–inorganic TE nanostructural composites have been paid increasing attention in recent years. The TE composites generally comprise an inorganic component, for example, nanostructures with a high ZT value, which are integrated into a polymer matrix so that the overall TE performance of the resulting composites are collectively better than the pristine polymer by enhancing TE characteristic parameters such as the Seebeck coefficient and electrical conductivity.

In this chapter, we aim to provide a comprehensive summary of the fundamental aspects of hybrid TE materials and related processing. Recent state-of-the-art TE devices based on hybrid materials that demonstrate high TE performance are discussed. First, the progress of the main preparation approaches of hybrid TE materials is discussed. Second, recent state-of-the-art of TE materials based on conducting polymers, inorganic/polymer nanocomposites as well as carbon-based/polymer nanocomposites and their TE properties are discussed. Finally, a short summary and future outlook are presented.

6.2 Preparation and processing of TE composites

Mixing inorganic materials with polymers is opening a way for engineering flexible composites that show advantageous electrical, magnetic, mechanical, and optical properties. Researchers have been keeping on looking for novel approaches to engineer materials that associate the desirable properties of polymers and inorganic materials for the formation of inorganic–polymer composites as a portion of this

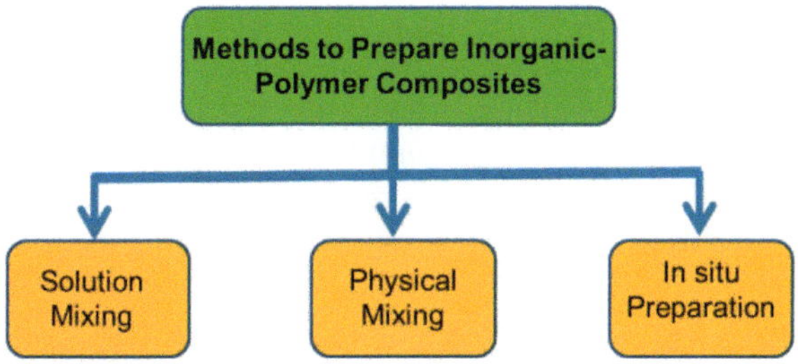

Scheme 6.1 General methods to prepare TE composites.

rehabilitated interest in composites. There have been several well-established methods for the synthesis of inorganic–polymer composites to render superior performance. These approaches can be can be classified under three major categories as solution mixing, physical mixing, and in-situ preparation (Scheme 6.1) (Balazs et al., 2006). Solution mixing is a method that mixes an inorganic filler and polymer as in a solution. In this technique, the inorganic fillers are predispersed in a solution of the polymer, followed by evaporation of the solvent from the polymer-filler solution or suspension, which sometimes permits the fillers to be homogeneously dispersed in the polymer matrix (Carotenuto et al., 1996; Kalaitzidou et al., 2007). Casting methods, however, use a polymer solution as a dispersant, and solvent evaporation yields the inorganic–polymer composites. Physical mixing preparation methods for inorganic–polymer composites are based on liquid-particle dispersions, but different from the type of the continuous phase. In melt processing, inorganic particles are dispersed into a melt polymer phase, and then, inorganic–polymer composites are obtained by extrusion (Althues et al., 2007). Finally, the in-situ technique includes dispersing the inorganic nanoparticles directly in the monomer solution prior to applying polymerization. The issue is that inorganic particles may tend to separate and sediment rapidly from the organic polymers (Althues et al., 2007). In general, solution mixing, physical mixing, and in-situ preparation have been employed to develop polymer and/or inorganic composites. This part will summarize some common methods to prepare and process TE composite materials in terms of solution mixing, physical mixing, and in-situ preparation.

The composites formed between organic semiconductors, particularly for various conducting conjugated polymers, and different inorganic nanostructures have shown advantageous characteristics that are not available with a single material component, just by merging beneficial optical and conducting properties with robustness and flexibility. Composite approaches have already been demonstrated in the areas of optoelectronic devices such as hybrid organic–inorganic light-emitting diodes and solar cells (Holder et al., 2008), and it is expected that similar composites approaches making use of the advantages of semiconducting polymers and inorganic component would offer improved properties for TE applications. Some representative examples are given below.

Powder composites formed by incorporating both n- and p-type Bi_2Te_3 ball-milled powders into the conducting polymer PEDOT:PSS led to PF enhancements for both p- and n-polymer composites. It was also demonstrated that n-type TE polymer composites could be obtained by loading n-type powders with high conductivity and high Seebeck coefficient into a low Seebeck coefficient p-type conducting polymer. For example, n-type samples with PEDOT volume percent from 10% to 30% show a PF from 80 to 60 $\mu W\ m^{-1}\ K^{-2}$ (Zhang et al., 2010).

Naoki et al. used both physical and solution mixing methods to prepare TE composites of PANi and bismuth(III) telluride nanoparticles (PANi/Bi_2Te_3) (Toshima et al., 2011). The TE performance of two types of PANI/Bi_2Te_3 composites prepared by physical and solution mixing is compared, showing the Bi_2Te_3/PANi composite film prepared by physical mixture has a much higher PF, leading to better TE performance with a calculated ZT value of about 0.18 at 350 K, larger than that of PANi/Bi_2Te_3 films prepared by solution mixture (0.003–0.009). In another PANI-based composite, PANi with 1–7 wt% and $Bi_{0.5}Sb_{1.5}Te_3$ dry powders were cold pressed by a pressure of about one GPA after physically mixed and the composite resulted a Seebeck coefficient about 10% smaller than $Bi_{0.5}Sb_{1.5}Te_3$. However, ~30%–70% of its original value of the electrical conductivity significantly declined and the material developed a notable reduction in PF. (Zhao et al., 2002).

Segalman et al. described the synthesis of composite nanocrystals comprising a tellurium core functionalized with the conducting polymer PEDOT:PSS (See et al., 2010). The Te nanorods are prepared in situ with a water-soluble conducting polymer, PEDOT:PSS, affording a continuous, two-component composite material. The resulting Te nanowires/PEDOT:PSS hybrid materials exhibited much higher electrical conductivity without significantly compromising with the Seebeck coefficient (reduced from 408 to 163 $\mu V\ K^{-1}$), and meanwhile, the thermal conductivity is almost unchanged, leading to a room temperature ZT ~ 0.1. The mechanism underlying this high-performance of hybrid materials was also studied by the same research group (Coates et al., 2013). The unusual electrical behavior of this composite is interpreted by a series-connected model where carrier transport occurs mainly through a highly conductive volume of polymer that exists at the nanowire polymer interface. The result suggests how to control interfacial property in composite material systems is of significance and therefore may offer a general approach to improve carrier transport in composite materials. Further study also demonstrated that the effect of Te nanowire morphology on carriers' transport properties of the same Te composites with PEDOT:PSS:longer nanowires results in a larger thermopower but a lower electrical conductivity (Yee et al., 2013).

In another example, PANi/PbTe composite powders composed of PbTe nanoparticles, PANi/PbTe core–shell nanostructure, and PbTe/PANi/PbTe three-layer sphere-like nanostructures were produced using in-situ interfacial polymerization at room temperature. These nanostructures resulted in enhanced electrical conductivity starting 0.019 to 0.022 $S\ cm^{-1}$, whilst the Seebeck coefficient reduced starting 626 to 578 $\mu V\ K^{-1}$ when the temperature was enlarged starting 293 to 373 K (Wang et al., 2010). Moreover, carbon nanotubes (CNTs) and PANi nanocomposites were prepared by in-situ polymerization using single-walled nanotubes

(SWNTs) as template. In this study, the strong $\pi-\pi$ interactions between PANi and CNT nanotubes create additional well-organized structure of composite than the pure PANi and thus allow enhanced carrier mobility so that their electrical conductivity and Seebeck coefficient are improved compared to pure PANi (Yao et al., 2010).

6.3 Conductive polymers for TE materials

The conductive polymers, including PA, PTH, PCZ, PANi, PEDOT, and others, are one type of organic materials that have been used for TE applications. Among these conducting polymer materials, PEDOT and PANi are the two major conducting polymers that have attracted much attention due to some unique characteristics, for example, conductivity, environmental stability, thermal conductivity, flexibility, and thermal stability. For instance, PEDOT has been extensively studied by doping a variety of additives, including organic acids, miner acids, and others, to improve its TE performance. As this chapter is mainly focused on the conducting polymer–inorganic TE composites materials, we will give a summary account of conducting polymer composites instead of concentrating on their doped forms for TE application, although a large number of research papers have been published in this area.

6.4 Polymer–inorganic hybrid TE materials

Organic conducting polymers have displayed great potential for the TE application, due to their advantages such as low density, low cost and easy synthesis, good processability, and low toxicity. However, the high-performance TE devices composed of organic conducting polymers for practical applications are still not available, which is limited by the defects of organic conducting polymers, such as low electrical conductivity, low Seebeck coefficient, and low PF. For example, the PF of organic conducting polymer TE materials is in the range of $10^{-6}-10^{-10}$ W m^{-1} K^{-2}, which is three order of magnitude lower than that of inorganic TE materials. In order to solve this problem, the TE composite materials formed by organic conducting polymer and inorganic additive have been received more attentions (Du et al., 2012b; Yu et al., 2008), which exhibited potentially better TE performance than pure organic conducting polymers.

6.4.1 PANi–inorganic TE nanocomposites

Anno et al. successfully prepared the PANi/Bi composites by planetary ball-milling technique, in which PANi/Bi powder was doped by (±)-10-camphorsulfonic acid (CSA) in *m*-cresol (Fig. 6.2). PANi effectively inhabited the aggregation and oxidation of Bi nanoparticles in final CSA-doped PANi/Bi composites.

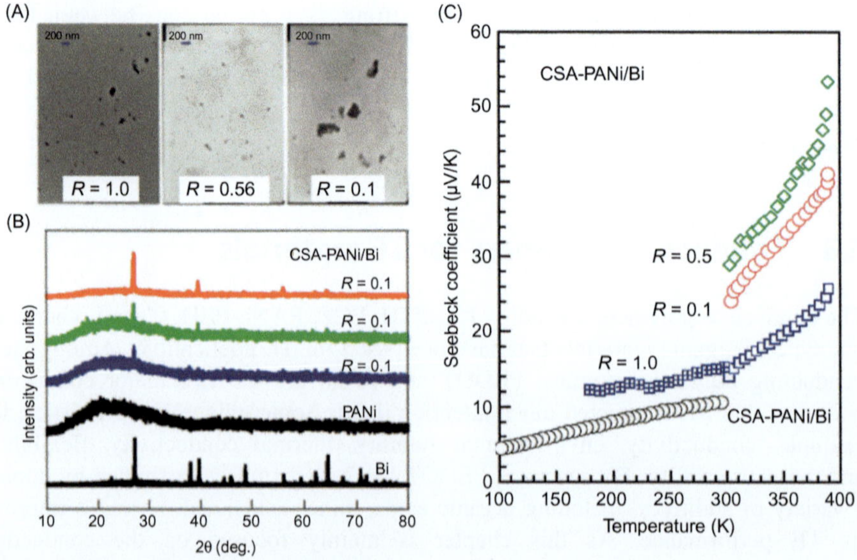

Figure 6.2 (A) TEM images of CSA-doped PANi/Bi nanocomposites with different molar ratios (R); (B) X-ray diffraction patterns of Bi source powder, PANi, and CSA-doped PANi/Bi nanocomposites with different molar ratios; (C) temperature dependence of the Seeback coefficient for PANi and CSA-doped PANi/Bi nanocomposites with different molar ratios (Anno et al., 2009).

Nevertheless, the composites exhibited a good crystalline morphology and thermal stability. The Seebeck coefficient and the electrical conductivity of CSA-doped PANi/Bi composites were measured in the range from room temperature to ~400 K, which displayed obviously improved the Seebeck coefficient than that of CSA-doped PANi without Bi particles (Anno et al., 2009). Li et al. prepared PANi/Bi_2Te_3 composites by mechanical blending of rice-like PANi particles and Bi_2Te_3 flakes, which were synthesized by chemical oxidation and hydrothermal reaction respectively. The PANi/Bi_2Te_3 composites exhibited an *n*-type conduction with Seebeck coefficient of $-50\,\mu V\,K^{-1}$, which is similar to that of Bi_2Te_3. Nevertheless, the PANi/Bi_2Te_3 composites showed a similar electrical conductivity to PANi. The mechanical blended composites have a lower PF than both of Bi_2Te_3 and PANi, which remained almost unchanged with temperature. Thus, the electrical properties of PANi/Bi_2Te_3 composites cannot be improved by mechanical blending (Li et al., 2011).

6.4.2 PTH-Inorganic TE Nano composites

Pintér et al. prepared poly(3-octylthiophene)/silver nanocomposites (P3OT/Ag) though impregnating P3OT powder in a silver perchlorate solution, demonstrating a simple method to produce Ag nanoparticles in P3OT matrix. Ag is predominantly in the metallic form rather than in the form of AgCl salt due to the redox interaction of Ag^+ with neural P3OT, which was combined by the affinity of Ag and the sulfur sites of P3OT. Nevertheless, the stabilization effect of Ag nanoparticles on the partially oxidized P3OT increased its electrical conductivity by more than five orders of magnitude. And the large Seebeck coefficient of 1283 $\mu V\ K^{-1}$ for P3OT/Ag composites indicated its potential for TE application (Pintér et al., 2007).

He et al. demonstrated a method to successfully enhance the Seebeck coefficient and PF of the poly(3-hexylthiophene) (P3HT) with addition of Bi_2Te_3 nanowires, in which the Bi_2Te_3 nanowires substantially scatter low-energy carriers by applying the energy-filtering effect on the interface of P3HT–Bi_2Te_3. Fig. 4.2A–C shows the SEM images of $FeCl_3$-doped P3HT and P3HT–Bi_2Te_3 films, which revealed that Bi_2Te_3 nanowires would be homogeneously dispersed in the P3HT matrix and microphase separation between P3HT and Bi_2Te_3 nanowires was not observed. The P3HT–Bi_2Te_3 composites had a higher PF of 13.6 $\mu W\ K^{-2}\ m^{-1}$ at room temperature compared to 3.9 $\mu W\ K^{-2}\ m^{-1}$ of P3HT, because the Seebeck coefficient of P3HT was increased after mixing with Bi_2Te_3 particles without greatly decreasing electrical conductivity (Fig. 6.3). The improved TE performance may be caused by the energy-filtering effect on the P3HT–Bi_2Te_3 interface, in which low-energy carriers were scattered more strongly than high-energy carriers, leading significant increase in Seebeck coefficient and PF. This method, via rationally modifying the interface of organic–inorganic semiconductor to largely improve TE performance, may provide promising route to high-performance, large-area, and flexible polymer TE materials (He et al., 2012).

6.4.3 PEDOT:PSS–Inorganic TE nanocomposites

See et al. prepared PEDOT:PSS/Te composite with high TE performance through aqueous solution process, which was composed of a tellurium (Te) core functionalized with PEDOT:PSS (Fig. 6.4). The PEDOT:PSS/Te film was smooth and nanoporous, in which the uniformly distributed rods bridged by conducting polymers (Fig. 4.3B–D). The composite film exhibited PF of 70.9 $\mu W\ K^{-2}\ m^{-1}$, which is much higher than that of PEDOT:PSS (0.05 $\mu W\ K^{-2}\ m^{-1}$) and Te nanowires (2.7 $\mu W\ K^{-2}\ m^{-1}$), leading a ZT value of ~0.1 at room temperature. This development demonstrated a method to form TE devices with a high ZT by compositional optimization and variation of quantum confined inorganic cores and also presented a new platform to study TE and charge transport in nanoscale heterostructures (See et al., 2010). Zhang et al. tested the TE performance of two newly commercialized PEDOT:PSS products CLEVIOS PH1000 and FE-T, which showed promising PF of 47 $\mu W\ K^{-2}\ m^{-1}$ and 30 $\mu W\ K^{-2}\ m^{-1}$, respectively. By incorporating both *n*- and *p*-type Bi_2Te_3 ball-milled powders into these PEDOT:PSS products, PF enhancements for both *n*- and *p*-composites are achieved as high as 70 and 40 $\mu W\ K^{-2}\ m^{-1}$, respectively. This demonstration provides a new fabrication option to form all-solution-processed TE devices on flexible substrates

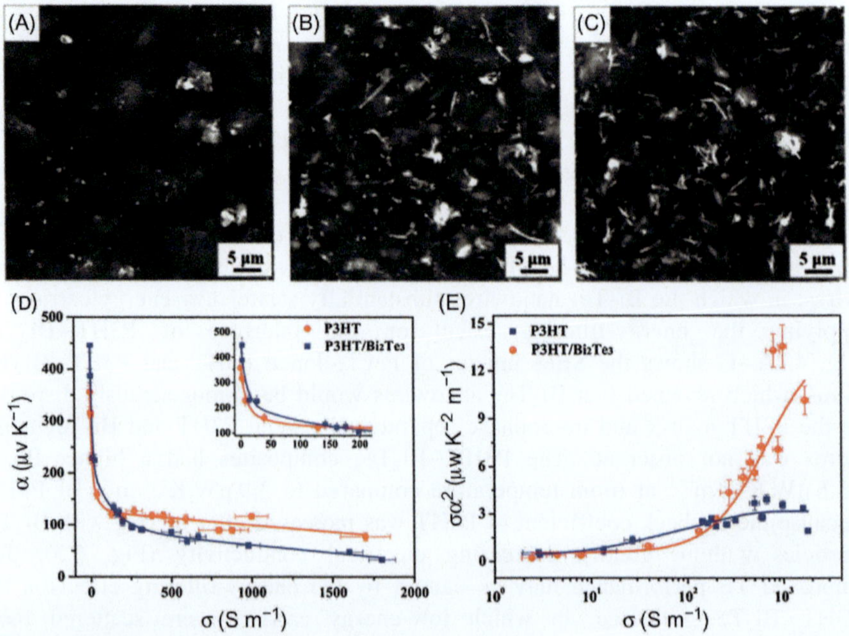

Figure 6.3 SEM images of P3HT and P3HT–Bi_2Te_3 films: (A) $FeCl_3$-doped P3HT solely; (B) $FeCl_3$-doped P3HT with 10 wt% Bi_2Te_3 nanowires; (C) $FeCl_3$-doped P3HT with 20 wt% Bi_2Te_3 nanowires. (D) The connection between the Seebeck coefficient and electrical conductivity in P3HT–Bi_2Te_3 composites and P3HT; the inset demonstrates the lock up in the variety of small conductivity ($\sigma < 200$ S m^{-1}). (E) The connection involving the PF and the electrical conductivity in P3HT–Bi_2Te_3 composites and P3HT (He et al., 2012). Reproduced from He, M., et al., Thermopower enhancement in conducting polymer nanocomposites via carrier energy scattering at the organic–inorganic semiconductor interface. Energy Environ. Sci., 2012. 5(8): p. 8351, with permission of The Royal Society of Chemistry.

(Zhang et al., 2010). Liu et al. prepared free-standing PEDOT:PSS/$Ca_3Co_4O_9$ composite films by casting PEDOT:PSS/$Ca_3Co_4O_9$ solution on polypropylene (PP) film. The $Ca_3Co_4O_9$ particles were composited well with PEDOT:PSS and in shape of sheets. The Seebeck coefficient would be enhanced by increasing $Ca_3Co_4O_9$ content in composite and reached the maximum value at $Ca_3Co_4O_9$ content of 24.8%. However, PF of composite film decreased with $Ca_3Co_4O_9$ content increasing due to the decline of electrical conductivity and limited increase in Seebeck coefficient (Liu et al., 2011). Wang et al. developed a facile route to form composites of PbTe-modified PEDOT nanotubes through adding PbTe nanoparticles to PEDOT nanotubes, which was fabricated in situ by an interfacial polymerization. The PF of composite film could be achieved to 1.44 μW K^{-2} m^{-1} by adjusting the PbTe content in composite to 28.7 wt% (Wang et al., 2011).

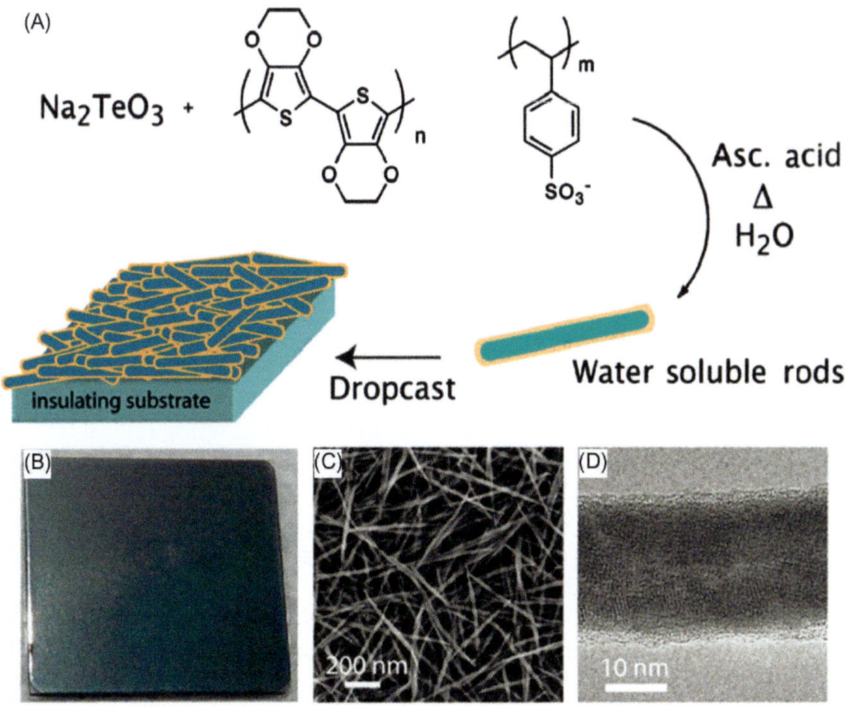

Figure 6.4 (A) Synthesis of PEDOT:PSS/Te composite film. (B) Image of a drop-cast composite film on a quart substrate. (C) SEM image of a drop-cast composite nanorod film. (D) TEM image illustrating the crystalline Te nanorod passivated with PEDOT:PSS (See et al., 2010).
Reproduced with permission from Yu, C., et al., Thermoelectric behavior of segregated-network polymer nanocomposites. Nano Lett., 2008. 8(12): p. 4428–4432, Copyright (2016) American Chemical Society.

6.5 Carbon-based nanostructures/polymer nanocomposites TE materials

6.5.1 CNTs and polymer composites TE materials

CNTs have been widely used in material engineering due to their wonderful physical, mechanical, and electronic characteristics. The use of pure CNTs for TE materials is greatly limited by their high thermal conductivity of 3000 W m^{-1} K^{-1} at room temperature (Kim et al., 2001). In contrast, polymers are intrinsically poor thermal conductors, which are ideal materials for TEs, but low electrical conductivity and Seebeck coefficient have excluded them as feasible candidates for TEs. However, recent progress in polymer and CNT has made it possible to formulate them into CNTs/polymer composites, in which the thermal conductivity, electrical conductivity, and thermopower can be adjusted efficiently, leading to better TE performance.

Yu et al. prepared segregated-network CNT–polymer composites and investigated their TE properties. As shown in Fig. 6.5A–D, the SEM images indicated that CNTs wrapped around the emulsion particles in a network fashion rather than randomly distributed in the composite with different CNT contents. The electrical conductivity of composites were found to be dramatically improved by incorporating CNTs in a network fashion, while the thermal conductivity and Seebeck coefficient remain almost no change on CNT concentration (Fig. 6.5E). This behavior results from thermally disconnected, but electrically connected, junctions in the nanotube network, which makes it feasible to tune the properties for a higher ZT. With 20 wt% CNT content, these composites exhibit an electrical conductivity of 4800 S m^{-1}, thermal conductivity of 0.34 W m^{-1} K^{-1} and a ZT of 0.006 at room temperature. This study suggests that polymeric TEs are possible and provide a basis for further development of lightweight, low cost, and nontoxic polymer composites for TE applications in the future (Yu et al., 2008).

Chen et al. systematically studied the electrical and thermal transportation properties of composites with 3D-CNT network, which exhibited extremely low thermal conductivity of 0.035 W m^{-1} K^{-1} in standard atmosphere at room temperature. The electrical transportation of composites would be easily converted by gas-fuming doping, for example, the original p-type composites can be converted to n-type by potassium (K) doping. The composites with 3D-CNT network is a good template for polymer composition due to its self-sustainable homogeneous network structure, which has been composed with PANi to form composites with significantly improved TE performance and good flexibility (Chen et al., 2011). Kim et al.

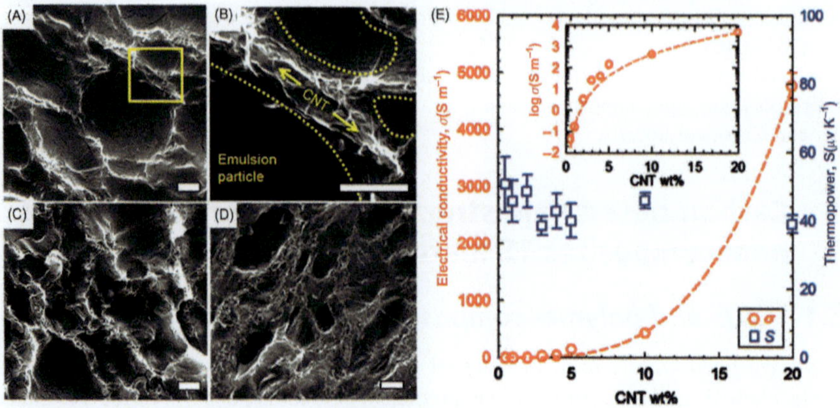

Figure 6.5 SEM images of CNT–polymer composites with different CNT contents: (A) 5%; (B) the high-magnification SEM of composites with 5% CNT content; (C) 10%; (D) 20%. (E) Electrical conductivity and Seebeck coefficient of CNT composites with different CNT contents at room temperature (Yu et al., 2008).
Reproduced with permission from Yu, C., et al., Thermoelectric behavior of segregated-network polymer nanocomposites. Nano Lett., 2008. 8(12): p. 4428–4432, Copyright (2016) American Chemical Society.

reported the preparation of segregated-network CNT—polymer composites, and systematically investigated the influence of components and processing parameters on the TE properties, including CNT type, CNT concentration, stabilizer, and drying temperature. It was found that PEDOT:PSS is presumably attached to CNTs and bridged tube—tube junctions. In this case, electrically conducting PEDOT:PSS facilitates electrons to travel more efficiently in the composites, resulting in high electrical conductivity. While the thermal transport across the tube—tube is impeded due to mismatches in vibrational spectra between CNT and PEDOT:PSS. The thermal conductivities of composites were effectively controlled in the range of typical polymeric materials (0.2—0.4 W m^{-1} K^{-1}) even the CNT concentration as high as 15 wt%. And the electrical conductivity of composite could be raised to ~400 S cm^{-1}, while Seebeck coefficient and thermal conductivity remained relatively constant. Finally, the CNT—polymer composite showed optimum ZT value of ~0.02 in the formulation of 35 wt% SWCNT and 35 wt% PEDOT:PSS (Kim et al., 2010). Yu et al. fabricated CNT/PEDOT:PSS composites without using heavy and/or toxic inorganic materials, which exhibited very high electrical conductivities and relatively constant Seebeck coefficient. Through adjusting the CNT concentration in composites, the electrical conductivity of composites could be enhanced from ~48 to ~1350 S cm^{-1}, whilst the Seebeck coefficient and thermal conductivity were kept at ~30 μV K^{-1} and ~0.4 W m^{-1} K^{-1}, respectively. Finally, the composites with 60% CNT concentration exhibited the highest electrical conductivity (1.35×10^3 S cm^{-1}) and thermopower (41 μV K^{-1}) at room temperature. The composites exhibited large PF of ~160 μW m^{-1} K^{-2}, which are orders of magnitude better than those of typical polymer composites (Yu et al., 2011).

Bounioux et al. reported several p-doped composites with improved TE performance, which were composited by P3HT and MW—CNT as well as SW—CNT. The use of SW—CNT consistently resulted in better TE performance (Fig. 6.6). The P3HT/SW—CNT composites displayed a promising PF of ~95 ± 12 μW m^{-1} K^{-2} when the SW—CNT content ranging from 42 to 81 wt%. In contrast, P3HT/MW—CNT composites offered their maximum PF of only ~6 ± 2 μW m^{-1} K^{-2} when MW—CNT composition in range of 10—40 wt%. Moreover, a CNT content of 8—10 wt% does not compromise with the low bulk thermal conductivity of the polymer matrix, which promises a high ZT value of 0.01 at room temperature. All samples are cast on plastic substrates, emphasizing their suitability for large-area, flexible TE applications (Fig. 6.6) (Bounioux et al., 2013). The PTH/MW—CNT composites prepared through in-situ polymerization method were firstly reported by Wang et al. The ZT of composites could be significantly improved through increasing MW—CNT content, due to the large increase in electrical conductivity whilst the almost constant in Seebeck coefficient and thermal conductivity. The highest ZT of composites was achieved as 8.71×10^{-4} at 120°C when MW—NT content is 80 wt% (Wang et al., 2013).

Liu et al. reported an effective method to improve the TE performance of PANi by mixing with PPy-functionalized MW—CNTs. The surface-functionalized PPy nanolayer on the MW—CNTs was found to yield a homogeneous dispersion of MW—CNTs and strong interfacial adhesion. The resulting composites demonstrated

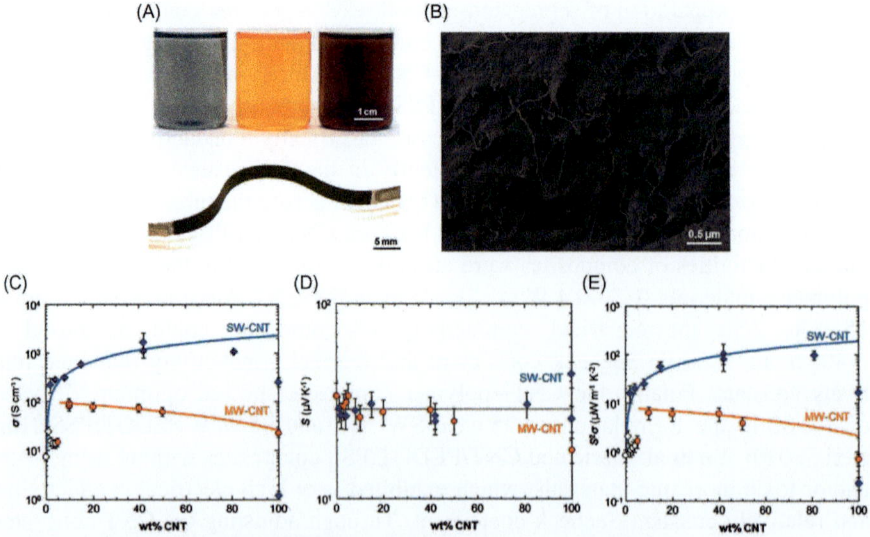

Figure 6.6 (A) Processing of P3HT/CNT composites: dispersion of MW−CNTs (left), solution of P3HT (center) and dispersion of MW−CNTs in P3HT solution (right). A long flexible tape composed of a composite film with 20 wt% MW−CNT on PET foil (bottom). (B) SME image of a cleaved film with SW−CNT content of 42 wt%. Thermoelectric properties of $FeCl_3$-doped P3HT/CNT composites: (C) electrical conductivity, (D) Seebeck coefficient, and (E) power factor (Bounioux et al., 2013).
Reproduced from Bounioux, C., et al., Thermoelectric composites of poly(3-hexylthiophene) and carbon nanotubes with a large power factor. Energy Environ. Sci., 2013. 6(3): p. 918−925, with permission of The Royal Society of Chemistry.

a remarkable enhancement in both electrical conductivity and Seebeck coefficient and exhibited a high PF of 3.1 $\mu W\ m^{-1}\ K^{-2}$, which was comparable with 0.006 $\mu W\ m^{-1}\ K^{-2}$ for PANi and 0.1 $\mu W\ m^{-1}\ K^{-2}$ for MW−CNT/PANi composite at 28.6 wt% MW−CNT content (Liu and Yu, 2014). Wang et al. prepared MW−CNT/PPy nanocomposites by an in-situ polymerization method using TSA as a dopant and $FeCl_3$ as an oxidant. The electrical transport of the PPy was greatly affected by MW−CNT. The carrier mobility increased parabolically with increase of MW−CNT content, while the carrier concentration is almost constant when the MW−CNT content is less than 15 wt%, and decreased dramatically when the MW−CNT content reaches 20 wt%. The PF of composites was significantly enhanced by the addition of MW−CNT and reached to the maximum PF of 2.079 $\mu W\ m^{-1}\ K^{-2}$ when MW−CNT content is 20%, which is almost 26-fold higher than that of pure PPy (Wang et al., 2014a). Yao et al. prepared SW−CNT/PANi composites through an in-situ polymerization reaction of SW−CNT as template and aniline as reactant. The PANi in the composites exhibited more ordered molecular structure than pure PANi due to the strong $\pi-\pi$ interaction between PANi and CNT, leading to improved carrier mobility. The composites provided significantly increased electrical conductivity and Seebeck coefficient, while restricted thermal conductivity, resulting optimum PF of 20 $\mu W\ m^{-1}\ K^{-2}$ and ZT value of 0.004 at room temperature (Yao et al., 2010).

6.5.2 Conducting polymer/graphene composites

In order to improve the TE performance of polymer composites, numerous efforts have been made to enhance the carriers' mobility whilst keeping their concentration intact. CNTs, including SW–CNTs and MW–CNTs, can induce polymers to form self-assembly alignments through π–π interactions, and thus, they are widely used to improve the TE performance of PEDOT:PSS. However, a large amount of CNTs (30–40 wt%) is always needed to achieve a reasonable ZT value. In contrast, high TE performance PEDOT:PSS/graphene composite can be fabricated with using a small amount of graphene, due to its larger surface area and higher level of carrier mobility because graphene can bridge the carrier transport via the strong π–π interaction with the rings in PEDOT:PSS (Yu et al., 2012; Zhao et al., 2012; Kim and Pipe, 2012).

Zhang et al. successfully integrated modified graphene into PEDOT:PSS, in which graphene was functionalized with fullerene by π–π stacking in a liquid–liquid interface. By incorporation of fullerene–graphene nanohybrids, the electrical conductivity of resulting composites could be increased from ~10.0 to ~70.0 $S\ cm^{-1}$, the thermal conductivity changed from 0.2 to 2.0 $W\ m^{-1}\ K^{-1}$, leading to a fourfold-enhanced Seebeck coefficient (Fig. 6.7). This phenomenon indicated that tuning the fullerene and graphene ratio helps to increase the composite electrical conductivity much higher than the thermal conductivity, due to the significant interfacial phonon scattering. Finally, the composites had a significantly improved ZT of 6.7×10^{-2}, which is more than one order of magnitude higher than that of single-phase filler-based polymer composites (10^{-3}) (Zhang et al., 2013). Kim et al. fabricated high TE performance PEDOT:PSS/graphene composite films, which has the maximum ZT of 2.1×10^{-2} at 300 K and a thermal conductivity of 0.14 $W\ m^{-1}\ K^{-1}$. By incorporating 2 wt% of graphene, the ZT value of PEDOT:PSS thin film could be increased about 10 fold. This enhancement arises from the facilitated carrier transfer between PEDOT:PSS and graphene, as well as the high electron mobility of graphene (200,000 $cm^2\ V^{-1}\ s^{-1}$) (Kim et al., 2012).

Xu et al. developed a convenient method to prepare PEDOT/rGO nanocomposites with a pie-like structure, in which the uniform PEDOT layers were formed on surfaces of rGO nanosheet. The nanocomposite exhibited a significantly enhanced TE performance at room temperature with a PF of $5.2 \pm 0.9\ \mu W\ m^{-1}\ K^{-2}$, which is 13.3 fold higher than that of PEDOT (Xu et al., 2013). Ju et al. investigated the TE properties of benzenesulfonate-doped PEDOT/graphene composites, which were prepared in five different solvents, including 1-hexanol, 1-propanol, ethanol, *n*-butanol, and methanol. Among them, PEDOT–MeOH/graphene composite with 75 wt% graphene showed the maximum ZT value of 1.9×10^{-2}. The improved TE properties was resulted by the increased carrier mobility and electrical conductivity, which may be due to the enhancement of PEDOT chain stacking by shorter-chain

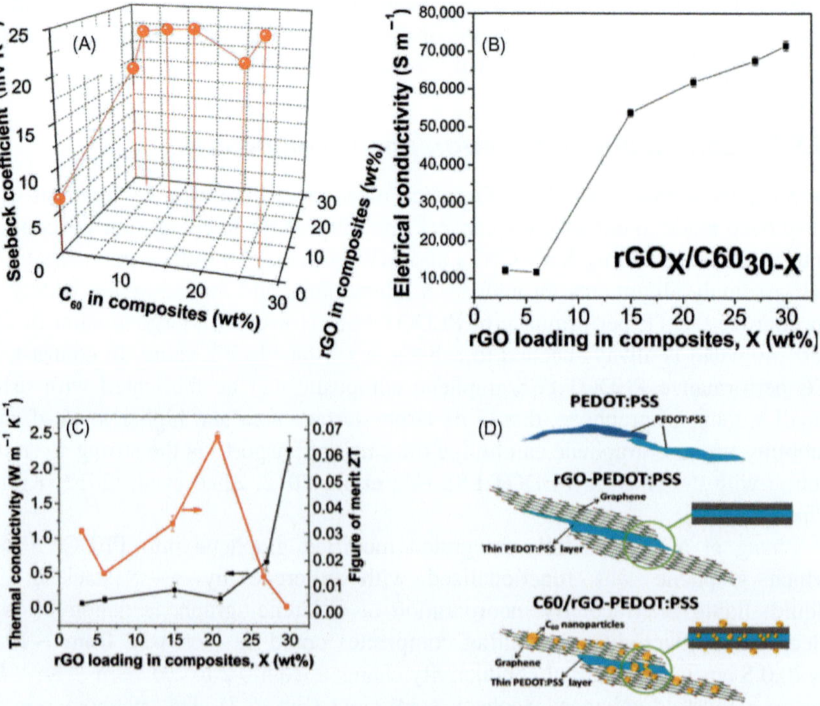

Figure 6.7 (A) Three-dimensional plot of the Seebeck coefficient of composites as a function of C_{60} and rGO weight loading in composites, indicating more than fourfold enhancement. (B) The electrical conductivity of nanohybrids filled polymer composites. (C) Thermal conductivity and ZT of nanohybrids filled polymer composites. (D) Schematic diagram of transport of carrier in the pure film polymer, polymer–rGO and rGO/C_{60} nanocomposites–polymer-filled hybrid (Zhang et al., 2013).

alcoholic solvents (Ju et al., 2015). Li et al. reported a facile method to prepare r-GO/PEDOT:PSS composite films by direct reduction of GO/PEDOT:PSS films, which were prepared via simple wet-chemical route. The r-GO/PEDOT:PSS composite films displayed the optimum PF of ~32.6 $\mu W\ m^{-1}\ K^{-2}$, which is about 1.5 fold higher than that of PEDOT:PSS film (Li et al., 2014).

Han et al. prepared rGO/PPy composites with the assistant of sodium dodecyl sulfate, in which uniform PPy coatings were conveniently grown on both sides of rGO nanosheet surfaces via a template-directed in-situ polymerization. The rGO/PPy composites exhibited greatly enhanced TE performance with a PF of 3.01 $\mu W\ m^{-1}\ K^{-2}$ at room temperature, which is 84 times greater than that of the pure PPy (Han et al., 2014). Wang et al. prepared the PPy/graphene nanosheets (PPy/GNs) composites by a convenient chemical polymerization method, in which PPy grew along the surface of the GNs to form a crystalline structure. The Seebeck coefficient of composites could be significantly enhanced through increasing GNs content, while a large increase in the thermal conductivity was not found. PPy/GNs composites exhibited its maximum ZT value of 2.8×10^{-3} when PPy:graphene nanoplatelet (GNP) weight ratio is 1:0.4, which is almost 210 fold higher than that of pure PPy (Wang et al., 2014b).

Lu et al. reported a PANi/GNs composite and investigated their TE properties in the temperature range from 323 to 453 K. The composites with 30 wt% GNs exhibited the highest ZT value of 1.95×10^{-3} at 453 K, which is about 70 fold of that obtained from the PANi (Lu et al., 2013). Xiang et al. prepared PANi/exfoliated GNP nanocomposites by in-situ polymerization of aniline monomer in the presence of GNP. PANi has a strong affinity for GNP due to π−electron interactions, forming a uniform nanofibril coating (Fig. 6.8). A paper-like nanocomposite was prepared by controlled vacuum filtration of an aqueous dispersion of PANi decorated GNP. By increasing the PANi content to 40 wt%, the composites had the optimum electrical conductivity and Seebeck coefficient as $59\ S\ cm^{-1}$ and $33\ \mu V\ K^{-1}$, respectively. Thus, the ZT value is two orders of magnitude higher than either of the constituents (Xiang and Drzal, 2012). Du et al. prepared GNs/P3HT composites by oxidative polymerization of three hexylthiophene in a GNs dispersed chloroform solution. The electrical conductivity of the composites was dramatically increased from $\sim 10^{-6}$ to $1.2\ S\ cm^{-1}$ as the GNs content increasing to 30 wt%, and the Seebeck coefficient slightly enhanced from 33.15 to $35.46\ \mu V\ K^{-1}$. The optimum PF was achieved as $\sim 0.16\ \mu W\ m^{-1}\ K^{-2}$ in composite with 30 wt% GNs (Du et al., 2012a).

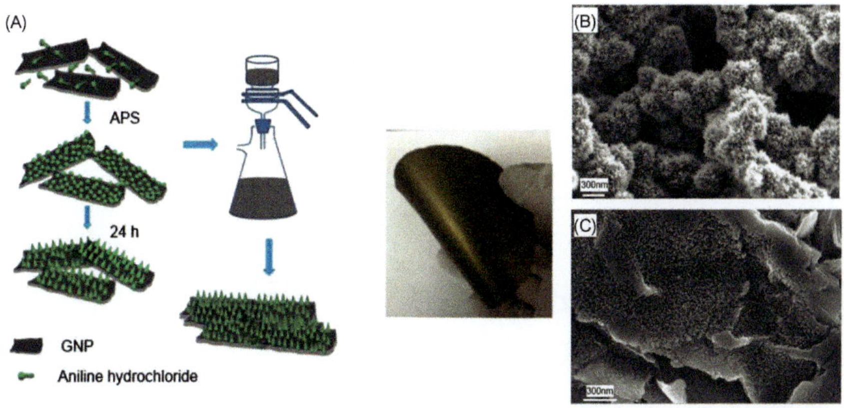

Figure 6.8 (A) Schematic demonstration of aniline in-situ polymerization in the occurrence of particles of GNP and a photograph displaying the final film of nanocomposite flexibility. (B) SEM image of neat polyaniline salt. (C) SEM image of PANi/GNP particles (Xiang and Drzal, 2012).
Reproduced from Xiang, J. and L.T. Drzal, Templated growth of polyaniline on exfoliated graphene nanoplatelets (GNP) and its thermoelectric properties. Polymer, 2012. 53(19): p. 4202–4210, with permission from Elsevier.

6.5.3 Conducting composites of polymer/graphite

Graphite is extensively used in electrodes for batteries and electrolysis reactions due to its high electrical conductivity of $\sim 10^4$ S cm^{-1}. Graphite is not ideal material for TE applications, due to its low Seebeck coefficient, high thermal conductivity, and low ZT value at room temperature. However, the TE properties of graphite are expected to be significantly modified by exfoliation. Recently, in order to generate materials with better TE performance, a lot of composites have been well composited by graphite and conducting polymers through improving the electrical conductivity and Seebeck coefficient, while restricting the thermal conductivity.

Zhao et al. prepared graphite oxide (GO)/ordered PANi composites through an in-situ polymerization, in which PANi grew along the surface of exfoliated GO to form high crystalline layers. Composites retained the thermal conductivity of pure PANi, whilst exhibited significantly improved electrical conductivity and Seebeck coefficient as 0.751 S cm^{-1} and 28.31 $\mu V\ K^{-1}$, respectively, leading PF of 0.602 $\mu W\ m^{-1}\ K^{-2}$ at 388 K Consequently, the composite provided a ZT value of 4.86×10^{-4} which is higher than that of the pure PANi by magnitude two orders (Zhao et al., 2012). Piao et al. applied expanded graphite (ExG) as filler in both p- and n-type organic materials for TE energy conversion (Fig. 6.9). The ExG/PEDOT:PSS composite displayed high electrical conductivity of 10^2 S cm^{-1} and enhanced Seebeck coefficient and could be used as a promising p-type material. The exfoliated graphitic sheets can be efficiently n-doped with polyethyleneimine (PEI) in polyvinyl alcohol (PVA), the resulting n-type ExG/PVA/PEI composite thin films exhibited high Seebeck coefficient of $-25.3\ \mu V\ K^{-1}$. A TE device was fabricated by using ExG/PEDOT:PSS composite as p-type components and ExG/PVA/PEI composite as n-type components, which produced an output voltage of $\sim$4 mV at a temperature gradient of 50 K and generated 1.7 nW power at load resistance of 1 kΩ (Piao et al., 2013). Wang et al. prepared $HClO_4$-doped PANi/graphite through mechanical ball milling and cold pressing. By adjusting the concentration of graphite, the composite with graphite concentration of 50 wt% had the optimum ZT value of 1.37×10^{-3} at 393 K, which is more 10,000 fold greater than that of the $HClO_4$-doped PANi without graphite (Wang et al., 2011).

6.6 Conclusions and outlook

In this chapter, we have summarized the recent advances in TE composite materials composed of organic conducting polymers and inorganic materials (Table 6.1). Some of the concepts for inorganic TE materials, such as phonon scattering, carrier-energy-filtering, and carrier-pocket engineering, have been used to design and construct TE composite materials. Nevertheless, several key principles for constructing energy-filtering interfaces in composites would be summarized as follows: (1) efficient interaction between polymers and inorganic particles to form a controllable organic–inorganic interface; (2) compatibility of different components to facilitate high-energy-carriers transferring on the interface; (3) interfacial barrier of

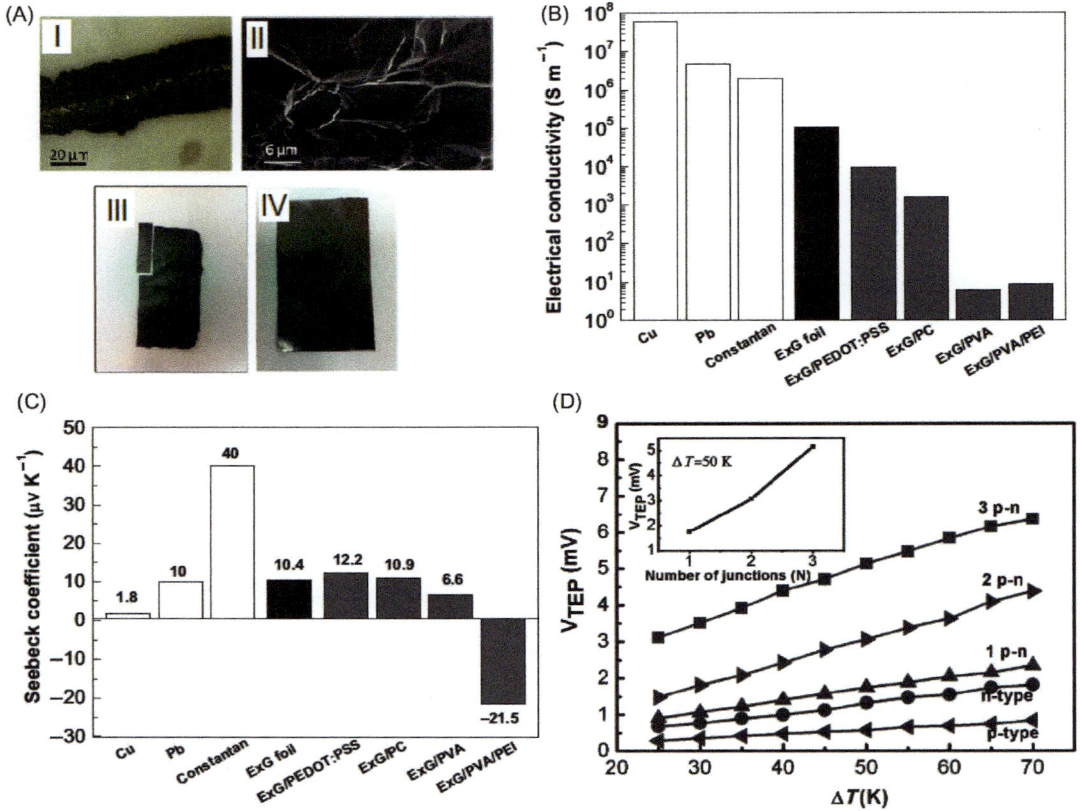

Figure 6.9 (A) Optical micrograph of ExG (I), SEM image of ExG (II), photograph of an ExG/PEI composite film (III), photography of a graphite foil obtained by lamination of ExG (IV). (B) Electrical conductivity of metals, ExG foil, ExG thin composite film with a 1:1 weight ratio between ExG and PEDOT:PSS, composite films with PC, PVA containing 10 wt% of ExG, and doped with PEI with a ExG:PEI weigh ratio 1:1. (C) Seebeck coefficients of the corresponding samples. (D) TE voltage generated from p−n junctions as a function of temperature gradient. One, two, and three p−n junctions were connected in series for testing (Piao et al., 2013).

Table 6.1 The electrical conductivity (σ), thermal conductivity (S), thermal conductivity (K), power factor (PF), and thermoelectric figure of merit (ZT) of TE materials covered in this chapter

Materials	σ (S cm^{-1})	S (μV K^{-1})	K (W m^{-1} K^{-2})	PF (W m^{-1} K^{-2})	ZT	Ref.
PEDOT	6.9–40.5 (310 K)	−122 to −35 (310 K)	–	(1.2–12) × 10^{-6} (310 K)	–	Taggart et al. (2011)
PEDOT:PSS/Bi_2Te_3	~152.5	~ −62.5	–	~80 × 10^{-6}	–	Taggart et al. (2011)
PANI/Bi_2Te_3	~2 (350 K)	~130 (350 K)	–	2.6 × 10^{-6} (350 K)	0.009 (350 K)	Toshima et al. (2010)
PANi-doped by naphthalene sulfonic acid	0.0077	212.4	0.21	–	4.86 × 10^{-5} (300 K)	Sun et al. (2010)
CSA-doped-PANI	10^2 (50–300 K)	~5–10 (100–300 K)	–	–	–	Anno et al. (2009)
CSA-doped-PANI/Bi	10^{-2}–10^1 (300–400 K)	~12–54 (180–400 K)	–	–	–	Anno et al. (2009)
PANI/Bi_2Te_3	~2 (300–473 K)	~ −50 (300–473 K)	–	–	–	Li et al. (2011)
P3HT	1283	–	–	–	–	Pintér et al. (2007)
Bi_2Te_3/PTH	~0.02–~17	–	–	13.6	–	He et al. (2012)
PEDOT:PSS/Te	19.3 (±2.3) (300 K)	163 (±4) (300 K)	0.22–0.30 (300 K)	70.9 × 10^{-6}	0.10	See et al. (2010)
PEDOT:PSS/Bi_2Te_3	~105	~105	–	~95.1 × 10^{-6}	–	Zhang et al. (2010)

PEDOT:PSS/$Ca_3Co_4O_9$)	~50–135 (100–300 K)	~1–18 (100–300 K)	–	~0.1–4 (100–300 K)	–	Liu et al. (2011)
PEDOT/PbTe	0.064–0.616 (300 K)	1205–4088 (300 K)	–	1.07–1.44	–	Wang et al. (2011)
Polymer/CNT	~0–48 (300 K)	~40–50 (300 K)	0.18–0.34 (300 K)	–	0.006 (300 K)	Yu et al. (2008)
PEDOT:PSS/SWCNT	~280–400 (300 K)	~21–25 (300 K)	0.4	~14–25 (300 K)	0.02 (300 K)	Kim et al. (2010)
PEDOT:PSS/CNT	~0–124 (300 K)	~17–40 (300 K)	0.26–0.38 (300 K)	1–11 (300 K)	–	Kim et al. (2010)
SWNT/PEDOT:PSS	1.35×10^5	28–33	0.2–0.4	160	–	Yu et al. (2011)
PANI/SWNT	~10–125 (300 K)	11–40 (300 K)	~0.5–1 (300 K)	–	–	Yao et al. (2010)
P3HT/CNT	1×10^3	12–32	0.13–0.16	24–25	–	Bounioux et al. (2013)
PTH/MW–CNT	0.0234–29.82	27.7–22.7	–	5–47	8.71×10^{-4} (393 K)	Wang et al. (2013)
PPy-MWNT/PANi	30.34	4.7–31.2	0.34	3.1	–	Liu and Yu (2014)
MWCNT/PPy	72	–	–	2.079	–	Wang et al. (2014a)
Graphene/PEDOT:PSS	700	32	2	–	6.7×10^{-2}	Zhang et al. (2013)
PEDOT:PSS/graphene	0.74–31.7	165.82–44.75	2.03–6.34	0.24–0.3	–	Kim et al. (2012)

(*Continued*)

Table 6.1 **(Continued)**

Materials	σ (S cm^{-1})	S (μV K^{-1})	K (W m^{-1} K^{-2})	PF (W m^{-1} K^{-2})	ZT	Ref.
PEDOT-rGO	50.8	31.8	–	5.2	–	Xu et al. (2013)
Graphene/PPy	29.4–41.6	21.8–26.9	–	3.01	–	Han et al. (2014)
PPy/GNs	101.68	31.74	–	10.24	24×10^{-2}	Wang et al. (2014b)
PANi/GNs	54	40	–	2.6	1.95×10^{-3}	Lu et al. (2013)
ANi/GNP	59	33	13	4.6	1.51×10^{-4}	Xiang and Drzal (2012)
GNs/P3HT	10^{-6}–1.2	33.15–35.46	–	0.16	–	Du et al. (2012a)
ExG	10	25	~0.3	–	6×10^{-4}	Piao et al. (2013)
$HClO_4$-doped PANI/ Graphite	120	18.66	–	4.18	1.37×10^{-4}	Wang et al. (2011)
GO/PANI	7.51	28.31	–	20	4.86	Zhao et al. (2012)

less than 0.1 eV to selectively scatter low-energy carriers. However, some critical issues in composites need to be resolved before being ideal TE materials for real application: (1) lack of straight reliable characterization methods duc to anisotropy of composite films; (2) need feasible strategies and theoretical models to design suitable composites and hence optimize the TE performance; (3) need efficient protocol to enhance the crystallinity, control the morphology of hybrid materials, and optimize the doping level.

Ultimately, with the rapid progress being made in organic chemistry, polymer engineering, nanomaterials science and theoretical modeling, a large number of novel polymers and nanostructure materials with high TE performance are expected to be produced. More importantly, the new development fabrication methods, such as screen printing, ink-jet printing, and roll-to-roll printing technologies, have shown potential for a large-scale production of solution processable TE composites. Thus, TE composites would be an extraordinarily active area for academic exploration as well as industry applications in the near future.

Acknowledgment

The authors acknowledge support from the A*STAR, Industry Alignment Fund, Pharos "Hybrid Thermoelectric Materials for Ambient Applications" program.

References

Aïch, R.B., et al., 2009. Electrical and thermoelectric properties of poly(2,7-carbazole) derivatives. Chem. Mater. 21 (4), 751–757.

Alam, H., Ramakrishna, S., 2013. A review on the enhancement of figure of merit from bulk to nano-thermoelectric materials. Nano Energy. 2 (2), 190–212.

Althues, H., Henle, J., Kaskel, S., 2007. Functional inorganic nanofillers for transparent polymers. Chem. Soc. Rev. 36 (9), 1454–1465.

Anno, H., et al., 2009. Preparation of conducting polyaniline–bismuth nanoparticle composites by planetary ball milling. J. Electr. Mater. 38 (7), 1443–1449.

Balazs, A.C., Emrick, T., Russell, T.P., 2006. Nanoparticle polymer composites: where two small worlds meet. Science. 314 (5802), 1107–1110.

Bell, L.E., 2008. Cooling, heating, generating power, and recovering waste heat with thermoelectric systems. Science. 321 (5895), 1457–1461.

Bhadra, S., et al., 2009. Progress in preparation, processing and applications of polyaniline. Progr. Polym. Sci. 34 (8), 783–810.

Biswas, K., et al., 2012. High-performance bulk thermoelectrics with all-scale hierarchical architectures. Nature. 489 (7416), 414–418.

Bounioux, C., et al., 2013. Thermoelectric composites of poly(3-hexylthiophene) and carbon nanotubes with a large power factor. Energy Environ. Sci. 6 (3), 918–925.

Carotenuto, G., Her, Y.-S., Matijevic, E., 1996. Preparation and characterization of nanocomposite thin films for optical devices. Ind. Eng. Chem. Res. 35 (9), 2929–2932.

Chen, J., et al., 2011. Superlow thermal conductivity 3D carbon nanotube network for thermoelectric applications. ACS Appl. Mater. Interfaces. 4 (1), 81–86.
Chen, Z.-G., et al., 2012. Nanostructured thermoelectric materials: Current research and future challenge. Progr. Nat. Sci.: Mater. Int. 22 (6), 535–549.
Choi, E., et al., 1997. The effect of high magnetic field on thermoelectric power of doped polyacetylene. Synth. Metals. 84 (1), 685–689.
Coates, N.E., et al., 2013. Effect of interfacial properties on polymer–nanocrystal thermoelectric transport. Adv. Mater. 25 (11), 1629–1633.
DiSalvo, F.J., 1999. Thermoelectric cooling and power generation. Science. 285 (5428), 703–706.
Du, Y., et al., 2012a. Preparation and characterization of graphene nanosheets/poly (3-hexylthiophene) thermoelectric composite materials. Synth. Metals. 162 (23), 2102–2106.
Du, Y., et al., 2012b. Research progress on polymer–inorganic thermoelectric nanocomposite materials. Progr. Polym. Sci. 37 (6), 820–841.
Han, S., et al., 2014. Morphology and thermoelectric properties of graphene nanosheets enwrapped with polypyrrole. RSC Adv. 4 (55), 29281–29285.
He, M., et al., 2012. Thermopower enhancement in conducting polymer nanocomposites via carrier energy scattering at the organic–inorganic semiconductor interface. Energy Environ. Sci. 5 (8), 8351.
Heremans, J.P., et al., 2013. When thermoelectrics reached the nanoscale. Nat. Nanotechnol. 8 (7), 471–473.
Holder, E., Tessler, N., Rogach, A.L., 2008. Hybrid nanocomposite materials with organic and inorganic components for opto-electronic devices. J. Mater. Chem. 18 (10), 1064–1078.
Holland, E., Monkman, A., 1995. Thermoelectric power measurements in highly conductive stretch-oriented polyaniline films. Synth. Metals. 74 (1), 75–79.
Ju, H., Kim, M., Kim, J., 2015. Enhanced thermoelectric performance by alcoholic solvents effects in highly conductive benzenesulfonate-doped poly(3,4-ethylenedioxythiophene)/graphene composites. J. Appl. Polym. Sci. 132, 42107.
Kalaitzidou, K., Fukushima, H., Drzal, L.T., 2007. A new compounding method for exfoliated graphite–polypropylene nanocomposites with enhanced flexural properties and lower percolation threshold. Compos. Sci. Technol. 67 (10), 2045–2051.
Kanatzidis, M.G., 2009. Nanostructured thermoelectrics: the new paradigm? Chem. Mater. 22 (3), 648–659.
Kim, D., et al., 2010. Improved thermoelectric behavior of nanotube-filled polymer composites with poly(3,4-ethylenedioxythiophene) poly(styrenesulfonate). ACS Nano. 4 (1), 513–523.
Kim, G., Pipe, K.P., 2012. Thermoelectric model to characterize carrier transport in organic semiconductors. Phys. Rev. B. 86 (8), 085208.
Kim, G.H., Hwang, D.H., Woo, S.I., 2012. Thermoelectric properties of nanocomposite thin films prepared with poly(3,4-ethylenedioxythiophene) poly(styrenesulfonate) and graphene. Phys. Chem. Chem. Phys. 14 (10), 3530–3536.
Kim, P., et al., 2001. Thermal transport measurements of individual multiwalled nanotubes. Phys. Rev. Lett. 87 (21), 215502.
Li, F., et al., 2014. Preparation and thermoelectric properties of reduced graphene oxide/ PEDOT: PSS composite films. Synth. Metals. 197, 58–61.
Li, J., et al., 2010. Synthesis and thermoelectric properties of hydrochloric acid-doped polyaniline. Synth. Metals. 160 (11–12), 1153–1158.

Li, Y., et al., 2011. Synthesis and characterization of Bi_2 Te_3/polyaniline composites. Mater. Sci. Semicond. Proc. 14 (3–4), 219–222.

Liu, C., et al., 2011. Free-standing PEDOT-PSS/Ca3Co4O9 composite films as novel thermoelectric materials. J. Electr. Mater. 40 (5), 948–952.

Liu, J., Yu, H.-Q., 2014. Thermoelectric enhancement in polyaniline composites with polypyrrole-functionalized multiwall carbon nanotubes. J. Electr. Mater. 43 (4), 1181–1187.

Lu, Y., Song, Y., Wang, F., 2013. Thermoelectric properties of graphene nanosheets-modified polyaniline hybrid nanocomposites by an in situ chemical polymerization. Mater. Chem. Phys. 138 (1), 238–244.

MacDonald, D.K.C., 1962. Thermoelectricity: An Introduction to The Principles. Wiley, New York.

Maddison, D., Unsworth, J., Roberts, R., 1988. Electrical conductivity and thermoelectric power of polypyrrole with different doping levels. Synth. Metals. 26 (1), 99–108.

Maddison, D., Roberts, R., Unsworth, J., 1989. Thermoelectric power of polypyrrole. Synth. Metals. 33 (3), 281–287.

Martín-González, M., Caballero-Calero, O., Díaz-Chao, P., 2013. Nanoengineering thermoelectrics for 21st century: energy harvesting and other trends in the field. Renewable Sustainable Energy Rev. 24, 288–305.

Moliton, A., et al., 1999. Thermoelectric power stability of the polyparaphenylene implanted with Cesium ions. Synth. Metals. 101 (1), 351–352.

Moreau, C., et al., 1997. Sensitive thermoelectric power and conductivity measurements on implanted polyparaphenylene thin films. Adv. Mater. Opt. Electr. 7 (6), 281–293.

Ohta, H., 2007. Thermoelectrics based on strontium titanate. Mater. Today. 10 (10), 44–49.

Pei, Y., et al., 2011. Convergence of electronic bands for high performance bulk thermoelectrics. Nature. 473 (7345), 66–69.

Piao, M., et al., 2013. Preparation and characterization of expanded graphite polymer composite films for thermoelectric applications. Phys. Status Solidi (B). 250 (12), 2529–2534.

Pintér, E., et al., 2007. Characterization of poly(3-octylthiophene)/silver nanocomposites prepared by solution doping. J. Phys. Chem. C. 111 (32), 11872–11878.

Rowe, D.M., 1999. Thermoelectrics, an environmentally-friendly source of electrical power. Renewable Energy. 16 (1), 1251–1256.

Rowe, D.M., 2005. Thermoelectrics Handbook Macro to Nano. CRC Press, New York.

See, K.C., et al., 2010. Water-processable polymer–nanocrystal hybrids for thermoelectrics. Nano Lett. 10 (11), 4664–4667.

Shinohara, Y., et al. The effect of carrier conduction between main chains on thermoelectric properties of polythiophene. in Thermoelectrics, 2007. ICT 2007. 26th International Conference on. 2007. IEEE.

Snyder, G.J., Toberer, E.S., 2008. Complex thermoelectric materials. Nat. Mater. 7 (2), 105–114.

Sootsman, J.R., Chung, D.Y., Kanatzidis, M.G., 2009. New and old concepts in thermoelectric materials. Angew. Chem. Int. Ed. Engl. 48 (46), 8616–8639.

Sun, Y., et al., 2010. A three-in-one improvement in thermoelectric properties of polyaniline brought by nanostructures. Synth. Metals. 160 (21–22), 2371–2376.

Szczech, J.R., Higgins, J.M., Jin, S., 2011. Enhancement of the thermoelectric properties in nanoscale and nanostructured materials. J. Mater. Chem. 21 (12), 4037–4055.

Taggart, D.K., et al., 2011. Enhanced thermoelectric metrics in ultra-long electrodeposited PEDOT nanowires. Nano Lett. 11 (1), 125–131.

Toshima, N., Imai, M., Ichikawa, S., 2010. Organic–inorganic nanohybrids as novel thermoelectric materials: hybrids of polyaniline and bismuth(III) telluride nanoparticles. J. Electr. Mater. 40 (5), 898–902.

Toshima, N., Imai, M., Ichikawa, S., 2011. Organic–inorganic nanohybrids as novel thermoelectric materials: hybrids of polyaniline and bismuth(III) telluride nanoparticles. J. Electr. Mater. 40 (5), 898–902.

Tritt, T.M., 1999. Holey and unholey semiconductors. Science. 283 (5403), 804–805.

Tritt, T.M., Subramanian, M., 2006. Thermoelectric materials, phenomena, and applications: a bird's eye view. MRS Bull. 31 (03), 188–198.

Wang, L., et al., 2011. Thermoelectric properties of conducting polyaniline/graphite composites. Mater. Lett. 65 (7), 1086–1088.

Wang, L., et al., 2013. Preparation and thermoelectric properties of polythiophene/multiwalled carbon nanotube composites. Synth. Metals. 181, 79–85.

Wang, J., et al., 2014a. Preparation and thermoelectric properties of multi-walled carbon nanotubes/polypyrrole composites. Synth. Metals. 195, 132–136.

Wang, L., et al., 2014b. Preparation of polypyrrole/graphene nanosheets composites with enhanced thermoelectric properties. RSC Adv. 4 (86), 46187–46193.

Wang, Y., Cai, K., Yao, X., 2011. Facile fabrication and thermoelectric properties of PbTe-modified poly(3,4-ethylenedioxythiophene) nanotubes. ACS Appl. Mater. Interfaces. 3 (4), 1163–1166.

Wang, Y.Y., et al., 2010. In situ fabrication and thermoelectric properties of PbTe–polyaniline composite nanostructures. J. Nanopart. Res. 13 (2), 533–539.

Wei, Q., et al., 2015. Recent progress on PEDOT-based thermoelectric materials. Materials. 8 (2), 732–750.

Xiang, J., Drzal, L.T., 2012. Templated growth of polyaniline on exfoliated graphene nanoplatelets (GNP) and its thermoelectric properties. Polymer. 53 (19), 4202–4210.

Xu, K., Chen, G., Qiu, D., 2013. Convenient construction of poly(3,4-ethylenedioxythiophene)–graphene pie-like structure with enhanced thermoelectric performance. J. Mater. Chem. A. 1 (40), 12395–12399.

Yao, Q., et al., 2010. Enhanced thermoelectric performance of single-walled carbon nanotubes/polyaniline hybrid nanocomposites. ACS Nano. 4 (4), 2445–2451.

Yee, S.K., et al., 2013. Thermoelectric power factor optimization in PEDOT: PSS tellurium nanowire hybrid composites. Phys. Chem. Chem. Phys. 15 (11), 4024–4032.

Yu, C., et al., 2008. Thermoelectric behavior of segregated-network polymer nanocomposites. Nano Lett. 8 (12), 4428–4432.

Yu, C., et al., 2011. Light-weight flexible carbon nanotube based organic composites with large thermoelectric power factors. ACS Nano. 5 (10), 7885–7892.

Yu, H., et al., 2012. Graphene/polyaniline nanorod arrays: synthesis and excellent electromagnetic absorption properties. J. Mater. Chem. 22 (40), 21679–21685.

Zebarjadi, M., et al., 2012. Perspectives on thermoelectrics: from fundamentals to device applications. Energy Environ. Sci. 5 (1), 5147–5162.

Zhang, B., et al., 2010. Promising thermoelectric properties of commercial PEDOT: PSS materials and their Bi_2Te_3 powder composites. ACS Appl. Mater. Interfaces. 2 (11), 3170–3178.

Zhang, K., Zhang, Y., Wang, S., 2013. Enhancing thermoelectric properties of organic composites through hierarchical nanostructures. Sci. Rep. 3, 3448.

Zhao, D., Tan, G., 2014. A review of thermoelectric cooling: materials, modeling and applications. Appl. Thermal Eng. 66 (1), 15–24.

Zhao, L.-D., et al., 2014. Ultralow thermal conductivity and high thermoelectric figure of merit in SnSe crystals. Nature. 508 (7496), 373–377.

Zhao, L.-D., Dravid, V.P., Kanatzidis, M.G., 2014. The panoscopic approach to high performance thermoelectrics. Energy Environ. Sci. 7 (1), 251–268.

Zhao, X., et al., 2002. Thermoelectric properties of $Bi_{0.5}Sb_{1.5}Te_3$/polyaniline hybrids prepared by mechanical blending. Mater. Lett. 52 (3), 147–149.

Zhao, Y., et al., 2012. The effect of graphite oxide on the thermoelectric properties of polyaniline. Carbon. 50 (8), 3064–3073.

Using recycled polymers for the preparation of polymer nanocomposites: properties and applications

7

Shadpour Mallakpour[1,2,3] *and Vajiheh Behranvand*[1]
[1]Organic Polymer Chemistry Research Laboratory, Department of Chemistry, Isfahan University of Technology, Isfahan, Islamic Republic of Iran, [2]Nanotechnology and Advanced Materials Institute, Isfahan University of Technology, Isfahan, Islamic Republic of Iran, [3]Center of Excellence in Sensors and Green Chemistry, Department of Chemistry, Isfahan University of Technology, Isfahan, Islamic Republic of Iran

Chapter Outline

7.1 Introduction 197
7.1.1 Why polymer recycling? 197
7.1.2 Solid-waste management 198
7.2 NCs obtained from recycled polymers 199
7.2.1 NCs obtained from recycled polyethylene (PE) 200
7.2.2 NCs obtained from polypropylene (PP) 204
7.2.3 NCs obtained from polyethylene terephthalate (PET) 207
7.2.4 NCs obtained from polystyrene (PS) 212
7.2.5 NCs obtained from other recycled polymers 215
7.3 Summary 217
7.4 Conclusions and challenges 222
Acknowledgments 222
References 222

7.1 Introduction

7.1.1 Why polymer recycling?

Polymers have been created a very important class of materials without which the life appears very difficult. They are all around us in everyday usage; in plastic, in rubber, in resins, and in adhesives. Polymer materials have valuable features such as

Hybrid Polymer Composite Materials: Applications. DOI: http://dx.doi.org/10.1016/B978-0-08-100785-3.00007-3

Figure 7.1 Plastic recycling as a waste-management strategy.

good mechanical properties, low density, rather low cost, and also ease of processing. They have main advantages over other conventional metallic components due to their specific strength properties with weight saving of 20%–40%, capability to meet stringent dimensional stability, potential for quick process cycles, recyclability, low cost, manufacturability, lower thermal expansion properties, and excellent fatigue and fracture resistance (Majka et al., 2016; Madi, 2013; Njuguna, 2013; Mallakpour and Behranvand 2016a). These plastics are synthetic materials resulting mainly from petroleum or natural gas. They can remain on both sea and land for years causing environmental pollution because they have high degradation temperature, high resistance to UV radiation and are mostly not biodegradable (Gürüa et al., 2014). Over the recent years, the amount of plastic waste randomly left in nature has increased considerably. These materials unlike natural polymers are not biodegradable in the natural environment and resists for hundreds of years, leading to environmental pollution (Park and Kim, 2014; Radu and Christiana, 2011).

Plastic recycling is clearly a waste-management strategy which is receiving increasing attention that can reduce environmental impact and resource depletion. It can decrease energy and material use per unit of output and therefore yield improved eco-efficiency (Fig. 7.1) (Hopewell et al., 2009). Different routes for recycling may be applied: chemical recycling, reuse and mechanical recycling (physical recycling) (Hamad et al., 2013). In chemical recycling, the polymer backbone under the recycling process is decomposed into monomer units or randomly breaded into larger chain fragments with associated generation of gaseous products (Al-sabagh et al., 2016). Under mechanical recycling, the plastic is ground down and then processed again and compounded to generate a novel component that may or may not be the same as its original usage. The most common examples of reuse are drinks bottles that are returned to be cleaned and used again (Hamad et al., 2013).

7.1.2 Solid-waste management

The solid-waste management tasks introduce complex technical challenges. They also pose an extensive variety of economic, administrative, and social problems that must be managed and solved. Solid-waste management is one among the basic essential services provided by municipal authorities in the country to keep urban centers clean (http://recycling.omicsgroup.com/events-list/solid-waste-management).

A key component of waste management and resource efficiency strategy, both for municipalities and for industrial processes is recycling. Recycling operations depend profoundly on production and consumption. A shift in plastics production has occurred from the West to Asia: in 2014, 40% by weight of world production is in Asia, with 20% each in Europe and North America—China is the largest individual country at 24%. Corresponding to its size and rapid financial development, China has become a main player in the global recycling market, particularly for plastics, paper and metals. The Chinese Government published the notice of the establishment of a complete and advanced recycling system of waste commodities on October 31, 2011. Plastic recycling in China is divided into three categories: industrial recycling, agricultural recycling, and municipal (household) recycling (Velis, 2014).

The Islamic Republic of Iran is the second largest oil producer in Organization of the Petroleum Exporting Countries (OPEC) and the second largest natural gas reserves. The history of municipal solid-waste management systems in the Islamic Republic of Iran goes back to 1911 when the first municipality was established. Even though solid waste is one of the most troublesome environmental problems in the Islamic Republic of Iran, there has so far been no well-defined public authority with an all-embracing responsibility for waste. However over the past years, substantial progress has been attained in some of the largest cities (Abduli, 2007).

7.2 NCs obtained from recycled polymers

For mechanical recycling only polymeric materials that may be remelted and reprocessed into products via techniques such as injection molding or extrusion are of interest. Thermosets may be chemically recycled back to feedstock or used as a carrier such as cement kilns. In spite of the implementation of new methods of recycling, always a certain amount of waste remains, which is not appropriate for recycling. Moreover, after several processing cycles, the structure of polymer is decomposed presenting the poorer mechanical properties than those of a virgin one. To overwhelm these limitations, it seems that the easiest way to recycle the waste plastics is development of blends and composites. Actually, an innovative idea of using plastic waste would produce repolymer composites (Majka and Pielichowski, 2011).

NCs are as multiphase materials, where one of the phases has nanoscale additives. Almost all kinds of polymers, such as thermosets, thermoplastics, and elastomers, have been applied to make polymer NCs (Anandhan and Bandyopadhyay, 2011; Jeon and Baek, 2010; Mallakpour and Adnany Sadaty, 2016).

Many research groups have incorporated nanofillers into the recycled waste polymers or used from recycled polymers as fillers and additives. In this chapter, it has been tried to focus on properties and applications of these recycled polymer NCs which are extensively discussed in the following sections. Application examples of packaging, building and construction, lubricating greases, water treatment, foundation and roadbed, transportation, construction, energy, defense, photocatalyst, and others will be presented by selection of recent related results from other groups as well as our own research.

7.2.1 NCs obtained from recycled polyethylene (PE)

One of the most common packaging materials is high density polyethylene (HDPE) from polyolefins family due to its excellent flexibility, good processability, and low cost. Though, its use is limited because of its lower strength, photodegradation, and so on (Li and Li, 2010). Polyolefin plastics can be mechanically recycled, regenerated, or incinerated, and they do not decompose under natural conditions which cause considerable environmental anxieties. So, manufacturers of plastics have started to recycle their waste products in an attempt to solve this problem (Alzerreca et al., 2015).

Navarro et al. formulated and characterized a novel composite, a recycled HDPE/graphite (Cg) mixture, for medium temperature thermal energy storage application (Yang et al., 2016). They mentioned that for decreasing CO_2 emissions and balancing energy, supply and request across the electricity grid, energy storage has become an important topic. Their aim was to improve the energy efficiency in established processes by recovering and storing heat. Also, using of recycled polymer reduces at the same time the overall cost. So, graphite content was added in different mass fractions into the recycled HDPE to improve thermal conductivities. The resulting 50-mm composites can be observed in Fig. 7.2.

The thermal conductivity of different samples with different graphite content at different temperatures was shown in Fig. 7.3. As can be observed when the graphite content is increased the thermal conductivity rises getting values of $1.31 \pm 0.04\ W\ m^{-1}\ K^{-1}$ at room temperature when the Cg is 20% that this showed increment of 160% compared to the recycled HDPE ($0.51\ W\ m^{-1}\ K^{-1}$). Also, the

Figure 7.2 HDPE/Cg, composite heat storage material with 4% and 20% Cg (50 mm diameter). Adapted from Yang, C., Navarro, M.E., Zhao, B., Leng, G., Xu, G., Wang, L., et al. (2016). Thermal conductivity enhancement of recycled high density polyethylene as a storage media for latent heat thermal energy storage. Solar Energy Mater. Solar Cells, 152, 103–110, with kind permission of Elsevier.

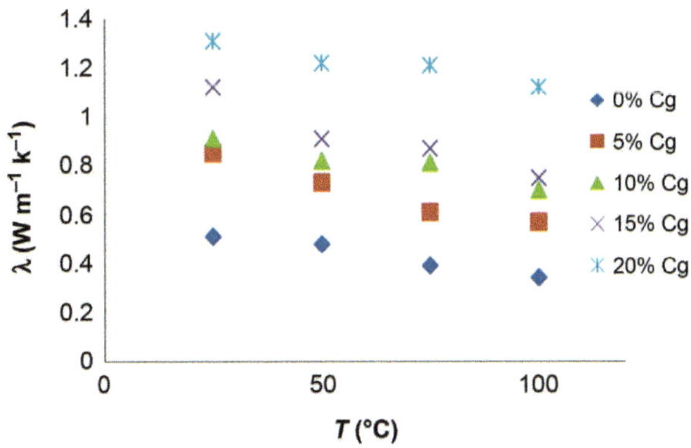

Figure 7.3 Effect of graphite content in HDPE sample as function of temperature.
Adapted from Yang, C., Navarro, M.E., Zhao, B., Leng, G., Xu, G., Wang, L., et al. (2016). Thermal conductivity enhancement of recycled high density polyethylene as a storage media for latent heat thermal energy storage. Solar Energy Mater. Solar Cells, 152, 103–110, with kind permission of Elsevier.

obtained composite presented a good chemical stability when it was working in nitrogen environment while in air the polymer started to be decomposed at lower temperatures.

Oblak et al. (2016) investigated the effect of extensive mechanical recycling on mechanical properties of HDPE. They found more decrease of hardness and modulus after 10th extrusion cycle. However, crystallinity didn't show significant changes through the first 20 extrusion cycles. They concluded that mechanical recycling causes deterioration of HDPE mechanical properties that gets apparent after 10th reprocessing cycle. This information should be taken into account when designing products from recycled HDPE. Nevertheless, it should be emphasized that even after 100 recycling, material retained 80% of its initial mechanical properties and durability was decreased for about 20%! Therefore, if these facts are taken into account through the product-designing phase recycled, HDPE may be often employed without any limitations. With the aim of improvement of mechanical and tribological properties of recycled HDPE, Brostow et al. (2016) used from wood sawdust (wood flour) to prepare composites. Since polymers are often hydrophobic whereas wood and wood products are hydrophilic, sawdust was modified. Traditional silane coupling was used as a first approach to modify the sawdust by 3-methacryloxypropyl-trimethoxysilane (3MPS) and thus forming C–O–Si bonds. As a second method, they decided to combine sol–gel process with 3MPS treatment to create Si–O–Si and C–O–Si linkages. They observed that the sol–gel modification process had positive effects on the bond and strength of the recycled HDPE/wood sawdust materials. Addition of unmodified wood caused increasing dynamic friction, while chemical modification of wood particles resulted lowering friction. Because the market share of wood–plastic composites (WPC) expected to develop sharply in Europe, Sommerhuber et al. (2015) decided to introduce WPC

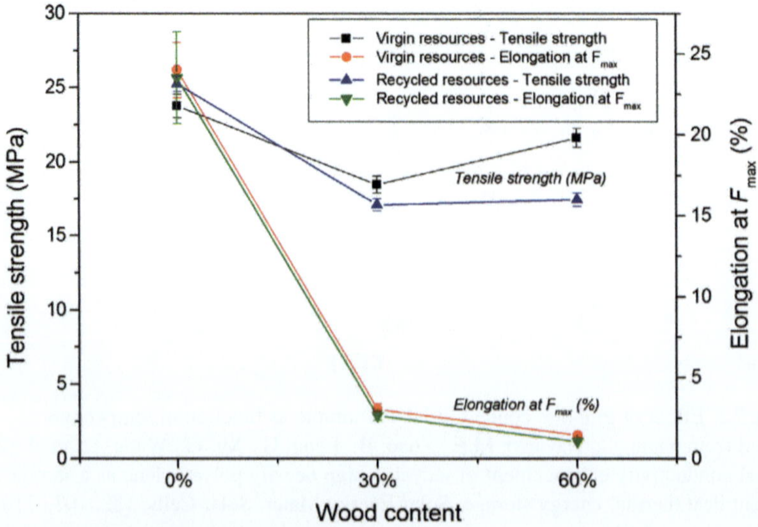

Figure 7.4 Tensile strength and elongation at F_{max}.
Adapted from Sommerhuber, P.F., Welling, J., Krause, A. (2015). Substitution potentials of recycled HDPE and wood particles from post-consumer packaging waste in wood–plastic composites. Waste Manage., 46, 76–85, with kind permission of Elsevier.

made from postconsumer recycled wood and HDPE and compared it to WPC made from virgin resources. To enhance the compounding of wood particles and HDPE and for better mechanical strength of WPC, esterification reaction was done between hydroxyl groups in wood with maleic anhydride polyethylene. Fig. 7.4 presents tensile strength on the left side and elongation at F_{max} on the right. The drop in tensile strength by the incorporation of wood in the polymer matrix is a known drawback of WPC regardless whether virgin or recycled HDPE is used. A proposition to attain better tensile properties of WPC could be the incorporation of wood fibers to the polymer matrix, as examined by Butylina et al. (2011) in a PP matrix. HDPE is a ductile material, as can be observed in the elongation at F_{max} of specimens A and B. Virgin and recycled specimens were not significantly different in elongation.

The cement-based materials, by their performance in terms of durability and mechanical strength, dominate the market of construction materials. However, cementing materials must be improved in their properties such as modulus, strength and ductility, decrease in permeability to chloride ions and liquids, and others (Chang, 2003; Ghernouti et al., 2009). The effect of foamed HDPE/montmorillonite (MMT) (Cloisite 15A) (0%–1.5%) in mortar compressive strength was studied by Ramírez-arreola et al. (2015). Siliceous sand was replaced by different amounts of foamed HDPE NCs (18%–100%). Foamed materials were obtained by extrusion of HDPE with azodicarbonamide as blowing agent and MMT as nucleation agent and also for improvement of the interfacial adhesion between the polymer foam and cement. The homogeneous size distribution and position of the bubbles resulting

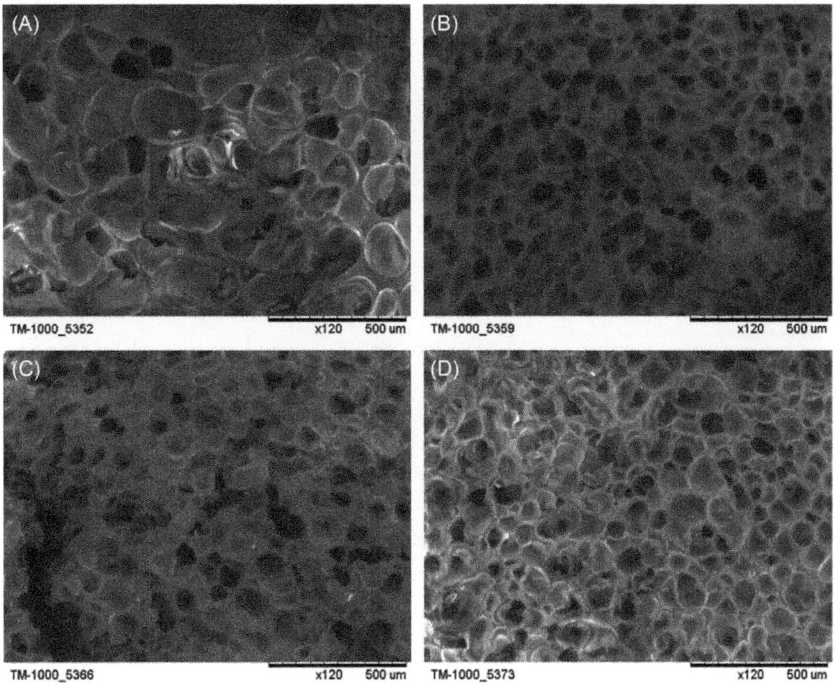

Figure 7.5 SEM micrographs of the foams to (A) 0%, (B) 0.5%, (C) 1.0%, and (D) 1.5% of nanoclay content.
Adapted from Ramírez-arreola, D.E., Sedano-de la Rosa, C., Haro-mares, N.B., Ramírez-morán, J.A., Pérez-fonseca, A.A., Robledo-ortíz, J.R. (2015). Compressive strength study of cement mortars lightened with foamed HDPE nanocomposites. Mater. Des., 74, 119–124, with kind permission of Elsevier.

from the use of a MMT was observed in scanning electron microscope (SEM) (Fig. 7.5).

The results showed a decrease in the resistance to compression as the amount of polymer added to the mortar was increased. They concluded that the mortars gained by this method can be used according to its mechanical resistance from structural concrete (18% of polymer) to building elements (100% replacement of sand by polymer). In 2014, hemp fiber reinforced PE (virgin and recycled) composite was manufactured by injection molding technique for varying fiber contents from 10% to 30% (Singh et al., 2014). Tensile strength and flexural strength of the composite decreased with the increase in hemp content from 10% to 30% in comparison with specimen made of 50–50 mixture of virgin and recycled HDPE. Samariha et al. (Hemmasi et al., 2011) studied the influence of nanoclay (0, 2, and 4 wt, phc) loading on the physical properties of NCs based on the recycled HDPE, bagasse flour and nanoclay using a counter-rotating twin-screw extruder and then injection molding. With the addition of nanoclay content, the water absorption decreased significantly and thickness swelling was lowered.

Figure 7.6 Flow chart of the experiment includes (A) materials preparation, (B) composite fabrication, and (C) composite characterization.
Adapted from Youssef, A.M., El-gendy, A., Kamel, S. (2015). Evaluation of corn husk fibers reinforced recycled low density polyethylene composites. Mater. Chem. Phys., 152, 26–33, with kind permission of Elsevier.

With the goal of disposal of environmental problematic agricultural and polymer waste, Youssef et al. (2015) designed NC sheets based on recycled low-density polyethylene (LDPE) as matrix and 5%, 10%, 15%, and 20% of corn husk fibers as filler. Fig. 7.6 shows a flow chart of their experiment. They proposed that the obtained composites materials can be used in packaging applications.

According to Table 7.1, increasing in fiber loading directed to increased moduli and tensile strength, whereas hardness was decreased. It is clear from these results that the maximum mechanical properties of recycled LDPE loading with corn husk fiber were attained with 10%–15% fiber loading and with increasing fiber loading the mechanical properties were slightly reduced.

7.2.2 NCs obtained from polypropylene (PP)

Filled PPs are commercial materials extensively used in large quantities in different application fields such as packaging, carpeting automotive industries, and other applications. PP has relatively low modulus, yield strength, and resistance to creeping which these disadvantages have limited its applications (Costantino et al., 2013; Majka et al., 2016; Madi, 2013). Manufacture of composite with high mechanical properties can be accomplished by adding of a clay or fiber as reinforcement in polymer matrix. Zaaba et al. (2016) investigated the performance of recycled

Table 7.1 Comparison of the influence of fiber loading rate on (A) Young's modulus, (B) tensile strength, and (C) hardness of resultant composites

Recycled LDPE	Fiber loading (%)	Young's modulus (MPa)	Load at maximum (N)	Tensile strength (MPa)	Hardness (KP mm^{-2})
100	00	166 ± 0.60	171 ± 0.40	13.5 ± 2.00	8.13 ± 0.18
95	5	327 ± 0.30	339 ± 1.50	24.7 ± 0.98	4.54 ± 0.60
90	10	456 ± 0.42	363 ± 0.62	24.3 ± 0.78	4.32 ± 0.20
85	15	456 ± 0.34	261 ± 0.74	17.9 ± 0.40	3.42 ± 2.00
80	20	410 ± 0.32	259 ± 0.80	17.7 ± 0.20	3.02 ± 0.30

Adapted from Youssef, A.M., El-gendy, A., Kamel, S. (2015). Evaluation of corn husk fibers reinforced recycled low density polyethylene composites. Mater. Chem. Phys., 152, 26–33, with kind permission of Elsevier.

PP/peanut shell powder (PSP) composites under the influence of chemical modification of PSP using polyvinyl alcohol (PVOH) (0%–40% by weight). Recycled PP/PSP modified with PVOH composites had better interfacial adhesion between the matrix and the filler than recycled PP/unmodified PSP composites as shown by SEM micrographs (Fig. 7.7).

Their results showed that recycled PP/PSP modified with PVOH composites had higher values of tensile strength, elongation at break, and tensile modulus, but lower water resistance than RPP/unmodified PSP composites. This can be due to the strong chemical interaction (hydrogen bonding) between the PSP and PVA, thus increasing the degree of adhesion between fillers and matrix. The effect of wood acetylation on the mechanical properties and creep resistance of wood/recycled PP composites (WRPCs) was studied by Wu et al. (Hung et al., 2016). In this study, the moisture content of WRPCs decreased with increasing the extent of wood acetylation. They concluded that the addition of acetylated wood particles in a recycled PP matrix can improve not only its tensile and flexural performances but also its creep resistance. So, they offer these composites as a high-performance alternative to conventional WPCs for building and construction applications. Samat et al. (Zulkifli et al., 2015) grafted maleic anhydride as a coupling agent on the PP to improve the interfacial adhesion between the microcrystalline cellulose (MCC) fibers and recycled PP matrix. A series of recycled PP/MCC composites, with or without maleic anhydride grafted PP (MAPP) were synthesized. They observed that without MAPP, an increase in MCC loadings caused degradation in the tensile strength of recycled PP composites. While all mechanical properties showed improvement after MAPP was added, especially in tensile modulus. Low cost wood–flour-recycled PP composite (WPC) was prepared with recycled PP nonwoven and wood flour by Ren et al. (2015). The effects of zinc borate (ZB), MMT, manganese dioxide (MnO_2), and stannic oxide (SnO_2) as fire retardant (FR) synergistic agents on the mechanical properties, thermal degradation, and flame retardant performance of WPC comprised of intumescent FR

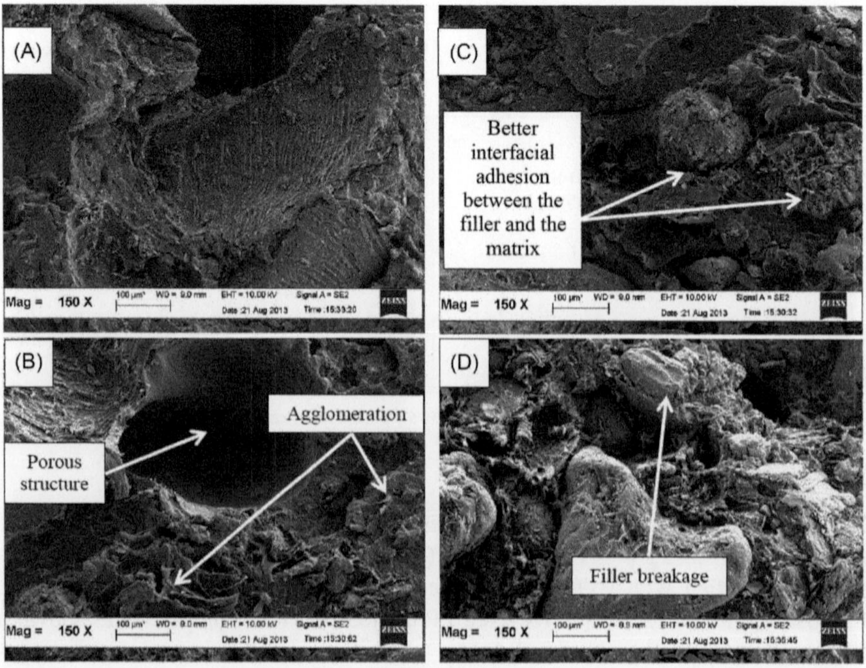

Figure 7.7 SEM micrographs of recycled PP/PSP composites with (A) 10%, (B) 40% filler loading by weight and SEM micrographs of recycled PP/PSP_{PVOH} composites with (C) 10%, and (D) 40% filler loading by weight.
Adapted from Zaaba, N.F., Ismail, H., Mariatti, M. (2016). Utilization of polyvinyl alcohol on properties of recycled polypropylene/peanut shell powder composites. Procedia Chem., 19, 763–769, with kind permission of Elsevier.

(IFR) were examined. The synergy between IFR and ZB or MnO_2 was greater than that of MMT or SnO_2. The addition of 5 wt% of ZB or MnO_2, MMT, and SnO_2 created fire retardant WPC which exhibited excellent fire retardancy. Ratanawilai's group investigated the effects of natural weathering on the physical and mechanical properties of PP/rubberwood flour (RWF) composites for various compositions, with different grades of plastic (virgin and recycled) and varied contents of wood flour and ultraviolet (UV) stabilizer (Homkhiew et al., 2014). Virgin PP had smaller relative changes of lightness and smaller relative loss of hardness, flexural strength (MOR) and modulus (MOE) than recycled PP, both in composites and as unfilled plastic. The recycled PP/RWF composites with 1 wt% UV stabilizer represented smaller changes in lightness and smaller relative loss of hardness, MOR, MOE, and maximum strain than without UV stabilizer, corresponding to reduced photodegradation of polymer. Martín-Alfonso et al. (2014) fabricated gel-like dispersions containing organo-bentonite (OBent)/recycled PP and mineral oil blends with the potential application as lubricating greases. Samples formulated with lower values of OBent/PP concentration ratio (0%, 25%,

50%, 75%, and 100%) yield gel-like formulations with suitable mechanical stabilities to be employed as lubricating greases. Wood derivatives like fiber, flour, and flakes are also agro wastes and have been widely used as reinforcement material in recycle thermoplastic composites (Almaadeed et al., 2012). Almaadeed et al. (2012) prepared recycled PP-based hybrid composites of date palm wood flour/glass fiber by different weight ratios of the two reinforcements. They mentioned that the prepared recycled hybrid composites will be more economical than pure glass fiber composites. Addition of as little 5 wt% glass fiber to wood flour reinforced recycled PP improved the tensile strength by about 18% relative to the wood flour reinforcement alone. They concluded that use of wood flour at a certain percentage as the filler for glass fiber reinforced hybrid composites will decrease the material cost without deteriorating the properties significantly.

7.2.3 NCs obtained from polyethylene terephthalate (PET)

PET is considered to be one of the most important thermoplastic polyesters which is widely used for various applications such as bottle containers, synthetic fibers, moldings, and sheets because of its excellent tensile and impact strength, low cost, clarity, processability, chemical resistance, and thermal stability (Mubarak, 2011; Park and Kim, 2014; Rodríguez-Uicab et al., 2013; Mallakpour and Behranvand, 2017; Mallakpour and Behranvand, 2016b). However, insufficient mechanical properties and thermal stability of PET have hindered its practical uses in a wide range of applications (Gorrasi et al., 2015). The universal amount of PET increased rapidly from about 14 to 60 million tons from the late 1990s to the year 2011. Congruently, equivalent amounts of PET waste are produced. The need of an suitable PET recycling is greater than ever because life cycle assessment studies has showed that reutilization of PET has a positive result on energy balance and the reduction of CO_2 emissions, for ecological reasons (Geyer et al., 2016; Reis et al., 2011). However, PET undergoes hydrolytic and thermal degradations during the recycling processes, which lead to a reduction in the molecular weight of the polymer and, therefore, its viscosity, melt strength, and mechanical properties. The property decline of recycled PET may be remunerated by the addition of reinforcing fillers and toughening modifiers (Makkam and Harnnarongchai, 2014). On the other hand, the use of chain extenders had been proposed as an effective method of increasing molecular weight (Makkam and Harnnarongchai, 2014; Duarte et al., 2016).

Mallakpour and Behranvand (2016c) recycled PET bottle waste through dissolution/reprecipitation method and obtained white and fine-recycled PET powders. Fig. 7.8 shows all stages for the preparation of PET powder.

Then, the effect of functionalized multiwalled carbon nanotubes (MWCNTs) with D-glucose (MWCNT-Gl) (1, 2, and 4 wt%) was examined on recycled PET properties. As can be seen from Fig. 7.9, there is hydrogen bonding interactions between hydroxyl groups on the surface of nanotubes and C = O groups in the recycled PET. Also, the aromatic rings in the recycled PET backbone have strong

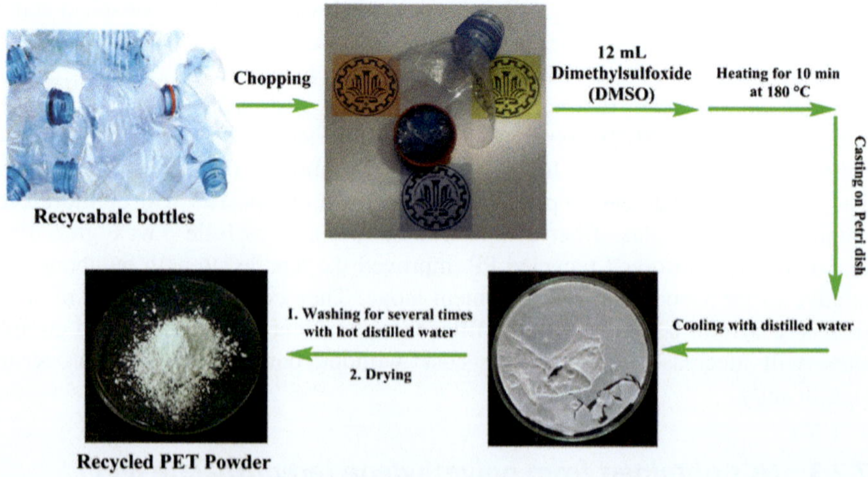

Figure 7.8 Schematic representation of PET recycling.
Adapted from Mallakpour, S., Behranvand, V. (2016). Manufacture and characterization of nanocomposite materials obtained from incorporation of d-glucose functionalized MWCNTs into the recycled poly(ethylene terephthalate). Des. Monomers Polym., 19 (4), 283–289.

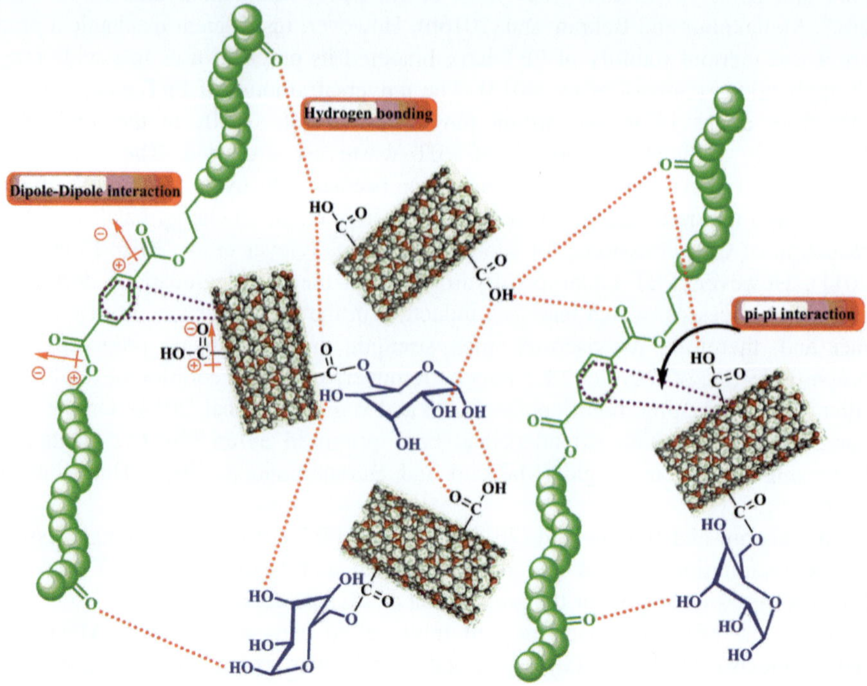

Figure 7.9 Possible interactions between recycled PET and MWCNTs-Gl.
Adapted from Mallakpour, S., Behranvand, V. (2016). Manufacture and characterization of nanocomposite materials obtained from incorporation of d-glucose functionalized MWCNTs into the recycled poly(ethylene terephthalate). Des. Monomers Polym., 19 (4), 283–289.

interactions with CNT surfaces through π−π interactions. The obtained NCs of 2 and 4 wt% exhibited higher char residue than recycled PET. The interaction of nanotubes with polar polymer matrix could weaken intermolecular interactions between the polymer chains and caused decreasing in melting point of the composites.

In another research work, Mallakpour and Javadpour (2016a) modified surface of ZnO NPs with biocompatible citric acid and then different percentages of surface-treated NPs (1, 3, and 5 wt%) were incorporated into the as-mentioned recycled PET. It was clear from transmission electron microscopy (TEM) images of NC 8 wt% (Fig. 7.10), that ZnO−citric acid nanoparticles (NPs) are rather well dispersed in the recycled PET matrix and the mean size of NPs was 21 nm. The obtained data by thermal analysis showed that prepared NCs can be categorized as self-extinguished materials.

Also, this group modified ZnO NPs with optically active diacid containing alanine amino acid and incorporated them into the recycled PET (Mallakpour and Javadpour, 2016b). The obtained data showed not only higher UV−Vis absorption, but also NCs exhibited higher char yield compared to the pure PET. In addition, burning behavior observations showed that the pure recycled PET burned fast after

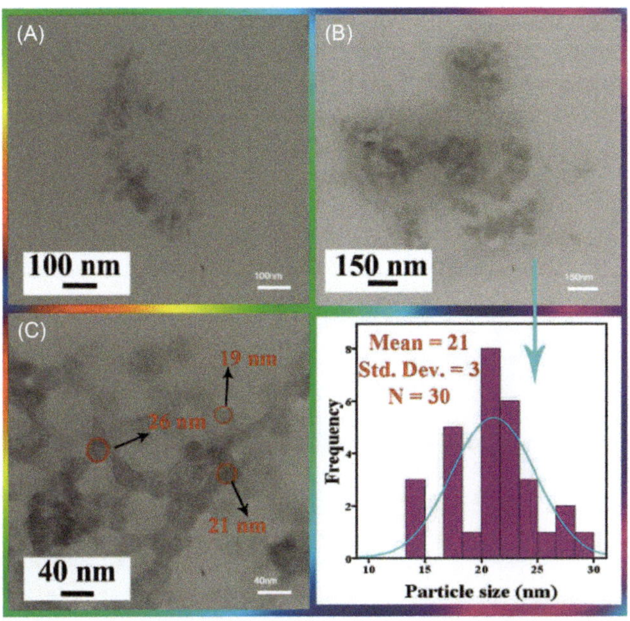

Figure 7.10 TEM images of PET/ZnO−citric acid NC 8 wt% at different magnifications and related histogram.
Adapted from Mallakpour, S., Javadpour, M. (2016a). The potential use of recycled PET bottle in nanocomposites manufacturing with modified ZnO nanoparticles capped with citric acid: preparation, thermal, and morphological characterization. RSC Adv., 6 (18), 15039−15047.

ignition, while modified ZnO NPs in matrix were resulted in delayed ignition, retarded rate of burning, and reduced smoke emission. Changizi and Haddad (2015) investigated the effects of recycled polyester fiber made from bottles waste, nano-SiO_2 and mixture of recycled PET fiber and nano-SiO_2 on the shear strength, and elastic modulus properties of clayey soil. The results showed that adding of nano-SiO_2 was more important than the addition of recycled PET fiber in increasing the stiffness of clay. On the other hand, the simultaneous use of recycled PET fiber and nano-SiO_2 has a significant influence on increasing the elastic modulus of clay. They observed narrower and shorter cracks on the surface of compacted soil, nano-SiO_2, and fiber mixture. They explained that the decrease in cracks could be ascribed to the fact that fiber acts as bridges between soil particles and cause matrix reinforcement. Consequently, the recycled material provides strength to crack propagation and keeps load transfer during tension.

Mats of ultrathin and nanoscale fibers were prepared via room temperature electrospinning of lignocellulosic sisal fibers (S) and recycled PET trifluoroacetic acid solutions by Ruvolo-Filho et al. (Santos et al., 2015). Materials obtained from PET, such as electrospun mats, due to its chemical nature, tend to possess a highly hydrophobic character which causes a disadvantage in certain applications, for example, in biological systems. The presence of high content of cellulose in lignocellulosic sisal fiber makes it an excellent reinforcing agent in polymeric composites (de Oliveira Santos et al., 2014). As can be observed in Fig. 7.11, the incorporation of sisal fibers into the PET created super hydrophilic surfaces, mainly seen for S/PET0.40, in which the water droplet was instantaneously absorbed upon contacting the surface. This change is because of the presence of sisal fibers polar groups on the surfaces of the mats.

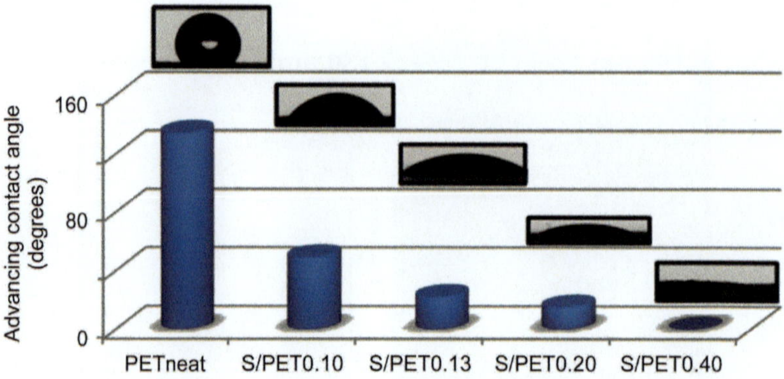

Figure 7.11 Advancing contact angle for neat PET and S/PET electrospun mats and respective snapshots taken after the first second of contact between a water droplet and their surfaces.

Adapted from de Oliveira Santos, R.P., Oliveira Castro, D., Ruvolo-Filho, A.C., Frollini, E. (2014). Processing and thermal properties of composites based on recycled PET, sisal fibers, and renewable plasticizers. J. Appl. Polym. Sci., 40386, 1–13, with kind permission of John Wiley and Sons.

Their results showed that the presence of sisal in the mat decreased the mobility of the segments of the PET chains and caused an increase in T_g by up to 20°C. They mentioned that intermolecular interactions between PET and the components of sisal, hindered the rotational movements of the covalent bonds present in the segments of polymer chains. Some interactions between these components were showed in Fig. 7.12.

Recycled PET NCs with various amounts of MMT clay were fabricated by using twin screw extrusion by Meri et al. (2014). Introduction of MMT in the recycled PET caused considerable increase of the modulus of elasticity *E*. Fig. 7.13 shows tensile stress–strain characteristics of the recycled PET NCs as functions of MMT weight contents W_f. At $W_f = 5\%$, value of *E* of the composite is 18% higher than that of neat recycled PET. Decrease in the increment of *E* of the composite along with increasing MMT content is usually caused by decreased exfoliation extent of layered nanofiller. Final deformation of the composites slowly decreased (from 375% at $W_f = 0\%$ to 280% at $W_f = 5\%$) along with growing MMT content, although materials maintained ductility in the entire MMT concentration range. Maximum ultimate tensile strength σ_b was obtained in 1 wt% MMT loading.

Generally, fillers materials such as carbon black, silica, or clay minerals are used in rubber composites to not only improve the mechanical properties but also reduce material costs, improve processing and decrease the weight (Rattanasom et al., 2007; Nabil and Ismail, 2013). From one hand, due to various problems such as pollution, dark color, and use of petroleum feedstock in synthesis of carbon black, researchers have been focused on the development of white fillers such as silica, calcium carbonate and clay. On the other hand, rubber is nonpolar polymer so poor interaction will be created between these fillers and the polymer matrix which leads

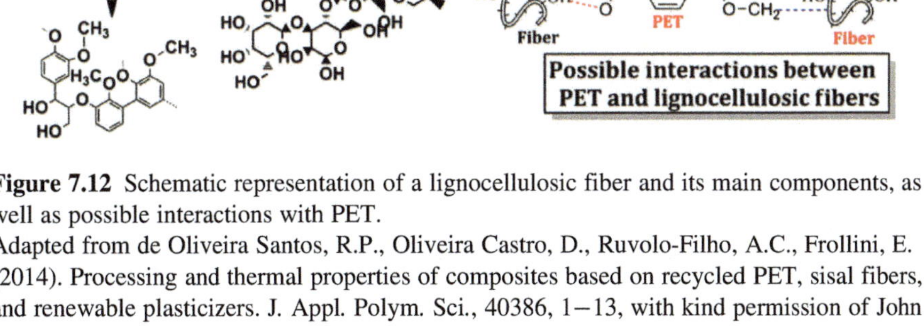

Figure 7.12 Schematic representation of a lignocellulosic fiber and its main components, as well as possible interactions with PET.
Adapted from de Oliveira Santos, R.P., Oliveira Castro, D., Ruvolo-Filho, A.C., Frollini, E. (2014). Processing and thermal properties of composites based on recycled PET, sisal fibers, and renewable plasticizers. J. Appl. Polym. Sci., 40386, 1–13, with kind permission of John Wiley and Sons.

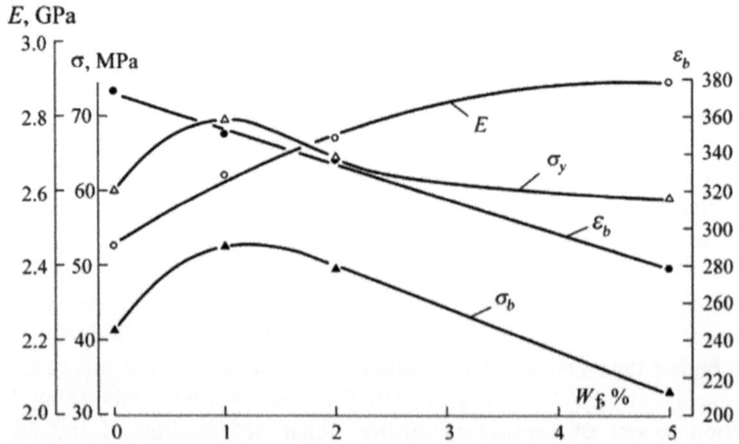

Figure 7.13 Yield strength σ_y, ultimate tensile strength σ_b, ultimate deformation ε_b, and elastic modulus E of recycled PET/MMT composites as functions of weight content of MMT W_f. Adapted from Meri, R.M., Zicans, J., Maksimovs, R., Ivanova, T., Kalnins, M., Berzina, R., et al. (2014). Elasticity and long-term behavior of recycled polyethylene terephthalate (rPET)/montmorillonite (MMT) composites. Composite Struct., 111, 453–458, with kind permission of Elsevier.

to the formation of aggregates. It needs the addition of coupling agents which complicates the processing method and consequently increases the production cost (Munusamy et al., 2009; Nabil and Ismail, 2013). In the lights of these facts, Nabil and Ismail (2013) partially replaced recycled PET with halloysite nanotubes (HNTs) as fillers in natural rubber composites. The addition of HNTs had noteworthy improvement on many properties, such as tensile modulus, tensile strength, elongation at break, fatigue life, and even better thermal stability. They concluded that recycled PET/HNTs could be employed as alternative filler in the rubber industry, and at special weight, recycled PET/HNTs can be used in applications where high strength is not essential.

7.2.4 NCs obtained from polystyrene (PS)

PS, an amorphous, thermoplastic material, is often selected as a host matrix because of its optically clarity. These plastic materials usually used in disposable cutlery, plasticmodels, CD and DVD cases, and smoke-detector housings. PS is very multipurpose but not biodegradable and its recycling is not easy (Jeeju et al., 2011).

Khan et al. (2014) encapsulated MWCNTs and NiZn ferrite ($Ni_{0.6}Zn_{0.4}Fe_2O_4$) NPs with recycled PS obtained from a local restaurant to produce NC fibers by means of electrospinning. The thermal conductivity results showed higher thermal conductivity in MWCNT-based NC fibers which is mainly due to the higher thermal conductivity of MWCNTs (1600 W m^{-1} K^{-1}) compared to NiZn ferrite NPs (6.3 W m^{-1} K^{-1}) (Fig. 7.14).

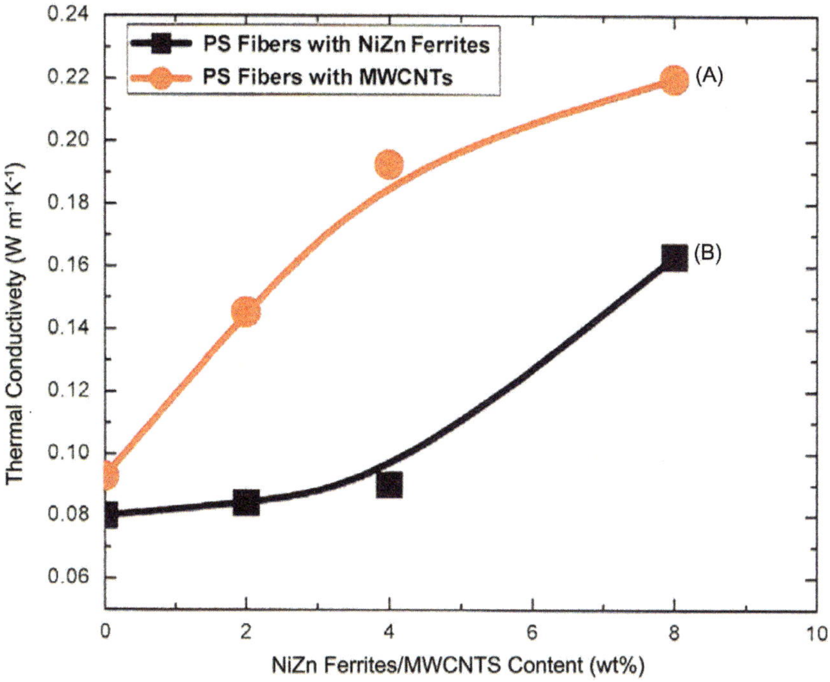

Figure 7.14 Thermal conductivity values of PS NC fibers as function of MWCNTs (A) and NiZn ferrite concentrations (B).
Adapted from Khan, W.S., Asmatulu, R., Davuluri, S., Dandin, V.K. (2014). Improving the economic values of the recycled plastics using nanotechnology associated studies. J. Mater. Sci. Technol., 30(9), 854–859, with kind permission of Elsevier.

Water contact angle values of the MWCNT- and NiZn-ferrite-based NC fibers were represented in Table 7.2. MWCNTs could increase both hydrophobicity and surface roughness, only with a small addition in the PS NC fibers compared to the ferrite-based NCs. While ferrite-based NC fibers got superhydrophobic in highest weight percentages. Overall, the authors proposed recycled plastics in different forms of nanomaterials have potential industrial applications, such as filtration, transportation, construction, energy, defense, and so on.

Mechanical properties of WPC obtained from wood flour, recycled PS, and nanoclay were investigated by Nemati et al. (2013). The wood flour was mixed with the recycled PS at three weight ratios of 40, 50, and 60%, whereas the nanoclay was used in 0, 3, and 5 phc. The introduction of wood fibers is a method employed to improve impact strength of PS due to some of the potential advantages it has. The obtained results showed that the tensile strength and flexural strength were enhanced by raising nanoclay content in the composites. Hernández-Rivera et al. (Herrera-Sandoval et al., 2013) reinforced recycled PS with titanium dioxide (TiO_2) NPs. The TEM images showed that TiO_2 with average particle sizes were

Table 7.2 **Water contact angle values of PS NC fibers**

MWCNTs (wt%)	Water contact angle (degree.)	Ferrites (wt%)	Water contact angle (degree.)
0	135	0	135.0
1	152.9	7.5	142.2
2	154.3	15	146.7
4	156.9	30	150.7

Adapted from Khan, W.S., Asmatulu, R., Davuluri, S., Dandin, V.K. (2014). Improving the economic values of the recycled plastics using nanotechnology associated studies. J. Mater. Sci. Technol., 30(9), 854–859, with kind permission of Elsevier.

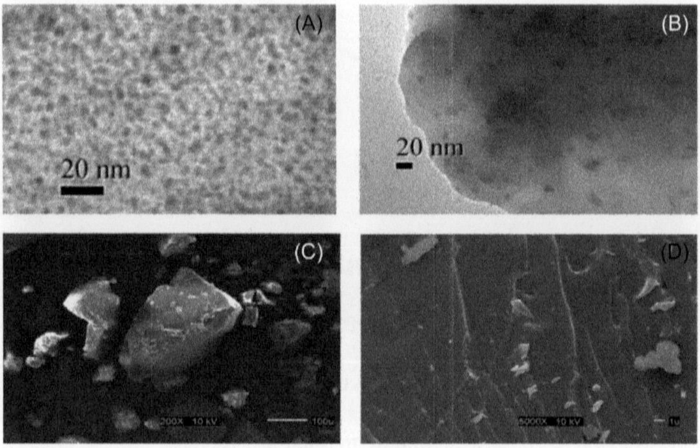

Figure 7.15 SEM and TEM images of the recycled PS/TiO_2 NC. Micrographs of the TEM images (A) and (B) and the SEM images, (C) and (D).
Adapted from Herrera-Sandoval, G.M., Baez-Angarita, D.B., Correa-Torres, S.N., Primera-Pedrozo, O.M., Hernández-Rivera, S.P. (2013). Novel EPS/TiO_2 nanocomposite prepared from recycled polystyrene. Mater. Sci. Appl., 4, 179–185.

between 5 and 15 nm occurred in the form of islands within the PS matrix and was agglomerated in some areas (Fig. 7.15A and B). The SEM images (Fig. 7.15C and D) indicated a granular morphology with a size of less than 100 μm. In Fig. 7.15D, the SEM image has focused on one recycled PS/TiO_2 grain. In this case, a smooth particle surface was observed.

The representative discoloration for MB with the PS/TiO_2 NC and TiO_2 photocatalysts was shown in Fig. 7.16. The best discoloration behavior was observed in the PS/TiO_2 NC. The percent discoloration of the MB solutions reached 98%. Poor discoloration of the MB solutions was observed when the photocatalysts were not added to the solutions (control runs), which confirmed the photocatalytic activity of the fabricated NC.

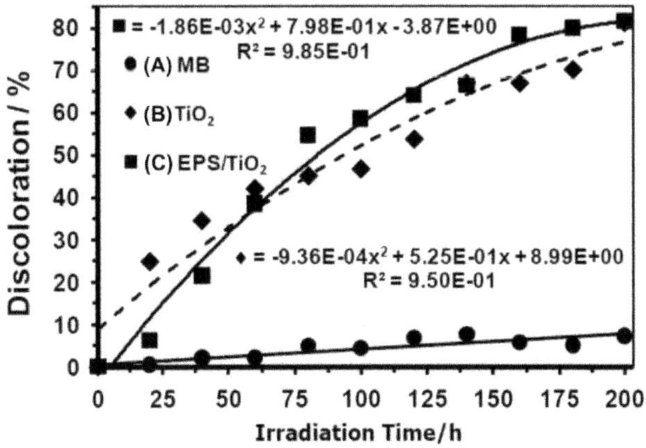

Figure 7.16 Discoloration of the MB solutions under UV light irradiation. (A) MB with recycled PS/TiO_2; (B) MB with TiO_2; and (C) MB without TiO_2 or recycled PS/TiO_2 (control) (Herrera-Sandoval et al., 2013).

7.2.5 NCs obtained from other recycled polymers

In 2015, Salleh et al. (Mohamed et al., 2015) designed an environmental-friendly recycled cellulose/N-TiO_2 NC thin film as a green portable photocatalyst. In this work, cellulose was extracted from recycled newspaper. There was good compatibility between recycled cellulose matrix and N-doped TiO_2 nanorods due to a strong interfacial interaction between polymer chain and N-doped TiO_2 nanorods which were confirmed from highly dispersed N-doped TiO_2 nanorods that are clear in TEM image (Fig. 7.17E). Well-dispersed NPs are essential for the better photocatalytic activities.

The recycled cellulose/N-TiO_2 NC thin film exhibited remarkable photocatalytic activity under UV and visible irradiation for the degradation of methylene blue solution, with degradation percentage of 96% and 78.8%, respectively. Nassar and Youssef (2012) coated the recycled carton paper by the PS NCs containing different concentration of silver NPs, namely 2%, 4%, 6%, and 8% based on PS. The mechanical properties of recycled carton sheet coated by PS/Ag NCs were increased by the coating process. The results showed that the incorporation of silver NPs leads to easier water vapor migration through carton sheet fibers. Also, the obtained product exhibited antibacterial activity toward different bacterial strains.

Polylactic acid (PLA) is a rigid thermoplastic and highly versatile biopolymer that has a special attention as a matrix in composite materials. PLA has been highlighted because it is derived from a renewable resource such as corn (Sujaritjun et al., 2013). However, there are some drawbacks include brittleness, low thermal resistance, and slow crystallization rate for PLA, which limit its applications. Wu et al., to overwhelm these disadvantages and reduce fossil fuel consumption, incorporated bamboo fibers into the recycled PLA to obtain fully bio-based composites (Yang et al., 2015). Although with increasing bamboo fiber loading, moisture content increased but the modulus of rupture and modulus of elasticity of bamboo–fiber-reinforced-recycled

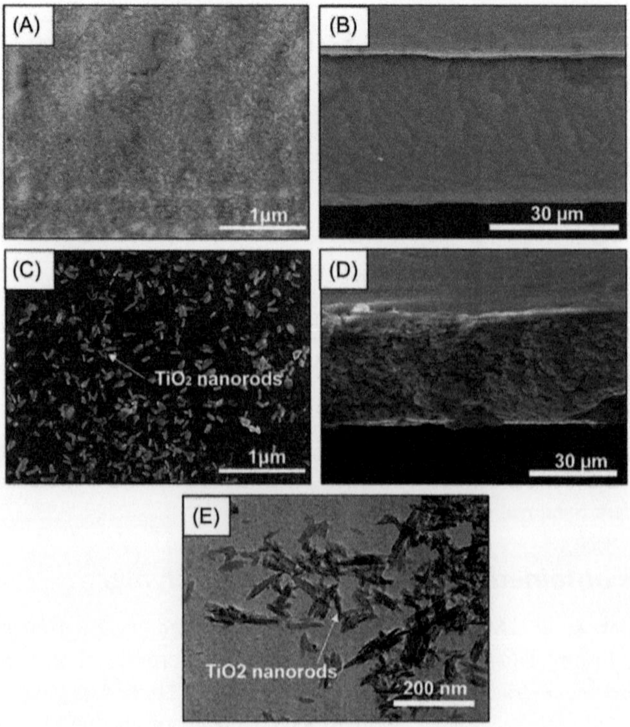

Figure 7.17 FE-SEM images of (A, B) recycled cellulose and (C, D) recycled cellulose/ N-TiO_2, (A) and (C) is surface, (B) and (D) is cross-sectional, (E) TEM image of N-doped TiO_2 nanorods.
Mohamed, M.A., Salleh, W.N.W., Jaafar, J., Ismail, A.F., Mutalib, M.A., Jamil, S.M. (2015). Incorporation of N-doped TiO_2 nanorods in regenerated cellulose thin films fabricated from recycled newspaper as a green portable photocatalyst. Carbohydr. Polym., 133, 429–437, with kind permission of Elsevier.

PLA composites were improved with increasing fiber loading up to 60 wt%. In 2015, the impact of recycling on gloss, color, and surface free energy properties of PLA neat polymer and its composite with MMT was studied by Leluk et al. (2015). The pure polymer and its NC were extruded once and press molded to provide the reference samples (PLA-0 cycle and PLA-MMT-0 cycle). PLA and PLA-MMT composite were recycled for 1, 5, and 10 times by means of extrusion. Recycling caused the most significant changes in PLA after the first and fifth cycles, while for PLA-MMT composite after the fifth and tenth cycles. For the PLA sample, the surface free energy, color, and gloss subsequently decreased because of polymer chain scission. A reverse result was seen for PLA–MMT, which was explained by the formation of mirror-like structures in the exfoliated or partially exfoliated composite.

Blending of different class of polymer was applied to manufacture new materials with unique properties. Polymer blending is a low-cost and rapid way to achieve desired properties from existing polymers. However, blending causes marginal improvement in physical properties which were still insufficient for engineering applications. Therefore, for the improvement in the strength and stiffness of polymer materials, different types of organic and inorganic fillers were used. It was observed that strength and stiffness of thermosetting polymer incorporated with long fibers is comparable to metals at a fraction of their weight (Hameed, 2012; Choudhary and Gupta, 2001). Khanam and Almaadeed (2014) investigated the study and preparation of date palm fiber reinforced recycled polymer ternary blends of (1) recycled LDPE, (2) recycled HDPE, and (3) recycled PP. 1% maleic anhydride was effective for chemical modification of the composites and endorsed dispersion and interfacial adhesion between the blend and the fibers. This statement was supported by data presenting increases in thermal stability, hardness, tensile strength, tensile modulus, and morphology of the composites. Also, the composites had good resistance to acids and alkalis. Martín-Alfonso et al. (2013) examined the effect of the PP concentration ratio (Wi) on the rheology, thermal, and some lubricant performance properties of polymer/oil blends to evaluate its applicability as a thickening agent for lubricating greases. Thermal degradations of these systems showed slightly higher decomposition temperatures than standard lubricating greases, and poorer mechanical stability than conventional lithium lubricating greases regardless of the polymer used. Gel-like dispersions containing higher recycled/amorphous PP concentration ratios ($Wi > 0.5$) exhibited an appropriate consistency for use as lubricating greases. Novel recycled PE/ground tire rubber/bitumen blends for use in roofing applications were designed by Navarro et al. (2010). Bitumen can be defined as a dark brown to black cementitious material. They studied the effect of recycled polymers, such as ground tire rubber (GTR) and recycled PE, on the thermal and rheological properties of modified bitumen blends. Remarkable changes in the rheological response of these materials were obtained by bitumen modification with recycled PE or GTR. However, the modification capabilities of recycled PE and GTR were different depending on thermal conditions. Results for binary blends showed that GTR was more effective in the low in-service temperature range, since the reduction of the mechanical glass transition temperature was more remarkable. By contrast, recycled PE improved the thermomechanical characteristics of the modified binder at high in-service temperatures. In ternary blends of recycled PE/GTR/bitumen blends increasing in deformation resistance at high temperatures due to the presence of recycled PE, better flexibility at low temperatures corresponding to the GTR addition and enhancing in fatigue resistance were observed. From an environmental point of view, recycled PE/GTR/bitumen blends, with 25 wt% of both recycled polymers, appeared to be an attractive alternative to roofing and waterproofing membranes made from virgin polymers, exhibiting similar or even improved properties.

7.3 Summary

The results for recycled polymer NCs which were examined in this chapter with regards to their properties and applications are summarized in Table 7.3.

Table 7.3 **Summary of the properties and applications of the recycled polymer NCs**

Composite type	Properties	Applications	References
Recycled HDPE/ graphite	Good chemical stability in nitrogen environment–high thermal conductivity	Medium temperature thermal energy storage application	Yang et al. (2016)
Recycled HDPE/ wood sawdust	Lowering of friction	–	Brostow et al. (2016)
Recycled HDPE/ waste wood	Decreasing of tensile strength- increasing of modulus- increasing water absorption	–	Sommerhuber et al. (2015)
Recycled HDPE/ Cloisite 15A	Decreasing of the resistance to compression- reducing mortar density	Lightweight cement mortars for the use in building	Ramírez-Arreola et al. (2015)
Recycled HDPE/ hemp fiber	Decreasing in tensile strength and flexural strength of the composite with increasing of hemp content	–	Singh et al. (2014)
Recycled HDPE/ bagasse flour and nanoclay	Decreasing in thickness swelling and water absorption	-	Hemmasi et al. (2011)
Recycled LDPE/corn husk fibers	Increasing of moduli and tensile strength- decreasing of hardness	Packaging applications	Youssef et al. (2015)
Recycled PP/peanut shell powder	Increasing of tensile strength, elongation at break, and tensile modulus- decreasing of water resistance	–	Zaaba et al. (2016)
Recycled PP/wood	Increasing of flexural and tensile strength and creep resistance	Building and construction applications	Hung et al. (2016)
Recycled PP/ cellulose fibers	Improving of all mechanical properties especially in tensile modulus	-	Zulkifli et al. (2015)

• Recycled PP/ wood flour/ZB • Recycled PP/ wood flour/MMT • Recycled PP/ wood flour/MnO_2 • Recycled PP/ wood flour/SnO_2	• Excellent fire retardancy • Excellent fire retardancy • Excellent fire retardancy • Excellent fire retardancy	–	Ren et al. (2015)
PP/rubberwood flour	Slower photobleaching of the wood component- improvement of the degradation resistance of plastic composites- decreasing of the relative loss of hardness- insignificant decreases of maximum strain- increasing of swelling	Composite materials for applying in exterior environments	Homkhiew et al. (2014)
Recycled PP/organo-bentonite	Decreasing of friction coefficient-lowering of viscoelastic functions	Lubricating greases	Martín-Alfonso et al. (2014)
Recycled/date palm wood flour/glass fiber	Increasing of tensile strength and Young's modulus	–	Almaadeed et al. (2012)
Recycled PET/ MWCNT	Higher thermal stability- decreasing of melting point	Potential application in water treatment	Mallakpour and Behranvand (2016)
Recycled PET/nano-ZnO	Increasing of UV-Visible absorption and crystallinity-	Potential application in packaging	Mallakpour and Javadpour (2016a,b)
Recycled PET fiber/ nano-SiO_2/soil	Increasing of the shear strength and elastic modulus of soil specimens	Foundation and roadbed (in civil engineering)	Changizi and Haddad (2015)

(*Continued*)

Table 7.3 (Continued)

Composite type	Properties	Applications	References
Recycled PET/ lignocellulosic sisal fibers	Increasing of T_g- increasing of hydrophilicity	–	Santos et al. (2015)
Recycled PET/MMT	Increasing of the modulus of elasticity E-decreasing of deformation	-	Meri et al. (2014)
Recycled PET/ halloysite nanotubes/ natural rubber	Improving of tensile modulus, tensile strength, elongation at break, fatigue life and thermal stability	-	Nabil and Ismail (2013)
• Recycled PS/ MWCNTs • Recycled PS/NiZn ferrite ($Ni_{0.6}Zn_{0.4}Fe_2O_4$) NPs	• MWCNTs could increase both hydrophobicity and surface roughness, only with a small addition in the PS NC fibers compared to the ferrite-based NCs • The thermal conductivity results showed higher thermal conductivity in MWCNT-based NC fibers compared to NiZn ferrite NPs	Potential industrial applications, such as filtration, transportation, construction, energy, defense, etc.	Khan et al. (2014)
Recycled PS/Wood flour/Nanoclay	Enhancing of tensile and flexural strength	–	Nemati et al. (2013)
Recycled PS/TiO_2	High discoloration efficiency for aqueous methylene blue	Photocatalyst	Herrera-Sandoval et al. (2013)

Recycled cellulose/N-dopedTiO_2	Remarkable photocatalytic activity under UV and visible irradiation for the degradation of methylene blue	Photocatalyst	Mohamed et al. (2015)
Recycled carton paper/PS/Ag	Increasing of mechanical properties—improving of water vapor permeability—antibacterial activity toward different bacterial strains	Packaging	Nassar and Youssef (2012)
Recycled PLA/bamboo fibers	Improving of thermal resistance, creep resistance and flexural properties	–	Yang et al. (2015)
Recycled LDPE/recycled HDPE/recycled PP/date palm fiber	Increasing of thermal stability, hardness, tensile strength and tensile modulus-good resistance to acids and alkalis	Building and construction materials, such as door and window frames, decking, railings for parapet walls and furniture sections, as well as door panels, setbacks, headlines, package trays, dashboards and trunk liners in the automobile industry	Khanam and Almaadeed (2014)
Amorphous/recycled PP	Higher decomposition temperatures than standard lubricating greases- poorer mechanical stability than conventional lithium lubricating greases	Lubricating greases	Martín-Alfonso et al. (2013)
recycled PE/ground tire rubber/bitumen	Increasing in deformation resistance at high temperatures—better flexibility at low temperatures—enhancing in fatigue resistance	Roofing and waterproofing membranes	Navarro et al. (2010)

7.4 Conclusions and challenges

The effective recycle of polymeric materials is an important issue and a challenge by considering sustainable development in modern society nowadays. The aim in recycling is to reduce the amount of waste by recirculation of raw materials and to improve the material utilization. Though, the major problem in this field includes the polymer structure degradation in reprocessing which cause much poorer properties. The addition of other components to the waste polymers seems to be the simplest way for reusing the recycled polymers. More excellent feature for improvement of all mechanical, thermal and barrier properties is presented by nanofillers. As it was summarized in Table 7.3, addition of different nanofillers to the recycled polymers, resulted not only better properties but also obtained novel NCs with various applications. Besides, in some cases, recycled polymers were employed as filler to produce lightweight composites. In other cases, blends of recycled polymers were prepared to obtain materials with potential applications in building or automobile industry. The key point in the NC fabrication is increasing of interfacial adhesion between the reinforcing agent and the polymer matrix. It was showed that an appropriate compatibilizer can significantly enhance the interfacial interactions and produce a NC with efficient characteristics.

Even though recycling is an environmentally attractive solution, but only a minor portion of plastics is recycled, and most of these wastes end up in municipal burial sites. In addition, it appears that the amounts of landfill sites are also limited and polymer containing toxic materials release to the environment. A promising development is the use of bio-based polymers rather than petroleum-based polymers.

Acknowledgments

We gratefully acknowledge the financial support from the Research Affairs Division Isfahan University of Technology (IUT) Isfahan and the National Elite Foundation (NEF).

References

Abduli, M.A., 2007. Solid Waste Management: Issues and Challenges in Asia. Part I. Environmental Management Centre, Mumbai (Ed.), *4. Islamic Republic Of Iran*. Tokyo: Asian Productivity Organization-1-2-10 Hirakawacho, Chiyoda-ku, Tokyo 102-0093, Japan.

Almaadeed, M.A., Kahraman, R., Khanam, P.N., Madi, N., 2012. Date palm wood flour/glass fibre reinforced hybrid composites of recycled polypropylene: mechanical and thermal properties. Mater. Des. 42, 289–294.

Al-sabagh, A.M., Yehia, F.Z., Eshaq, Gh, Rabie, A.M., Elmetwally, A.E., 2016. Greener routes for recycling of polyethylene terephthalate. Egypt. J. Petrol. 25 (1), 53–64.

Alzerreca, M., Paris, M., Boyron, O., Orditz, D., Louarn, G., Correc, O., 2015. Mechanical properties and molecular structures of virgin and recycled HDPE polymers used in gravity sewer systems. Polym. Test. 46, 1–8.
Anandhan, S., Bandyopadhyay, S., 2011. Polymer nanocomposites: from synthesis to applications. In: Cuppoletti, J. (Ed.), Nanocomposites and Polymers with Analytical Methods. InTech Europe, Slavka Krautzeka, pp. 1–28.
Brostow, W., Datashvili, T., Jiang, P., Miller, H., 2016. Recycled HDPE reinforced with sol–gel silica modified wood sawdust. Eur. Polym. J. 76, 28–39.
Butylina, S., Martikka, O., Kärki, T., 2011. Properties of wood fibre–polypropylene composites: effect of wood fibre source. Appl. Compos. Mater. 18, 101–111.
Chang, D.D.L., 2003. Cement-matrix structural nanocomposites. Metals Mater. Int. 10, 55–67.
Changizi, F., Haddad, A., 2015. Strength properties of soft clay treated with mixture of nano-SiO_2 and recycled polyester fiber. J. Rock Mech. Geotech. Eng. 7, 367–378.
Choudhary, V., Gupta, A., 2001. Polymer/carbon nanotube nanocomposites. In: Yellampalli, S. (Ed.), *Carbon Nanotubes-Polymer Nanocomposites*. InTech Europe, Slavka Krautzeka, pp. 65–90.
Costantino, A., Pettarin, V., Viana, J., Pontes, A., Pouzada, A., Frontini, P., 2013. Polypropylene/clay nanocomposites produced by shear controlled orientation in injection moulding: deformation and fracture properties. J. Mech. Eng. 59, 697–704.
de Oliveira Santos, R.P., Oliveira Castro, D., Ruvolo-Filho, A.C., Frollini, E., 2014. Processing and thermal properties of composites based on recycled PET, sisal fibers, and renewable plasticizers. J. Appl. Polym. Sci. 40386, 1–13.
Duarte, I.S., Tavares, A.A., Lima, P.S., Andrade, D.L.A.C.S., Carvalho, L.H., Canedo, E.L., et al., 2016. Chain extension of virgin and recycled poly(ethylene terephthalate): effect of processing conditions and reprocessing. Polym. Degrad. Stab. 124, 26–34.
Geyer, B., Lorenz, G., Kandelbauer, A., 2016. Recycling of poly(ethylene terephthalate)—a review focusing on chemical methods. eXPRESS Polym. Lett. 10 (7), 559–586.
Ghernouti, Y., Rabehi, B., Safi, B., Chaid, R., 2009. Use of recycled plastic bag waste in the concrete. J. Int. Sci. Publ.: Mater. Methods Technol. 8, 480–487.
Gorrasi, G., Milone, C., Piperopoulos, E., Pantani, R., 2015. Preparation, processing and analysis of physical properties of calcium. Compos. B. 81, 44–52.
Gürüa, M., Cubuk, M.K., Arslan, D., Farzanian, S.A., Bilici, I., 2014. An approach to the usage of polyethylene terephthalate (PET) waste as roadway pavement material. J. Hazard. Mater. 279, 302–310.
Hamad, K., Kaseem, M., Deri, F., 2013. Recycling of waste from polymer materials: an overview of the recent works. Polym. Degrad. Stab. 98, 2801–2812.
Hameed, T., 2012. Study of Reactions between Highly Functionalized Low Molecular Weight Polyethylene and Polyamines to Produce Thermoset Materials. Open Access Dissertations and Teses. Paper 6881.
Hemmasi, A.H., Ghasemi, I., Bazyar, B., Samariha, A., 2011. Influence of nanoclay on the physical properties of recycled high-density polyethylene/bagasse nanocomposite. Middle-East J. Sci. Res. 8 (3), 648–651.
Herrera-Sandoval, G.M., Baez-Angarita, D.B., Correa-Torres, S.N., Primera-Pedrozo, O.M., Hernández-Rivera, S.P., 2013. Novel EPS/TiO_2 nanocomposite prepared from recycled polystyrene. Mater. Sci. Appl. 4, 179–185.
Homkhiew, C., Ratanawilai, T., Thongruang, W., 2014. Effects of natural weathering on the properties of recycled polypropylene composites reinforced with rubberwood flour. Ind. Crops Prod. 56, 52–59.

Hopewell, J., Dvorak, R., Kosior, E., 2009. Plastics recycling: challenges and opportunities. Philos. Trans. R. Soc. 364, 2115–2126, http://recycling.omicsgroup.com/events-list/solid-waste-management.

Hung, K.C., Wu, T.L., Chen, Y.L., Wu, J., 2016. Assessing the effect of wood acetylation on mechanical properties and extended creep behavior of wood/recycled-polypropylene composites. Constr. Build. Mater. 108, 139–145.

Jeeju, P.P., Sajimol, A.M., Sreevalsa, V.G., Varma, S.J., Jayalekshmi, S., 2011. Size-dependent optical properties of transparent, spin-coated polystyrene/ZnO nanocomposite films. Polym. Int. 60, 1263–1268.

Jeon, I.Y., Baek, J.B., 2010. Nanocomposites derived from polymers and inorganic nanoparticles. Materials. 3, 3654–3674.

Khan, W.S., Asmatulu, R., Davuluri, S., Dandin, V.K., 2014. Improving the economic values of the recycled plastics using nanotechnology associated studies. J. Mater. Sci. Technol. 30 (9), 854–859.

Khanam, P.N., Almaadeed, M.A., 2014. Improvement of ternary recycled polymer blend reinforced with date palm fibre. Mater. Des. 60, 532–539.

Leluk, K., Iwanczuk, A., Kozłowski, M., 2015. Surface characterization of recycled PLA and its nanocomposite. Cellulose Chem. Technol. 49 (7–8), 653–657.

Li, S.C., Li, Y.N., 2010. Mechanical and antibacterial properties of modified nano-ZnO/high-density polyethylene composite films with a low doped content of nano-ZnO. J. Appl. Polym. Sci. 116, 2965–2969.

Madi, N.K., 2013. Thermal and mechanical properties of injection molded recycled high density polyethylene blends with virgin isotactic polypropylene. Mater. Des. 46, 435–441.

Majka, T.M., Pielichowski, K., 2011. Application of waste plastics for efficient flood protection systems. Environment. 4, 1–10.

Majka, T.M., Bartyzel, O., Raftopoulos, K.N., Pagacz, J., Leszczyńska, A., Pielichowski, K., 2016. Recycling of polypropylene/montmorillonite nanocomposites by pyrolysis. J. Anal. Appl. 119, 1–7.

Makkam, S., Harnnarongchai, W., 2014. Rheological and mechanical properties of recycled PET modified by reactive extrusion. Energy Procedia. 56, 547–553.

Mallakpour, S., Adnany Sadaty, M., 2016. Thiamine hydrochloride (vitamin B1) as modifier agent for TiO_2 nanoparticles and the optical, mechanical, and thermal properties of poly (vinyl chloride) composite films. RSC Adv. 6, 92596–92604.

Mallakpour, S., Behranvand, V., 2017. Application of recycled PET/carboxylated multi-walled carbon nanotube composites for Cd^{2+} adsorption from aqueous solution: a study of morphology, thermal stability, and electrical conductivity. Colloid Polym. Sci.http://dx.doi.org/10.1007/s00396-017-4022-z, Online.

Mallakpour, S., Behranvand, V., 2016a. Nanocomposites based on biosafe nano ZnO and different polymeric matrixes for antibacterial, optical, thermal and mechanical applications. Eur. Polym. J. 84, 377–403.

Mallakpour, S., Behranvand, V., 2016b. Recycled PET/MWCNT-ZnO quantum dot nanocomposites: Adsorption of Cd(II) ion, morphology, thermal and electrical conductivity properties. Chem. Eng. J.http://dx.doi.org/10.1016/j.cej.2016.10.129, Online.

Mallakpour, S., Behranvand, V., 2016c. Manufacture and characterization of nanocomposite materials obtained from incorporation of d-glucose functionalized MWCNTs into the recycled poly(ethylene terephthalate). Des. Monomers Polym. 19 (4), 283–289.

Mallakpour, S., Javadpour, M., 2016a. The potential use of recycled PET bottle in nanocomposites manufacturing with modified ZnO nanoparticles capped with citric acid: preparation, thermal, and morphological characterization. RSC Adv. 6 (18), 15039–15047.

Mallakpour, S., Javadpour, M., 2016b. The thermal, optical, flame retardant, and morphological consequence of embedding diacid-capped ZnO into the recycled PET matrix. J. Appl. Polym. Sci. 43433, 1–11.

Martín-Alfonso, J.E., Valencia, C., Franco, J.M., 2013. Effect of amorphous/recycled polypropylene ratio on thermo-mechanical properties of blends for lubricant applications. Polym. Test. 32, 516–524.

Martín-Alfonso, J.E., Valencia, C., Franco, J.M., 2014. Composition-property relationship of gel-like dispersions based on organo-bentonite, recycled polypropylene and mineral oil for lubricant purposes. Appl. Clay Sci. 87, 265–271.

Meri, R.M., Zicans, J., Maksimovs, R., Ivanova, T., Kalnins, M., Berzina, R., et al., 2014. Elasticity and long-term behavior of recycled polyethylene terephthalate (rPET)/montmorillonite (MMT) composites. Compos. Struct. 111, 453–458.

Mubarak, Y.A., 2011. Thermal and mechanical properties of polyethylene terephthalate/polycarbonate nanocomposites modified by lanthanum acetyl acetonate hydrate thermal and mechanical properties of polyethylene terephthalate/polycarbonate nanocomposites modified by lantha. Polym.-Plast. Technol. Eng. 50, 635–645.

Mohamed, M.A., Salleh, W.N.W., Jaafar, J., Ismail, A.F., Mutalib, M.A., Jamil, S.M., 2015. Incorporation of N-doped TiO_2 nanorods in regenerated cellulose thin films fabricated from recycled newspaper as a green portable photocatalyst. Carbohydr. Polym. 133, 429–437.

Munusamy, Y., Ismail, H., Mariatti, M., Ratnam, C.T., 2009. Effects of different preparation methods on the properties of poly[ethylene-*co*-(vinyl acetate)]/(Standard Malaysian Natural Rubber)/organoclay nanocomposites. J. Vinyl Add. Technol. 15, 244–251.

Nabil, H., Ismail, H., 2013. Preparation and properties of recycled poly(ethylene terephthalate) powder/halloysite nanotubes hybrid-filled natural rubber composites. J. Thermoplast. Compos. Mater. 28, 1–16.

Nassar, M.A., Youssef, A.M., 2012. Mechanical and antibacterial properties of recycled carton paper coated by PS/Ag nanocomposites for packaging. Carbohydr. Polym. 89, 269–274.

Navarro, F.J., Partal, P., Martínez-boza, F.J., Gallegos, C., 2010. Novel recycled polyethylene/ground tire rubber/bitumen blends for use in roofing applications: thermo-mechanical properties. Polym. Test. 29, 588–595.

Nemati, M., Khademieslam, H., Talaiepour, M., Ghasemi, I., Bazyar, B., 2013. Investigation on the mechanical properties of nanocomposite based on wood flour/recycle polystyrene and nanoclay. J. Basic Appl. Sci. Res. 3 (3), 688–692.

Njuguna, J., 2013. Part II: Polymers. In: Lehmhus, D., Busse, M., Herrmann, A.S., Kayvantash, K. (Eds.), Structural Materials and Processes in Transportation. John Wiley & Sons, Weinheim, p. 500.

Oblak, P., Gonzalez-Gutierrez, J., Zupančič, B., Aulova, A., Emriin, I., 2016. Mechanical properties of extensively recycled high density polyethylene (HDPE). Mater. Today: Proc. 3, 1097–1102.

Park, S.H., Kim, S.H., 2014. Poly(ethylene terephthalate) recycling for high value added textiles. Fashion Text. 1, 1–17.

Radu, M., Christiana, C., 2011. Using PET (polyethylene terephthalate) waste for buildings. Civil Eng. Installations. 1 (14), 73–80.

Ramírez-arreola, D.E., Sedano-de la Rosa, C., Haro-mares, N.B., Ramírez-morán, J.A., Pérez-fonseca, A.A., Robledo-ortíz, J.R., 2015. Compressive strength study of cement mortars lightened with foamed HDPE nanocomposites. Mater. Des. 74, 119–124.

Rattanasom, N., Saowapark, T., Deeprasertkul, C., 2007. Reinforcement of natural rubber with silica/carbon black hybrid filler. Polym. Test. 26, 369–377.

Reis, J.M.L., Chianelli-junior, R., Cardoso, J.L., Marinho, F.J.V., 2011. Effect of recycled PET in the fracture mechanics of polymer mortar. Constr. Build. Mater. 25, 2799–2804.

Ren, Y., Wang, Y., Wang, L., Liu, T., 2015. Evaluation of intumescent fire retardants and synergistic agents for use in wood flour/recycled polypropylene composites. Constr. Build. Mater. 76, 273–278.

Rodríguez-Uicab, O., May-Pat, A., Avilés, F., Toro, P., Yazdani-Pedram, M., 2013. Influence of processing method on the mechanical and electrical properties of MWCNT/PET composites. J. Mater. 2013, 1–10.

Santos, R.P.O., Rodrigues, B.V.M., Ramires, E.C., Ruvolo-filho, A.C., Frollini, E., 2015. Bio-based materials from the electrospinning of lignocellulosic sisal fibers and recycled PET. Ind. Crops Prod.1–8.

Singh, S., Deepak, D., Aggarwal, L., Gupta, V.K., 2014. Tensile and flexural behavior of hemp fiber reinforced virgin- recycled HDPE matrix composites. Procedia Mater. Sci. 6, 1696–1702.

Sommerhuber, P.F., Welling, J., Krause, A., 2015. Substitution potentials of recycled HDPE and wood particles from post-consumer packaging waste in wood–plastic composites. Waste Manage. 46, 76–85.

Sujaritjun, W., Uawongsuwan, P., Pivsa-art, W., Hamada, H., 2013. Mechanical property of surface modified natural fiber reinforced PLA biocomposites. Energy Procedia. 34, 664–672.

Velis C.A., 2014. Global recycling markets-plastic waste: A story for one player–China. Report Prepared by FUELogy and Formatted by D-waste on Behalf of International Solid Waste Association-Globalisation and Waste Management Task Force. ISWA, Vienna, September 2014.

Yang, C., Navarro, M.E., Zhao, B., Leng, G., Xu, G., Wang, L., et al., 2016. Thermal conductivity enhancement of recycled high density polyethylene as a storage media for latent heat thermal energy storage. Solar Energy Mater. Solar Cells. 152, 103–110.

Yang, T.C., Wu, T.L., Hung, K.C., Chen, Y.L., Wu, J., 2015. Mechanical properties and extended creep behavior of bamboo fiber reinforced recycled poly(lactic acid) composites using the time–temperature superposition principle. Constr. Build. Mater. 93, 558–563.

Youssef, A.M., El-gendy, A., Kamel, S., 2015. Evaluation of corn husk fibers reinforced recycled low density polyethylene composites. Mater. Chem. Phys. 152, 26–33.

Zaaba, N.F., Ismail, H., Mariatti, M., 2016. Utilization of polyvinyl alcohol on properties of recycled polypropylene/peanut shell powder composites. Procedia Chem. 19, 763–769.

Zulkifli, N.I., Samat, N., Anuar, H., Zainuddin, N., 2015. Mechanical properties and failure modes of recycled polypropylene/microcrystalline cellulose composites. Mater. Des. 69, 114–123.

Recent developments in the synthesis of hybrid polymer/clay nanocomposites: properties and applications

8

Shadpour Mallakpour[1,2,3] *and Shima Rashidimoghadam*[1]
[1]Organic Polymer Chemistry Research Laboratory, Department of Chemistry, Isfahan University of Technology, Isfahan, Islamic Republic of Iran, [2]Nanotechnology and Advanced Materials Institute, Isfahan University of Technology, Isfahan, Islamic Republic of Iran, [3]Center of Excellence in Sensors and Green Chemistry, Department of Chemistry, Isfahan University of Technology, Isfahan, Islamic Republic of Iran

Chapter Outline

8.1 Introduction 227
8.1.1 Clay 227
8.1.2 Structure of clay 228
8.1.3 Classification of clay minerals 229
8.1.4 Clay surface treatments 230
8.1.5 Polymer/clay nanocomposites 233
8.2 Preparation of PCNs and their applications 234
8.2.1 Food packaging 235
8.2.2 Biomedical applications 240
8.2.3 Wastewater pretreatment 246
8.2.4 Other applications 254
8.3 Conclusions 259
Acknowledgments 259
References 260

8.1 Introduction

8.1.1 Clay

The term "clay" refers to a class of materials generally made up of layered silicates or clay minerals with traces of metal oxides and organic matter. Clays minerals, which may be either natural or synthetic, are constitutes an important class of fillers,

Hybrid Polymer Composite Materials: Applications. DOI: http://dx.doi.org/10.1016/B978-0-08-100785-3.00008-5

such as mica, talc, kaolin, montmorillonite (MMT), and hectorite are layered silicates, often used as filler in polymers for many years. However, since clays in their pristine state are hydrophilic and do not have good compatibility with the hydrophobic polymer matrix to achieve good dispersion, it is necessary to chemically modify natural clay so that it can be compatible with a chosen polymer matrix. A very small amount of such modified clay can significantly enhance the mechanical, optical, thermal, and physicochemical properties when compared with virgin polymer or conventional microcomposites (Mittal, 2009a,b; Hundáková et al., 2015).

8.1.2 Structure of clay

The commonly used clays for the preparation of polymer–clay nanocomposites (PCNs) belong to the same general family of 2:1 layered or phyllosilicates (Tan and Thomas, 2016). Their crystal structure consists of layers made up of two silica tetrahedral fused to an edge-shared octahedral sheet of either aluminum or magnesium hydroxide (Fig. 8.1). The thickness of a layer is around 1 nm, whereas the lateral dimensions are in the range from 30 nm to several microns or even larger depending on the particular silicate, the source of the clay and the method of preparation. The ability of clay minerals to hold cations is named the cation exchange capacity, which is generally expressed as meq/100 g clay. Stacking of the layers leads to a regular van der Waals gap between the layers called the interlayer or gallery. Small molecules or macromolecules can be inserted (intercalated) between the layers under appropriate

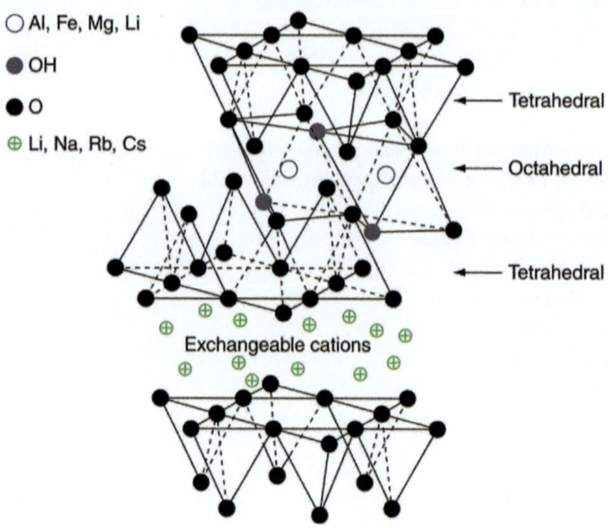

Figure 8.1 Structure of 2:1 layered silicates.
Adapted from Giannelis, E.P., Krishnamoorti, R., and Manias, E. (1999). Polymer-Silicate Nanocomposites: Model Systems for Confined Polymers and Polymer Brushes. In Polymers in Confined Environments (S. Granick, K. Binder, P.G. de Gennes, E.P. Giannelis, G.S. Grest, H. Hervet, R. Krishnamoorti, L. Léger, E. Manias, E. Raphaël and S.Q. Wang, eds.), pp. 107–147. Springer Berlin Heidelberg, Berlin, Heidelberg. With kind permission of Springer.

conditions. Isomorphous substitution within the layers can cause the formation of negative charges on clay layers, for example, in the substitution of Al^{3+} by Mg^{2+} and Fe^{2+} or Mg^{2+} by Li^{1+}. These charges are counterbalanced by alkali and alkalinc earth cations such as Na^{+} or Ca^{2+} situated inside the galleries. The majority of these cations are present in the interlayers between the sheets, but some percentages of them are present on the edges of the sheets. The interlayer spacing between the silicate layers is quite hydrophilic and normally hydrated unless heated to several hundred degrees celsius. Alkyl–ammonium or alkyl–phosphonium cations are used for the modification of interlayer spacing through the cation exchange reactions. This modification improves the hydrophobicity of the interlayer spacing; in addition, the organic modifiers may contain a variety of functional groups, thereby providing access to a wide range of chemistries and properties. Such functionalities can include polymerizable groups (e.g., vinyl) or initiating groups (Chen et al., 2016; Giannelis et al., 1999; Pavlidou and Papaspyrides, 2008; Mittal, 2009a,b).

8.1.3 Classification of clay minerals

Clay minerals are classified in number of ways. A committee on the terminology of the Clay Minerals Group of the Mineralogy Society of Great Britain categorized the crystalline clay minerals into two major groups: chain and layer structures which the layer structures are classified into 2:1 (TOT) and 1:1 (TO) families (T: tetrahedral, O: octahedral). The classification shown in Table 8.1 is mainly

Table 8.1 Classification of clay minerals

I. Amorphous
Allophane group
II. Crystalline
- A. Two-layer type (sheet structures composed of units of one layer of silica tetrahedrons and one layer of alumina octahedrons
  1. Equidimensional kaolinite group (kaolinite, nacrite, etc.)
  2. Elongate halloysite group
- B. Three-layer types (sheet structures composed of two layers of silica tetrahedrons and one central dioctahedral or trioctahedral layer)
  1. Expanding lattice
     - a. Equidimensional montmorillonite group (montmorillonite, sauconite, etc.), vermiculite
     - b. Elongate montmorillonite group (nontronite, saponite, hectorite)
  2. Nonexpanding lattice, illite group
- C. Regular mixed-layer types (ordered stacking of alternate layers of different types): Chlorite group
- D. Chain-structure types (hornblende-like chains of silica tetrahedrons linked together by octahedral groups of oxygens and hydroxyls containing Al and Mg atoms): Attapulgite, sepiolite, palygorskite

Adapted from Gurses, A., Ejder-Korucu, M., Karaca, S. (2011). Clay-Organoclay and Organoclay/Polymer Nanocomposites. In Clay: Types, Properties and Uses. Boyd D.E., Humphery J.P. (Eds.), pp. 155–192. Nova Science Publisher. Open access journal.

due to distinctions in the shape of clay minerals and the expandable or nonexpendable character of the 2:1 and 1:1 layer silicates.

The 1:1 layer type usually has no layer charge or a very small layer charge because the tetrahedral cation sites usually are all occupied by Si^{4+} and the octahedral sites by all Al^{3+} or all Mg^{2+} If there is substitution in one sheet of a 1:1 layer silicate, there almost always is a compensating substitution in the other sheet so that neutrality is maintained. There are dioctahedral and trioctahedral varieties of 1:1 layer silicates and within these subgroups there are individual species (Gurses et al., 2011).

8.1.4 Clay surface treatments

Layered silicate clays are naturally hydrophilic. This makes them poorly suited to mixing and interacting with most polymer matrices which are mostly hydrophobic (Jordan, 1949; Van Olphen, 1964; Giannelis, 1996). Moreover, clay layers are closely stacked and held together by electrostatic interaction in the interlayer space where hydrated alkaline or alkaline earth cations make it difficult for macromolecule diffusion (Carrado and Komadel, 2009). For these reasons, it is necessary to modify pristine clay before use in preparation of PCNs. Making a composite out of untreated clay would not be a very effective because most of the clay would be unable to interact with the matrix (Singla et al., 2012). There are several types of chemical treatments that can be grouped as follows.

8.1.4.1 Purification

The available commercial clay materials are often raw clays, and they usually contain embarrassing impurities such as carbonates, cristobalite, feldspars, quartz, organic matter, iron hydroxides so that the clays should be purified using sedimentation isolates of the smectite portion. Consequently, the purification process would result in a highly pure material with improved properties (Ludueña et al., 2015).

8.1.4.2 Activation

One of the most common chemical modifications of clays, used for both industrial and scientific purposes, is their acid activation. The method involves treatment of the clays with a mineral acid solution. These acids include hydrochloric acid, sulphuric acid, phosphoric acid, nitric acid, organic acids such as acetic, citric, oxalic, and lactic acids. Among these acids, hydrochloric acid and sulfuric acid are the most widely used in acid activation, because they gives good result regarding the specific surface area, porosity, and adsorption capacity of the activated clay. The first step of acid attack is to remove the exchangeable cations by protons. The second effect is the leaching of Al, Mg, and Fe from the octahedral and tetrahedral sheet, but the SiO_4 groups of the tetrahedral sheet stay largely intact. Thus, acid attack causes a corrosion of the octahedral sheet. Acid treatment is expected to improve the specific surface area and number of acid centers with respect to the parent clays. The acid-treated clays are composed of a mixture of nonattacked clay layers and a hydrous, amorphous, and partially protonated silica phase (Komadel and Madejova, 2006; Panda et al., 2010; Steudel et al., 2009; Aishat et al., 2015). Acid activated clay minerals find wide industrial applications as bleaching earth (Siddiqui, 1968; Kaufhold and Meyer, 2001), catalysts or catalyst supports (Rhodes et al., 1991; Rhodes and Brown, 1992), and in carbonless copying paper (Fahn and Fenderl, 1983).

8.1.4.3 Ion exchange reactions

This is an easy method of modifying the clay surface that allows the exchange of any of the interlaminar cations by a desirable cation (Lee and Tiwari, 2012). The surfactants used are mainly primary, secondary, tertiary, and quatenary alkylammonium although other "onium" salts can be used, such as sulfonium and phosphonium (Fig. 8.2) (Azeez et al., 2013).

The driving force behind the ion exchange reaction is linked to two aspects: (1) the hydration of interlayer cations in the aqueous solution leading to the swelling of nanoclay and (2) the tendency of the hydrophobic surfactants to be repelled by the aqueous solution and collected on the nanoclay surface. The longer are the organic tails of the surfactant, the stronger is the repelling force from the aqueous solution (Xia, 2014). The organic cations modify interlayer interactions by lowering the surface energy of the inorganic component and improve the wetting characteristics with the polymer or in some cases initiate the polymerization of monomers to improve the strength of the interface between the inorganic and polymer matrices (Nguyen and Baird, 2006; Joshi et al., 2006). Moreover, the long organic chains of such surfactants, with positively charged ends, are tethered to the surface of the negatively charged silicate layers, resulting in an increase of the gallery height (Kim et al., 2001). It then becomes possible for organic species (i.e., polymers or prepolymers) to diffuse between the layers and eventually separate them (Kornmann et al., 2001; Zerda and Lesser, 2001). Depending on the packing density, temperature and alkyl chain length, different arrangements of surfactants in the clay gallery are possible. The organic chains have been long thought to lie either parallel to the silicate layer, forming mono or bilayers or, depending on the chain

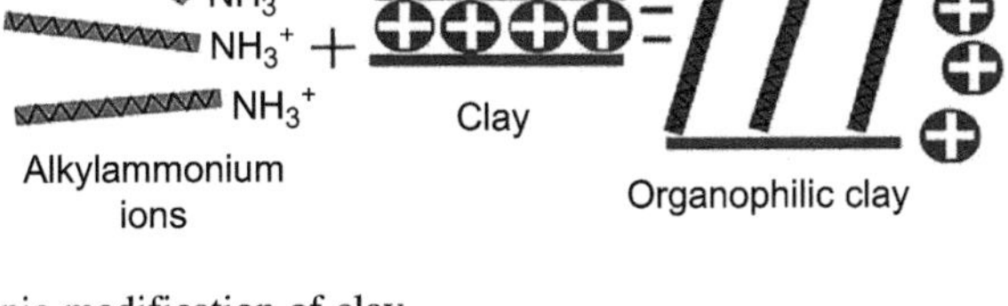

Figure 8.2 Organic modification of clay.
Adapted from Azeez, A.A., Rhee, K.Y., Park, S.J., and Hui, D. (2013). Epoxy clay nanocomposites—processing, properties and applications: a review. Compos. B: Eng. 45, 308–320. With kind permission of Elsevier.

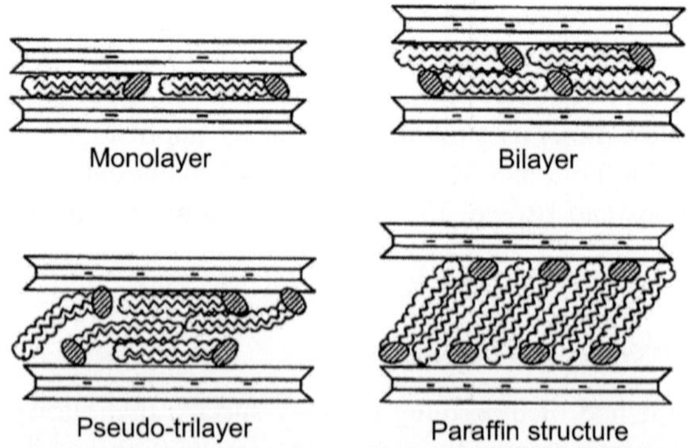

Figure 8.3 Schematic diagram of the idealized arrangements of alkylammonium cations between the clay layers: (A) parallel monolayer, (B) parallel bilayer, (C) radiated monolayer, and (D) radiated bilayer. The clay layers are expressed by the straight dark lines.
Adapted from Mittal, V. (2009a). Barrier Properties of Polymer Clay Nanocomposites, Nova Science Publisher; Mittal, V. (2009b). Polymer layered silicate nanocomposites: a review. Materials 2, 992. Open access journal.

length, to radiate away from the surface, forming pseudo trilayers or even tilted paraffinic arrangement (Fig. 8.3) (Vaia et al., 1994; Xia, 2014).

8.1.4.4 Grafting

The most common grafting process is silylation, also known as silane grafting, which consists of covalently bonding the silane molecules to the clay platelets. The Si–OH groups present on silane condense with the M-OH on the clay mineral surfaces (M is Si, Al, or other metals) and introduce special specific functionalities, such as amino, mercapto, glycidoxy, vinyl, or methacryloxy groups which interact with the polymer matrix or react with it producing a network between clay, silane and polymer via covalent bonding (He et al., 2013; Romanzini et al., 2015). This can significantly improve the mechanical property of the resultant PCNs and may lead to a breakthrough in synthesis of novel materials (Su et al., 2013).

8.1.4.5 Reactions with biomolecules

Clays have been modified with different kinds of biomolecules like proteins, enzymes, amino acids, or peptides. Among these, amino acids have the significant improvement of biodegradability, low toxicity, and various possible structures when compared with the chemically synthesized modifier. In this case, the reaction consists of an ion exchange process of the sodium cations of the protonated amino acid clay. It is remarkable that because these clays don't endanger human health, they are quite proper as reinforcing materials in the preparation of bionanocomposites. Moreover, they might contain curative properties and can also be employed in pharmaceutical formulations (Mallakpour and Shahangi, 2013; Mallakpour and Dinari, 2013; Ludueña et al., 2015).

8.1.5 Polymer/clay nanocomposites

The growing interest in polymer nanocomposites initiates from the improvement of fundamental properties and, also, the development of new materials to meet different applications. The significant changes in properties of polymer nanocomposites are occurred by incorporation of nanofillers such as nanoclay, graphene, carbon nanotubes, metal oxides, and layered double hydroxide (Azzam, 2014; Mallakpour and Khani, 2015, 2016; Ensafi et al., 2014; Mallakpour and Barati, 2014; Cailloux et al., 2016; Mallakpour and Soltanian, 2016; Banks-Sills et al., 2016; Mallakpour and Jarahiyan, 2016; Mallakpour and Javadpour, 2016; Mallakpour and Dinari, 2014). Among these, clays have long been used in polymer systems owing to its low cost and further scope to improve their interfacial properties by use of coupling agent in polymer composites. Low amounts of nanoclay fillers dispersed in a polymer can dramatically improve the material performance of polymers as reinforcements and minimize the utilization of polymers within composites (Nigam and Lal, 2008).

PCNs are a new class of materials which have deserved an increasing interest from both scientists and engineers in recent years because of their novel and unique properties like high dimensional stability, heat deflection temperature, gas barrier performance, reduced gas permeability, optical clarity, flame retardancy (Usuki et al., 1993; Kojima et al., 1993; LeBaron et al., 1999; Manias et al., 2001; Lan et al., 1995; Tien and Wei, 2001; Huang et al., 2001; Agag and Takeichi, 2000; Galgali et al., 2001; Fu and Qutubuddin, 2001; Okamoto et al., 2000, 2001; Messersmith and Giannelis, 1995; Kawasumi et al., 1997; Hasegawa et al., 1998). With the rapid development of material science, the molecular level incorporation of layered clays into polymer matrix had been achieved through the addition of modified clays, which can further improve the mechanical properties of composite materials (Feng et al., 2016). The structure and properties of PCNs are governed mostly by the interfacial interaction of layered silicates and the polymer matrix. Depending on the nature of the components (layered silicate, organic cation, and polymer matrix) used and the preparation method, three main types of composites can be thermodynamically achievable when layered clay is associated with a polymer (Jamshidi, 2014; Naveau, 2010; Alexandre and Dubois, 2000).

1. **Phase-separated structure:** When the organic polymer is interacted with inorganic clay (unmodified clay), polymer chains are not able to penetrate between the silicate layers gallery spacing and the clay is dispersed as aggregates or regular stacks of layers formed by stacking together within the polymer matrix. The resulting composite structure is considered "phase separated." The properties of phase-separated PCNs cannot be better than conventional microcomposites (Gurses, 2015).
2. **Intercalated nanocomposites:** The intercalated NCs are formed when the polymer chains are penetrated into the interlayer region of the clay, resulting in an ordered multiple layer structure with alternating polymer/inorganic layers at a repeated distance of a few nanometers (Mukhopadhyay and Dc, 2014).
3. **Exfoliated nanocomposites:** An exfoliated or delaminated NC is obtained when the clay layers are well dispersed in a continuous polymer matrix at random orientation by an average distance (d-spacing) larger than 8.8 nm which depends on the amount of nanoclay loading. In general, the clay content in this type of NC is much lower than that in intercalated NCs (Gurses, 2015).

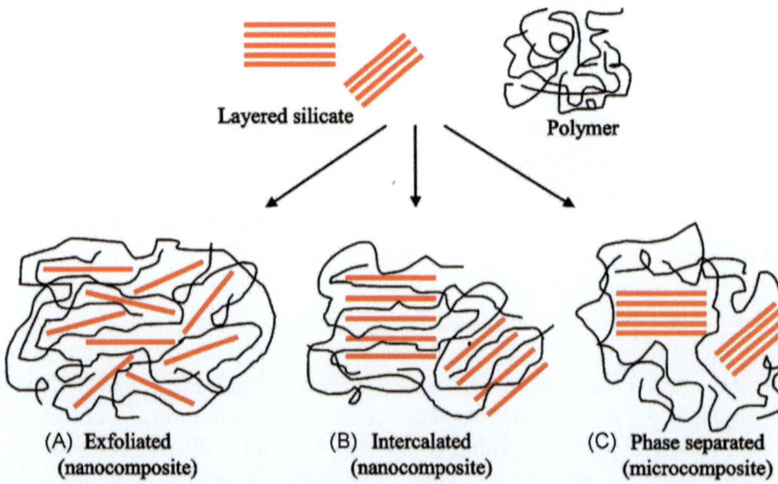

Figure 8.4 Schematic representation of different types of composites arising from interaction of layered silicate and polymer.
Adapted from Alexandre, M., and Dubois, P. (2000). Polymer-layered silicate nanocomposites: preparation, properties and uses of a new class of materials. Mater. Sci. Eng.: R: Rep. 28, 1–63. With kind permission of Elsevier.

The exfoliation configuration has considerable importance because it increases the interactions between polymer and clay and makes the whole surface of layers available for the polymers. This should lead to the most substantial changes in mechanical and physical properties. Actually, it is generally believed that exfoliated systems compared with intercalated ones, offer better mechanical properties. The complete dispersion of clay nanolayers in a polymer optimizes the number of available reinforcing elements for carrying an applied load and deflecting cracks. The coupling between the tremendous surface area of the clay and the polymer matrix facilitates stress transfer to the reinforcement phase, allowing for mechanical property improvements. Fig. 8.4 illustrates a schematic representation of the three morphologies of polymer layered silicates.

8.2 Preparation of PCNs and their applications

PCNs have attracted a great deal of attention because of their huge potential for industrial and technological applications, thus many efforts have been made for the preparation of PCNs with improved properties. The applications for PCNs are broad, and these materials have a promising commercial impact. The main application segments of them include: food packaging, biomedical applications, wastewater pretreatment, electricals/electronics, optoelectronics, sensors, and automobiles (Gurses, 2015).

Several chapters have been published recently on various aspects of PCNs, but this chapter has been written to present an update overview on the different synthetic methods, characterization and applications of various clay-based polymer nanocomposite systems reported from 2014 to 2016 with an emphasis on the various applications of this materials. We mainly studied the food packaging, biomedical, and wastewater pretreatment applications of these composites.

8.2.1 Food packaging

Packaging and the associated food industry are the biggest users of PCNs (Komadel and Madejova, 2006). The food package should hinder gain or loss of moisture, prevent microbial contamination, and act as a barrier against permeation of water vapor, oxygen, carbon dioxide, and other volatile compounds such as flavors and taints, in addition to the basic properties of packaging materials such as mechanical, optical, and thermal properties (Gurses, 2015). During the last decades, as a result of their advantages over other traditional materials, the polymers has widely employed as food packaging materials. The possibility to improve the performances of polymers for food packaging by adding NPs has led to the development of a variety of polymer nanomaterials (Silvestre et al., 2011). Although several NPs have been recognized as possible additives to enhance polymer performance, in the packaging industry, layered inorganic solids like clays and silicates have been the subject of increasing focus, due to their availability, low cost, significant enhancements, and relative simple process ability (Azeredo, 2009).

Yahiaoui et al. used the melt mixing process to produce antimicrobial poly(e-caprolactone) (PCL)/clay nanocomposites for food packaging application. Two kinds of quaternary ammonium surfactants: hexadecyl trimethylammonium chloride and hexadecyl pyridinium chloride were employed as organic modifiers for MMT (OMMT1 and OMMT2) with molar ratio clay/ammonium salt of 1:2 then these organoclays used as nanofillers at 3 wt% in the synthesis of NCs. Based on TEM and XRD observations, PCL/OMMTs NCs showed mainly intercalated structures. The addition of 3% of both OMMT1 and OMMT2 nanoclays to PCL matrix was found to be sufficient to: (1) inhibit about 90% of the growth of *Escherichia coli* and *Staphylococcus aureus*, while pure PCL does not show any antimicrobial activity and (2) greatly reduced water vapor permeability (WVP) values by about 56% for PCL/OMMT1 and 48% for PCL/OMMT2, indicating their barrier effects. These properties have led to the use of these NCs in food packaging applications application (Yahiaoui et al., 2015).

Savas et al. investigated the influence of organically modified clay on the antibacterial activity of low density polyethylene (LDPE)/clay NC against powerful gram negative *E. coli* bacteria. They first ion-exchanged the MMT with silver ions (Ag + Mt) then organically modified it with cetyltrimethylammoniumbromide, CTAB (Ag + OMt). PCNs were synthesized via embedding of various concentrations of Ag + OMt (1.25, 2.25, and 5%) in to LDPE using a high shear force corotating twin screw extruder. The SEM examination of nanocomposites with different clay loading indicated that with increasing MMT amount, particle agglomeration

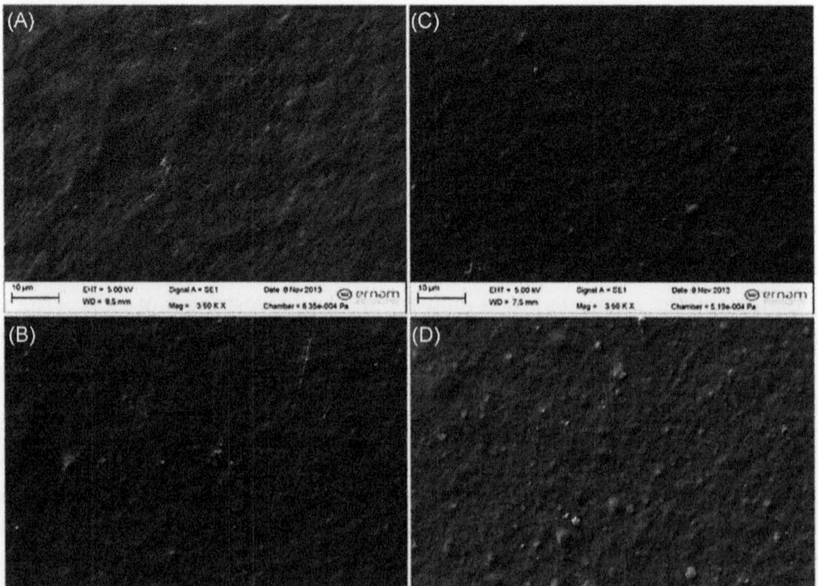

Figure 8.5 SEM micrographs of surface of (A) pure LDPE, (B) 1.25%, (C) 2.25%, and (D) 5% CPN.
Adapted from Savas, L.A., and Hancer, M. (2015). Montmorillonite reinforced polymer nanocomposite antibacterial film. Appl. Clay Sci. 108, 40–44. With kind permission of Elsevier.

becomes more pronounced as evidenced by the surface texture of the films (Fig. 8.5). The results of the antibacterial activity test indicated that among MMT, OMt, and Ag + OMt, only Ag + OMt showed inhibitory activity against bacteria due to presence of silver. It was further observed that the PCN films with 1.25 (30% reduction) and 2.25% (47% reduction) of silver content showed slight antibacterial activity, CPN film with 5% of Ag + OMt content was demonstrated a complete growth inhibition of *E. coli* (70% reduction) (Fig. 8.6). Thus, these NCs can be successfully employed in the food packaging (Savas and Hancer, 2015).

Jafarzadeh et al. found that the introduction of nanokaolin into semolina film allow bionanocomposites with enhanced barrier properties to be achieved, indicating the potential application of these bionanocomposites in food packaging industry especially cheeses, owing to its good antioxidant property. Kaolin was incorporated into biofilms at different concentrations (1%, 2%, 3%, 4%, and 5%, w/w total solid). All films were plasticized with 50% (w/w total solid) combination of sorbitol/glycerol at 3:1 ratio. Effects of the nanoclay were evaluated on oxygen permeability (OP), WVP, as well as antimicrobial activity of the NCs. WVP and OP of these NCs decreased significantly with increasing kaolin content. On the other hand, tensile strength (TS) of bionanocomposite films was improved by increasing

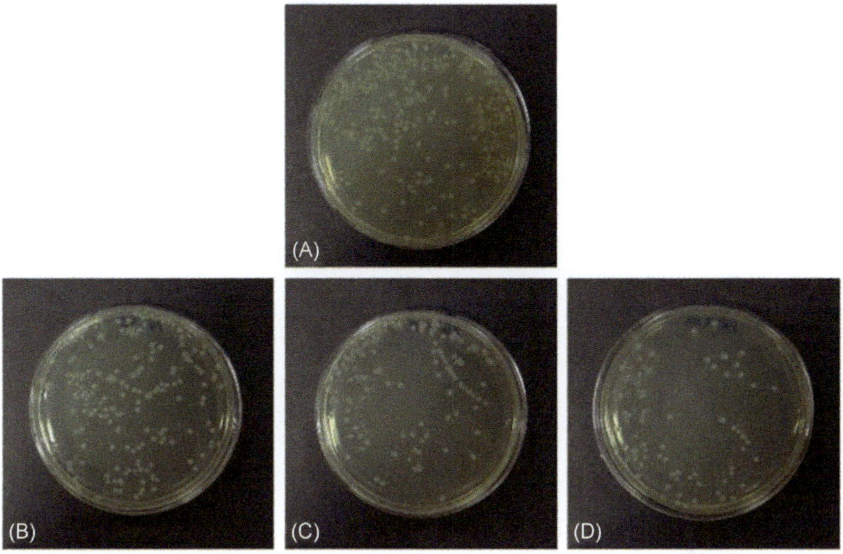

Figure 8.6 The photographs above show nutrient agar plates on *E. coli* containing different concentrations of CPN film: (A) pure LDPE, (B) 1.25% CPN, (C) 2.25% CPN, and (D) 5% CPN according to ASTM E-2149 standard.
Adapted from Savas, L.A., and Hancer, M. (2015). Montmorillonite reinforced polymer nanocomposite antibacterial film. *Appl. Clay Sci.* **108**, 40–44. With kind permission of Elsevier.

nanokaolin concentration. Improvement in TS plays an important role in food packaging applications. The incorporation of nanokaolin in semolina films did not exhibit any antimicrobial effect against Gram-negative and Gram-positive bacteria (Jafarzadeh et al., 2016).

Neves et al. examined the morphology, water vapor transmission (WVTR), WVP, and mechanical properties of composites obtained from bentonite nanoclay (0%, 1%, or 3%) and chitosan aqueous solutions (containing either 2 or 3 wt%) using a solvent-casting method. Tensile properties of composite films (containing either 1 or 3 wt% nanoclay) was increased compared to that of pure polymer matrix (either 2 or 3 wt%). Moreover, the highest TS among all samples (18 MPa) was observed when the nanoclay content was 1 wt% in the chitosan matrix (2 wt%), which has been attributed to the more uniform distribution of nanoclay on to the host polymer. As shown in Fig. 8.7, the addition of 1 or 3 wt% nanoclay to the chitosan premixture resulted in between 18% and 54% reduction on WVP respectively, in relation to the films prepared using 2 wt% chitosan alone. These results showed that various loadings of nanoclay significantly affect the WVP of biopolymer films at certain concentration of chitosan. This is believed to be due to the fulfilling of interspaces in the chitosan matrix by nanoclay particles, leading to enhancement of the gas barrier performances of polymer. In addition, the authors found that the lowest WVTR was observed in films consisting of 2 wt% chitosan and 3 wt%

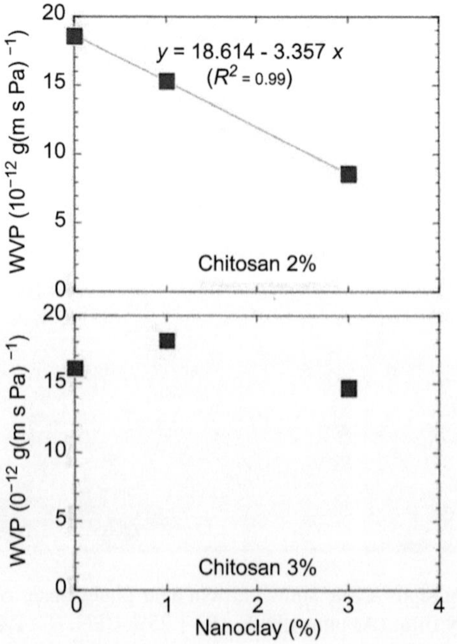

Figure 8.7 Water vapor permeability (WVP) of the composite biopolymer-based films. Adapted from M.A. Neves, J. Hashemi, T. Yoshino, K. Uemura, N. M. (2016). Development and characterization of chitosan-nanoclay composite films for enhanced gas barrier and mechanical properties. HSOA J. Food Sci. Nutr. 2, 007. Open access journal.

nanoclay. Thus, these nanocomposites with enhanced film properties have a high potential for being used as food packaging materials (Neves et al., 2016).

There are few studies on the effect of various composition of copolymer on the morphological, optical, and moisture sorption behavior of the final polymeric films.

Murima et al. evaluated the effect of clay content and copolymer composition on the optical properties and water vapor sorption behavior of PSBA–montmorillonite clay (PSBA–MMT) films. Surface modification of MMT by ion exchange was done with the use of vinylbenzyldodecyldimethylammonium chloride. The hybrid latexes containing MMT–VBDA clay concentrations of 10–30 wt% and copolymer (with various styrene–butyl acrylate comonomer molar ratios: 50:50, 40:60, and 30:70) were prepared using miniemulsion polymerization. TEM image indicates that the MMT particles are predominantly adhered onto the surface of the PSBA latex particles because MMT clay platelets have a larger size than the effective size of the copolymer particles (Fig. 8.8). Copolymer films with 30 mol% styrene present high WVP because of the unique morphological organization of MMT platelets in the matrix.

A direct utilization of the good barrier properties of these PCNs is in food packaging applications. The best optical properties were observed in the PSBA-30:70 films, in which the light transmittance only decreases from 85% (unfilled film) to 60% in the NC films containing 30 wt% clay (Murima et al., 2016).

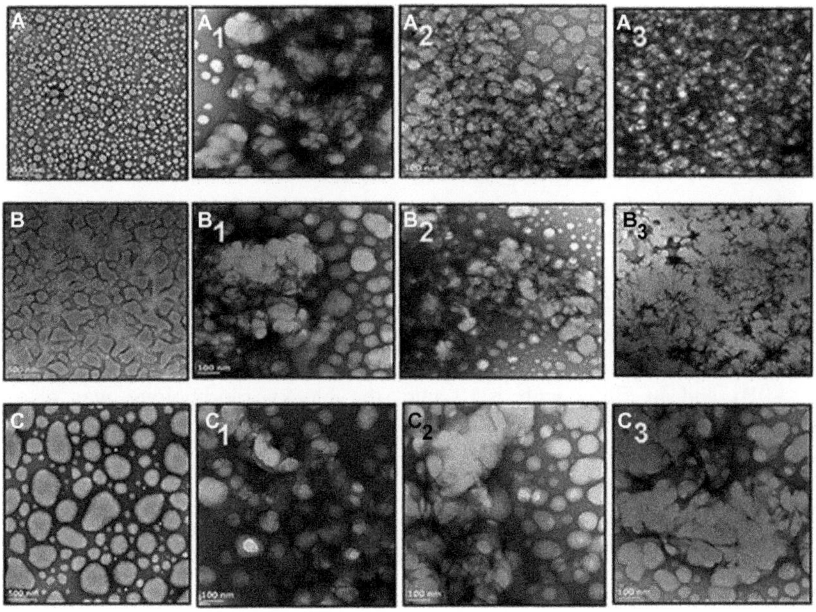

Figure 8.8 Micrographs obtained by TEM analysis of (A) neat PSBA-50:50 and its hybrid latexes [(A_1) PSBA-50:50-MMT10%, (A_2) PSBA-50:50-MMT20%, and (A_3) PSBA-50:50-MMT30%]; (B) neat PSBA-40:60 and its hybrid latexes [(B_1) PSBA-40:60-MMT10%, (B_2) PSBA-40:60-MMT20%, (B_3) PSBA-40:60-MMT30%]; and (3) neat PSBA-30:70 and its hybrid latexes [(C_1) PSBA-30:70-MMT10%, (C_2) PSBA-30:70-MMT20%, and (C_3) PSBA-30:70-MMT30%].

Adapted from Murima, D., Pfukwa, H., Tiggelman, I., Hartmann, P.C., and Pasch, H. (2016). Novel polymer clay-based nanocomposites: films with remarkable optical and water vapor barrier properties. Macromol. Mater. Eng. 301, 836–845. With kind permission of John Wiley and Sons.

Abreu et al. reported the synthesis of MMT modified with a quaternary ammonium salt C30B/starch NC (C30B/ST-NC), silver NPs/starch NC (Ag-NPs/ST-NC) and both silver NPs /C30B/starch NCs (Ag-NPs/C30B/ST-NC) films by solution casting method. They showed that incorporation of C30B clay (3 wt%) and different amounts of silver NPs (0.3, 0.5, 0.8, and 1.0 mM) in to starch showed improvement in the mechanical, barrier, and antimicrobial properties of polymer NCs. SEM results indicate that the incorporation of both NPs, Ag-NPs and C30B, had a synergetic effect, resulting in a material with higher homogeneity and better clay dispersion. The results of the antimicrobial activity test indicate that all films showed antimicrobial activity against *S. aureus*, *E. coli*, and *Candida albicans* without significant differences between Ag-NPs concentrations, while pure starch exhibited no antimicrobial activity against any of tested microorganisms. According to the water contact angle measurements which investigate the incorporation effect of Ag-NPs and C30B on films surface polarity, more hydrophilic surface (lower values of contact angles) was achieved when both NPs (Ag-NPs and C30B) were added. In terms of gas barrier properties, the best water and oxygen barrier properties were obtained for the NCs containing 1 mM Ag and C30B (34% permeability reduction) which can be linked to the morphology accomplished. The mentioned properties are useful in food packaging applications (Abreu et al., 2015).

To study the effect of talc morphology on structural characteristics and optical, thermal, barrier, and mechanical properties of thermoplastic corn starch (TPS), Castillo et al. incorporated 0 and 5% w/w of both Australian (A10) and Argentinean (SJ10) talc samples (respect to TPS) in to corn starch by melt mixing and thermocompression. SEM results indicated that talc crystalline character influenced on bionanocomposites microstructure. Fracture surface of bionanocomposites containing A10 talc exhibited some aggregates as a result of the onion-type structure of these particles (Fig. 8.9B) and in films with SJ10 talc, the blocky morphology was observed (Fig. 8.9C).

It was found that macrocrystalline character of SJ10 talc induced a higher WVP reduction (~85%) than A10 microcrystalline one, which decreased this property up to 55%. This reduction is a result of preferential spatial disposition of the filler within the matrix, hydrogen bond interactions between the hydroxyl groups of the TPS and talc hydrophilic sites on mineral edge surfaces and repelling of water vapor flow by hydrophobic talc basal surfaces. Also, SJ10 blocky morphology provides to TPS-based films a more effective physical barrier against UV and visible radiation than A10 platy one. These improved UV and water vapor barrier can encourage use of these NCs in packaging field (Castillo et al., 2015).

8.2.2 Biomedical applications

A variety of polymeric bionanocomposite materials are produced by the incorporation of clay into polymers of synthetic or natural origin. Applications of polymer–silicate NCs used for this purpose are shown schematically in Fig. 8.10 (Wu et al., 2010).

Saha et al. performed in-vitro transdermal drug release study in order to understand the availability of Ketorolac tromethamine (a nonsteroidal antiinflammatory drug) (KT) from drug loaded (MC)/pectin (PEC)/MMT NC films. They prepared films with different weight ratios of MC and PEC and found that only MC: PEC (90:10) (MP_{10}) gave a strong and tough film. Then, different concentrations of MMT (1, 3, and 5 wt%) were incorporated into the MP_{10} matrix via solution mixing process which has some advantages over a melt mixing process. Fig. 8.11 shows the cumulative amount of drug released from the pure MC and MMT loaded NC film samples in phosphate buffer saline media of pH 7.4 at 37°C. Results displayed that the rate of release of the drug decreases with the addition of PEC and MMT in MC-based NC formulations and 5 wt% MMT-loaded MC/PEC/MMT NC film shows better performance than the other one in terms of controlled release of a transdermal drug. Also, it was found that mechanical properties, moisture absorption, and WVP studies gave best result in the case of NC film based on 3 wt

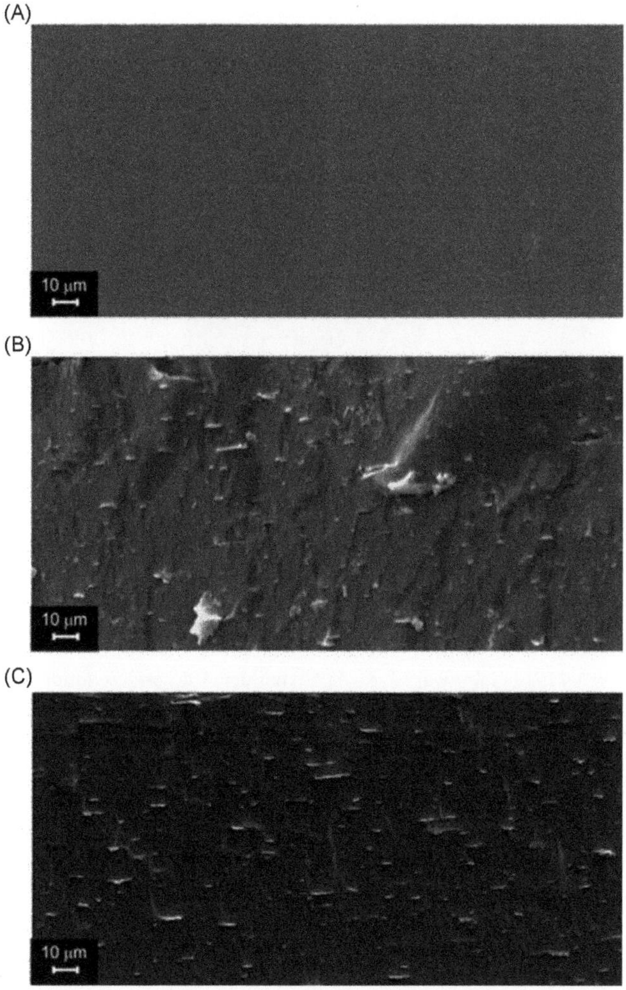

Figure 8.9 SEM micrographs of films based on: (A) TPS; (B) and (C) bionanocomposites of TPS with 5% w/w of A10 and SJ10 talc particles, respectively.
Adapted from Castillo, L.A., López, O.V., Ghilardi, J., Villar, M.A., Barbosa, S.E., and García, M.A. (2015). Thermoplastic starch/talc bionanocomposites. Influence of particle morphology on final properties. Food Hydrocoll. 51, 432–440. With kind permission of Elsevier.

% MMT-loaded MC/PEC/MMT, and it is suitable for packaging applications (Saha et al., 2016).

Othman et al. investigated the potential use of poly(D,L-lactide) (PLA)/MMT NC in biomedical applications as drug carriers. They fabricated paracetamol (PCM)-loaded composite NPs composed of a biodegradable poly (D,L-lactide) (PLA) polymer matrix filled with 2, 5, and 20 wt% of organically modified MMT NPs via

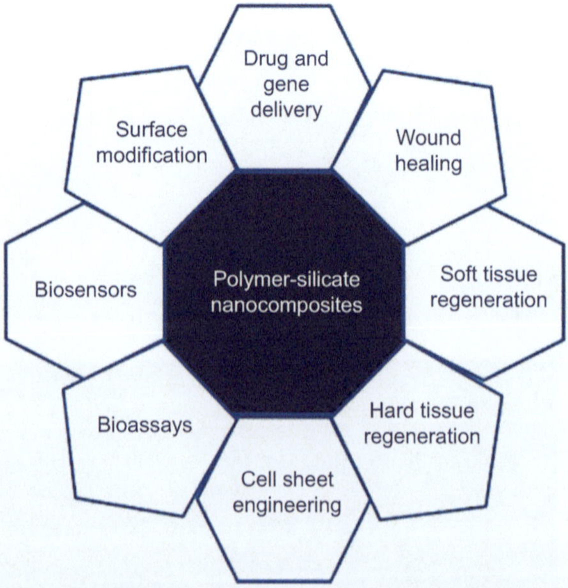

Figure 8.10 Schematic representation of biomedical applications of polymer silicate nanocomposites.
Adapted from Wu, C.-J., Gaharwar, A.K., Schexnailder, P.J., and Schmidt, G. (2010). Development of biomedical polymer-silicate nanocomposites: a materials science perspective. Materials 3, 2986.

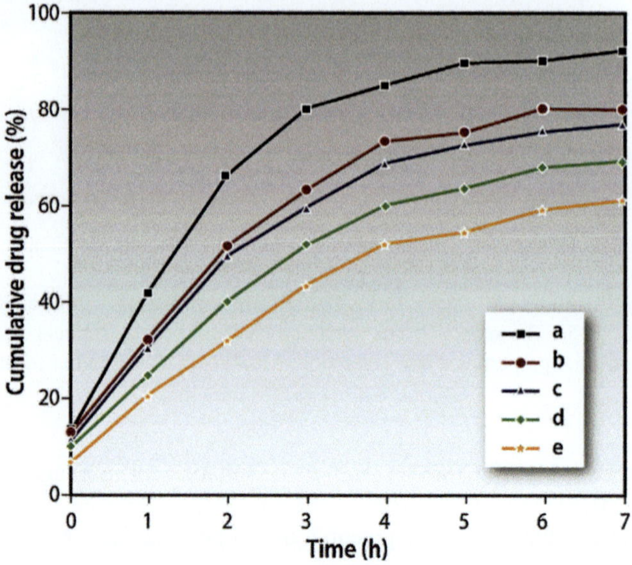

Figure 8.11 Cumulative drug (KT) release from drug loaded NC films (A) MK, (B) $MP_{10}K$, (C) $MP_{10}M(MMT)_1K$, (D) $MP_{10}M_3K$, and (E) $MP_{10}M_5K$.

antisolvent nanoprecipitation in a microfluidic coflow glass capillary device. The incorporation of nanoclay in the polymer matrix improved both the drug encapsulation efficiency and drug loading in the final formulation and extended the rate of drug release into simulated intestinal fluid. The results showed that the 2 wt% PLA/MMT composite film was found to be the most suitable drug carrier due to the spherical shape of the fabricated NPs and almost complete inclusion of the nanoclay platelets inside the host polymer (Othman et al., 2016).

Wang et al. described the formation of aspirin loaded MMT-polypyrrole (PPy) hybrids by in-situ electro polymerization of pyrrole in the dispersion of MMT in the presence of aspirin. The FESEM results confirmed the aggregated structure of MMT. After MMT was wrapped by PPy, compact cauliflower shaped morphology was observed for MMT-PPy (Fig. 8.12).

Also, aspirin release from the MMT-PPy hybrids and PPy without and with external electrical stimulation was investigated and compared, and the cumulative release of aspirin versus time was plotted in Fig. 8.13. It was found that, compared with PPy, the as-prepared MMT-PPy presented significantly enhanced electroresponsive drug release profile attributed to the relatively large SSA (small specific surface area) of Mt (>60 $m^2\ g^{-1}$) compared with that of PPy. This property makes as-prepared MMT-PPy hybrid a promising candidate for smart and intelligent drug release (Wang et al., 2016).

A complex of strontium ranelate (SRA), a drug for osteoporosis, with laponite (LS) was prepared by Nair et al. for the first time for bone tissue engineering applications, with a view to release the drug in a localized and controlled manner and benefit from the dual effect of laponite and SRA for bone tissue regeneration. The laponite–SRA complex was solution blended with 3, 6, and 12 wt% of PCL to obtain composite scaffolds (PCL-LS) for bone tissue engineering applications. TEM images of laponite and LS1 showed that the laponite consists of randomly distributed clay platelets, whereas good structural ordering was evident for LS1. The bulky SRA molecules which unable to intercalate in the clay interlayer through the cation exchange mechanism, interacted with the negatively charged surface of the clay tactoids, through electrostatic interaction (Fig. 8.14).

Release profile of strontium ranelate from the scaffolds exhibited an inverse relation on the drug–clay complex content, due to enhanced hydrophilicity and retention capacity of the composite scaffolds. The composite scaffolds with varying laponite–SRA complex content of 3–12 wt% were evaluated in vitro using human osteosarcoma cells. It was found that an optimum composition of the scaffold with 3 wt% laponite–SRA complex loading would be ideal for obtaining enhanced ALP (alkaline phosphatase) activity, by maintaining cell viability (Nair et al., 2016).

Different immobilization platforms using poly(methylmethacrylate) (PMMA)–clay NC and a conjugated polymer, poly(4-(2,3-dihydrothieno[3,4-*b*][1,4]dioxin-5yl)-7-(2,3-dihydro-thieno[3,4-*b*][1,4]dioxin-7-yl)-2-benzyl-1*H*-benzo[*d*]imidazole) (poly(BIPE)), were examined for the detection of glucose concentration by Emre et al. PMMA–clay NC was synthesized by grafting PMMA with 2-(methacryloyloxy) ethyl trimethylammonium chloride (MTMA)-modified laponite using

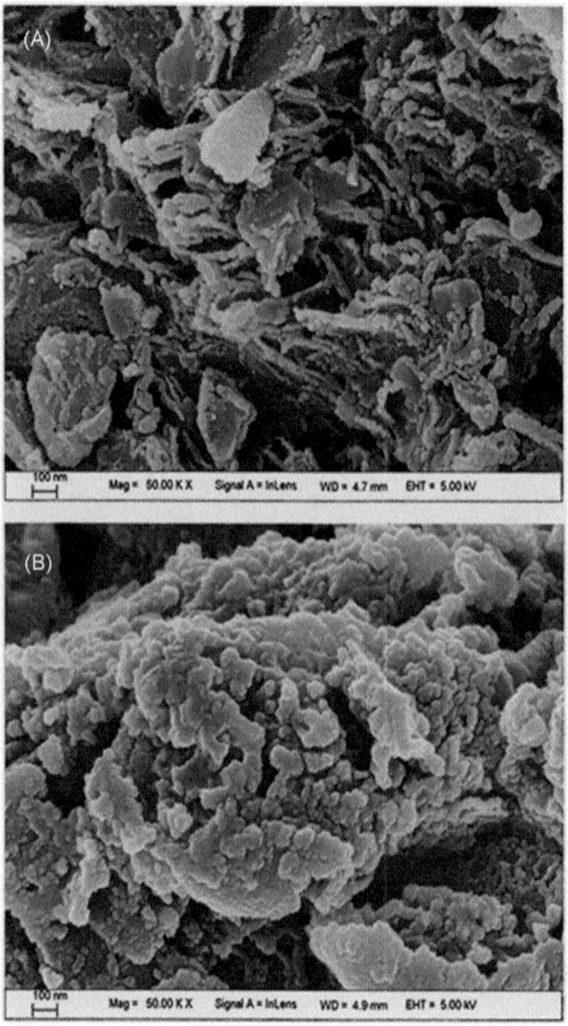

Figure 8.12 FE-SEM images of Mt (A) and Mt-PPy hybrids (B).
Adapted from Wang, F., Chang, P.R., Zheng, P., and Ma, X. (2015). Monolithic porous rectorite/starch composites: fabrication, modification and adsorption. Appl. Surface Sci. 349, 251–258. With kind permission of Elsevier.

suitable grafting agent via in-situ suspension polymerization. Glucose oxidase (GOx) was selected as the model enzyme which was immobilized onto the desired surface in the presence of glutaraldehyde which was used as the crosslinking agent to attain a more compact enzyme structure and to improve the stability of the biosensor. Owing to the presence of aromatic units in conjugated polymer, it hosts enzyme molecules and clay NC excellently. Three different electrodes were prepared; poly(BIPE)/GOx, PCN/GOx, and PCN/poly (BIPE)/GOx, and it was found

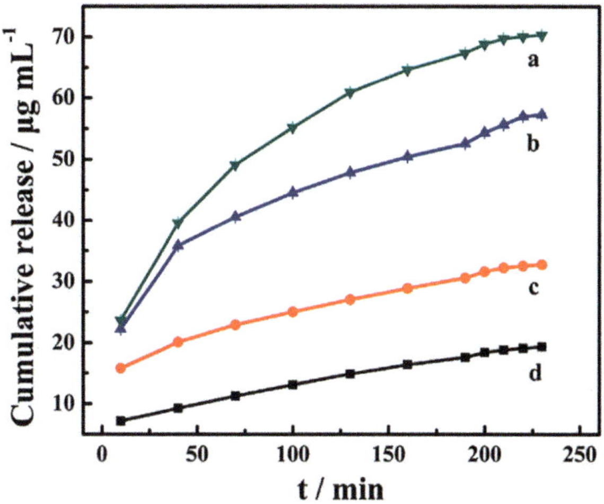

Figure 8.13 The plots of cumulative aspirin release versus time. Release of aspirin from Mt-PPy (A) and PPy (B) by an electrical stimulation at −0.6 V; spontaneous release of aspirin from Mt-PPy (C) and PPy (D) without an electrical stimulation.
Adapted from Wang, F., Chang, P.R., Zheng, P., and Ma, X. (2015). Monolithic porous rectorite/starch composites: fabrication, modification and adsorption. *Appl. Surface Sci.* **349**, 251−258. With kind permission of Elsevier.

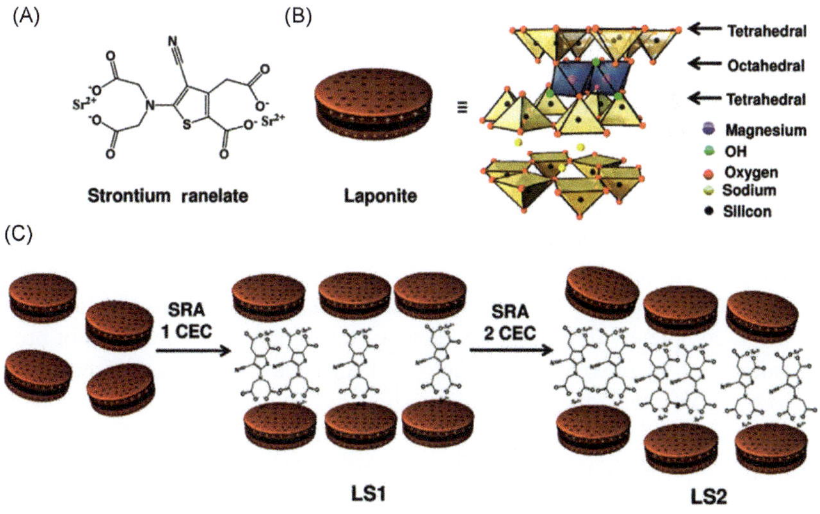

Figure 8.14 (A) Chemical structure of SRA, (B) schematic representation of laponite, and (C) scheme of interaction between laponite and SRA in LS1 and LS2.
Adopted from Nair, B.P., Sindhu, M., and Nair, P.D. (2016). Polycaprolactone-laponite composite scaffold releasing strontium ranelate for bone tissue engineering applications. Colloids Surf. B: Biointerfaces 143, 423−430. With kind permission of Elsevier.

that the enzyme electrode prepared with both poly (BIPE) and PCN exhibited the superior biosensor responses (Emre et al., 2015).

Novel bio-inspired methyl cellulose (MC)–sodium alginate (SA)–MMT-clay-based bionanocomposite films were manufactured by solvent-casting method for use as wound healing materials by Mishra et al. The MC solution was mixed with SA solution on a high speed homogenizer so as to get a blend ratio of 1:1, 1:2, 1:3, 2:1, and 3:1, and various concentrations of unmodified MMT (1%–5%, w/v) was added to polymer blends. The SEM results of bionanocomposite showed uniform aggregation of clay particles in the composite films but further increase in MMT lead to irregular morphology. It was also observed that the developed films were able to exhibit antimicrobial activity against four typical pathogenic bacteria found in the presence of wound. All the developed films were able to inhibit *E. faecium* and *P. aeruginosa* at high concentrations (10 mg mL^{-1}) with no significant differences between any of the films. In vitro scratch assay indicated potential wound closure activities of MC-2–4 bionanocomposite films at their respective highest subtoxic doses. The properties of these ternary bionanocomposite films make them suitable for wound healing applications (Mishra et al., 2014).

Mauro et al. evaluated the suitability of MMT-reinforced hydrogels, based on a peptidomimetic polyamidoamine carrying guanidine pendants (AGMA1), as scaffolds for osteoblast culture and bone repair. It was considered that AGMA1, being prevailingly cationic, was particularly apt to interact with MMT platelets in aqueous media, establishing strong and complex polyelectrolytic bridges, which may act as additional physical crosslinkers. Two series of MMT-reinforced AGMA1 composite hydrogels were synthesized (Fig. 8.15.):

1. the MMT content was maintained constant at 10%, while varying the degree of crosslinking, according to the different lengths of the starting bis-acrylamide determinate oligomers and
2. the degree of crosslinking was constantly 25%, whereas the MMT content varied in the range 1%–10%.

Hydrogels of the AGMA25–MMT series with MMT amounts in the range 0%–10% when evaluated as scaffolds for the osteogenic differentiation of mouse calvaria-derived preosteoblastic MC3T3-E1 cells, proved able to support cell adhesion and proliferation and clearly induced differentiation toward the osteoblastic phenotype, as indicated by different markers. In addition, AGMA1–MMT hydrogels proved completely degradable in aqueous media at pH 7.4 and did not provide any evidence of cytotoxicity. Morphological analysis by confocal laser scanning microscopy confirmed these findings, showing an excellent cellular network throughout the hydrogel composite (Figs. 8.16 and 8.17). The experimental evidence suggests that AGMA1–MMT composites definitely warrant potential as scaffolds for osteoblast culture and bone grafts (Mauro et al., 2016).

8.2.3 Wastewater pretreatment

The chemical contamination of water from a wide range of toxic substances such as heavy metals, organic compounds, and dyes is a serious problem owing to their

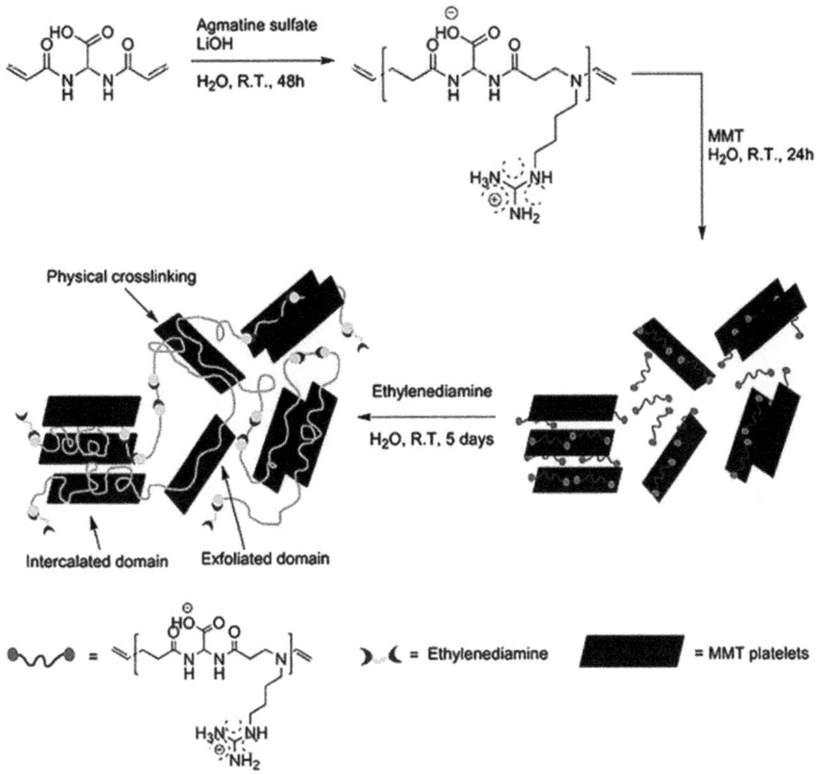

Figure 8.15 Synthesis of AGMA1−hydrogel composites.
Adapted from Mauro, N., Chiellini, F., Bartoli, C., Gazzarri, M., Laus, M., Antonioli, D., et al. (2016). RGD-mimic polyamidoamine−montmorillonite composites with tunable stiffness as scaffolds for bone tissue-engineering applications. J. Tissue Eng. Regenerative Med., n/a-n/a. With kind permission of John Wiley and Sons.

human toxicity. Therefore, numerous researchers have focused their attention on the developing techniques that can remove toxic pollutants from wastewater. The different techniques have been utilized to remove these and other micropollutants from water and industrial wastewater including chemical precipitation, conventional coagulation, reverse osmosis, ion-exchange, electrodialysis, electrolysis, and adsorption. Among these, adsorption process has become widely used for the removal of inorganic and organic micropollutants from aqueous solution and various adsorbents were prepared for water treatment over the years. In recent years, PCNs have been intensively used for the removal of organic pollutants from water, because of large specific surface area, cation exchange, low cost, and toxicity of clay. The exchangeable cations, the distance between the clay mineral layers, and the existence of water molecules between the layers are important factors affecting the adsorption of organic molecules to these minerals (Unuabonah and Taubert, 2014; Mishra, 2014; Hernández-Hernández et al., 2016).

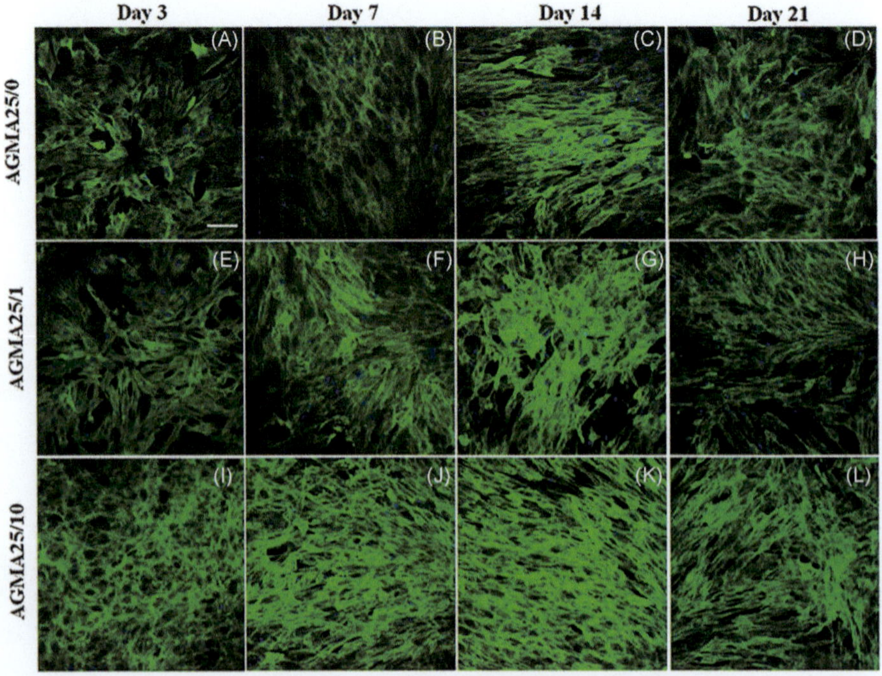

Figure 8.16 Confocal laser scanning micrographs of MC3T3-E1 cultured on AGMA25–MMT hydrogels; magnification = ×20 (scale bar = 100 μm).
Adapted from Mauro, N., Chiellini, F., Bartoli, C., Gazzarri, M., Laus, M., Antonioli, D., et al. (2016). RGD-mimic polyamidoamine–montmorillonite composites with tunable stiffness as scaffolds for bone tissue-engineering applications. *J. Tissue Eng. Regenerative Med.*, n/a-n/a. With kind permission of John Wiley and Sons.

Kara et al. showed that the sepiolite/poly(vinylimidazole) (sepiolite/PVI) composite is useful for Hg(II) removal from wastewater. They fabricated sepiolite/PVI composite by in-situ polymerization of vinyl imidazole and unmodified sepiolite, as shown in Fig. 8.18.

Sepiolite/PVI composite showed a dense and smooth surfaces and the dispersion of sepiolite in the polymer matrix was good, as confirmed from SEM results. Also, the effects of the pH, initial Hg(II) concentration, and temperature on the adsorption of Hg(II) by the sepiolite/PVI composite samples were investigated. It was found that the removal of mercury was increased with an increase in the initial Hg(II) concentration due to possibility of binding sites for the adsorption of Hg(II) ions and the adsorption capacity of sepiolite/PVI composite is higher in comparison to that of sepiolite under the same conditions. The effect of pH on the Hg(II) adsorption was investigated at different pH in the range of 2–6.5. The optimum pH value for Hg(II) adsorption is found at pH 6 because the decrease of the concentration of H^+ promoted the formation of the mercury–imidazole complex (Fig. 8.19) and increased the adsorption capacity of the sepiolite/PVI composite. In terms of

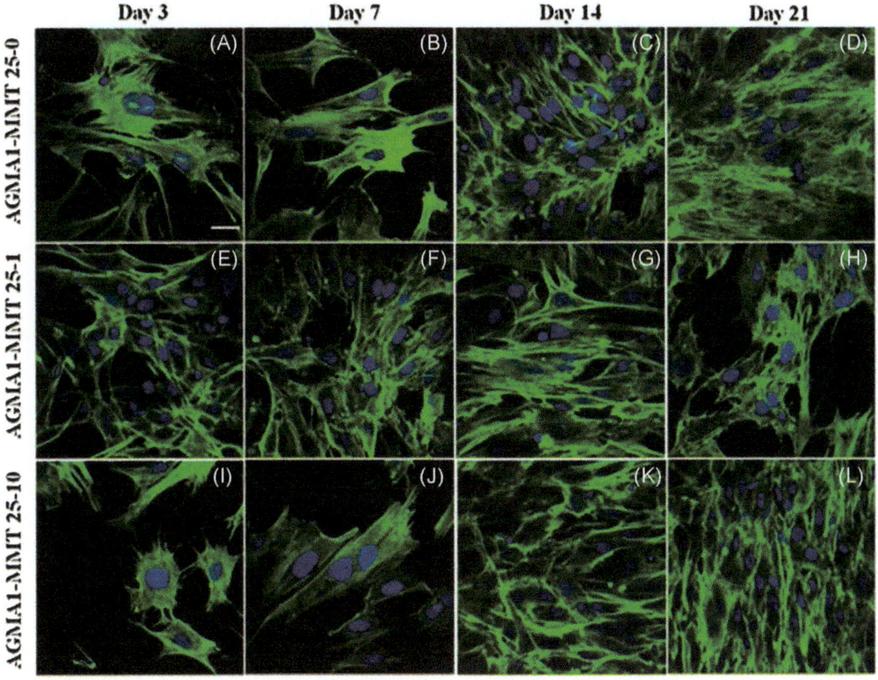

Figure 8.17 Confocal laser scanning micrographs of MC3T3-E1 cultured on AGMA25–MMT hydrogels; magnification = ×60 (scale bar = 30 μm).
Adapted from Mauro, N., Chiellini, F., Bartoli, C., Gazzarri, M., Laus, M., Antonioli, D., et al. (2016). RGD-mimic polyamidoamine–montmorillonite composites with tunable stiffness as scaffolds for bone tissue-engineering applications. *J. Tissue Eng. Regenerative Med.*, n/a-n/a. With kind permission of John Wiley and Sons.

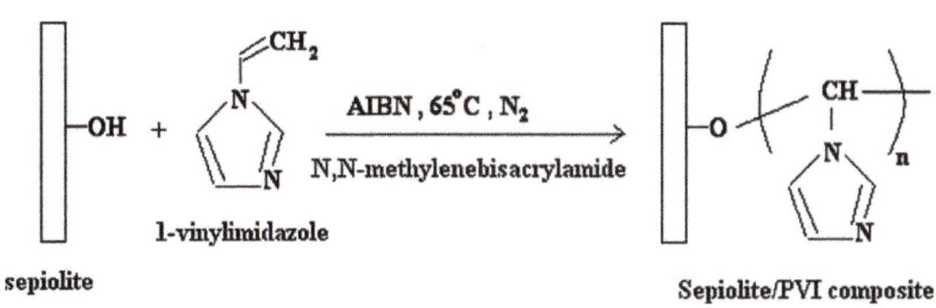

Figure 8.18 Proposed the formation process of sepiolite/PVI composite.
Adapted from Kara, A., Tekin, N., Alan, A., and Şafaklı, A. (2016). Physicochemical parameters of Hg(II) ions adsorption from aqueous solution by sepiolite/poly (vinylimidazole). J. Environ. Chem. Eng. 4, 1642–1652. With kind permission of Elsevier.

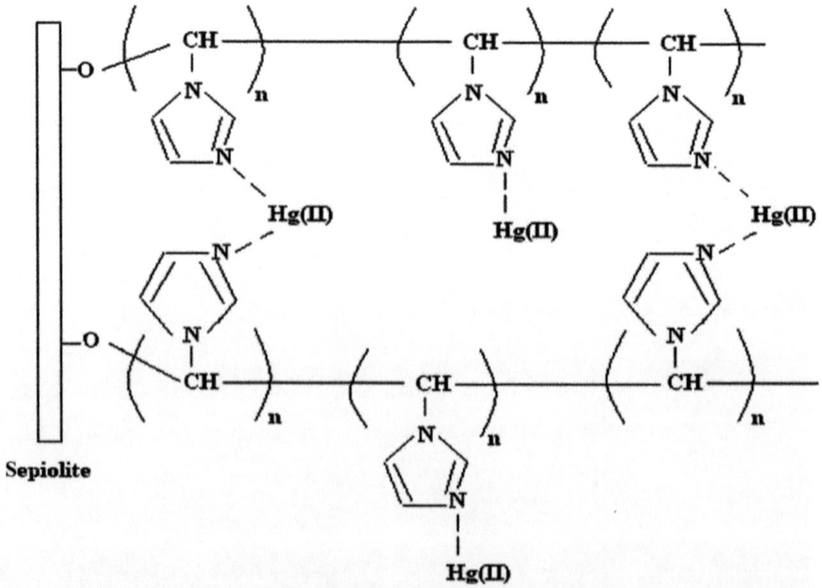

Figure 8.19 Proposed adsorption mechanism between imidazole groups in sepiolite/PVI composite and Hg(II).
Adapted from Kara, A., Tekin, N., Alan, A., and Şafaklı, A. (2016). Physicochemical parameters of Hg(II) ions adsorption from aqueous solution by sepiolite/poly (vinylimidazole). *J. Environ. Chem. Eng.* **4**, 1642–1652. With kind permission of Elsevier.

temperature, it is found that adsorption capacity values increased from 65.04 to 288.10 mg g^{-1} as temperature was increased from 277 to 338 K. This change could be due to higher collision frequencies at higher temperature and hence leading to more sorption (Kara et al., 2016).

Rusmin et al. synthesized chitosan–palygorskite composites at 1:1, 1:2, and 2:1 mass ratios of the components (CP1, CP2, and C2P, respectively) by using acetic acid (1%, v/v) as the solvent and sodium tripolyphosphate (3%, m/v) as the cross-linking agent. The interaction between chitosan and palygorskite was mainly through electrostatic attraction between the cationic polyelectrolyte and the negatively charged clay mineral also hydrogen bonding between palygorskite's silanol groups and chitosan's hydroxyl and amino groups (Fig. 8.20).

The composite beads' performance in removing Pb from aqueous solutions through adsorption was studied, and the results showed that higher adsorption capacities were obtained for chitosan–palygorskite composites than their pristine materials in the order of CP1 > C2P > CP2 > chitosan > palygorskite, which suggested that besides the availability of functional groups in chitosan and palygorskite, other factors like structural characteristics, porosity, and solute transportation mechanism might affect the adsorption process. Adsorption by C2P is best described by Langmuir model, while that by CP1 and CP2 fits to Freundlich model.

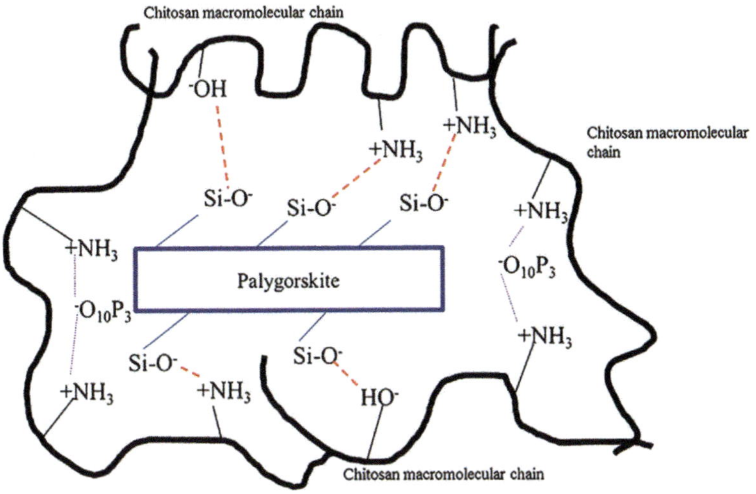

Figure 8.20 Proposed chemical structure of chitosan–palygorskite composite.
Adapted from Rusmin, R., Sarkar, B., Liu, Y., McClure, S., and Naidu, R. (2015). Structural evolution of chitosan–palygorskite composites and removal of aqueous lead by composite beads. Appl. Surface Sci. 353, 363–375. With kind permission of Elsevier.

Therefore, these composites have a great potential for application as an inexpensive adsorbent in removing toxic metals as an example of a pollutant from wastewater (Rusmin et al., 2015).

Wang et al. showed that the modification of monolithic rectorite/starch composites (PRSs) improved the adsorption of methylene blue (MB) and Pb^{+2} adsorption from the water. They fabricated these composites by freezing the composite gels and exchanging ice with ethanol to obtain the porous structures. The modification of PRSs was made (removal of starch by calcinating and the introduction of xanthate groups) without destroying the porous structures which were found to improve the adsorption capacities for MB dye and Pb^{+2}. The maximum adsorption capacities could reach 277.0 for MB by calcinated PRSs because when starch components were calcinated, more pores were opened, and the exposed rectorite adsorbed more MB dye. Porous rectorite/starch xanthate composites (PRSX) gave better adsorption efficiency (180.8 mg g^{-1}) for Pb(II) due to chelation between xanthate groups and Pb^{2+} ions which could effectively remove Pb^{2+} from the water (Wang et al., 2015).

Zeng et al. prepared the novel crosslinking chitosan/rectorie (CTS/REC) nanohybrid composite microsphere in which the mass ratios of chitosan to rectorite were 2:1 (CTS/REC-1), 3:1 (CTS/REC-2), and 4:1 (CTS/REC-3). They found that a large number of the microspheres were formed after crosslinking and SEM analysis revealed that the surfaces of the crosslinking CTS/REC nanohybrid composite microspheres with more molar ratios of REC were noted to be more coarsely (Fig. 8.21). The adsorption efficiency of CTS/REC-1, CTS/REC-2, and CTS/REC-3 for the removal of Cd(II), Cu(II), and Ni(II) from water was examined. The study found a decrease in adsorption capacity with the increasing proportion of chitosan

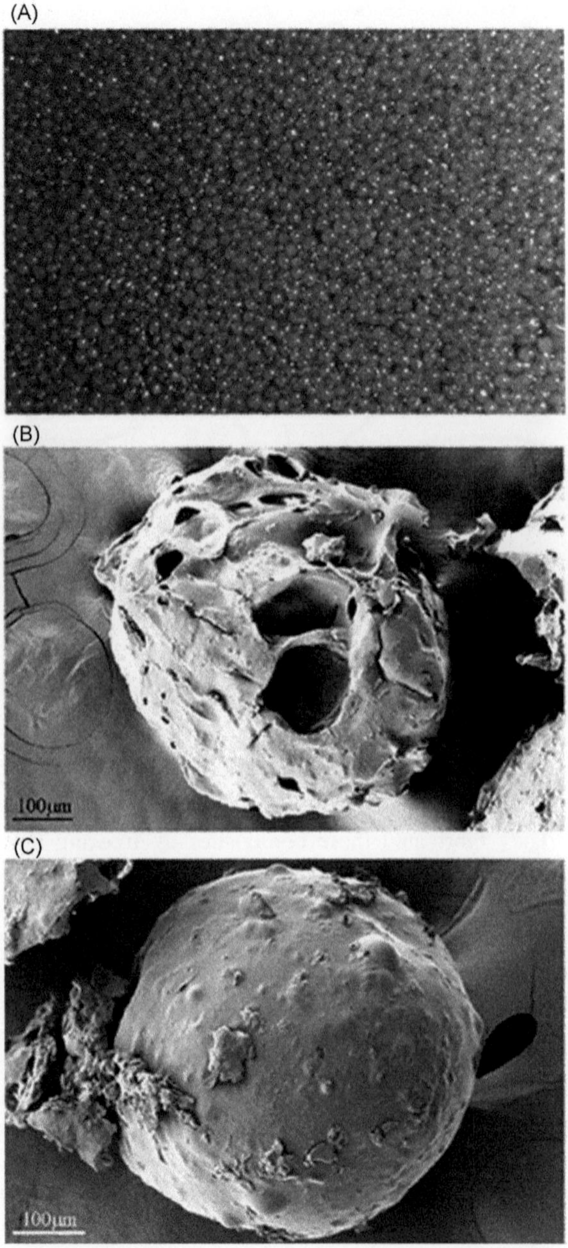

Figure 8.21 (A) The photograph of the wet crosslinked-CTS/REC nanohybrid 235 composite resins and SEM micrographs of, (B) CTS/REC-1, and (C) CTS/REC-2. Adapted from Zeng, L., Chen, Y., Zhang, Q., Guo, X., Peng, Y., Xiao, H., et al. (2015). Adsorption of Cd(II), Cu(II) and Ni(II) ions by cross-linking chitosan/rectorite nano-hybrid composite microspheres. Carbohydr. Polym. 130, 333–343. With kind permission of Elsevier.

which attributed to the porous surface of CTS/REC-1 samples with the maximum content of rectorite. Also, the adsorption capacity follows the order of Cu(II) > Cd (II) > Ni(II). The effect of pH on the adsorption was investigated, and the study found an increase in metal uptake with an increase in pH up to 5.0. Copper was more strongly attracted to the adsorbent, with higher removal percentages at the same pH than the other two metals due to the competition from H^+ ions and the protonation effect on amino group at low pH. It could be concluded that the crosslinking CTS/REC nanohybrid composite microspheres serve as suitable adsorbent in metal ions removal. These results indicated that the crosslinking CTS/REC nanohybrid composite microspheres were useful for metal ions removal from aqueous solutions (Zeng et al., 2015).

El-Zahhar reported the synthesis of organically modified kaolinite-doped polyacrylonitrile (PAN) NCs through grafting of PAN onto clay. The kaolinite was modified with different ratios of hexadecylethyldimethylammonium (HDEDMA) to kaolinite followed by the chemical grafting polymerization of acrylonitrile (AN) onto organophilic kaolinite. The resultant composite was studied for simultaneous adsorption of Cr(VI) and methylene blue dye (MB) from aqueous solutions and adsorption experiments showed that the adsorbed amount of Cr(VI) and MB increased with increasing the concentration of HDEDMA in the clay which is due to increasing the surface active sites on the composite adsorbent with increasing HDEDMA concentration. Also, it was found that an increase in temperature resulted in increase in the amount of Cr(VI) and MB adsorbed (El-Zahhar, 2015).

Yildiz et al. investigated the adsorption properties of PVI–clay composites for the removal of remazol black B (RB) from water for the first time. PVI was adsorbed on two clay minerals sodium bentonite (Na–Bt) and calcium bentonite (Ca–Bt) to form PCN. The SEM pictures of the clays and prepared clay composites with PVI indicated that interlayer spaces are decreased in Na–Bt–PVI and Ca–Bt–PVI which may be related to the impregnation of the PVI between the clay layers. RB adsorption speed was determined by using PVI (maximum adsorbed PVI), and different amounts of RB were added into 0.1% clay suspensions. Adsorption process progresses because of the electrostatic interaction between anionic group (sulfonic group) of the dye and the positively charged surface of organo clays (Yildiz and Senkal, 2016).

Ma et al. reported the fabrication of high performance and low cost composite superabsorbent (PAsp-*g*-PAA/PGS) by graft copolymerization of acrylic acid (PAA), polyaspartic acid (PAsp), and palygorskite (PGS) clay (0 and 2.5 wt%) in the presence of initiator and crosslinking agents (Fig. 8.22). The whole process includes chain initiation, graft and crosslinking. SEM micrographs of composite superabsorbent revealed that after introducing palygorskite clay, the surface of superabsorben becomes coarse and loose, and many creases can be observed. The composite superabsorbent showed high water absorbency (1405 $g\ g^{-1}$ in distilled water); pH-stability (pH 6–10), and salt tolerance (93 $g\ g^{-1}$ in 0.9 wt% salt solution) which revealed that the prepared composite could be potentially used as multivalent metal ions removal from wastewater (Ma et al., 2015).

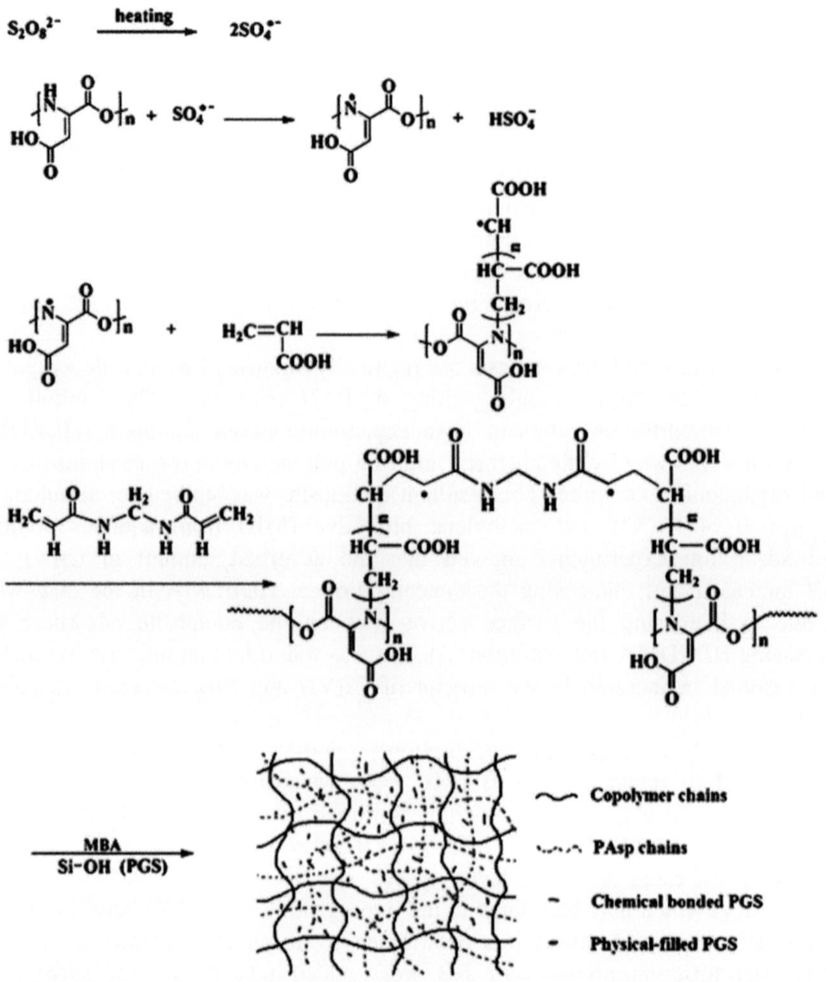

Figure 8.22 Proposed reaction mechanism for synthesis of PAsp-*g*-PAA/PGS composite superabsorbent.
Adapted from Ma, G., Yang, Q., Ran, F., Dong, Z., and Lei, Z. (2015). High performance and low cost composite superabsorbent based on polyaspartic acid and palygorskite clay. Appl. Clay Sci. 118, 21–28. With kind permission of Elsevier.

8.2.4 Other applications

Jain et al. reported the synthesis of polyacrylamide/clay NC (PANC) via free radical polymerization technique with potassium persulfate (KPS) as an initiator. The change in the morphology of the poly(acryl amide) after the formation of NC which was studied using FE-SEM analysis indicated the completely covered clay layers in NC and the compatibility between the dispersed clay layers and polymer matrix. Various ratios (0.05–0.25) of clay to monomer were used to get well dispersed polymer matrix with nanoclay and the effect of synthesized product on the rheological and filtration properties of the drilling fluid system was studied. A decrease in these properties was found at higher clay to monomer ratios. Also, the rheological properties of the water-based inhibitive mud system were increased with the increase in the concentration of the NC. In shale dispersion tests, the recovery rate of the shale cuttings increased with the increase in NC concentration. The study showed that NC could inhibit swelling and dispersion of the shales more effectively. The NC had better shale recovery rate than the partially hydrolyzed polyacrylamide (PHPA) as revealed by shale recovery tests on the Indian shale sample. For this reason, these NCs may be employed as adrilling fluid additive in inhibitive water based drilling fluid system (Jain and Mahto, 2015).

The effect of sepiolite clay on the thermal and fire behavior of an epoxy resin (bisphenol-A/epichlorohydrin derived liquid epoxy resin) was studied by Zotti et al. Sepiolite was used in different concentrations (from 2 to 10 wt%) as inorganic filler in two different forms: hydrated and dehydrated. According to the SEM and optical micrographs of nanocomposites, a good level of dispersion and homogeneity was found up to 10 wt%. Thermogravimetrical results showed that the effect of sepiolite, in its both forms (hydrated and dehydrated), on the degradation behavior of the final NC, is substantially negligible confirming that sepiolite act as inert filler without any chemical interaction with the hosting system. The various morphology of the char layer during the burning process can have relevant flame retardant effects acting on both condensate and vapor phase. According to Cone calorimeter analysis which investigated the fire behavior of the neat system and its filled mixture with sepiolite clay, a reduction of about 27% of the peak of heat release rate for the highest sepiolite percentage was measured and the burning total period was increased so confirming that sepiolite when pretreated represents a valid fire retardant inorganic filler for such a system (Zotti et al., 2015).

Fang et al. showed that the composite separators of poly(vinylidene fluoride) (PVDF) with different contents of MMT (0, 3, 5, and 7 wt%) are compatible for applications in Li-ion batteries. The morphology of composite membranes with different filler contents was investigated SEM (Fig. 8.23). The membranes consist of multilayered and three dimensional network structure with numerous randomly oriented continuous fibers.

The results of electrochemical measurements suggest that PVDF/MMT-5% composite membrane has maximum ionic conductivity of 4.2 mS cm^{-1} at room temperature, which is higher than the ionic conductivity of unfilled polymer (2.04 mS m^{-1}), lower interfacial resistance of 97 V, and excellent electrochemical stability. The mechanical properties of membranes, which are necessary for the safe handling and mechanical stability of batteries, were measured. It was noted that the tensile modulus and elongation-at-break of the PVDF fibrous membranes increased with increasing filler content which was due to the increased fiber diameter, the enhanced crystallinity, and the reinforcement effect of MMT particles. This experimental evidence suggests that the PVDF/MMT composite membranes are promising separators for high performance Li-ion batteries (Fang et al., 2016).

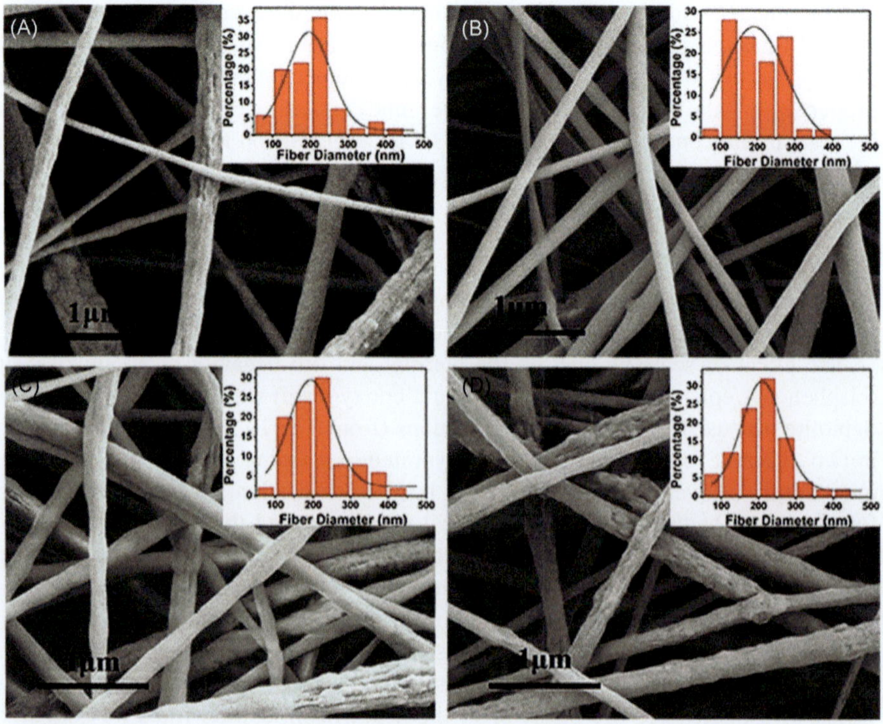

Figure 8.23 SEM image and histograms of fiber diameter distribution of composite membranes with different filler contents: (A) 0%, (B) 3%, (C) 5%, and (D) 7%.
Adapted from Fang, C., Yang, S., Zhao, X., Du, P., and Xiong, J. (2016). Electrospun montmorillonite modified poly(vinylidene fluoride) nanocomposite separators for lithium-ion batteries. Mater. Res. Bull. 79, 1–7. With kind permission of Elsevier.

Sahu et al. reported the fabrication of flame retardant poly(vinyl acetate) (PVAc)/MMT clay NCs via emulsifier-free emulsion technique. It was found that the PVAc was exfoliated into gallery structure of silicate by the catalytic action of Cu(II)/EDTA complex with help of the initiator potassium monopersulfate (KMPS) in microwave (MW) oven. The MMT layers are well dispersed in the PVAc matrix as evident from TEM studies. The fine dispersion of the MMT and the interactions between PVAc and clay created significant improvement of the flame retardancy which has been studied using cone calorimeter. Flame retardant mechanism of NCs is owing to the formation of Si–O–C (carbonaceous-silicate) char, which acts as insulating barrier for flow of heat and mass. To conclude, as the synthesized NCs are ecofriendly environmental benign green synthesis by MW, they may open the door for future prospects as commercial flame retardants (Sahu et al., 2014).

Essabir et al. prepared the hybrid composites by addition of oil palm fibers as an organic filler and clay as inorganic filler in to thermoplastic matrix (high density polyethylene (HDPE)). The SEM images of the HDPE/clay display that clay particles were well dispersed into the composites with no evidence of aggregate which was

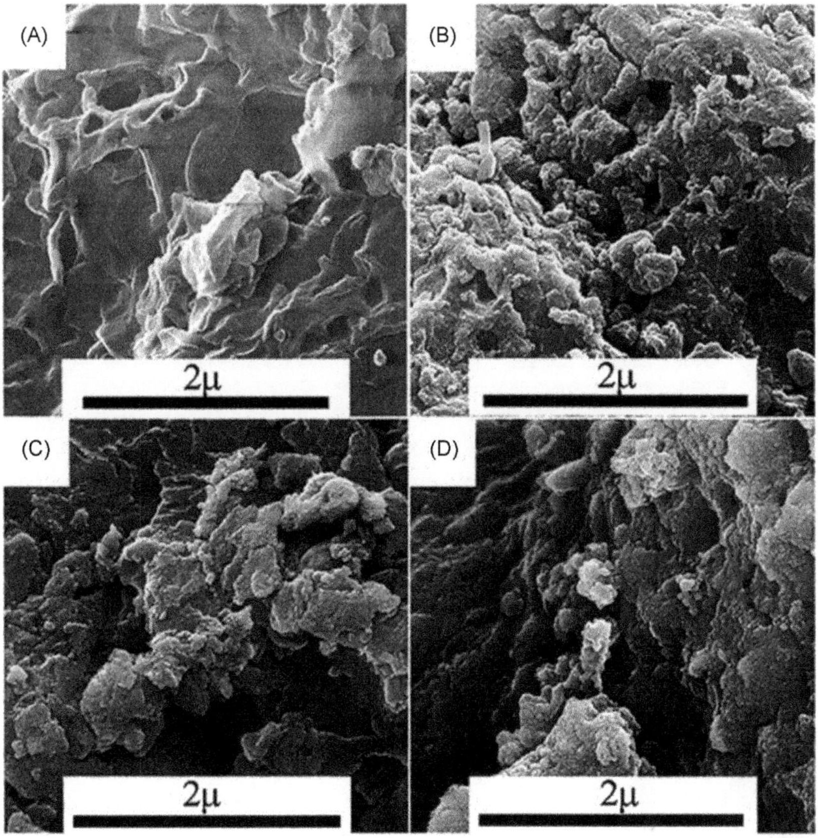

Figure 8.24 SEM images of (A) acid-treated bentonite; (B) PAA—75% ABent; (C) PAA—66% ABent; and (D) PAA—50% ABent.
Adapted from Liu, E., Sarkar, B., Wang, L., and Naidu, R. (2016). Copper-complexed clay/poly-acrylic acid composites: extremely efficient adsorbents of ammonia gas. Appl. Clay Sci. 121–122, 154–161. With kind permission of Elsevier.

due to the high affinity between the particles and the polymer by using the compatibilizer combined with optimum extrusion conditions. The tensile properties results showed that the hybrids composite with 12.5:12.5 clay/fiber content had the best tensile properties with a gain of 49% in Young's modulus and 11% in TS. The thermal analyses showed that the use of clay enhanced the degradation temperature of the composites giving better thermal stability to the fiber composites. In conclusion, the hybrid composites demonstrated higher thermal, tensile, and rheological properties which are the main properties required for plastics applications (Essabir et al., 2016).

Liu et al. prepared the copper-complexed clay/poly-acrylic acid (PAA) composites for ammonia gas (NH_3) adsorption with two kinds of clays: acid-treated bentonite and untreated palygorskite and studied the effect of clay types on the adsorption capacities of NH_3. The clay to polymer proportion was varied from 50% to 75% to find the

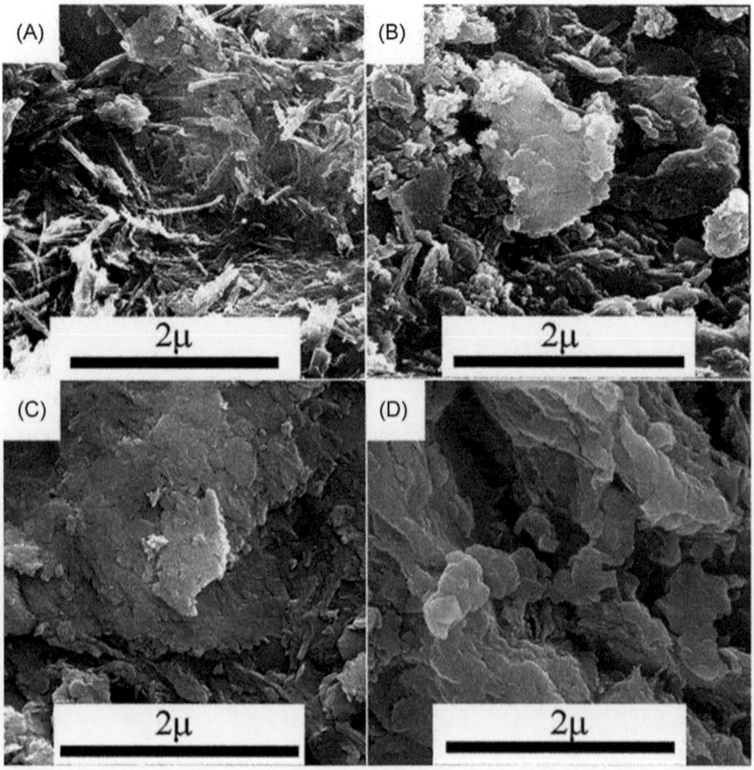

Figure 8.25 SEM images of (A) palygorskite; (B) PAA—75% Paly; (C) PAA—66% Paly; and (D) PAA—50% Paly.
Adapted from Liu, E., Sarkar, B., Wang, L., and Naidu, R. (2016). Copper-complexed clay/poly-acrylic acid composites: extremely efficient adsorbents of ammonia gas. *Appl. Clay Sci.* **121–122**, 154–161. With kind permission of Elsevier.

optimum adsorption performance for NH_3. The morphology of the synthesized composites was investigated by scanning electron microscopy (Figs. 8.24 and 8.25).

The results indicated that bentonite/PAA composites showed an extremely rough surface and the polymer was tightly attached onto the clay layers but in palygorskite/PAA composites the fibrous morphology became difficult to see mainly because all the particles were embedded in the polymer matrix. Also it was found that, in both composites, the more clay added into the composites resulted in a greater porosity of the resultant material.

Adding the clay resulted in more functional groups being accessible to NH_3 gas. For the copper clay/PAA composites, the highest adsorption capacity for NH_3 was achieved by adding 75% of the palygorskite (with capacities of 65.8 mg g^{-1}) and 66% of the acid-treated bentonite (with capacities of 80.0 mg g^{-1}). The key mechanism of NH_3 retention on the copper clay/PAA composites was a Lewis acid/base interaction between NH_3 and the Cu ions (Fig. 8.26). Trapping NH_3 on such adsorbents can lead

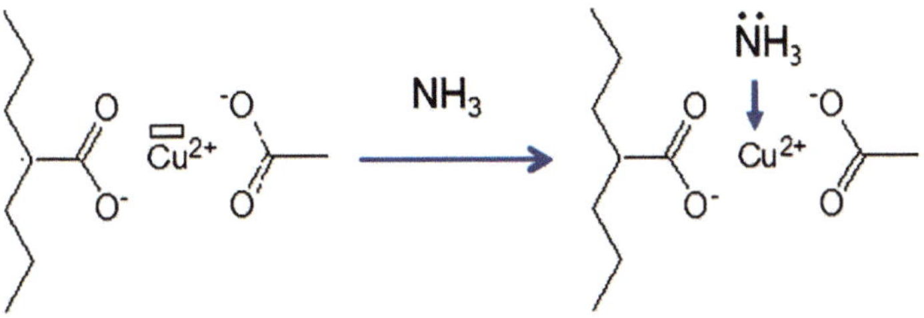

Figure 8.26 Mechanism of NH3 adsorption by Cu ion through Lewis acid/base interaction. Adapted from Liu, E., Sarkar, B., Wang, L., and Naidu, R. (2016). Copper-complexed clay/poly-acrylic acid composites: extremely efficient adsorbents of ammonia gas. *Appl. Clay Sci.* **121–122**, 154–161. With kind permission of Elsevier.

to color change, and this makes it possible to predict the lifetime of the adsorption bed visually, which is extremely important for real applications. Therefore, these composites can potentially be utilized in air filters. They may provide an effective and cheap way for removing NH_3 from contaminated air (Liu et al., 2016).

8.3 Conclusions

PCNs have attracted a great deal of attention, because of their huge potential for industrial and technological applications including, food packaging, biomedical applications, wastewater pretreatment, electricals/electronics, optoelectronics, sensors, automobiles, and so on. This chapter focuses on the synthesis of PCNs and aims to provide an overview of its current applications. We studied the food packaging, biomedical, and wastewater pretreatment applications of PCNs mainly. The incorporation of nanoclays into the polymer matrix to improve the physicochemical (mechanical, optical, gas, and vapor barrier) properties of the polymers with a view to be used in food packaging would be of great interest and have great potential. Clay have been widely used for curative and protective purposes by humans since ancient times due to their physicochemical characteristics such as high surface reactivity, good rheological behavior, high acid absorbing capacity, and high dispersibility in water, which renders them suitable for various biomedical applications. Also, PCNs have been intensively used for the removal of organic pollutants from water because of large specific surface area, cation exchange, low cost, and toxicity of clay. They are lost work could be done in this area in terms of synthesis and applications, and we are sure there will be extensive research work in the future to obtain wide verity PCNs for many applications.

Acknowledgments

The authors gratefully acknowledge the financial support provided by Research Affairs Division Isfahan University of Technology (IUT) Isfahan and the National Elite Foundation (NEF).

References

Abreu, A.S., Oliveira, M., de Sá, A., Rodrigues, R.M., Cerqueira, M.A., Vicente, A.A., et al., 2015. Antimicrobial nanostructured starch based films for packaging. Carbohydr. Polym. 129, 127–134.

Agag, T., Takeichi, T., 2000. Polybenzoxazine–montmorillonite hybrid nanocomposites: synthesis and characterization. Polymer. 41, 7083–7090.

Aishat, A.B., Olalekan, S.T., Arinkoola, A.O., Omolola, J.M., 2015. Effect of activation on clays and carbonaceous materials in vegetable oil bleaching: state of art review. Br. J. Appl. Sci. Technol. 5, 130–141.

Alexandre, M., Dubois, P., 2000. Polymer-layered silicate nanocomposites: preparation, properties and uses of a new class of materials. Mater. Sci. Eng.: R: Rep. 28, 1–63.

Azeez, A.A., Rhee, K.Y., Park, S.J., Hui, D., 2013. Epoxy clay nanocomposites—processing, properties and applications: a review. Compos. B: Eng. 45, 308–320.

Azeredo, H. M. C. d, 2009. Nanocomposites for food packaging applications. Food Res. Int. 42, 1240–1253.

Azzam, W.R., 2014. Behavior of modified clay microstructure using polymer nanocomposites technique. Alexandria Eng. J. 53, 143–150.

Banks-Sills, L., Shiber, D.G., Fourman, V., Eliasi, R., Shlayer, A., 2016. Experimental determination of mechanical properties of PMMA reinforced with functionalized CNTs. Compos. B: Eng. 95, 335–345.

Cailloux, J., Hakim, R.N., Santana, O.O., Bou, J., Abt, T., Sánchez-Soto, M., et al., 2016. Reactive extrusion: a useful process to manufacture structurally modified PLA/o-MMT composites. Compos. A: Appl. Sci. Manuf. 88, 106–115.

Carrado, K.A., Komadel, P., 2009. Acid activation of bentonites and polymer–clay nanocomposites. Elements. 5, 111–116.

Castillo, L.A., López, O.V., Ghilardi, J., Villar, M.A., Barbosa, S.E., García, M.A., 2015. Thermoplastic starch/talc bionanocomposites. Influence of particle morphology on final properties. Food Hydrocoll. 51, 432–440.

Chen, L., Zhou, C.H., Fiore, S., Tong, D.S., Zhang, H., Li, C.S., et al., 2016. Functional magnetic nanoparticle/clay mineral nanocomposites: preparation, magnetism and versatile applications. Appl. Clay Sci. 127–128, 143–163.

El-Zahhar, A.A., 2015. A polymer–organoclay nanocomposite for simultaneous removal of chromium(VI) and organic dyes. Eur. Chem. Bull. 4, 493–497.

Emre, F.B., Kesik, M., Kanik, F.E., Akpinar, H.Z., Aslan-Gurel, E., Rossi, R.M., et al., 2015. A benzimidazole-based conducting polymer and a PMMA–clay nanocomposite containing biosensor platform for glucose sensing. Synth. Metals. 207, 102–109.

Ensafi, A., Heydari-Bafrooei, E., Dinari, M., Mallakpour, S., 2014. Improved immobilization of DNA to graphite surfaces, using amino acid modified clays. J. Mater. Chem. B. 2, 3022–3028.

Essabir, H., Boujmal, R., Bensalah, M.O., Rodrigue, D., Bouhfid, R., Qaiss, A.K., 2016. Mechanical and thermal properties of hybrid composites: oil-palm fiber/clay reinforced high density polyethylene. Mech. Mater. 98, 38–43.

Fahn, R., Fenderl, K., 1983. Reaction products of organic dye molecules with acid-treated montmorillonite. Clay Miner. 18, 447–458.

Fang, C., Yang, S., Zhao, X., Du, P., Xiong, J., 2016. Electrospun montmorillonite modified poly(vinylidene fluoride) nanocomposite separators for lithium-ion batteries. Mater. Res. Bull. 79, 1–7.

Feng, X., Zhang, G., Zhuo, S., Jiang, H., Shi, J., Li, F., et al., 2016. Dual responsive shape memory polymer/clay nanocomposites. Compos. Sci. Technol. 129, 53–60.

Fu, X., Qutubuddin, S., 2001. Polymer–clay nanocomposites: exfoliation of organophilic montmorillonite nanolayers in polystyrene. Polymer. 42, 807–813.

Galgali, G., Ramesh, C., Lele, A., 2001. A rheological study on the kinetics of hybrid formation in polypropylene nanocomposites. Macromolecules. 34, 852–858.

Giannelis, E.P., 1996. Polymer layered silicate nanocomposites. Adv. Mater. 8, 29–35.

Giannelis, E.P., Krishnamoorti, R., Manias, E., 1999. Polymer-Silicate Nanocomposites: Model Systems for Confined Polymers and Polymer Brushes. In: Granick, S., Binder, K., de Gennes, P.G., Giannelis, E.P., Grest, G.S., Hervet, H., Krishnamoorti, R., Léger, L., Manias, E., Raphaël, E., Wang, S.Q. (Eds.), Polymers in Confined Environments. Springer Berlin Heidelberg, Berlin, Heidelberg, pp. 107–147.

Gurses, A., 2015. Introduction to Polymer–Clay Nanocomposites. Pan Stanford, Singapore.

Gurses, A., Ejder-Korucu, M., Karaca, S., 2011. Clay-Organoclay and Organoclay/Polymer Nanocomposites. In: Boyd, D.E., Humphery, J.P. (Eds.), Clay: Types, Properties and Uses. Nova Science Publisher, Hauppauge, NY, pp. 155–192.

Hasegawa, N., Kawasumi, M., Kato, M., Usuki, A., Okada, A., 1998. Preparation and mechanical properties of polypropylene-clay hybrids using a maleic anhydride-modified polypropylene oligomer. J. Appl. Polym. Sci. 67, 87–92.

He, H., Tao, Q., Zhu, J., Yuan, P., Shen, W., Yang, S., 2013. Silylation of clay mineral surfaces. Appl. Clay Sci. 71, 15–20.

Hernández-Hernández, K.A., Illescas, J., Díaz-Nava, M.C., Muro-Urista, C.R., Martínez-Gallegos, S., Ortega-Aguilar, R.E., 2016. Polymer-clay nanocomposites and composites: structures, characteristics, and their applications in the removal of organic compounds of environmental interest. Med. Chem. 6, 201–210.

Huang, J., Zhu, Z., Yin, J., Qian, X., Sun, Y., 2001. Poly(etherimide)/montmorillonite nanocomposites prepared by melt intercalation: morphology, solvent resistance properties and thermal properties. Polymer. 42, 873–877.

Hundáková, M., Tokarský, J., Valášková, M., Slobodian, P., Pazdziora, E., Kimmer, D., 2015. Structure and antibacterial properties of polyethylene/organo-vermiculite composites. Solid State Sci. 48, 197–204.

Jafarzadeh, S., Alias, A.K., Ariffin, F., Mahmud, S., Najafi, A., 2016. Preparation and characterization of bionanocomposite films reinforced with nano kaolin. J. Food Sci. Technol. 53, 1111–1119.

Jain, R., Mahto, V., 2015. Evaluation of polyacrylamide/clay composite as a potential drilling fluid additive in inhibitive water based drilling fluid system. J. Petrol. Sci. Eng. 133, 612–621.

Jamshidi, S., 2014. Mechanical, Rheological and Thermal Properties of Polyethylene (PE)/Clay Nanocomposite for Rotomolded Containers, CALGARY. ALBERTA.

Jordan, J.W., 1949. Organophilic bentonites. I. Swelling in organic liquids. J. Phys. Colloid Chem. 53, 294–306.

Joshi, M., Banerjee, K., Prasanth, R., Thakare, V., 2006. Polymer/clay nanocomposite based coatings for enhanced gas barrier property. Indian J. Fiber Text. Res. 31, 202–214.

Kara, A., Tekin, N., Alan, A., Şafaklı, A., 2016. Physicochemical parameters of Hg(II) ions adsorption from aqueous solution by sepiolite/poly(vinylimidazole). J. Environ. Chem. Eng. 4, 1642–1652.

Kaufhold, S., Meyer, F.M., 2001. Untersuchungen zur Eignung von natürlich alterierten sowie mit Oxalsäure aktivierten Bentoniten als Bleicherde für Pflanzenöle. Publikationsserver der RWTH Aachen University, Aachen, Germany.

Kawasumi, M., Hasegawa, N., Kato, M., Usuki, A., Okada, A., 1997. Preparation and mechanical properties of polypropylene – clay hybrids. Macromolecules. 30, 6333–6338.

Kim, G.M., Lee, D.H., Hoffmann, B., Kressler, J., Stöppelmann, G., 2001. Influence of nanofillers on the deformation process in layered silicate/polyamide-12 nanocomposites. Polymer. 42, 1095–1100.

Kojima, Y., Usuki, A., Kawasumi, M., Okada, A., Fukushima, Y., Kurauchi, T., et al., 1993. Mechanical properties of nylon 6-clay hybrid. J. Mater. Res. 8, 1185–1189.

Komadel, P., Madejova, J., 2006. Acid Activation of clay minerals. In: Bergaya, F., Theng, B.K.G., Lagaly, G. (Eds.), Handbook of Clay Science, Vol. 1. Elsevier Ltd, Amsterdam, Netherlands, pp. 263–287.

Kornmann, X., Lindberg, H., Berglund, L.A., 2001. Synthesis of epoxy–clay nanocomposites. Influence of the nature of the curing agent on structure. Polymer. 42, 4493–4499.

Lan, T., Kaviratna, P.D., Pinnavaia, T.J., 1995. Mechanism of clay tactoid exfoliation in epoxy-clay nanocomposites. Chem. Mater. 7, 2144–2150.

LeBaron, P.C., Wang, Z., Pinnavaia, T.J., 1999. Polymer-layered silicate nanocomposites: an overview. Appl. Clay Sci. 15, 11–29.

Lee, S.M., Tiwari, D., 2012. Organo and inorgano–organo-modified clays in the remediation of aqueous solutions: an overview. Appl. Clay Sci. 59–60, 84–102.

Liu, E., Sarkar, B., Wang, L., Naidu, R., 2016. Copper-complexed clay/poly-acrylic acid composites: extremely efficient adsorbents of ammonia gas. Appl. Clay Sci. 121–122, 154–161.

Ludueña, L., Morán, J., Alvarez, V., 2015. Biodegradable Polymer/Clay Nanocomposites. In: Thakur, K.V., Thakur, K.M. (Eds.), Eco-friendly Polymer Nanocomposites: Processing and Properties. Springer India, New Delhi, pp. 109–135.

Ma, G., Yang, Q., Ran, F., Dong, Z., Lei, Z., 2015. High performance and low cost composite superabsorbent based on polyaspartic acid and palygorskite clay. Appl. Clay Sci. 118, 21–28.

Mallakpour, S., Barati, A., 2014. A straightforward preparation and characterization of novel poly(vinyl alcohol)/organoclay/silver tricomponent nanocomposite films. Progr. Org. Coat. 77, 1629–1634.

Mallakpour, S., Dinari, M., 2013. Preparation, characterization, and thermal properties of organoclay hybrids based on trifunctional natural amino acids. J. Therm. Anal. Calorim. 111, 611–618.

Mallakpour, S., Dinari, M., 2014. Bionanocomposite materials from layered double hydroxide/Ntrimellitylimido-L-isoleucine hybrid and poly(vinyl alcohol): structural and morphological study. J. Thermoplast. Compos. Mater. 29, 623–637.

Mallakpour, S., Jarahiyan, A., 2016. An eco-friendly approach for the synthesis of biocompatible poly(vinyl alcohol) nanocomposite with aid of modified CuO nanoparticles with citric acid and vitamin C: mechanical, thermal and optical properties. J. Iran. Chem. Soc. 13, 509–518.

Mallakpour, S., Javadpour, M., 2016. An efficient preparation and characterization of nanocomposite films based on poly(vinyl chloride) and modified ZnO quantum dot with an optically active diacid containing amino acid as coupling agent. Polym.-Plast. Technol. Eng. 55, 498–509.

Mallakpour, S., Khani, M., 2015. Composites of semiaromatic poly(amide-ester-imide) based on bioactive diacid and oragnomodified nanoclay produced by solution intercalation method: thermal and morphological study. Polym.-Plast. Technol. Eng. 54, 541–547.

Mallakpour, S., Khani, M., 2016. Thermal and morphological studies of poly(vinyl alcohol)/poly(vinyl pyrrolidone)/organoclay nanocomposites containing L-leucine moiety. Colloid Polym. Sci. 294, 583–590.

Mallakpour, S., Shahangi, V., 2013. Bio-modification of cloisite Na + with chiral L-leucine and preparation of new poly(vinyl alcohol)/organo-nanoclay bionanocomposite films. Synth. React. Inorg. Metal-Org. Nano-Metal Chem. 43, 996–971.

Mallakpour, S., Soltanian, S., 2016. Chemical surface coating of MWCNTs with riboflavin and its application for the production of poly(ester-imide)/MWCNTs composites containing 4,4′-thiobis(2-tert-butyl-5-methylphenol) linkages: thermal and morphological properties. J. Appl. Polym. Sci. 133, n/a-n/a.

Manias, E., Touny, A., Wu, L., Strawhecker, K., Lu, B., Chung, T.C., 2001. Polypropylene/montmorillonite nanocomposites. Review of the synthetic routes and materials properties. Chem. Mater. 13, 3516–3523.

Mauro, N., Chiellini, F., Bartoli, C., Gazzarri, M., Laus, M., Antonioli, D., et al., 2016. RGD-mimic polyamidoamine–montmorillonite composites with tunable stiffness as scaffolds for bone tissue-engineering applications. J. Tissue Eng. Regener. Med. http://dx.doi.org/10.1002/term.2115, online.

Messersmith, P.B., Giannelis, E.P., 1995. Synthesis and barrier properties of poly(ε-caprolactone)-layered silicate nanocomposites. J. Polym. Sci. A: Polym. Chem. 33, 1047–1057.

Mishra, A.K., 2014. Nanocomposites in Wastewater Treatment. Pan Stanford, Singapore.

Mishra, R.K., Ramasamy, K., Lim, S.M., Ismail, M.F., Majeed, A.B.A., 2014. Antimicrobial and in vitro wound healing properties of novel clay based bionanocomposite films. J. Mater. Sci.: Mater. Med. 25, 1925–1939.

Mittal, V., 2009a. Barrier Properties of Polymer Clay Nanocomposites. Nova Science Publisher, Hauppauge.

Mittal, V., 2009b. Polymer layered silicate nanocomposites: a review. Materials. 2, 992.

Mukhopadhyay, R., Dc, N., 2014. Nano clay polymer composite: synthesis, characterization, properties and application in Rainfed Agriculture. Global J. Bio-Sci. Biotechnol. 3, 133–138.

Murima, D., Pfukwa, H., Tiggelman, I., Hartmann, P.C., Pasch, H., 2016. Novel polymer clay-based nanocomposites: films with remarkable optical and water vapor barrier properties. Macromol. Mater. Eng. 301, 836–845.

Nair, B.P., Sindhu, M., Nair, P.D., 2016. Polycaprolactone-laponite composite scaffold releasing strontium ranelate for bone tissue engineering applications. Colloids Surf. B: Biointerfaces. 143, 423–430.

Naveau, E., 2010. Preparation of New Organoclays in Supercritical Carbon Dioxide and Their Industrial Potential in Polymer Nanocomposites. Serveur institutionnel des thèses de doctorat, Universite de Liege, Liège, Belgium.

Neves, M.A., Hashemi, J., Yoshino, T., Uemura, K., N. M., 2016. Development and characterization of chitosan-nanoclay composite films for enhanced gas barrier and mechanical properties. HSOA J. Food Sci. Nutr. 2, 007.

Nguyen, Q.T., Baird, D.G., 2006. Preparation of polymer–clay nanocomposites and their properties. Adv. Polym. Technol. 25, 270–285.

Nigam, V., Lal, G., 2008. Review on recent trends in polymer layered clay nanocomposites. Proc. Indian Natl. Sci. Acad. 74, 87–96.

Okamoto, M., Morita, S., Taguchi, H., Kim, Y.H., Kotaka, T., Tateyama, H., 2000. Synthesis and structure of smectic clay/poly(methyl methacrylate) and clay/polystyrene nanocomposites via in situ intercalative polymerization. Polymer. 41, 3887–3890.

Okamoto, M., Morita, S., Kotaka, T., 2001. Dispersed structure and ionic conductivity of smectic clay/polymer nanocomposites. Polymer. 42, 2685–2688.

Othman, R., Vladisavljević, G.T., Thomas, N.L., Nagy, Z.K., 2016. Fabrication of composite poly(d,l-lactide)/montmorillonite nanoparticles for controlled delivery of acetaminophen by solvent-displacement method using glass capillary microfluidics. Colloids Surf. B: Biointerfaces. 141, 187–195.

Panda, A.K., Mishra, B.G., Mishra, D.K., Singh, R.K., 2010. Effect of sulphuric acid treatment on the physico-chemical characteristics of kaolin clay. Colloids Surf. A: Physicochem. Eng. Aspects. 363, 98–104.

Pavlidou, S., Papaspyrides, C.D., 2008. A review on polymer–layered silicate nanocomposites. Prog. Polym. Sci. 33, 1119–1198.

Rhodes, C.N., Brown, D.R., 1992. Structural characterisation and optimisation of acid-treated montmorillonite and high-porosity silica supports for ZnCl2 alkylation catalysts. J. Chem. Soc. Faraday Trans. 88, 2269–2274.

Rhodes, C.N., Franks, M., Parkes, G.M.B., Brown, D.R., 1991. The effect of acid treatment on the activity of clay supports for $ZnCl_2$ alkylation catalysts. J. Chem. Soc., Chem. Commun. 1991, 804–807.

Romanzini, D., Frache, A., Zattera, A.J., Amico, S.C., 2015. Effect of clay silylation on curing and mechanical and thermal properties of unsaturated polyester/montmorillonite nanocomposites. J. Phys. Chem. Solids. 87, 9–15.

Rusmin, R., Sarkar, B., Liu, Y., McClure, S., Naidu, R., 2015. Structural evolution of chitosan–palygorskite composites and removal of aqueous lead by composite beads. Appl. Surface Sci. 353, 363–375.

Saha, N.R., Sarkar, G., Roy, I., Rana, D., Bhattacharyya, A., Adhikari, A., et al., 2016. Studies on methylcellulose/pectin/montmorillonite nanocomposite films and their application possibilities. Carbohydr. Polym. 136, 1218–1227.

Sahu, M., Samal, R., Biswal, T., Sahoo, P.K., 2014. Synthesis and characterization of poly (vinyl acetate)/MMT nanocomposite flame retardant. Int. J. Mater. Sci. 4, 91–98.

Savas, L.A., Hancer, M., 2015. Montmorillonite reinforced polymer nanocomposite antibacterial film. Appl. Clay Sci. 108, 40–44.

Siddiqui, M.K.H., 1968. Bleaching Earths. 1st/Ed. Pergamon Press, Oxford.

Silvestre, C., Duraccio, D., Cimmino, S., 2011. Food packaging based on polymer nanomaterials. Prog. Polym. Sci. 36, 1766–1782.

Singla, P., Mehta, R., Upadhyay, S., 2012. Clay modification by the use of organic cations. Sci. Res. 2, 21–25.

Steudel, A., Batenburg, L.F., Fischer, H.R., Weidler, P.G., Emmerich, K., 2009. Alteration of swelling clay minerals by acid activation. Appl. Clay Sci. 44, 105–115.

Su, L., Tao, Q., He, H., Zhu, J., Yuan, P., Zhu, R., 2013. Silylation of montmorillonite surfaces: dependence on solvent nature. J. Colloid Interface Sci. 391, 16–20.

Tan, B., Thomas, N.L., 2016. A review of the water barrier properties of polymer/clay and polymer/graphene nanocomposites. J. Membr. Sci. 514, 595–612.

Tien, Y.I., Wei, K.H., 2001. Hydrogen bonding and mechanical properties in segmented montmorillonite/polyurethane nanocomposites of different hard segment ratios. Polymer. 42, 3213–3221.

Unuabonah, E.I., Taubert, A., 2014. Clay–polymer nanocomposites (CPNs): adsorbents of the future for water treatment. Appl. Clay Sci. 99, 83–92.

Usuki, A., Kawasumi, Y., Kawasumi, M., Okada, A., Fukushima, Y., Kurauchi, T., et al., 1993. Swelling behavior of montmorillonite cation exchanged for ω-amino acids by E-caprolactam. J. Mater. Res. 8, 1174–1178.

Vaia, R.A., Teukolsky, R.K., Giannelis, E.P., 1994. Interlayer structure and molecular environment of alkylammonium layered silicates. Chem. Mater. 6, 1017–1022.

Van Olphen, H. (1964). An introduction to clay colloid chemistry. By H van Olphen. Interscience Publishers, Div. of John Wiley & Sons, 605 Third Ave., New York 16,

N. Y, 1963. xvi + 301 pp. 15.5 × 23 cm. Price $10. Journal of Pharmaceutical Sciences 53, 230–230.

Wang, F., Chang, P.R., Zheng, P., Ma, X., 2015. Monolithic porous rectorite/starch composites: fabrication, modification and adsorption. Appl. Surface Sci. 349, 251–258.

Wang, R., Peng, Y., Zhou, M., Shou, D., 2016. Smart montmorillonite-polypyrrole scaffolds for electro-responsive drug release. Appl. Clay Sci. 54, 50–54.

Wu, C.-J., Gaharwar, A.K., Schexnailder, P.J., Schmidt, G., 2010. Development of biomedical polymer-silicate nanocomposites: a materials science perspective. Materials. 3, 2986.

Xia, Y., 2014. Release of Nanoclay and Surfactant from Polymer-Clay Composite-System. Michigan State Uiversity, USA.

Yahiaoui, F., Benhacine, F., Ferfera-Harrar, H., Habi, A., Hadj-Hamou, A.S., Grohens, Y., 2015. Development of antimicrobial PCL/nanoclay nanocomposite films with enhanced mechanical and water vapor barrier properties for packaging applications. Polym. Bull. 72, 235–254.

Yildiz, G., Senkal, B.F., 2016. Formation of composites between polyvinylimidazole and bentonites and their use for removal of remazol black B from water. Sep. Sci. Technol. 1–8.

Zeng, L., Chen, Y., Zhang, Q., Guo, X., Peng, Y., Xiao, H., et al., 2015. Adsorption of Cd(II), Cu(II) and Ni(II) ions by cross-linking chitosan/rectorite nano-hybrid composite microspheres. Carbohydr. Polym. 130, 333–343.

Zerda, A.S., Lesser, A.J., 2001. Intercalated clay nanocomposites: morphology, mechanics, and fracture behavior. J. Polym. Sci. B: Polym. Phys. 39, 1137–1146.

Zotti, A., Borriello, A., Ricciardi, M., Antonucci, V., Giordano, M., Zarrelli, M., 2015. Effects of sepiolite clay on degradation and fire behaviour of a bisphenol A-based epoxy. Compos. B: Eng. 73, 139–148.

Graphene-based materials and their potential applications: a theoretical study

9

Hong-ping Zhang[1] *and Youhong Tang*[2]
[1]School of Materials Science and Engineering, Southwest University of Science and Technology, Mianyang, China, [2]Centre for NanoScale Science and Technology and School of Computer Science, Engineering and Mathematics, Flinders University, Adelaide, SA, Australia

Chapter Outline

9.1 Introduction 267
9.2 Properties of different graphene, including pristine graphene, defective graphene, graphene oxide, and metal-doped graphene 268
9.3 Theoretical study methodologies 268
9.4 Graphene-based materials with applications 269
9.4.1 Interaction between functionalized graphene and chitosan for tissue engineering 269
9.4.2 Metal-doped graphene for gas sensing 272
9.4.3 Interaction between MgH_2 and graphene for hydrogen storage 280
9.5 Conclusion and remarks 282
References 283

9.1 Introduction

Graphene, which is the monolayer of graphite, currently attracts much research attention by virtue of its excellent properties, for which reason graphene and its derivatives are widely studied (Novoselov et al., 2004, 2005; Wang et al., 2008; Park, 2008; Shao et al., 2010). Among the varieties, defective graphene, graphene oxide (that can be regarded as a monolayer of graphite oxide), and metal-doped graphene are the main objects of interest (Tan and Thomas, 2016; Zhao et al., 2016; Ji et al., 2016; Rasheed et al., 2016; Chen et al., 2016). Using various forms, graphene-based materials can be fabricated, such as graphene-based gas sensors, graphene-based tissue engineering scaffolds, and graphene-based advanced energy materials (Ke and Wang, 2016; Marmolejo-Tejada and Velasco-Medina, 2016; Putria et al., 2015; Chatterjee et al., 2015; Liu et al., 2015; Toda et al., 2015;

Hybrid Polymer Composite Materials: Applications. DOI: http://dx.doi.org/10.1016/B978-0-08-100785-3.00009-7

Kuttappan et al., 2016). Although many studies have investigated the effects of multiple factors, such as temperature, content, and pH, on the properties of graphene-based materials, the interactions between graphene-based derivatives and other materials are very important, and these interactions should be considered at atomic scale. For this purpose, common experimental methods can do little, but computer simulation can be an effective way to study the interactions between graphene-based materials and other components, providing information about interactions at the atomic level that cannot be obtained directly from experiments.

9.2 Properties of different graphene, including pristine graphene, defective graphene, graphene oxide, and metal-doped graphene

As is known by researchers, the properties of materials are greatly related to their structure that entails multiscale information. Some of those properties, such as electron conductivity, microstress, and chemical activity, can be very sensitive to the microstructure of materials. Graphene is famous for its real 2D microstructures, the C–C bonds, and the excellent mechanical properties induced by its bonds. The zero band gap is an apt illustration of the superconducting property of graphene; the good biocompatibility exhibited by graphene has also been reported (Liu et al., 2015; Kuttappan et al., 2016). If the microstructure of graphene is changed to a very small degree, different kinds of graphene-based materials can be obtained, such as defective graphene, graphene oxide, and metal-doped graphene (Xiong et al., 2015; Botello-Méndez et al., 2013; Tan et al., 2014). Among the various defective graphenes, the vacancy defect and the Stone–Wales defect have attracted much attention (Wang et al., 2011). Both of these are closely related to adjustments of the band gap of intrinsic graphene. Stone–Wales defects in graphene are regarded as very helpful for the transition of graphene to nanotubes and fullerenes. Compared to intrinsic graphene, defective graphenes are more common, and Stone–Wales defective graphene is more likely to be present under nonequilibrium conditions. Meanwhile, graphene oxide is of high interest to researchers because of the abundant chemical functional groups on it, including –OH, –COOH, and phenolic hydroxyl groups. It is of great interest to chemists and materials scientists that several kinds of graphene derivatives and materials can be fabricated based on graphene oxide. It has been reported that metal dopants can regulate the band gap of graphene and improve the chemical activities of the graphene (Ji et al., 2016; Putria et al., 2015; Chatterjee et al., 2015).

9.3 Theoretical study methodologies

Due to the limitations of experimental methods, graphene and its derivatives have been studied broadly through density functional theory (DFT) and molecular dynamics (MD) methods. In accordance with the time and length scale rules of

DFT and MD theoretical methods, researchers can study or predict the different properties of graphene and its derivatives (Jian et al., 2015). Quantum mechanics (QM) is a widely used molecular modeling method (Friesner, 2005), based on the solving of the Schrodinger equation. Hartree–Fock and DFT are two classic QM methods. The Hartree–Fock equation (Hartree, 1928), utilizing a wide variety of approximations when solving the Schrodinger equation, is an effective and successful method of solving the Schrodinger equation. DFT takes the total energy of the system as a function of electron density, which is not necessary for solving the complex Schrodinger equation, and therefore, DFT is considered to be more efficient than the Hartree–Fock equation.

The DFT method, which is based on QM, can deal with the interactions of electrons. As the MD method is based on classical mechanics, it cannot deal with the interactions of electrons but can calculate the interactions between atoms according to the force field parameters. Thus, for structural and electronic details, studies of graphene and its derivatives, the DFT method is more suitable. Several commercial packages exist that can be used for DFT calculations, such as Gaussian, VASP (The Vienna Ab-initio Simulation Package), CASTEP, and $Dmol^3$. $Dmol^3$ produces highly accurate results while keeping computational cost fairly low. The double numerical plus polarization (Delley, 2000), which is comparable to the 6–31G** basis set, has been utilized during theoretical study (Delley, 1990, 2002). The core electrons were treated with DFT semicore pseudopotentials. The exchange–correlation energy was calculated using the Perdew–Burke–Ernzerhof-generalized gradient approximation (Perdew et al., 1996). Special point sampling integration over the Brillouin zone has been employed using the Monkhorst–Pack schemes (Monkhorst and Pack, 1976). A Fermi smearing of 0.005 Ha (1 Ha = 27.211 eV) and a global orbital cutoff of 5.2 Å were employed. The convergence criteria for the geometric optimization and energy calculation were set as follows: (1) a self-consistent field tolerance of 1.0×10^{-6} Ha atom^{-1}, (2) an energy tolerance of 1.0×10^{-5} Ha atom^{-1}, (3) a maximum force tolerance of 0.002 Ha Å^{-1}, and (4) a maximum displacement tolerance of 0.005 Å.

9.4 Graphene-based materials with applications

9.4.1 *Interaction between functionalized graphene and chitosan for tissue engineering*

Graphene-based materials are attractive for their excellent mechanical features, thermal conductivity, and good biocompatibility. It is difficult for intrinsic graphene to combine intimately with biomolecules due to its chemical inertness. However, graphene oxide and reduced graphene oxide, with abundant oxygen-containing groups, provide a good foundation for graphene to interact with biomolecules at atomic scale. Generally, functional groups such as –COOH, $–NH_2$, and –OH can be used to modify graphene. These are chemically active groups, and they are also abundant in biomolecules such as peptides, proteins, polysaccharides, and biodegradable polymers. Chitosan has attracted much attention for its biodegradability, biocompatibility, and antibacterial ability (Chieu and Tamutsiwa, 2015; Anna et al., 2015; Yang et al., 2015; Tsaia et al., 2011; Margarita et al., 2006; Majeti and Ravi, 2000; Rinaudo, 2006; Yang et al., 2010). Graphene oxide/chitosan is regarded as an excellent candidate for tissue engineering scaffolds because it can combine the desirable properties of both graphene and chitosan. The interfacial interactions between graphene and chitosan are highly related to their solubility, processability, and mechanical properties. In our previous study (Zhang et al., 2016), interfacial interactions between chitosan and functionalized graphene sheets with carboxylization (COOH−), amination (NH_2−), and hydroxylation (OH−) groups were systematically studied at the electronic level using the method of ab-initio simulations based on QM theory.

Fig. 9.1 shows the initial configurations between chitosan and functionalized graphene sheet interaction systems. Different chitosan initial positions were considered in the chitosan/graphene interactions to ensure accuracy of the simulation. Meanwhile, the ratios of the different kinds of functional groups were also considered during the simulation. Fig. 9.2 shows the different combinations of the functional group modified graphenes used to investigate the former simulation results of the single functional group modified graphene, shown in Fig. 9.1. It was found that covalent linkages existed between COOH-modified graphene and the chitosan unit. The combination of multifunctionalization on graphene could regulate the interfacial interactions between the graphene and the chitosan. The interfacial interactions between chitosan and properly functionalized graphene are critical for the preparation of graphene and graphene-based composites for tissue engineering scaffolds and other applications. The mechanical properties of graphene/chitosan composites can be controlled by adjusting the interfacial interactions between them. The different chemical functional groups and their ratios might be the two key factors in that study (Zhang et al., 2016).

The details of interactions between chitosan and functionalized graphene can be intuitively discovered by electron density difference analysis. Fig. 9.3 shows the results of the electron density difference analysis in the aforementioned study. The blue area represents electron accumulation, and the yellow area represents electron depletion. It can be found that apparent electron transformations occur between chitosan and COOH- or OH-modified graphene sheets. However, NH_2-modified graphene seems to prefer not to interact with chitosan (Zhang et al., 2016).

In order to demonstrate the results of the electron density difference, PDOS (partial density of state) analysis was carried out. The results are shown in Fig. 9.4. The results of PDOS analysis indicated that the number of interactions between chitosan and COOH- or OH-modified graphene was greater than those between chitosan and NH_2-modified graphene.

Thus, computer simulation can be an effective way to support the design of specific interfacial structures. Among COOH-, NH_2-, and OH-modified graphenes, COOH− and OH− can be useful for enhancing interfacial interactions between chitosan and graphene.

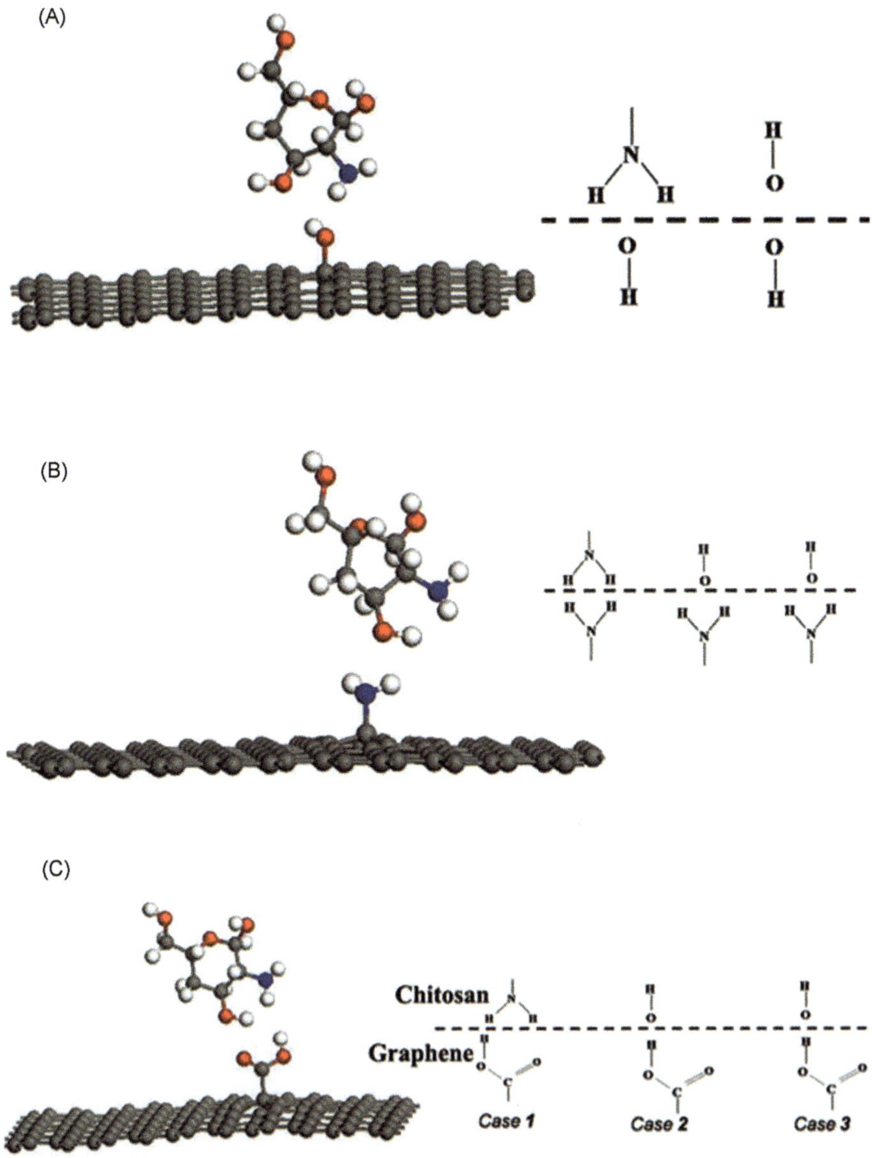

Figure 9.1 Chitosan interacts with functionalized graphene with (A) hydroxylation (OH−), (B) amination (NH_2−), and (C) carboxylization (COOH−) groups (Zhang et al., 2016). Red color: Oxygen atom; White color: Hydrogen atom; Gray color: Carbon atom; Blue color: Nitrogen atom.
Reproduced with permission from Elsevier.

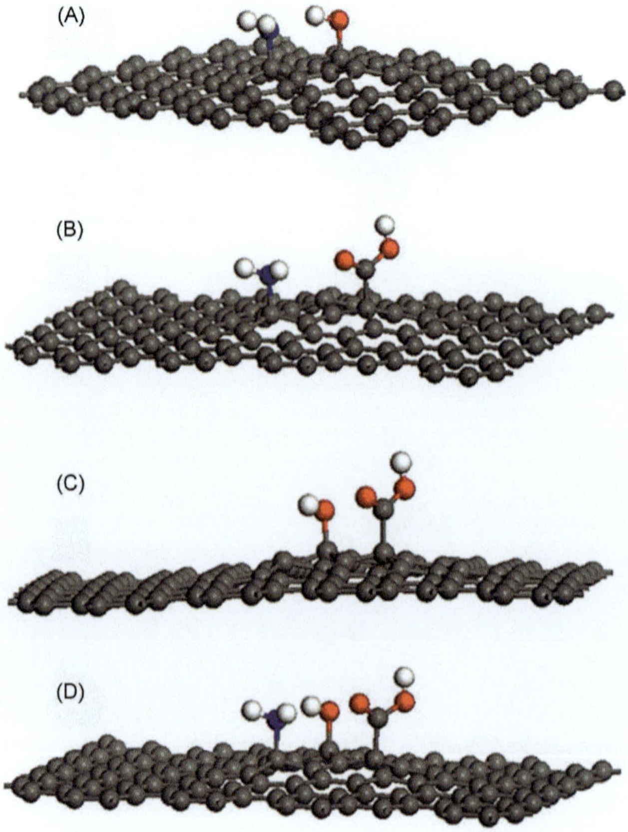

Figure 9.2 Different combinations between functional groups on a graphene sheet: (A) NH_2- and OH-modified graphene, (B) NH_2- and COOH-modified graphene, (C) OH- and COOH-modified graphene, and (D) NH_2-, COOH-, and OH-modified graphene (Zhang et al., 2016). Red color: Oxygen atom; White color: Hydrogen atom; Gray color: Carbon atom; Blue color: Nitrogen atom.
Reproduced with permission from Elsevier.

9.4.2 Metal-doped graphene for gas sensing

Graphene-based materials have also been widely studied as sensor materials. Among the various kinds of sensors, gas sensors are very important (Banerjee and Bhattacharyya, 2008; Basu and Bhattacharyya, 2012; Chen et al., 2012; Avramov et al., 2011; Gao et al., 2011; Grande et al., 2012; Liao and Duan, 2010). It has been widely demonstrated that intrinsic graphene, graphene obtained by CVD, and reduced graphene oxide can be excellent candidates for gas sensing (Banerjee and Bhattacharyya, 2008; Basu and Bhattacharyya, 2012; Chen et al., 2012). All graphene-based gas sensors are based on the mechanism whereby their resistance changes obviously after gas adsorption. Thus, these graphene materials are very

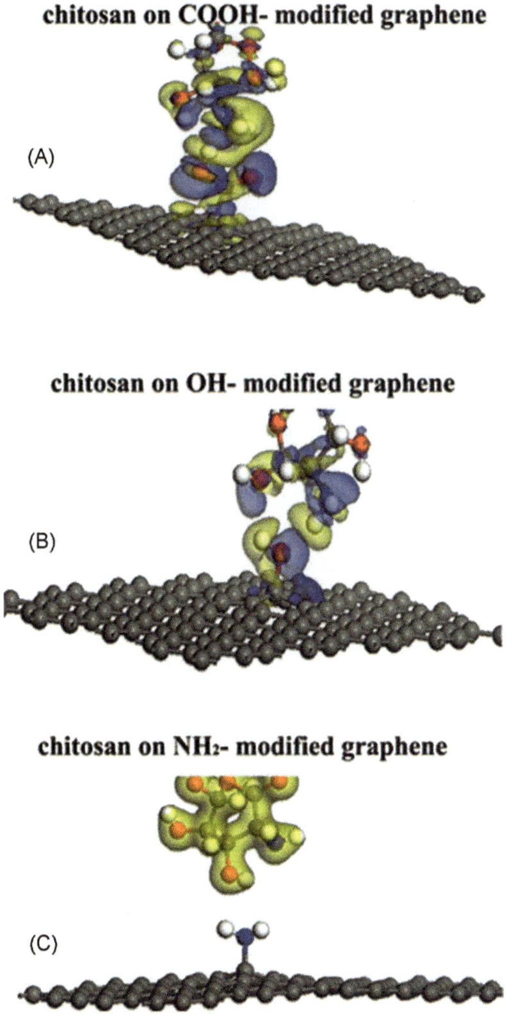

Figure 9.3 The electron density difference analysis between functionalized graphene and chitosan (Zhang et al., 2016); Red color: Oxygen atom; White color: Hydrogen atom; Gray color: Carbon atom; Blue color: Nitrogen atom.
Reproduced with permission from Elsevier.

suitable for the testing of low concentrations of toxic gases, such as NH_3, NO_X, and COx (Choong and Tiroumalechetty, 2008; Prashant et al., 2008; Cole et al., 2010; Duong et al., 2011). There are, however, certain difficulties with these types of gas sensor. First, the desorption process of adsorbed gas molecules can be a problem because of the intense adhesion between gas molecules and graphene substrates. Meanwhile, poor reproducibility can also be a notable problem for the application of graphene-based gas sensors. Thus, graphene has always been used together with

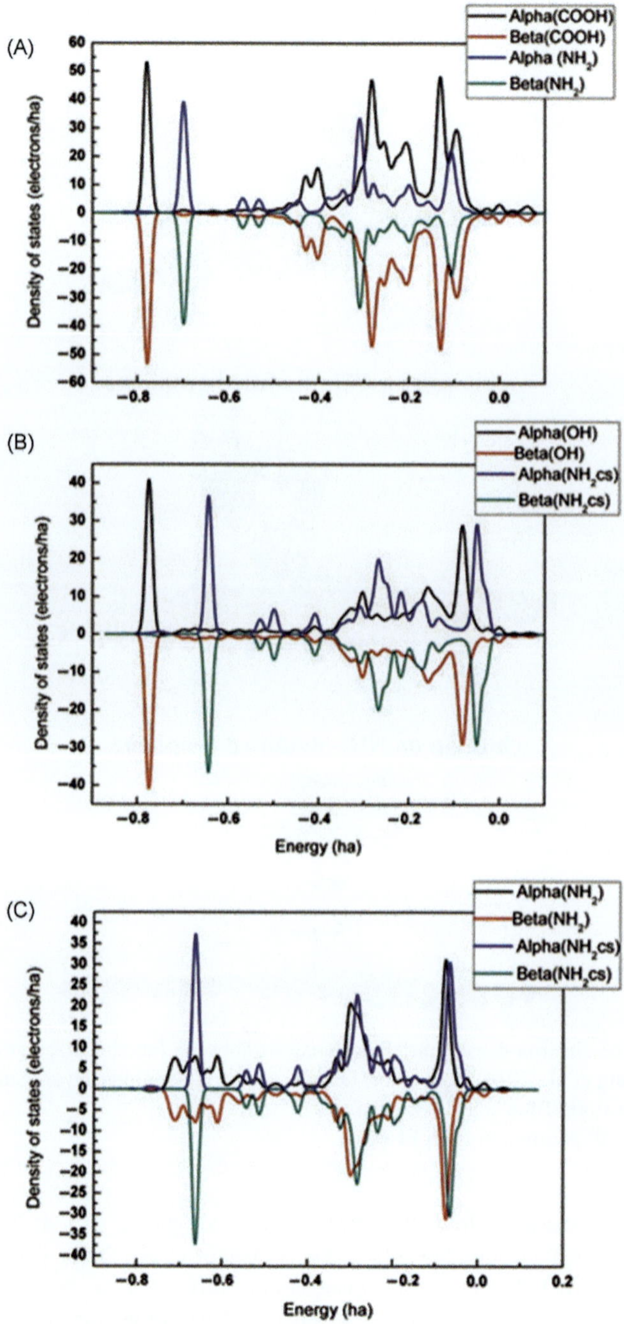

Figure 9.4 PDOS of the different graphene-chitosan systems (Zhang et al., 2016). Reproduced with permission from Elsevier.

a semiconductor. Studies have shown that the excellent sensing behavior of graphene is mainly related to its band gap, and the zero band gap of graphene causes the obvious change of conductivity of graphene.

Hydrogen storage has attracted much attention because of its close relationship with hydrogen energy, and carbon derivatives such as carbon nanotubes, graphene, and carbon fiber exhibit outstanding behavior in hydrogen storage (Kothari et al., 2012; Midilli et al., 2005; Sherif et al., 2005; Jain et al., 2001, 2010; Lopez-Corral et al., 2012; Luxembourg et al., 2007; Mou et al., 2011; Zini and Tartarini, 2010; Sealy, 2009). To broaden the application of graphene-based gas sensors, the influence of dopants on the gas sensing properties of graphene-based materials has been studied. In 2013, Ti-, Zn-, Zr-, Al-, and N-doped graphenes were investigated, and the effect of these dopants on the interactions between graphene and hydrogen molecules were studied (Zhang et al., 2013a). Fig. 9.5 shows an optimized ball and stick model of intrinsic and doped graphenes by Ti−, Zn−, Zr−, Al−, and N, respectively. It was found that the strength of interactions between hydrogen molecules and doped graphene was greatly increased compared with that between hydrogen molecules and intrinsic graphene.

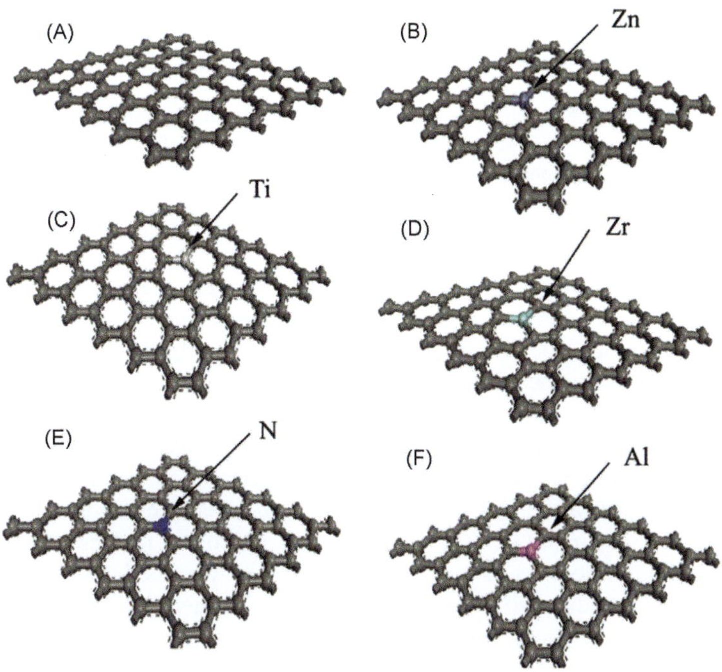

Figure 9.5 DFT-optimized graphene sheet and doped graphene sheet: (A) intrinsic graphene, (B) Zn-doped graphene, (C) Ti-doped graphene, (D) Zr-doped graphene, (E) N-doped graphene, and (F) Al-doped graphene (Zhang et al., 2013a).
Reproduced with permission from Elsevier.

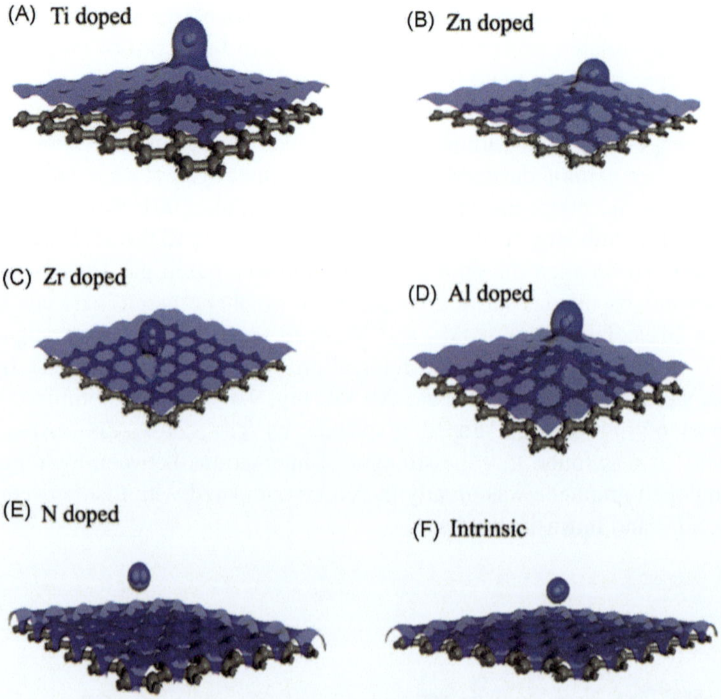

Figure 9.6 Isosurface of the electron density of graphene sheet and doped graphene sheet: (A) Ti-doped graphene, (B) Zn-doped graphene, (C) Zr-doped graphene, (D) Al-doped graphene, (E) N-doped graphene, and (F) intrinsic graphene (Zhang et al., 2013a). Reproduced with permission from Elsevier.

The different effects of the different dopants are evident in Fig. 9.6, which shows the isosurface of the electron density of hydrogen/graphene systems. The metallic dopants, such as Ti, Al, Zn, and Zr, could be very effective in enhancing the strength of interactions between hydrogen and graphene, whereas the N-doped graphene did not have the same effect.

Details of electron transfer can be studied via analysis of the differences in electron density. Fig. 9.7 shows that the metallic dopants could enhance the interactions between graphene and hydrogen. There are clearly electron accumulation areas between dopants (Ti, Zn, and Zr) and hydrogen molecules. However, the contribution of dopants Al and N could be weak. Also the interaction between the toxic tetrachlorodibenzo-*p*-dioxin (TCDD) and graphene has been studied, and it was found that it could be regulated by Ti, Ag, or N doping (Zhang et al., 2013a).

TCDD is one of the most toxic gases to threaten human health (Schectera et al., 2006; Gaylor and Aylward, 2004; Ishii and Furuichi, 2007; Peltier et al., 2013). Fig. 9.8 shows the interaction systems of TCDD and graphene. To explore the effect of different chemical regions on the interaction, three different initial

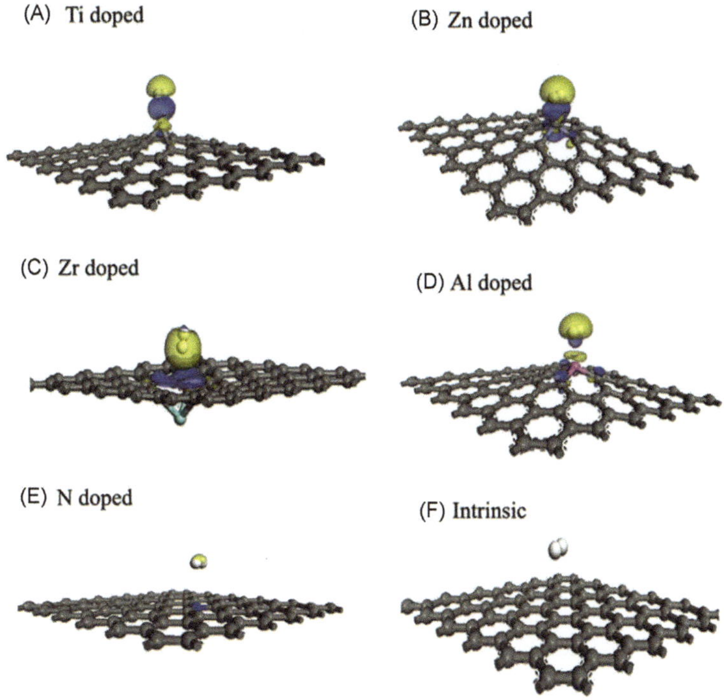

Figure 9.7 Electron density difference analysis of graphene sheet and doped graphene sheet: (A) Ti-doped graphene, (B) Zn-doped graphene, (C) Zr-doped graphene, (D) Al-doped graphene, (E) N-doped graphene, and (F) intrinsic graphene (Zhang et al., 2013a). Reproduced with permission from Elsevier.

configurations were studied (Zhang et al., 2014a). Fig. 9.9 shows analyses of the electron density differences in TCDD/graphene systems. It can be seen that although intrinsic graphene interacted with a TCDD molecule, the dopant Ti-captured TCDD through electron interactions. The configurations of the TCDD molecule had also been changed, as shown in Fig. 9.9.

Another toxic gas molecule, formaldehyde (HCHO), has also attracted much attention, because of its close relationship with human health (Tunga et al., 2010; Wang et al., 2009). The interaction between HCHO and Ti-doped graphene was studied by the DFT method (Zhang et al., 2013b). Fig. 9.10 shows the electron density difference analysis of Ti-doped graphene. It can be seen that electrons tend to transfer from the C atoms to the doped Ti atom. Thus, doping with Ti could break the electron equilibrium of the graphene, and the HCHO molecule could be captured by Ti-doped graphene. Fig. 9.11 shows the results of the electron density and electron density difference. The interactions between electrons and transformation of the electrons are clearly evident in Fig. 9.11.

H_2S adsorption behavior on Fe-doped graphene has also been studied by the DFT method, and it was found that H_2S could dissociate into S and H_2

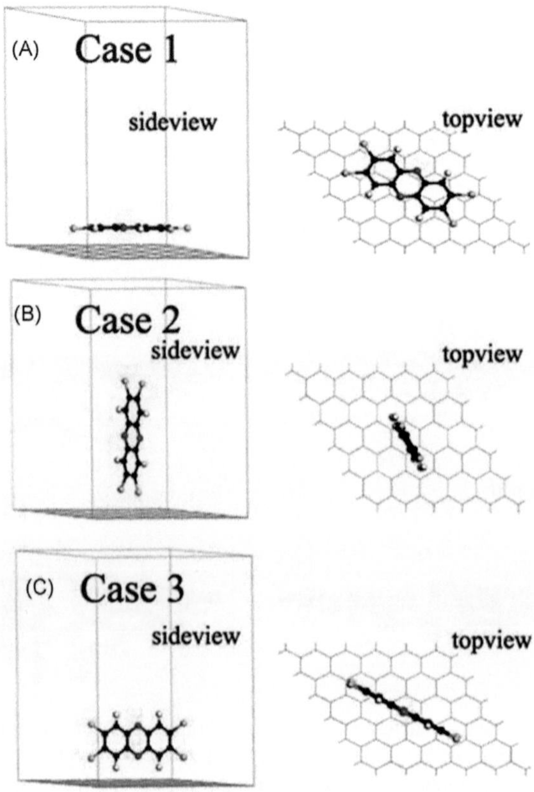

Figure 9.8 Ball and stick model of the TCDD-graphene interaction systems (Zhang et al., 2014a); Red color: Oxygen atom; White color: Hydrogen atom; Black color: Carbon atom; Green color: Chlorine atom.
Reproduced with permission from Springer.

(Zhang et al., 2014b). Fig. 9.12 shows the model of intrinsic graphene, Pt- and Fe-doped graphene, and the H_2S molecules. Through the DFT simulation, it was found that H_2S could be captured by Fe-doped graphene, and the doped Fe could be very helpful for the transformation of H_2S (from H_2S to H_2 and S). Fig. 9.13 shows the reaction process of the transformation of H_2S on Fe-doped graphene.

Table 9.1 shows the binding energy results of the H_2S/Fe, Pt-doped graphene systems. From the binding energy, it is found that, doped with Fe, graphene has the ability to capture H_2S similar to that of the Pt-doped graphene. This finding indicates that Fe doping could be used to replace Pt doping in dealing with H_2S. Meanwhile, the different energy dispersion correction methods could affect the results of binding energy; however, the trend of the values was not changed by these different correction methods.

Given the former results, through the DFT calculation, the electron structures of graphene, the effect of the dopants, and the interactions between gas molecules

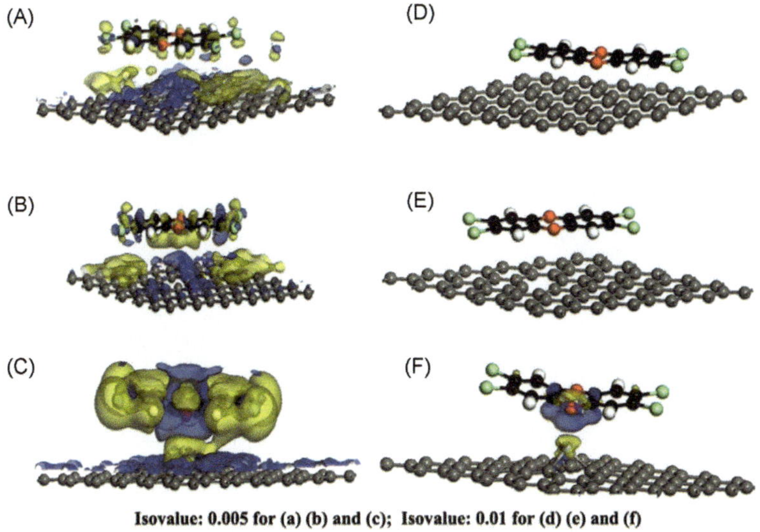

Figure 9.9 Electron density difference analysis of TCDD-graphene and TCDD-doped graphene (Zhang et al., 2014a); Red color: Oxygen atom; White color: Hydrogen atom; Black color: Carbon atom; Green color: Chlorine atom.
Reproduced with permission from Springer.

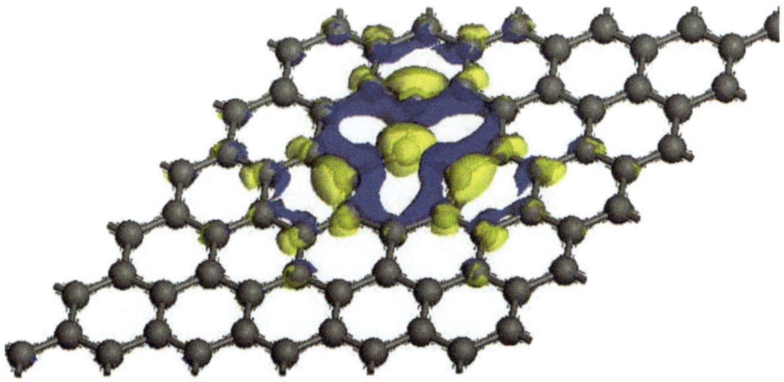

Figure 9.10 Electron density difference analysis of a Ti-doped graphene sheet (Zhang et al., 2013b).
Reproduced with permission from Elsevier.

and different graphene sheets could be studied at atomic scale. Many phenomena could be comprehended by understanding the transfer of electrons. The results indicate that study of the influence of dopants on the interactions between graphene and different gases is very significant for the design of graphene-based sensors.

Figure 9.11 Isosurface of electron density and electron density difference analysis of HCHO/Ti-doped graphene sheet interaction system (Zhang et al., 2013b).
Reproduced with permission from Elsevier.

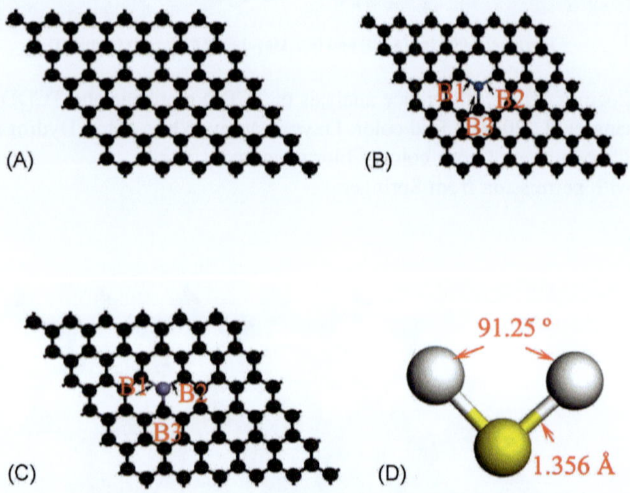

Figure 9.12 Ball and stick models of intrinsic, Pt- and Fe-doped graphenes and H_2S molecule (Zhang et al., 2014b).
Reproduced with permission from Elsevier.

9.4.3 Interaction between MgH_2 and graphene for hydrogen storage

Hydrogen energy, as a clean energy, is one of the most promising candidates for replacing fossil energy. MgH_2 has attracted much attention because of the high volumetric and gravimetric density of H_2. However, the high desorption temperature (about 400°C) of hydrogen and the low desorption rate are two important negative factors for the application of MgH_2 in hydrogen energy. Finding a proper substrate to load MgH_2 that can carry much MgH_2 and increase the desorption rate of MgH_2 would be very profitable. It is well known that graphene is very attractive by virtue

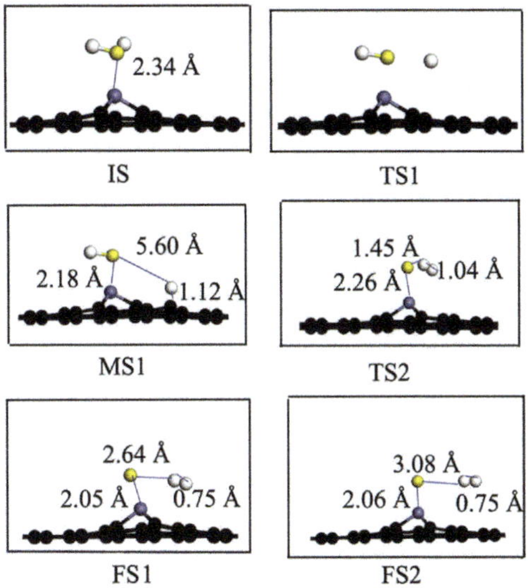

Figure 9.13 Ball and stick model of the H_2S–Fe-doped graphene interaction system in the initial and transition states (Zhang et al., 2014b).
Reproduced with permission from Elsevier.

of its large specific surface area, which implies that it can load many MgH_2 molecules. If a method can be found to induce good catalytic properties to graphene, that can help the dissociation process of MgH_2, it would be very favorable for the application of MgH_2 in hydrogen energy engineering. Interactions with Pt-, Fe-, Ti-, Al-, F-, Cl-, B-, and N-doped graphene have been studied by the theoretical method (Zhang et al., 2015).

Fig. 9.14 shows the interaction models of MgH_2 with different graphene sheets. It can be seen that the configurations of MgH_2 changed in different ways after interaction with the doped graphene sheets. From the results, it was found that different dopants could regulate the interactions between graphene and MgH_2. MgH_2 could be physically absorbed on intrinsic, B- and N-doped graphenes, whereas it was chemically absorbed on Pt-, Fe-, Ti-, Al-, F-, and Cl-doped graphenes. To clarify the results, an analysis of electron density difference was performed. Fig. 9.15 shows the transfer of electrons between an MgH_2 molecule and different kinds of graphene sheet. The effects of the metallic dopants are demonstrated.

It is interesting that the chemical absorption process of MgH_2 induces a change in the molecule configuration, indicating that these specific dopants might be helpful for the dissociation of MgH_2 on graphene sheets. Thus, graphene-based materials can be promising candidates for MgH_2 loading, and the substrate can be regulated to help the dissociation of MgH_2. However, many details still need to be confirmed. The density of dopants, the amount of MgH_2 molecules, and the surface functional groups on graphene sheets are key issues regarding the control of the loading of MgH_2 and its dissociation.

Table 9.1 **Adsorption energy of H_2S–graphene systems (unit: eV) (Zhang et al., 2014b)**

Graphene	Intrinsic			Fe-doped			Pt-doped		
Case	1	2	3	1	2	3	1	2	3
Adsorption energy	0.2320	0.0530	0.017	−1.023	−1.022	−0.131	−1.023	−1.022	−0.131

Case 1: lying; Case 2: S standing; Case 3: H standing.
Reproduced with permission from Elsevier.

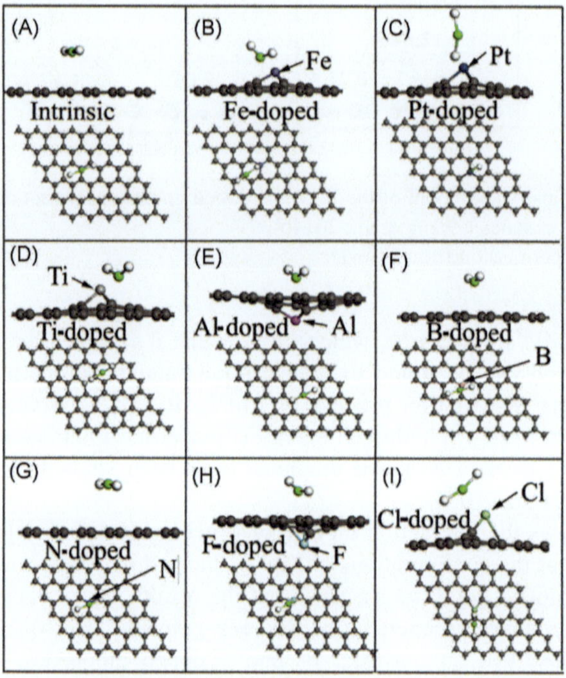

Figure 9.14 MgH_2 interacting with different doped graphene sheets (Zhang et al., 2015). Reproduced with permission from Elsevier.

9.5 Conclusion and remarks

As shown in our previous studies, the DFT method can be very effective for investigating the basic relationships between the microstructures of graphene-based materials and their physical/chemical properties. These materials can be adjusted to meet the demands of specific applications. Graphene/chitosan composites have been regarded as excellent candidates for tissue engineering scaffolds due to their outstanding mechanical properties, biocompatibility, and biodegradability.

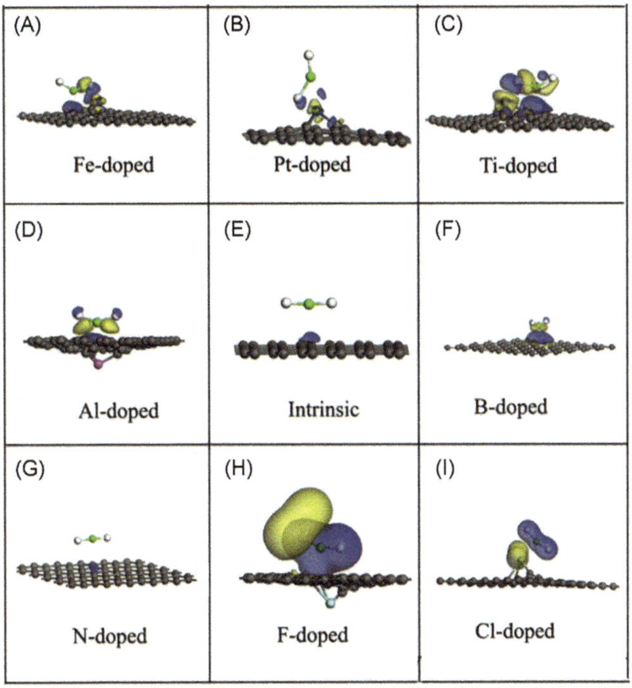

Figure 9.15 Isosurface of electron density difference analysis of MgH_2/doped graphene sheet interaction system (Zhang et al., 2015).
Reproduced with permission from Elsevier.

The specific behavior of graphene/chitosan composites is strongly related to the interfacial interactions between graphene and chitosan. The DFT method can also be used to select appropriate chemical functional groups. As knowledge of graphene has developed, band gap engineering has attracted much attention in relation to the aim of replacing silicon by graphene in the future. Studies indicate that doping with metal can be very helpful for opening the band gap of graphene. These metal-doped graphenes can be used in hydrogen storage and gas sensing, as shown in this chapter.

References

Anna, R.F., Małgorzata, K.L., Victor, S., Silvia, I., Manuel, A., Grazyna, S., et al., 2015. Development of noncytotoxic chitosan–gold nanocomposites as efficient antibacterial materials. ACS Appl.. Mater. Interfaces. 7, 1087–1099.

Avramov, P.V., Sakai, S., Entani, S., Matsumoto, Y., Naramoto, H., 2011. Ab initio LC-DFT study of graphene, multilayer graphenes and graphite. Chem. Phys. Lett. 508, 86–89.

Banerjee, S., Bhattacharyya, D., 2008. Electronic properties of nano-graphene sheets calculated using quantum chemical DFT. Comput. Mater. Sci. 44, 41–45.

Basu, S., Bhattacharyya, P., 2012. Recent developments on graphene and graphene oxide based solid state gas sensors. Sens. Actuators B: Chem. 173, 1–21.
Botello-Méndez, A.R., Lherbier, A., Charlier, J.C., 2013. Modeling electronic properties and quantum transport in doped and defective graphene. Solid State Commun. 175–176, 90–100.
Chatterjee, S.G., Chatterjee, S., Ray, A.K., Chakraborty, A.K., 2015. Graphene-metal oxide nanohybrids for toxic gas sensor: a review. Sens. Actuators B. 221, 1170–1181.
Chen, X.W., Hai, X., Wang, J.H., 2016. Graphene/graphene oxide and their derivatives in the separation/isolation and preconcentration of protein species: a review. Anal. Chim. Acta. 922, 1–10.
Chen, Y., Wang, J., Liu, Z.-M., 2012. Graphene and its derivative-based biosensing systems. Chin. J. Anal. Chem. 40, 1772–1779.
Chieu, D.T., Tamutsiwa, M.M., 2015. Cellulose, chitosan, and keratin composite materials. Controlled drug release. Langmuir. 31, 1516–1526.
Choong, K.Y.N.S., Tiroumalechetty, M., 2008. Dioxin levels in fly ash coming from the combustion of bagasse. J. Hazard. Mater. 155, 179–182.
Cole, P., Adami, H.O., Trichopoulos, D., Mandel, J., 2010. Formaldehyde in the indoor environment. Regul. Toxicol. Pharmacol. 58, 161–166.
Delley, B., 1990. An all-electron numerical method for solving the local density functional for polyatomic molecules. J. Chem. Phys. 92, 508–517.
Delley, B., 2000. From molecules to solids with the $DMol^3$ approach. J. Chem. Phys. 113, 7756–7764.
Delley, B., 2002. Hardness conserving semilocal pseudopotentials. Phys. Rev. B: Condens. Matter Mater. Phys. 66, 155125–155133.
Duong, A., Steinmaus, C., McHale, C.M., Vaughan, C.P., Zhang, L., 2011. Reproductive and developmental toxicity of formaldehyde: a systematic review. Mutat. Res./Rev. Mutat. Res. 728, 118–138.
Friesner, R.A., 2005. Ab initio quantum chemistry: methodology and applications. Proc. Natl. Acad. Sci. U. S. A. 102, 6648–6653.
Gao, S., Ren, Z., Wan, L., Zheng, J., Guo, P., Zhou, Y., 2011. Density functional theory prediction for diffusion of lithium on boron-doped graphene surface. Appl. Surf. Sci. 257, 7443–7446.
Gaylor, D.W., Aylward, L.L., 2004. An evaluation of benchmark dose methodology for non-cancer continuous-data health effects in animals due to exposures to dioxin (TCDD). Regul. Toxicol. Pharmacol. 40 (1), 9–17.
Grande, L., Chundi, V.T., Wei, D., Bower, C., Andrew, P., Ryh(a)nen, T., 2012. Graphene for energy harvesting/storage devices and printed electronics. Particuology. 10, 1–8.
Hartree, D.R., 1928. The wave mechanics of an atom with a non-Coulomb central field. Part I. Theory and methods Mathematical Proceeding of the Cambridge Philosophical Society. Cambridge University Press, Cambridge, pp. 89–110.
Ishii, K., Furuichi, T., 2007. Development of bioreactor system for treatment of dioxin-contaminated soil using *Pseudallescheria boydii*. J. Hazard. Mater. 148 (3), 693–700.
Jain, I.P., Jain, P., Jain, A., 2001. Novel hydrogen storage materials: a review of lightweight complex hydrides. J. Alloys Compd. 503 (2), 303–339.
Jain, I.P., Lal, C., Jain, A., 2010. Hydrogen storage in Mg: a most promising material. Int. J. Hydrogen Energy. 35 (10), 5133–5144.
Ji, X.Q., Xu, Y.H., Zhang, W.L., Cui, L., Liu, J.Q., 2016. Review of functionalization, structure and properties of graphene/polymer composite fibers. Composites: A. 87, 29–45.

Jian, N., Han, J.Y., Wang, H., Zhu, X.L., Ge, Q.F., 2015. A DFT study of oxygen reduction reaction mechanism over O-doped graphene-supported Pt4, Pt3Fe and Pt3V alloy catalysts. Int. J. Hydrogen Energy. 40, 5126–5134.

Ke, Q.Q., Wang, J., 2016. Graphene-based materials for supercapacitor electrodes: a review. J. Materiomics. 2, 37–54.

Kothari, R., Singh, D.P., Tyagi, V.V., Tyagi, V.V., 2012. Fermentative hydrogen production —an alternative clean energy source. Renewable Sustainable Energy Rev. 16 (4), 2337–2346.

Kuttappan, S., Mathew, D., Nair, M.B., 2016. Biomimetic composite scaffolds containing bioceramics and collagen/gelatin for bone tissue engineering—a mini review. Int. J. Biol. Macromol.. Available from: http://dx.doi.org/10.1016/j.ijbiomac.2016.06.043.

Liao, L., Duan, X., 2010. Graphene-dielectric integration for graphene transistors. Mater. Sci. Eng.: R: Rep. 70, 354–370.

Liu, L.L., Qing, M.Q., Wang, Y.B., Chen, S.M., 2015. Defects in graphene: generation, healing, and their effects on the properties of graphene: a review. J. Mater. Sci. Technol. 31, 599–606.

Lopez-Corral, I., German, E., Juan, A., Volpe, M.A., Brizuela, G.P., 2012. Hydrogen adsorption on palladium dimer decorated graphene: A bonding study. Int. J. Hydrogen Energy. 37 (8), 6653–6665.

Luxembourg, D., Flamant, G., Beche, E., Sans, J.L., Giral, J., Goetz, V., 2007. Hydrogen storage capacity at high pressure of raw and purified single wall carbon nanotubes produced with a solar reactor. Int. J. Hydrogen Energy. 32 (8), 1016–1023.

Majeti, N.V., Ravi, K., 2000. A review of chitin and chitosan applications. React. Funct. Polym. 46, 1–27.

Margarita, D., Mar, L.B., Pilar, A., Antonio, J.A., Julio, B., Eduardo, R.H., 2006. Microfibrous chitosan-sepiolite nanocomposites. Chem. Mater. 18, 1602–1610.

Marmolejo-Tejada, J.M., Velasco-Medina, J., 2016. Review on graphene nanoribbon devices for logic applications. Microelectr. J. 48, 18–38.

Midilli, A., Ay, M., Dincer, I., Rosen, M.A., 2005. On hydrogen and hydrogen energy strategies: I: Current status and needs. Renewable Sustainable Energy Rev. 9 (3), 255–271.

Monkhorst, H.J., Pack, J.D., 1976. Special points for Brillouin-zone integrations. Phys. Rev. B: Condens. Matter Mater. Phys. 13, 5188–5192.

Mou, Z., Dong, Y., Li, S., Du, Y., Wang, X., Yang, P., et al., 2011. Eosin Y functionalized graphene for photocatalytic hydrogen production from water. Int. J. Hydrogen Energy. 36 (15), 8885–8893.

Novoselov, K.S., Geim, A.K., Morozov, S.V., Jiang, D., Zhang, Y., Dubonos, S.V., et al., 2004. Electric field effect in atomically thin carbon films. Science. 306, 666–669.

Novoselov, K.S., Geim, A.K., Morozov, S.V., Jiang, D., Katsnelson, M.I., Grigorieva, I.V., et al., 2005. Two-dimensional gas of massless dirac fermions in graphene. Nature. 438, 197–200.

Park, S.J., 2008. Aqueous suspension and characterization of chemically modified graphene sheets. Chem. Mater. 20, 6592–6594.

Peltier, M.R., Arita, Y., Klimova, N.G., Gurzenda, E.M., Koo, H.C., Murthy, A., et al., 2013. 2,3,7,8-Tetrachlorodibenzo-*p*-dioxin (TCDD) enhances placental inflammation. J. Reprod. Immunol. 98 (1–2), 10–20.

Perdew, J.P., Burke, K., Ernzerhof, M., 1996. Generalized gradient approximation made simple. Phys. Rev. Lett. 77, 3865–3868.

Prashant, S.K., João, G.C., Afonso, C.A.M., 2008. Dioxins sources and current remediation technologies—a review. Environ. Int. 34 (1), 139–153.

Putria, L.K., Ong, W.J., Chang, W.S., Chai, S.P., 2015. Heteroatom doped graphene in photocatalysis: a review. Appl. Surf. Sci. 358, 2–14.
Rasheed, A.K., Khalid, M., Rashmi, W., Gupta, T.C.S.M., Chan, A., 2016. Graphene based nanofluids and nanolubricants—review of recent developments. Renewable Sustainable Energy Rev. 63, 346–362.
Rinaudo, M., 2006. Chitin and chitosan: properties and applications. Prog. Polym. Sci. 31, 603–632.
Schectera, A., Birnbaumb, L., Ryanc, J.J., Constabled, J.D., 2006. Dioxins: an overview. Environ. Res. 101 (3), 419–428.
Sealy, C., 2009. Graphene-CNT superstructure holds promise for hydrogen storage. Nano Today. 4 (1), 6–6.
Shao, Y., Wang, J., Wu, H., Liu, J., Aksay, I.A., Lin, Y., 2010. Graphene based electrochemical sensors and biosensors: a review. Electroanalysis. 22, 1027–1036.
Sherif, S.A., Barbir, F., Veziroglu, T.N., 2005. Wind energy and the hydrogen economy-review of the technology. Solar Energy. 78 (5), 647–660.
Tan, B., Thomas, N.L., 2016. A review of the water barrier properties of polymer/clay and polymer/graphene nanocomposites. J. Mem. Sci. 514, 595–612.
Tan, S.H., Tang, L.M., Chen, K.Q., 2014. Band gap opening in zigzag graphene nanoribbon modulated with magnetic atoms. Curr. Appl. Phys. 14, 1509–1513.
Toda, K., Furue, R., Hayami, S., 2015. Recent progress in applications of graphene oxide for gassensing: a review. Anal. Chim. Acta. 878, 43–53.
Tsaia, W.B., Chen, Y.R., Liu, H.L., Lai, J.Y., 2011. Fabrication of UV-crosslinked chitosan scaffolds with conjugation of RGD peptides for bone tissue engineering. Carbohydr. Polym. 85, 129–137.
Tunga, S., Sibel, M., Marutzky, R., 2010. Formaldehyde in the indoor environment. Chem. Rev. 110, 2536–2572.
Wang, G., Yang, J., Park, J., Gou, X., Wang, B., Liu, H., et al., 2008. Facile synthesis and characterization of graphene nanosheets. J. Phys. Chem. C. 112, 8192–8195.
Wang, J., Zhang, P., Qi, J.Q., Yao, P.J., 2009. Silicon-based micro-gas sensors for detecting formaldehyde. Sens. Actuators B: Chem. 136, 399–404.
Wang, S.F., Yao, Y., Zhang, H.L., Wang, R., 2011. Stone-Wales defect as a dipole of dislocation and anti-dislocation. Phys. Lett. A. 375, 4109–4112.
Xiong, D.B., Li, X.F., Shan, H., Zhao, Y., Dong, L., Xu, H., et al., 2015. Oxygen-containing functional groups enhancing electrochemical performance of porous reduced graphene oxide cathode in lithium ion batteries. Electrochim. Acta. 174, 762–769.
Yang, X.M., Tu, Y.F., Li, L., Shang, S.M., Tao, X., 2010. Well-dispersed chitosan/graphene oxide nanocomposites. ACS Appl. Mater. Interfaces. 2, 1707–1713.
Yang, Y., Yuan, S.X., Zhao, L.H., Wang, C., Ni, J.S., Wang, Z.G., et al., 2015. Ligand-directed stearic acid grafted chitosan micelles to increase therapeutic efficacy in hepatic cancer. Mol. Pharm. 12, 644–652.
Zhang, H.P., Luo, X.G., Lin, X.Y., Lu, X., Tang, Y., 2013a. Density functional theory calculations of hydrogen adsorption on Ti-, Zn-, Zr-, Al-, and N-doped and intrinsic graphene sheets. Int. J. Hydrogen Energy. 38, 14269–14275.
Zhang, H.P., Luo, X.G., Lin, X.Y., Lu, X., Leng, Y., Song, H.T., 2013b. Density functional theory calculations on the adsorption of formaldehyde and other harmful gases on pure, Ti-doped, or N-doped graphene sheets. Appl. Surf. Sci. 283, 559–565.
Zhang, H.P., He, W.D., Luo, X.G., Lin, X.Y., Lu, X., 2014a. Adsorption of 2,3,7,8-tetrochlorodibenzo-p-dioxins on intrinsic, defected, and Ti (N, Ag) doped graphene: a DFT study. J. Mol. Model. 20, 2238–2244.

Zhang, H.P., Luo, X.G., Lin, X.Y., Lu, X., 2014b. DFT study of adsorption and dissociation behavior of H_2S on Fe-doped graphene. Appl. Surf. Sci. 317, 511–516.

Zhang, H.P., Luo, X.G., Lin, X.Y., Lu, X., Tang, Y., 2015. Modulating the interactions between MgH_2 and graphene using different dopants. Chem. Phys. Lett. 623, 82–88.

Zhang, H.P., Luo, X.G., Lin, X.Y., Lu, X., Tang, Y., 2016. The molecular understanding of interfacial interactions of functionalized graphene and chitosan. Appl. Surf. Sci. 360, 715–721.

Zhao, Y., Li, X.G., Zhou, X., Zhang, Y.N., 2016. Review on the graphene based optical fiber chemical and biological sensors. Sens. Actuators B. 231, 324–340.

Zini, G., Tartarini, P., 2010. Wind-hydrogen energy stand-alone system with carbon storage: modeling and simulation. Renewable Energy. 35 (11), 2461–2467.

Synthesis and applications of cellulose nanohybrid materials

10

Nathaniel T. Garland[1], *Eric S. McLamore*[2], *Carmen Gomes*[3], *Ethan A. Marrow*[4], *Michael A. Daniele*[4,5], *Scott Walper*[6], *Igor L. Medintz*[6] *and Jonathan C. Claussen*[1]

[1]Department of Mechanical Engineering, Iowa State University, Ames, IA, United States, [2]Agricultural and Biological Engineering Department, Institute of Food and Agricultural Sciences, University of Florida, Gainesville, FL, United States, [3]Department of Biological and Agricultural Engineering, Texas A&M, College Station, TX, United States, [4]Joint Department of Biomedical Engineering, UNC-Chapel Hill/NC State University, Raleigh, NC, United States, [5]Department of Electrical & Computer Engineering, NC State University, Raleigh, NC, United States, [6]Center for Biomolecular Science and Engineering, U.S. Naval Research Laboratory, Washington, DC, United States

Chapter Outline

10.1 Introduction 289
10.2 Cellulose 291
10.3 Regenerated cellulose 291
10.4 Nanocrystalline cellulose 293
10.5 Functionalization 295
10.6 Applications 295
- 10.6.1 Propulsion 296
- 10.6.2 Cellulose-based electronics 298
- 10.6.3 Biosensing 302
- 10.6.4 Packaging materials and drug delivery systems 308

10.7 Conclusion 310
References 311

10.1 Introduction

Cellulose, a polymer comprised of repeated glucose monomers, is the most abundant biopolymer on earth. Cellulose contains straight, rod-like chain structures with side hydroxyl groups that form myriad hydrogen bonds with neighboring cellulose strands—a network structure that is found ubiquitously in plant cell walls.

Hybrid Polymer Composite Materials: Applications. DOI: http://dx.doi.org/10.1016/B978-0-08-100785-3.00010-3

These cellulose networks provide a tremendous mechanical support structure that allows plants to reach surprising heights (Sequoia sempervirens (redwoods) can reach heights of over 115 m, Sillett et al., 2015). The concomitance of strength, abundance, and biorenewability makes cellulose an appealing low-cost material for use in a wide variety of applications. Until recently, these applications were primarily limited to paper and textiles, but recent efforts to structure these materials on the micro and nanoscale have opened the door for their use in emerging technologies such as flexible/disposable sensors and electronics.

Cellulose has been incorporated into man-made inventions since the Chinese first developed paper in A. D. 100. However, the polymer cellulose wasn't isolated and identified until the French chemist Anselme Payen purified the carbohydrate in 1838 (Klemm et al., 2005). Since its discovery, numerous industrial processes have been developed to extract cellulose from plant materials such as wood (40%–50% of wood comprises cellulose, Pettersen, 1984). To turn plant material into cellulose for paper, raw plant material is mechanically and chemically broken down into smaller, soluble fibers that can be recast as paper, fibers, or membranes (Klemm et al., 2005). Likewise cotton, another important source of cellulose has been twisted and spun into fibers to form clothing for thousands of years (Klemm et al., 2005). More recently, synthetic fibers such as rayon have been synthesized from cellulose fibers via chemical processing. Furthermore, cellulose fibers can be converted into guncotton, an explosive first discovered by Henri Braconnot in 1832 (Braconnot, 1833), when mixed with nitric acid. Even early motion-picture films used a flexible, moldable film called celluloid that was made by mixing nitrocellulose with camphor. Although many of the historical uses of cellulose have now been replaced by other materials, cellulose-based materials still hold a promising future for integration into advanced technologies.

The applications of cellulose have been broadened in recent years due in part to the merging or hybridization with advanced materials such as carbon nanotubes (CNTs), graphene, and metallic nanoparticles. These novel cellulose hybrid materials have enabled diverse technical applications, from bendable and disposable conductive films for biosensors to degradable but robust catalyst for UUV propulsion (Claussen et al., 2014; Daniele et al., 2015; Burrs et al., 2016; Mahadeva et al., 2015). These emerging technologies have been fueled by the unique fabrication and material properties of the cellulose hybrids including the ability to form these materials into fibers, sheets, membranes, and matrixes while leveraging high mechanical strength, chemical stability, and biocompatibility. This chapter will extensively review these emerging cellulose nanohybrid-based technologies in context with their unique material properties. In fact, this chapter is divided into the following two main components: First, a description of cellulose properties and material manufacturing is reviewed, followed by an overview of emerging cellulose nanohybrid materials and their subsequent applications. Other reviews on cellulose have focused on synthesis, fabrication, or applications of solely cellulose materials. For example, other review articles detail the use of cellulose in sensors (Mahadeva et al., 2015; Joo-Hyung et al., 2014) and nanocellulose printed electronics (Hoeng et al., 2016). This chapter is a departure from these reviews as it highlights the broad range of applications possible by incorporating nanoscale materials with cellulose.

10.2 Cellulose

The diverse applications discussed in this chapter require a broad spectrum of cellulose forms. Depending on the desired material properties, cellulose can be changed into fibers (Lee et al., 2013; Cai et al., 2007), nanowhiskers (Azizi Samir et al., 2005), paper (Klemm et al., 2005), membranes (Klemm et al., 2005), films (Zhang et al., 2005), or be functionalized (Klemm et al., 2005; Ribeiro-Viana et al., 2016; Hu et al., 2014) with chemical groups. The form that cellulose takes on is determined by its molecular and intermolecular structure. Cellulose found in nature in plants and bacteria is made up of chains that exhibit both highly crystalline domains and noncrystalline or amorphous domains, the ratio of which varies between species and conditions during synthesis and directly affect the physical properties of the material Park et al. (2010).

The structure of cellulose is defined at the molecular and crystalline scale. Cellulose is a linear polysaccharide chain comprised of β(1−>4) linked D-glucose monomers (Jones and Wegner, 2009). Four crystalline allotropes of cellulose exist due to the interaction of intermolecular hydrogen bonds. Cellulose I is the native form of cellulose that is found in plants, while cellulose II can be made from cellulose I through either alkali treatment (mercerization) or solubilization and recrystallization (regeneration), the process used for synthetic fiber and film production (Park et al., 2010). These two forms can be reversibly converted to celluloses IIII and IIIII, respectively, with liquid ammonia treatment (Hayashi et al., 1975).

These different crystalline forms are characterized by the microfibril helix angle (orientation of microfibrils). In plants and man-made cellulosic materials, module of elasticity and elongation are tuned by varying helix angle (Park et al., 2010). With higher helix angles (low orientation), the material is of lower strength and higher elongation at breakage, whereas lower helix angles (high orientation) give rise to higher strengths (Klemm et al., 2005). As the amount of material processing increases, so does the elastic modulus. For example, the modulus of wood increases from 10 to 40 GPa after chemical pulping and to 70 GPa after hydrolysis and mechanical disintegration. In man-made cellulose fibers, the greater the degree of polymerization (longer molecular chains), the higher the tensile strength of the material (Fink et al., 1994).

We will now explore the methods of cellulose material production. Regenerated cellulose is pure, soluble, and chemically prepared from native sources. The most often used subset of regenerated cellulose in nanohybrid materials is nanocrystalline cellulose, which comprises fibers less than 100 nm in length. These materials' properties depend on the molecular structure discussed (Fig. 10.1).

10.3 Regenerated cellulose

Regenerated cellulose is the material produced by mechanical and chemical breakdown of plant material to a soluble pure cellulose solution, with less

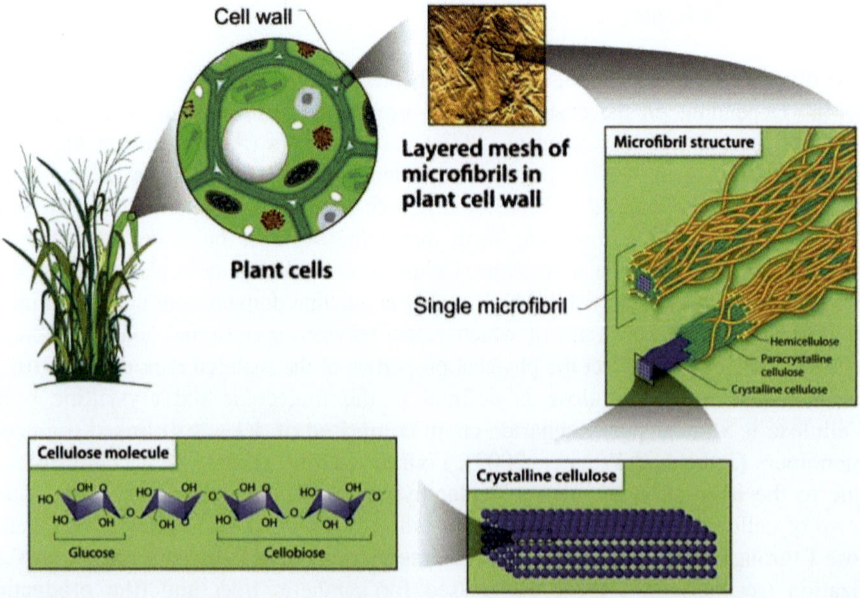

Figure 10.1 Macro to microscale of cellulose.
Reprinted with permission by U.S. Department of Energy Office of Science.

crystallinity and smaller particles as processing increases from the natural state. The process of creating materials from regenerated cellulose begins with the natural source. Wood and other plant matter from cotton, jute, flax, sisal, and other sources can all be harvested for their cellulose-based fibers (Bledzki and Gassan, 1999).

The chemical processing required to produce regenerated cellulose must overcome the intermolecular hydrogen bonding between strands that makes cellulose hydrophobic. Strong dissolving chemicals are required, such as copper(II) hydroxide, aqueous ammonia, strong acids, and alkali solutions used with catalysts (Zhang et al., 2010; Ke et al., 2006; Wang et al., 2008; Qi et al., 2009). Recently, *N*-methyl-morpholine-*N*-oxide and cadmium chloride ($CdCl_2 \cdot 2.5H_2O$) have been identified as some of the most powerful solvents to form regenerated cellulose with favorable properties (Fink et al., 2001; Ruan et al., 2005; Cai et al., 2007; Walther et al., 2011). Once prepared, regenerated cellulose can be spun into fibers, or spread into films and membranes.

Fibers made from regenerated cellulose were first produced in the 1860s, when Courtaulds developed a chemical method to block the reactive hydroxyl groups of cellulose, resulting in a spinnable solution called rayon (Borbély, 2008). The degree of polymerization, fiber angle, and cellulose content (weight%)—all affect the resulting mechanical properties (Young's modulus, Tensile strength).

10.4 Nanocrystalline cellulose

Nanocrystalline cellulose (NCC) comprises needle-like morphologies of less than 100 nm and is formed by removing lignin and hemicellulose from plant/wood pulp. NCC has high tensile strength and can form solid transparent films as thin as a few microns. NCC can be uniformly dispersed in water and is biocompatible/biodegradable (Jones and Wegner, 2009). Although significant efforts have been made, removal of lignin and hemicellulose is notoriously difficult, expensive, and environmentally unfavorable. A review of the process can be found in Ng et al. (2015) and is summarized briefly here. Typically, once the biomass has reduced to a manageable form, the plant material is treated with alkali and bleaching solutions to remove the lignin and hemicellulose components of the cellulose bundle. Mercerization is typically performed with concentrated sodium or potassium hydroxide solutions that allow the displacement of hemicellulose from the cell wall material. The subsequent bleaching steps remove residual contaminants and lignin. This is typically done under high heat and acidic conditions. Once other contaminants of the plant cell wall are removed, the amorphous regions of the cellulose microfibril are removed through acid hydrolysis. A number of different acids have been employed for this stage of the processing that results in the hydrolytic cleavage of glycosidic bonds (Sirvio et al., 2016; Dufresne, 2010). Any residual hemicellulose and pectin are also removed in this aggressive stage of the purification. Final mechanical treatments are often included to ensure uniformity of the NCC.

NCC can also be produced in certain species of bacteria. These organisms produce cellulose that is excreted from the cell into the environment or extracellular space often with the goal of generating a protective environment in which the bacterial community can grow and thrive. Though chemically identical, bacterial cellulose exhibits several unique properties as a material compared to plant-derived cellulose. Bacterial cellulose exhibits a greater crystallinity, higher tensile strength, and improved capacity for water retention compared to plant materials. The unique properties of this material in both its hydrated and dehydrated forms have led to interest in using the material in wound dressings (Czaja et al., 2006; Rouabhia et al., 2014), artificial vasculature (Klemm et al., 2001) and tissue engineering (Svensson et al., 2005), cosmetics (Ullah et al., 2016; Amnuaikit et al., 2011), textiles, and numerous other areas. Industrial applications of bacterial cellulose have thus far been limited due to difficulties with large-scale manufacture; however, research in alternate bioreactors and fermenters has improved the process and may eventually lead to the commercialization of this material.

To date, the most studied bacterial cellulose producer is *Gluconacetobacter xylinus* (formerly *Acetobacter*), though cellulose synthesis has been observed in a broad range of genera including *Agrobacterium*, *Aerobacter*, *Salmonella*, *Rhizobium*, and *Azotobacter*. The Gram-negative, α-proteobacterium *G. xylinus,* is strict aerobe that utilizes fixed carbon sources such as sugars and alcohols to produce a cellulose. Cellulose generally forms from static cell cultures as a thin film on solid nutrient sources such as rotting fruit, as a thick pellicle that forms at the air–media interface in liquid cultures, or as spherical particles if the culture is sufficiently agitated during growth. Although glucose is the preferred carbon source for cellulose production in *G. xylinus*, the bacterium is able to utilize a variety of carbon sources including other mono-, di-, and polysaccharides as well as some alcohols and other organic molecules (Masaoka et al., 1993). Generally, properties of the macromolecular pellicle change slightly with variation in growth conditions and carbon sources.

The morphology of the bacterial cellulose pellicle can vary dramatically on the basis of the conditions under which it is grown. The ability to control the physical dimensions of the pellicle through the choice of growth chamber is one of the more unique features of the cellulose pellicle; however, the very specific requirements for cellulose production have proven limiting for industrial-scale manufacture of this material. Aside from the dimensions of the growth chamber itself, the most influential growth condition and one that has proven most limiting is whether the culture is grown statically or with agitation. Grown statically, pellicles can be shaped to form thin films, thick mats, tubes, or foils using methods that have been optimized over the years in the food industry for the production of Nata de Coco or SCOBY pellicles. Although successful for many applications, bacterial growth, and pellicle formation under static conditions is notoriously slow, often taking weeks for a suitable pellicle to be formed. In contrast to the uniform pellicles formed under static conditions, agitated cultures produce small aggregates, elongated ribbons, or spheres during culture maintenance. Both methods of cellulose production have found success in the small-scale manufacture of cellulose-based materials.

The formation of flat pellicles and foils under static conditions has been utilized by the food industry in East Asian countries for decades. The properties (fibril size, density, etc.) of the cellulose pellicle grown under static conditions can be well controlled through choice of carbon source and basic physical conditions such as temperature, pH, and media composition. This method of cellulose production, however, is very slow and costly in more developed countries where labor costs are significantly higher. As static cultures will rapidly experience a depletion of nutrients within proximity of the pellicle, researchers have investigated alternative methods of growing static cultures typically referred to as Fed-Batch cultivation. Here, fresh media is supplied to the culture either beneath the culture (Bae and Shoda, 2004) or to the surface of the pellicle itself using jets to aerosolize the media (Hornung et al., 2007). These methods have led to formation of a significantly thicker pellicle compared to the standard static growth method.

A number of different reactors are currently utilized for the production of cellulose under agitated conditions. Reactor types include traditional shake flasks (Czaja et al., 2004), stirred culture flasks (Zhou et al., 2007), and airlift reactors (Chao et al., 2000). Although not fully understood, in agitated cultures, it is believed that cultures form at the interface with air bubbles trapped within the media. As bacteria proliferate cellulose ribbons move outward forming compact particles (Czaja et al., 2004). Though agitated cultures can be grown in large-scale industrial fermenters and bioreactors, great care must be taken when maintaining these cultures for extended periods. Under such conditions, a subpopulation of the

G. xylinus culture will spontaneously develop a Cel-phenotype (inability to produce cellulose) (Hestrin and Schramm, 1954). This population possesses a growth advantage which allows them to rapidly overtake their cellulose producing relatives.

Regardless of the method of production, pellicles are typically treated with either detergents or alkali solutions to decellularize them and remove bacterial protein and media contaminants. The cellulose materials can then be stored in any number of ways determined by downstream application. If free of fungal contaminants that would consume the cellulose pellicles, they can effectively be stored indefinitely.

10.5 Functionalization

Once prepared, these cellulose structures have further utility in their ability to be functionalized with chemical groups and nanomaterials. Chemical functionalization takes advantage of cellulose's strongly reactive hydroxyl groups to alter solubility, increase metal binding (Diekmann et al., 2003; Kowalik et al., 2000), create new selective protecting groups for organic synthesis (Diekmann et al., 2003; Wuts and Greene, 2006), and realize new structural forms of cellulose (Kowalik et al., 2000; Jaehne et al., 2002). The many hydroxyl side groups of cellulose are advantageous toward functionalization with other chemical moieties including biological agents (Tanaka and Sackmann, 2006). Subsequently, cellulose can be carboxymethylated (Cheng et al., 2013), succinylated (Ribeiro-Viana et al., 2016), acetylated, and phosphorylated (Hu et al., 2014). The aim of these various chemical functionalization strategies is to improve the surface properties for the desired application while maintaining the useful three-dimensional (3D) structure of the cellulose.

Cellulose can also be functionalized with other carbon-based materials such as CNTs (Mandal et al., 2016; Zhao et al., 2010), graphene (Burrs et al., 2015, 2016; Cheng et al., 2013; Cao et al., 2016; Weng et al., 2011; Xiong et al., 2016), and nanometals (Claussen et al., 2014; Pinto et al., 2007). In these nanohybrid materials, cellulose acts as a ubiquitous and biofriendly substrate for modern applications including catalysts for small-scale propulsion, printable and flexible electronics, disposable sensors and biosensors, and biocompatible drug delivery applications. In the following section, we will investigate these specific applications and explain the synthesis of the nanohybrid cellulose materials used in each case.

10.6 Applications

This section will discuss the use of nanohybrid cellulosic materials in small-scale UUV propulsion, flexible electronics, biosensors, and drug delivery. Material synthesis, imaging, current results, and future directions will be investigated. These exciting fields offer new and useful applications for the versatility of cellulose used in current research and future technologies.

10.6.1 Propulsion

Cellulose nanohybrid materials are beginning to be used for micropropulsion, such as to power UUVs (Marr et al., 2015) that can be deployed for various missions including chemical sensing, payload delivery, and underwater surveillance. Such cellulose-based catalysts provide a low-cost and degradable pathway for the decomposition of hydrogen peroxide that produces oxygen and water/steam that can drive UUVs through pulsation-based propulsion. Such hydrogen peroxide propelled UUVs eliminate the need for complex electromechanical systems (e.g., mechanical propellers), provide fine maneuvering through complex features (e.g., interrogating sunken vessels or operating in dense mine fields), and emit environment-friendly byproducts (oxygen and water/steam). These cellulose-based catalysts are effective for hydrogen peroxide decomposition when combined with nanomaterials (i.e., nanostructured metal or oxide) such as platinum nanoparticles. The decomposition of H_2O_2 degradation by platinum catalyst as a method of propulsion has been used in many UUV applications, such as janus motors (Baraban et al., 2012; Ke et al., 2010; Gibbs and Zhao, 2009), conical-shaped bubble thrusters (Campuzano et al., 2011; Gao et al., 2011; Mirkovic et al., 2010; Manesh et al., 2010), catalytic nano/micromotors (Laocharoensuk et al., 2008; Soler et al., 2013; Wang et al., 2006), and a parallel arrangement of microtubular bubble thrusters (Mei et al., 2008; Solovev et al., 2009). An additional advantage of these H_2O_2-based systems is their scalability. They have been used in a number of applications such as platinum-loaded stomatocytes (Watt, 2014), platinum and catalase conical-shaped or tubular bubble thrusters (Baraban et al., 2012; Ke et al., 2010; Gibbs and Zhao, 2009; Campuzano et al., 2011; Gao et al., 2011; Mirkovic et al., 2010; Manesh et al., 2010; Wernimont, 2006), chemotaxis-driven silica-manganese "matchstick" particles and bimetallic nanorods (Laocharoensuk et al., 2008; Soler et al., 2013; Wang et al., 2006; Mei et al., 2008; Solovev et al., 2009), catalytic janus motors (Hutchison et al., 2010; Song et al., 2011; Cho et al., 2001; Zhang et al., 2011; Tasaltin and Basarir, 2014), and microelectromechanical systems based thrusters (Claussen et al., 2009; Maruyama and Xiang, 2012).

Recently, researchers synthesized platinum nanourchins (PNUs) via the reduction of hexachloroplatinic acid (H_2PtCl_6) with formic acid (HCOOH) onto cellulose-based papers (Claussen et al., 2014). The PNUs were deposited on microfibrillated cellulose films (MFC) through the process of submerging the MFC in the hexachloroplatinic acid solution. Upon completion of the reaction, a PNU-MFC nanohybrid material with a porous structure was formed. These PNU-MFC nanohybrid materials or catalytic Pt-papers were evaluated and characterized using SEM and TEM. SEM images revealed that the Pt-papers had a highly dense tall and coral-like structures on the surface of the MFC similar to Pt nanowires that have been fabricated on CNTs carbon spheres, and 3D graphene (Meng et al., 2011; Sun et al., 2008, 2010; Sattayasamitsathit et al., 2013). High-resolution transmission electron microscopy also showed that the PNUs form (1 1 1) planes with a lattice spacing of 0.23 nm. These observations satisfy the requirements of tight size distribution, high proportion of (1 1 1) faces, and high metal loading and surface area

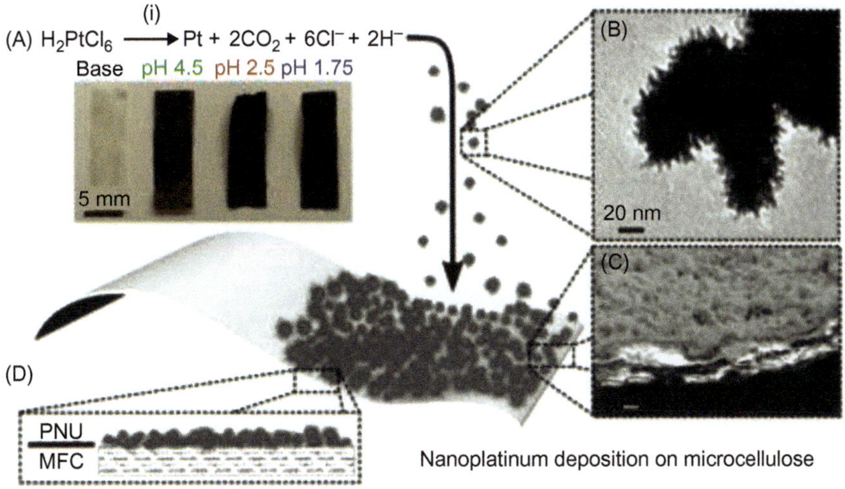

Figure 10.2 Scanning electron microscopy images of PNUs deposited on MFCs at different pHs and magnifications.
Reprinted with permission by American Chemical Society.

that is essential to optimize the use of platinum as a monopropellant catalyst (Fig. 10.2).

The resulting Pt-catalyst paper was evaluated by testing its performance in hydrogen peroxide decomposition. The performance of the Pt-catalyst was characterized by measuring the differential pressures between two sealed round-bottom flasks containing H_2O_2; one containing H_2O_2 and a bare MFC that acts as a control and another containing H_2O_2 and the PNU-MFC. From experimental results, it was apparent that the presence of the PNU-MFC greatly increases the decomposition rate of H_2O_2, compared to the flask containing bare MFC or even MFC with a planar evaporated platinum. This observation may be attributed to the fact that PNU-MFC papers have the ability to lower the activation energy required for H_2O_2 decomposition by over a 50% as compared to traditional catalytic material, such as metal oxides, thus accelerating the rate of decomposition. The robustness and improved efficiency of this Pt-nanohybrid material can thus be exploited to generate sustainable thrust for micro-UAVs from the size of a few centimeters up to half a meter. To further prove the feasibility of the concept, a tiny reaction chamber was filled with the PNU-MFC strips and attached to a strain gauge underwater. Next, various flow rates of H_2O_2 were supplied to the chamber, and the measurements on the strain gauge were recorded. It was found that for a reaction chamber of 20 mL filled with 10 strips of PNU-MFCs and supplied with 30% w/w of H_2O_2, the reaction process was capable of providing 0.6 N of thrust, or an equivalent of 50 m at an average velocity of 4.8 m s^{-1}. In comparison, a micro-UAV propelled by compressed CO_2 of the same volume only moves approximately 37 m at an average velocity of 2.7 m s^{-1} (Fig. 10.3).

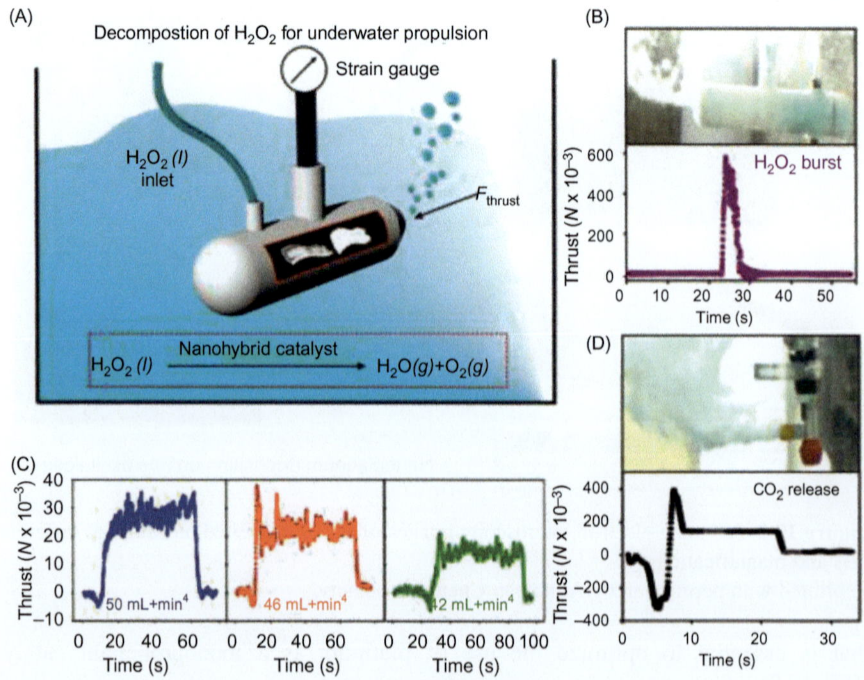

Figure 10.3 Test submersible assembly (A). Thrust generated by H_2O_2 (B). Thrust generated by different volumes of H_2O_2 (C). Thrust generated by CO_2 release (D).
Reprinted with permission by American Chemical Society.

Other groups have also explored cellulose as a substrate for UUV propulsion. One method of particular interest is artificial flagella made using cellulose. In a recent review (Majumdar et al., 2013), cellulose acetate was identified as a viable material to construct flagella due to its flexural modulus range, biocompatibility, and nanostructured size. Another group successfully immobilized flagellated bacteria in a bacterial NCC matrix, resulting in a microbial-robot that was capable of traveling at 4.8 $\mu m\ s^{-1}$ (Higashi and Miki, 2014). Dhar et al. (2015) also used cellulose as a platform to immobilize a method of propulsion, instead using zerovalent iron nanoparticles supported on cellulose nanocrystals whose trajectory and speed can be remotely tuned using pH and chemical gradients. These intriguing new technologies offer new approaches to drug delivery, bioremediation, and remote sensing by harnessing the unique material properties of cellulose (Fig. 10.4).

10.6.2 Cellulose-based electronics

Silicon is a reliable substrate for printing electronic devices and has revolutionized the electronics industry. Silicon-based central processing units can be found throughout the electronic devices we use today from computers to smartphones.

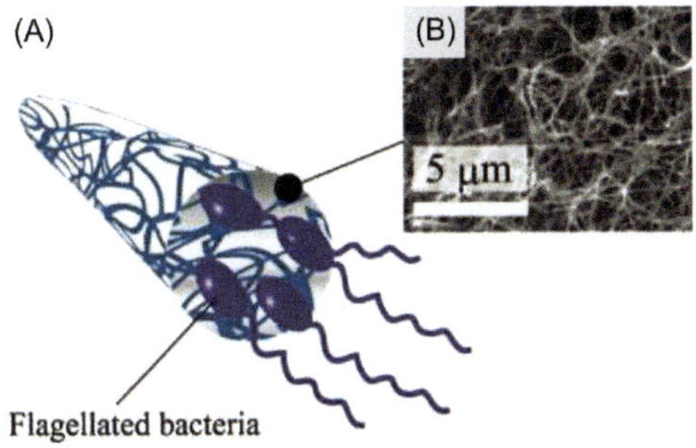

Figure 10.4 Flagellated microbial-robot (A) utilizing a cellulose matrix (B) to immobilize bacteria. Reprinted with permission by Elsevier.

However, one of the major drawbacks to silicon-based electronics is their brittleness. Silicon does not bend, stretch, or flex and is a hindrance to engineers trying to develop flexible and wearable electronics. Therefore, cellulose is becoming a material of interest as an engineering substrate for flexible and wearable electronics. For example, flexible conductive substrates would be useful in developing a wearable biosensor that must maintain solid contact with tissue over time to make accurate readings. The inherent strength of cellulose, its chemically inert nature, and biological abundance make it an attractive material with similar properties to silicon, with the addition of flexibility.

One of the potential drawbacks for using cellulose in electronic devices is its resistivity. Cellulose can easily dampen electronic signals and must be modified in instances in which substrate conductivity is required. Graphene is a highly conductive allomorph of carbon and has recently been the focus of many materials and electrical engineers due to its conductivity and mechanical toughness. Cellulose nanocrystals have been shown to be a powerful reinforcing agent in polymer chemistry, and so they can also be a reinforcing agent in flexible, wearable electronics. According to a recent publication, a thin-film graphene oxide film was created with layers of cellulose nanocrystals sandwiched between layers of graphene oxide. At $\sim$40% NCC and 60% graphene oxide, the thin film remained highly conductive (5000 S m^{-1}), transparent, and flexible (Xiong et al., 2016). The method of layer-by-layer incorporation of NCCs into a graphene thin film is very labor intensive and, although useful, may not be necessary for many applications. Other groups have shown that cellulose can be coated with graphene sheets by simple filtration of a graphene solution through a cellulose membrane to yield cellulose-based capacitors (Weng et al., 2011; Beeran et al., 2016). This simple method yields graphene-coated cellulose networks that remain flexible. The graphene adsorbs to the cellulose through electrostatic interactions. According to the authors, the only limitation to this method lies in the varying pore sizes within the network; as long as the graphene can penetrate the pores, it will coat the fibers.

Cellulose is an exceptionally versatile polymer with more applications than graphene-coated capacitors. Bacterial cellulose membranes are very thin membranes of cellulose produced by certain strains of bacteria. The cellulose comprising these membranes have higher crystallinity and fiber length than typical plant cellulose and make a suitable substrate for printed electronics for wearable applications. As cellulose is a biocompatible, conformable and does not occlude skin, it is an exciting substrate material for wearable electronics. Physically, cellulose is also a suitable substrate on which to print other electrical components due to a combination of the abundance of hydroxyl groups and porous structure which enable a strong adhesion between metals and substrate. Cellulose is also insulating and does detract from electrical current.

Printed circuit wearable biosensors are a growing class of products utilizing cellulose substrates. As cellulose does not occlude skin and transports water exceptionally well, it has gained interest as a sweat wicking substrate for detecting sweat-borne biosignals. Daniele et al. (2015) have produced cellulose-based circuits on microbial cellulose that adhere to skin through van der Waals forces with high conformality. This type of engineering is relatively new, but if further developed, could yield "temporary tattoo-like" sensors that could be produced cheaply.

In a world needing more cost-effective health care and cheaper sensor technology, cellulose-based substrates are continuing to build interest. Cellulose-based paper was first used as the dielectric layer for a flexible transistor in 2008 (Fink et al., 1994). The transistor, despite its ease of production, remained stable over a long period of time and rivaled its silicon and glass-based counterparts. The ability to rapidly generate flexible, paper-based electronics could be of use for point of care diagnostics, smart labels, smart packaging, and more. The flexibility and durability of paper electronics could ensure continued functionality through transportation as their ability to be damaged is significantly less than that of rigid substrates. Das et al. (2016) have demonstrated the ability to inkjet print graphene onto cellulose-based substrates for printable electronics and sensors. This research details the use of a postprint laser process that anneals the printed graphene [reduced graphene oxide (RGO) flakes] and welds or stiches individual graphene flakes; the postprint laser annealing process graphitizes the printed RGO flakes. The laser process reduces the sheet resistance of the printed graphene to below $1\ k\Omega\ sq^{-1}$ (a sheet resistance lower than previous reports of printed graphene) while nanostructuring the surface to produce 3D nanopetals with high surface area and high electroreactivity that is well-suited for electrochemical sensing (Fig. 10.5).

Another interesting use of cellulose in electronics is found in transparent cellulose films. As discussed before, cellulose can form networks of microfibers with proper preparation. Under the proper conditions, cellulosic microfiber networks can be made such that they are transparent to visible light. Transparent paper can be processed to tune light transmission from nearly 100% down to typical opaque paper. As transparent paper is made of the same robust microfibrillar cellulose as mentioned before, it remains a very durable material. Transparent paper could be a viable alternative to plastic films as an electrical device substrate in applications where flexible circuits are common, like conductive screens, OLED substrates, or solar cells. Interestingly, transparent paper could have a very promising future in conductive screen technology due to the lack of glare associated with cellulose films that seems always present in other film materials (Zhu et al., 2014).

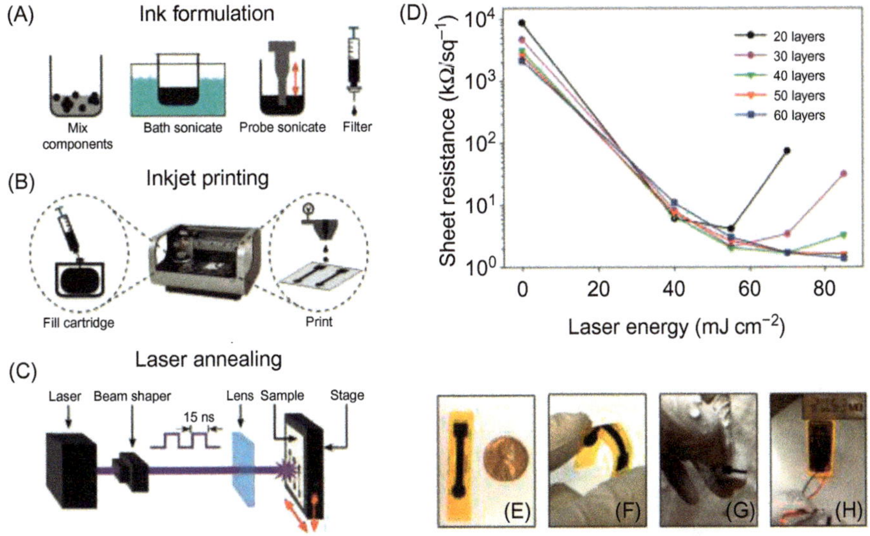

Figure 10.5 Inkjet-printed graphene on cellulose paper with subsequent laser annealing to increase conductivity. Graphene powder is mixed with other ink components, sonicated, and filtered (A). Using a specialized printer, graphene ink is printed onto a substrate (B). Laser annealing decreases the printed ink's sheet resistance and increases electroactive surface area (C). Minimization of sheet resistance can be achieved by increasing laser energy and number of printed layers (D). Resulting electrode can be miniaturized, and is flexible, water resistant, and conductive (E–H).
Reprinted with permission by Royal Society of Chemistry.

One interesting process to create conductive transparent cellulose resembles the layer-by-layer method as previously mentioned, but instead of adding NCCs to a graphene matrix, TEMPO-oxidized cellulose microfiber paper was coated with graphene oxide in a layer-by-layer fashion using divalent copper as a crosslinking agent (Gao et al., 2013). The resulting film was able to have resistivity and conductivity changed by varying layers and was shown to have little change in conductivity after 200 flexion cycles.

The inherent flexibility of cellulose films makes them well-suited for thin wire electronics. As the flexibility of a metal is proportional to its thickness, thin metal wires can be printed, deposited, or impregnated into thin cellulose membranes to create flexible circuitry. In a 2013 article, this principle was put to the test with the formation of highly foldable silver-nanowire printed antennas that outperformed commercial plastics (Nogi et al., 2013). The folding of the material-induced shifts in antenna receiving frequency, which means that flexible antennas could be made to transmit data at variable frequencies.

In the coming years, there will likely be a continued interest in flexible and biorenewable resources and a distancing from petrochemicals. In the electronic device industry, flexible substrates are based on expensive petrochemical polymers that cost significantly more to produce than typical silicon or glass substrates. Cellulose is a widely abundant polymer and has been shown to rival many of its flexible polymer industrial standards. Due to the expense of microfibrillation, it is difficult to industrialize the formation of microfibrillar paper. As the desire to use more renewable resources grows, the demand for cellulose-based electronics could increase in order to meet the growing demand for flexible, durable, and cheap electronic systems.

Although research into cellulosic materials has a long history, the use of cellulose in flexible electronic represents a bold new direction. Between fibrillation and device preparation methods, there is still a great deal of work to be done before large-scale industrialization can turn cellulose electronics into common devices. Despite the need for further research, it has been demonstrated that microbial cellulose films are durable and wearable materials that conform and adhere to skin and can be used to create flexible conductive films for use in wearable electronics.

10.6.3 Biosensing

Many hybrid materials have been formed with NCC that combines the physicochemical properties of cellulose with polymers, metals, or nanocarbon, to name a few (Idumah and Hassan, 2016). For example, conductive NCC has been fabricated by creating composites with polypyrrole (PPY) (Dubitsky and Zhubanov, 1993; Abthagir, 2010), nanoplatinum/grapheme (Burrs et al., 2016 (Fig. 10.6); Campuzano et al., 2011), polyaniline (PANI) (Jang et al., 2011), and poly ethylene dioxythiophene (Kaihovirta et al., 2010). During cellulose-polymer synthesis, oxidized monomers form polymer chains that entangle and penetrate the cellulose fibers, creating a strong bond between the base cellulose and the conducting polymer (e.g., PPY).

Applications included fabrics that can discharge static electricity (Frenning et al., 2009), artificial muscles by coating Cladophora algae (Mihranyan et al., 2008), and gas/temperature sensors (Li et al., 2011). Although not covered here in detail, hybrid composites have also been created using biomaterials such as proteins, peptides, and other biomaterials (Salas et al., 2015). NCC and hybrid composite materials have been used in food packaging (Khan et al., 2014), development of autonomous micropropulsion systems (Campuzano et al., 2011), medical (Lin and Dufresne, 2014), and sensing. In sensing, NCC has been used for detection of glucose (Esmaeili et al., 2015), esterase (Derikvand et al., 2016), alcohol oxidase (Burrs et al., 2015), pathogens such as shiga toxin–producing *Escherichia coli* (Burrs et al., 2016), elastase (Edwards et al., 2013, 2016), silver nanoparticles (Ruiz-Palomero et al., 2016), and ions (Gao et al., 2015; Schyrr et al., 2014; Zeinali et al., 2014), among others.

One of the most promising applications of NCC is in the development of high strength and low-cost paper (Zheng et al., 2013; Zhu et al., 2015; Morales-Narváez et al., 2015). This has been extended for the development of electrically conductive paper for biosensing (Shi et al., 2013). For example, Burrs et al. (2016) demonstrated development of low cost, biocompatible sensors that are durable and have a relatively long shelf life based on conductive paper fabricated with NCC and

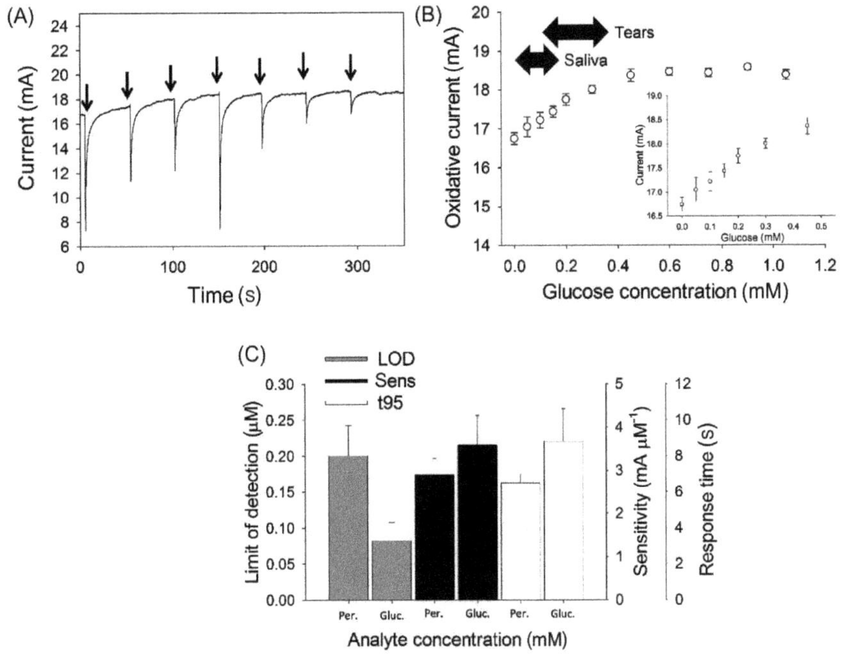

Figure 10.6 Glucose biosensor developed using the graphene-NCC paper. DC potential amperometry was used to calculate the sensitivity, limit of detection (LOD), and response time (t95). (A) Representative oxidative current response to injections of 50 μM glucose (indicated by vertical arrows). (B) Average calibration curves for glucose derived from replicate DCPA tests ($n = 3$). Inset shows linear range between 0 and 500 μM. (C) Average LOD, sensitivity, and t_{95} for peroxide and glucose biosensors tested in buffer at 25°C; Per. = hydrogen peroxide data, Gluc. = Glucose data. Error bars represent the standard error of the mean. Reproduced from Burrs, S.; Bhargava, M.; Sidhu, R.; Kiernan-Lewis, J.; Gomes, C.; Claussen, J., et al., A paper based graphene-nanocauliflower hybrid composite for point of care biosensing. *Biosens. Bioelectron.* **2016**, 85, 479–487. Reprinted with permission by Elsevier.

graphene. This conductive NCC paper was used as a platform for developing biosensors to detect small molecules (glucose), or pathogenic bacteria (Burrs et al., 2016) in a proof of concept study. In their study, Burrs et al. demonstrated the development of NCC-graphene paper functionalized with fractal platinum nanocauliflower for use in electrochemical biosensing of small molecules (glucose) or detection of pathogenic bacteria (*E. coli* O157:H7). Furthermore, Das et al. (2016) demonstrated the ability to inkjet-printed graphene on cellulose-based films for electrochemical sensing. A postprint laser process significantly increased the electrical conductivity and electrochemical reactivity of the graphene (while not damaging the underlying cellulose) via graphene flake welding and nanostructuring. The resultant inkjet-printed and laser-processed graphene displays favorable electrochemical sensing characteristics—ferricyanide cyclic voltammetry with a redox peak separation (ΔEp) ≈ 0.7 V as well as hydrogen peroxide (H_2O_2) amperometry with a sensitivity of 3.32 μA mM^{-1} and a response time of <5 s.

10.6.3.1 Small molecules detection

Raman spectroscopy, scanning electron microscopy, and energy-dispersive spectroscopy show that graphene oxide-coated NCC was partially reduced by both thermal treatment and further reduced by chemical treatment (ascorbic acid). The average sensitivity (2.9 ± 0.4 mA mM^{-1}), limit of detection (LOD) (0.2 ± 0.1 μM), and response time (6.5 ± 0.5 s) toward hydrogen peroxide were within the range of laboratory electrodes and biochips functionalized with graphene–nanometal hybrid materials for detection of peroxide (Chaturvedi et al., 2014; Claussen et al., 2011; Vanegas et al., 2014). The average sensitivity for enzymatic glucose biosensors prepared on the graphene-NCC paper was 3.59 ± 0.68 mA mM^{-1}, which is lower or within the range of laboratory electrochemical devices fabricated on glass, silicon, Katpon tape, and other materials. The average LOD (0.08 ± 0.02 μM) and response time (8.8 ± 1.8 s) toward glucose are applicable for POC testing of glucose levels in saliva, tears, or diluted blood (approximately $10\times$ dilution). The response time toward glucose changed significantly within the testing range but was less than 10 s, and as low as 6 s (Fig. 10.7).

In another recent study, NCC was compared against other three different hydrogels as protein encapsulants in a mediator-free biosensor based on graphene–nanometal–enzyme composites (Burrs et al., 2015). Alcohol oxidase

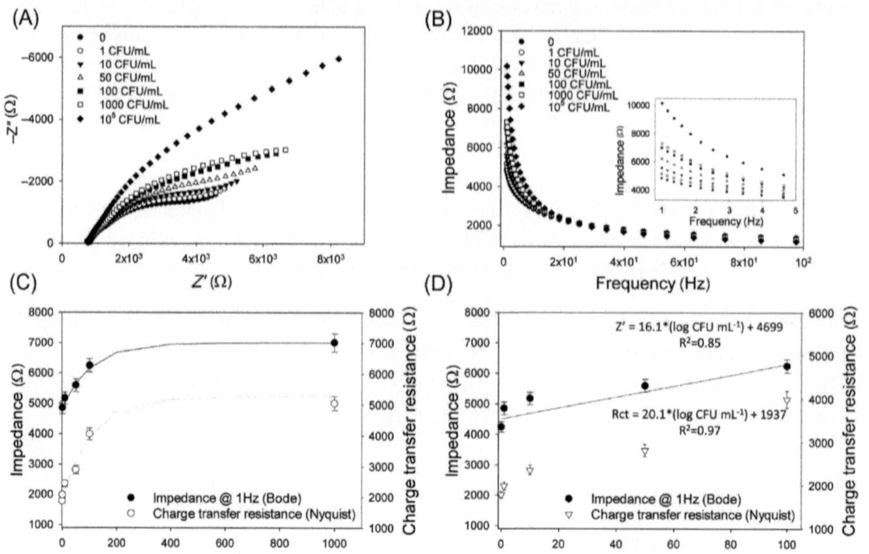

Figure 10.7 Representatives (A) Nyquist and (B) Bode magnitude plot (inset shows exploded view of lower frequency range). (C) Calibration curve using direct impedance measurement and charge transfer resistance ($n = 3$). (D) Exploded view of linear sensing region from panel.

Reprinted with permission by Elsevier.

was encapsulated in chitosan poly-*N*-isopropylacrylamide (PNIPAAM), silk fibroin or NCC hydrogels, and then spin coated onto a nanoplatinum–graphene modified electrode. Electroactive surface area (ESA), electrochemical impedance spectroscopy (EIS), sensitivity to methanol, response time, LOD, and shelf life were measured for each bionanocomposite. Chitosan and PNIPAAM had the highest sensitivity (0.46 ± 0.2 and 0.3 ± 0.1 μA mM^{-1}, respectively) and ESA (0.2 ± 0.06 and 0.2 ± 0.02 cm^2, respectively), as well as the fastest response time (4.3 ± 0.8 and 4.8 ± 1.1 s, respectively). Silk and NCC demonstrated lower sensitivity (0.09 ± 0.02 and 0.15 ± 0.03 μA mM^{-1}, respectively), lower ESA (0.12 ± 0.02 and 0.09 ± 0.03 cm^2, respectively), and longer response time (8.9 ± 2.1 and 6.3 ± 0.8 s, respectively). The high porosity of chitosan, PNIPAAM, and silk gels led to excellent transport, which was significantly better than NCC bionanocomposites. Electrochemical performances of NCC bionanocomposites were relatively poor, which may be linked to poor gel stability under the testing conditions (pH 7.1 at 25°C). Each of these composites was within the range of other published devices in the literature, whereas some attributes were significantly improved (namely, response time, and shelf life).

10.6.3.2 Bacteria detection

Burrs et al. (2016) also developed an aptasensor based on NCC-graphene paper for measuring *E. coli* O157:H7. Fig. 10.8 shows a representative Nyquist plot (Fig. 10.8A) and Bode plot (Fig. 10.8B). Calibration curves obtained from the charge transfer resistance (R_{ct}, derived using a Randles equivalent circuit) and the impedance at 1 Hz (Z', from Bode plot) are shown in Fig. 10.8C and D (Olsson et al., 2010). The calibration plots for both measurements (Fig. 10.8C) follow a nonlinear saturation behavior, and the rate constant for each plot is nearly identical (≈ 0.01 mL CFU^{-1}), as expected.

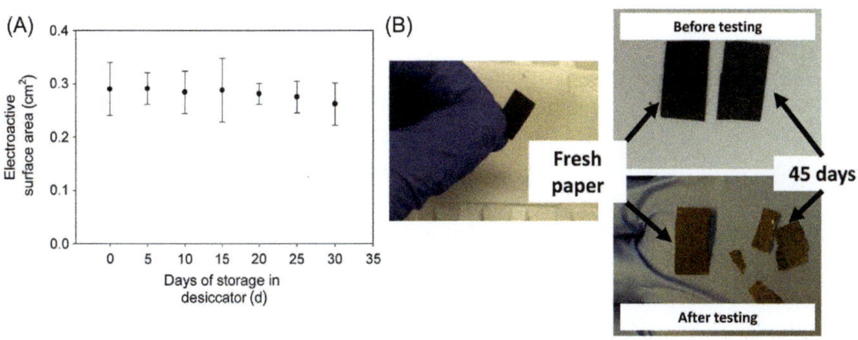

Figure 10.8 (A) Change in electroactive surface area for Pt nanocauliflower-decorated graphene paper after 45 days of storage in a desiccator at room temperature. (B) Photos of fresh Pt nanocauliflower-decorated graphene paper and a sample after 45 days of storage. Reprinted with permission by Elsevier.

Based on these two plots, the LOD was 3.5 ± 1.1 CFU mL^{-1} for the impedance plot and 4.2 ± 0.8 for the charge transfer resistance (S/N ratio = 3). For both measurements, the signal at 1 CFU mL^{-1} was significantly higher than baseline ($P < 0.001$, $\alpha = 0.05$), although calculation of LOD using the 3δ method indicated that the minimum detectable value in vegetable broth may be closer to ≈ 4 CFU mL^{-1}. Using the calculated LOD above, the linear range using both Z' and R_{ct} as the output was from 4 to 10^5 CFU mL^{-1} with an average response time of 12 min (including 10 min for bacteria capture and 2 min for EIS). This LOD is relevant for food safety and public health since the infectious dose for *E. coli* O157:H7, which is estimated to be 10 to 100 cells. The shiga toxin-producing serotype is implicated in the most incidents of illness worldwide (USDFA Bacteriological Analytical Manual (BAM)). These infections are mostly food and water borne and have implicated undercooked chicken and ground beef, raw milk, cold sandwiches, water, unpasteurized apple juice, and sprouts and vegetables.

The performance values obtained in this study (LOD, response time, and sensitivity) are comparable to, if not superior to, aptasensors published for the detection of *E. coli* O157:H7. Most recently, Ning et al. (2014) developed an impedance-based immunosensor for O157:H7 based on self-assembled lectins with a LOD of 100 CFU mL^{-1}, a linear range from 10^2 to 10^7 CFU mL^{-1}, and a total detection time of 1 h. Fang et al. (2014) developed a lateral flow biosensor based on aptamer-mediated strand displacement amplification with a detection limit of 10 CFU mL^{-1}; however, it required preenrichment steps and isothermal amplification prior to lateral flow test.

A brief shelf life study indicated that the Pt nanocauliflower-decorated paper maintained 90% of the reported ESA when stored in a desiccator at room temperature for at least 30 days but completely deteriorated after 30 days; the paper was extremely brittle, and testing of ESA was not possible using the methods herein after 30 days so no data can be shown (Fig. 10.9).

10.6.3.3 In vivo detection and other sensing applications

NCC offers potential for detection and biosensing in vivo. Contrary to CNTs, whose acute toxicity precludes their utilization in vivo, NCCs appear to be rather biocompatible. In addition, cellulose's strong hydrophilicity promises reduced nonselective adsorption of biological materials that often causes degradation of transducer surfaces (Lam et al., 2006). For example, Edwards et al. (2013) have reported biosensing based on NCC by conjugation of a fluorogenic peptide for detection of human neutrophil elastase. The NCCs were also used for the immobilization of enzymes or labeled with fluorescent molecules to study cellular update (Hassan et al., 2012).

Hybrid NCC materials have been used for the preparation of sensitive materials for bioimaging or biosensing applications. For example, metallic nanoparticles adsorbed on carboxylated NCCs were prepared for electrochemical detection for DNA hybridization (Liu et al., 2011). The self-assembly of NCC by shear alignment into highly ordered crystalline structures have been used to produce hybrid films with gold nanorods, which exhibit strong surface plasmon resonance

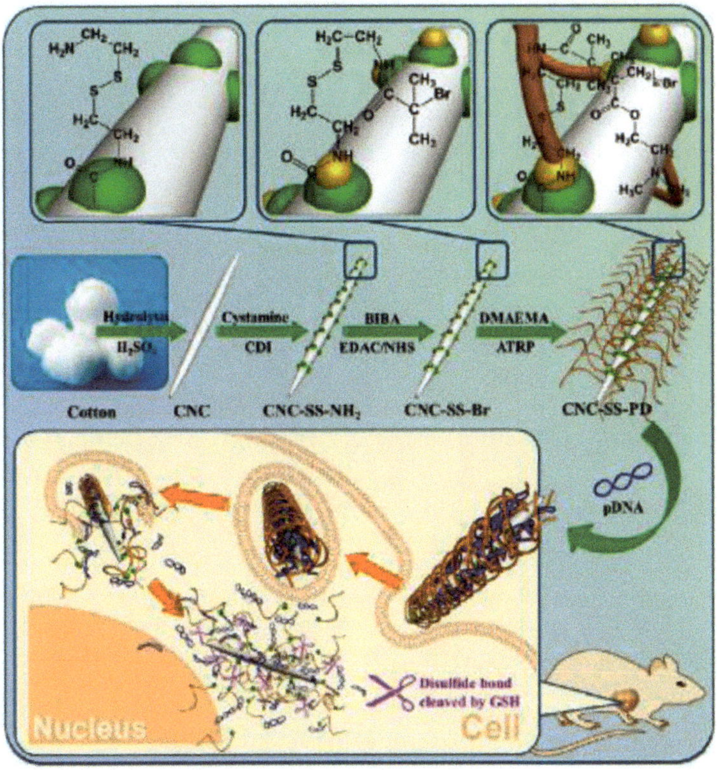

Figure 10.9 Gene theraphy delivery process using a cellulose substrate.
Reprinted with permission by American Chemical Society.

(Campbell et al., 2014). NCCs have also been reported in composite electronics with tin oxide layers for flexible organic field effect transistor (Valentini et al., 2014). Hybrid luminescent single-walled CNT–NCC films were prepared using layer-by-layer technique. The NCC facilitated the dispersion of CNTs (up to 24%) in aqueous media. These dispersions were used to produce films that exhibited near-infrared luminescence (Olivier et al., 2012).

Bacterial cellulose nanofibril (CNF) has also been used as templates for growing cobalt-ferrite nanoparticles to produce magnetic hybrid aerogels with adjustable response depending on the concentration of precursors salts ($FeSO_4$/$CoCl_2$) (Olsson et al., 2010), which could be used for preconcentration step in sensing applications. Furthermore, it was shown that carbon fibers derived from CNF have superior reversibility, the rate capability required for fast charging and excellent cycling capacity, important properties in electrochemical sensing. Preparation of CNF-based electroactive composites by coating cellulose with fibrils with PPY (conductive polymer) has been demonstrated by Momtazi et al. (2014). The resulting composite was conductive, electroactive and suitable for energy storage and electrochemically controlled separation.

10.6.4 Packaging materials and drug delivery systems

NCC is often pressed into thin transparent films referred to as nanopapers (Olivier et al., 2012) and has been identified as potential packaging material, especially if it is further modified for barrier properties. In addition to passive packaging, NCC films have been shown to be suitable for loading active substances and for controlled release (Lavoine et al., 2014). The highly crystalline NCC provides a rigid surface with tunable functional groups available for modification and grafting. Typical NCC grades show no cytotoxicity. Hybrid NCC composites have been proposed in targeted delivery of chemotherapeutical drugs (Dong et al., 2014).

Hybrids of NCC with different clays have been explored for packaging applications. Some examples include CNF-montmorillonite cast films (Liu et al., 2011). CNF-montmorillonite and poly(vinyl) alcohol (PVA) films with polyacrylic acid as crosslinker (Lavoine et al., 2014), vermiculite-CNF hybrid films obtained by solvent casting (Aulin et al., 2010), and others. Due to its excellent film-forming properties and numerous positive interactions through hydroxyl groups, PVA constitutes a good polymer choice for deposition of NCC films (Schyrr et al., 2014). The cationization of NCC has been applied and evaluated in preparation of hybrids with different layered silicates (vermiculite, kaolin, talc, smectite, and mica). The results indicate a dependence of the barrier properties with the amount of silicate in the composites (Ho et al., 2012).

NCC hybrid composites have been shown to improve packaging properties. For example, to form hybrid nanopapers, a genetically modified protein (class II hydrophobin) was used to bind graphene flakes and CNF. The mechanical properties of nanopapers improved after addition of graphene, furthermore, for a constant amount of graphene, the Young modulus increased linearly with the amount of engineered proteins (Laaksonen et al., 2011). Similarly, hybrid graphene–cellulose–PVA organic aerogels were prepared using TEMPO (tetramethylpiperidinyloxy) oxidized CNF, and the results indicated materials with ultralow density, good mechanical and thermal properties (Javadi et al., 2013). Another example, George et al. (2012) used NCC from bacterial cellulose were used to reinforce PVA composites containing silver nanoparticles for improved thermal and mechanical properties. Starch-based coating formulations have been filled with CNF and zinc oxide in order to prepare coating in antibacterial paper. The coated papers showed bactericidal activity against Gram-positive and Gram-negative bacteria (Martins et al., 2013). The above text describes advances in packaging systems and highlights NCC hybrids as a supporting or active material as well as a coating for flexible packaging (Li et al., 2013).

The high surface area-to-volume ratio, low cost, nontoxicity, biodegradability, and their ease of chemical modification lends NCC materials to potentially be used as excipients or matrices for the binding and controlled release of active pharmaceutical ingredients. However, when evaluating the overall scientific literature on NCCs, there are relatively few publications to date focused specifically on their use as a vehicle for drug delivery. Of interest is the possibility to use NCC to provide controlled and targeted release of both hydrophobic and hydrophilic compounds, consequently providing truly versatile materials regarding drug delivery. Recent trends have focused on incorporating nanoparticles such as polymeric, metallic, and carbon-based nanomaterials, within the hydrogel network to obtain nanocomposite hydrogels/matrices (Merino et al., 2015) with reinforced properties for various biomedical applications.

Many types of nanoparticles such as inorganic/ceramic nanoparticles (hydroxyapatite, silica, silicates, and calcium phosphate), polymeric nanoparticles (hyperbranched polyesters, cyclodextrins, etc.), metal/metal–oxide nanoparticles (gold, silver, and iron oxide), and carbon-based nanomaterials (CNTs and graphene) are combined with the polymeric network to create reinforced polymeric hydrogels, developing nanocomposites with tailored physical properties and custom-made functionalities. Recently, hybrid nanocomposite hydrogels have opened a new avenue and are able to improve the mechanical strength of hydrogels. In principle, the nanocomposite hydrogels are prepared by incorporation of nanofillers onto the hydrogel matrix (Trakakis et al., 2015). For example, Mandal et al. (2016) have developed carboxymethyl cellulose (CMC) and acid functionalized multiwalled CNT (MWCNT) base hybrid nanocomposite hydrogel by ultrasonication at room temperature. The nanocomposite was biodegradable and noncytotoxic toward rat fibroblasts. It has sufficient gel strength for intended application of transdermal patch. The transdermal release study revealed that CMC–MWCNT nanocomposite released the drug (diclofenac sodium) in a controlled way. Finally, the stability study predicted that the drug remains stable up to 3 months in presence of composite matrix.

Another use of NCC materials is to assist nano-sized particles such as gold (Nasrolahi Shirazi et al., 2013) and silver nanoparticles (Hebeish et al., 2013), quantum dots (Probst et al., 2013), and magnetic particles (Momtazi et al., 2014) of different shapes in drug delivery applications. These drug delivery systems encapsulate the drug to manage the poor distribution and stability of the therapeutic agents (Pissuwan et al., 2011). Encapsulation efficiency can be optimized by enlargement of nanoparticles into the polymer networks that increases the therapeutic efficiency by many folds (Pissuwan et al., 2011). A recent example, Das et al. (2015) synthesized a nanocomposite derived from crosslinked hydroxypropyl methyl cellulose and Au–nanoparticles where crosslinked hydrogel acted as reducing agent as well as stabilizing agent. Results showed that this nanocomposite has stimuli-responsive swelling behavior, biodegradability, noncytotoxicity, proficient drug stability, and controlled release behavior of colon-targeted drugs.

Similarly, for delivery of genes, some nanoparticles with specific morphologies, such as mesoporous silica nanoparticles (Lin et al., 2015), CNTs (Nunes et al., 2010), iron oxide (Kievit et al., 2009), and layered double hydroxides nanoparticles (Choy et al., 2000), have been investigated for gene/drug delivery systems by surface modification using NCC for improved transfection and stability. These biocompatible solid vectors show similar or better transfection performances than soft polycationic vectors (Hu et al., 2016). Because of their excellent physicochemical properties, biocompatible rod-like cellulose nanocrystals (NCCs) were reportedly expected to replace CNTs and similar particles. Hu et al. (2016) prepared Au NP-conjugated heterogeneous polymer brush-coated NCCs as multifunctional delivery vectors via different controllable polymerization techniques using poly(2-(di-methylamino)ethyl methacrylate) (PDMAEMA) side chains and were able to improve the transfection performance and imaging via computed tomography. Another example on gene therapy demonstrated that NCCs functionalized with redox-responsive polycations was an effective method for gene-delivery systems (Hu et al., 2015). NCCs from natural cotton wool were functionalized with disulfide bond-linked PDMAEMA brushes. The needle-like shape of NCCs had an important effect on enhancing transfection efficiency (Hu et al., 2015).

The most challenging delivery systems are the ones with reversible on-off switching capability. Nanoparticles can promote the response of hydrogels to a new stimulus, which can be modulated in a versatile way, with the modification of the nanoparticle. Moreover, the hybrid systems could enable routes of drug administration with limited systemic absorption but that can be useful in long-term delivery systems. Furthermore, therapies that promise minimally invasive localized treatment such as on cancerous tissue with little-to-no side effects are desirable. Irreversible electroporation (IRE), a nonthermal focal ablation technique, exposes cells to electric pulses to increase the permeability of the plasma membrane past the point of recovery with no damage to the supporting stroma in the vicinity (Thomson et al., 2011; Davalos et al., 2005; Maor et al., 2007; Li et al., 2011). A recent study showed the potential use of folic acid-conjugated NCC in the potentiation of IRE-induced cell death in folate receptor (FR)-positive cancers. The NCC-FA was selective toward cancer cells and did not potentiate IRE-induced cytotoxicity in an FR-negative cancer cell type (Colacino et al., 2015).

Similarly, NCC in conjunction with a conducting polymer, such as PANI, has the potential to act as an electrically stimulated drug delivery device. Considering this, iontophoresis (drug delivery by electric potential), bacterial NCC was combined with MWCNTs and electroconductivity was increased up to 1.4×10^{-1} S cm^{-1} (Shi et al., 2013; Yoon et al., 2006). Transdermal delivery of buprenorphine induced by electric stimuli was improved when using CMC hydrogels. Iontophoresis improved permeation by a factor of 14.3 compared to passive diffusion (Fang et al., 2002).

10.7 Conclusion

Cellulose is an abundant and unique polymer with nanoscale-dimensional structures and unique properties. Cellulose forms very mechanically strong nanocrystals due to the tight inter- and intramolecular hydrogen bonding between glucose units in a cellulose crystal. The native fibrillar form of cellulose is made of noncrystalline and crystalline regions. Theses alternating regions enable the tough crystalline regions to have flexible sections that intermingle to create tough but flexible fiber networks with a vast array of uses. Manipulations of the fiber diameter make-up of a filter have also been shown to affect filter properties due to changes in surface roughness and porosity. These cellulose networks have been shown to have moduli close to that of steel with high ductility and low electrical and thermal conductivity. Furthermore, cellulose is biodegradable and biocompatible, so it can provide eco-friendly strength to other biodegradable polymers that could be used in vivo, or in any external application where a strong and biodegradable material is needed.

Cellulose is truly a broad application polymer. Its mechanical strength, flexibility, abundance, and biocompatibility make it a uniquely capable material to combine with nanomaterials such as graphene, CNTs, metallic nanoparticles, or biological agents for use in a diverse range of applications such as packaging barriers, reinforcing agents, wound repair, flexible substrates, capacitors, filters, biosensors/sensors, and drug delivery systems. With the increasing global need for renewable and ecofriendly technology, there is no doubt that cellulose could be a vital component in products that meet that demand. Future commercial electronics could integrate seamlessly into daily life by utilizing the flexibility and conductivity of graphene-NCC electronic circuits. Swarms of autonomous microsubmersibles may one day explore the oceans of Titan powered by cellulose-platinum catalysts. Soon, patients will potentially be able to receive targeted release of drugs thanks to the biocompatibility of NCC-silica nanoparticle films. The growth of cellulose from plant to textile to nanohybrid material has been an exciting story, and with so many interesting applications on the horizon, its end seems far to come.

References

Abthagir, S., 2010. Study of conducting nanocrystalline cellulose composites. In: Wood & Biofiber Plastic Composites and Cellulose Nanocomposites Symposium.

Amnuaikit, T., Chusuit, T., Raknam, P., Boonme, P., 2011. Effects of a cellulose mask synthesized by a bacterium on facial skin characteristics and user satisfaction. Med. Devices (Auckl). 4, 77–81.

Aulin, C., Netrval, J., Wågberg, L., Lindström, T., 2010. Aerogels from nanofibrillated cellulose with tunable oleophobicity. Soft Matter. 6 (14), 3298–3305.

Azizi Samir, M.A.S., Alloin, F., Dufresne, A., 2005. Review of recent research into cellulosic whiskers, their properties and their application in nanocomposite field. Biomacromolecules. 6 (2), 612–626.

Bae, S., Shoda, M., 2004. Bacterial cellulose production by Fed-Batch fermentation in molasses medium. Biotechnol. Prog. 20, 1366–1371.

Baraban, L., Makarov, D., Streubel, R., Moönch, I., Grimm, D., Sanchez, S., et al., 2012. Catalytic janus motors on microfluidic chip: deterministic motion for targeted cargo delivery. ACS Nano. 6 (4), 3383–3389.

Beeran, Y., Bobnar, V., Gorgieva, S., Grohens, Y., Finšgar, M., Thomas, S., et al., 2016. Mechanically strong, flexible and thermally stable graphene oxide/nanocellulosic films with enhanced dielectric properties. RSC Adv. 6 (54), 49138–49149.

Bledzki, A.K., Gassan, J., 1999. Composites reinforced with cellulose based fibres. Prog. Polym. Sci. 24 (2), 221–274.

Borbély, É., 2008. Lyocell, the new generation of regenerated cellulose. Acta Polytech. Hung. 5 (3), 11–18.

Braconnot, H., 1833. De la transformation de plusieurs substances végétales en un principe nouveau. Ann. Chim. Phys. 52, 290–294.

Burrs, S., Bhargava, M., Sidhu, R., Kiernan-Lewis, J., Gomes, C., Claussen, J., et al., 2016. A paper based graphene-nanocauliflower hybrid composite for point of care biosensing. Biosens. Bioelectron. 85, 479–487.

Burrs, S., Vanegas, D., Bhargava, M., Mechulan, N., Hendershot, P., Yamaguchi, H., et al., 2015. A comparative study of graphene–hydrogel hybrid bionanocomposites for biosensing. Analyst. 140 (5), 1466–1476.

Cai, J., Zhang, L., Zhou, J., Qi, H., Chen, H., Kondo, T., et al., 2007. Multifilament fibers based on dissolution of cellulose in NaOH/urea aqueous solution: structure and properties. Adv. Mater. 19 (6), 821–825.

Campbell, M.G., Liu, Q., Sanders, A., Evans, J.S., Smalyukh, I.I., 2014. Preparation of nanocomposite plasmonic films made from cellulose nanocrystals or mesoporous silica decorated with unidirectionally aligned gold nanorods. Materials. 7 (4), 3021–3033.

Campuzano, S., Orozco, J., Kagan, D., Guix, M., Gao, W., Sattayasamitsathit, S., et al., 2011. Bacterial isolation by lectin-modified microengines. Nano Lett. 12 (1), 396–401.

Cao, J., Zhang, X., Wu, X., Wang, S., Lu, C., 2016. Cellulose nanocrystals mediated assembly of graphene in rubber composites for chemical sensing applications. Carbohydr. Polym. 140, 88–95.

Chao, Y., Ishida, T., Sugano, Y., Shoda, M., 2000. Bacterial cellulose production by *Acetobacter xylinum* in 50-L internal-loop airlift reactor. Biotechnol. Bioeng. 68 (3), 345–352.

Chaturvedi, P., Vanegas, D., Taguchi, M., Burrs, S., Sharma, P., McLamore, E., 2014. A nanoceria–platinum–graphene nanocomposite for electrochemical biosensing. Biosens. Bioelectron. 58, 179–185.

Cheng, Y., Feng, B., Yang, X., Yang, P., Ding, Y., Chen, Y., et al., 2013. Electrochemical biosensing platform based on carboxymethyl cellulose functionalized reduced graphene oxide and hemoglobin hybrid nanocomposite film. Sens. Actuators, B: Chem. 182, 288–293.

Cho, Y.-R., Lee, J.H., Song, Y.-H., Kang, S.-Y., Hwang, C.-S., Jung, M.-Y., et al., 2001. Photolithography-based carbon nanotubes patterning for field emission displays. Mater. Sci. Eng., B. 79 (2), 128–132.

Choy, J.-H., Kwak, S.-Y., Jeong, Y.-J., Park, J.-S., 2000. Inorganic layered double hydroxides as nonviral vectors. Angew. Chem. 39 (22), 4041–4045.

Claussen, J.C., Artiles, M.S., McLamore, E.S., Mohanty, S., Shi, J., Rickus, J.L., et al., 2011. Electrochemical glutamate biosensing with nanocube and nanosphere augmented single-walled carbon nanotube networks: a comparative study. J. Mater. Chem. 21 (30), 11224–11231.

Claussen, J.C., Daniele, M.A., Geder, J., Pruessner, M., Makinen, A.J., Melde, B.J., et al., 2014. Platinum-paper micromotors: an urchin-like nanohybrid catalyst for green monopropellant bubble-thrusters. ACS Appl. Mater. Interfaces. 6 (20), 17837–17847.

Claussen, J.C., Franklin, A.D., ul Haque, A., Porterfield, D.M., Fisher, T.S., 2009. Electrochemical biosensor of nanocube-augmented carbon nanotube networks. ACS Nano. 3 (1), 37–44.

Colacino, K.R., Arena, C.B., Dong, S., Roman, M., Davalos, R.V., Lee, Y.W., 2015. Folate conjugated cellulose nanocrystals potentiate irreversible electroporation-induced cytotoxicity for the selective treatment of cancer cells. Technol. Cancer Res. Treat. 14 (6), 757–766.

Czaja, W., Krystynowicz, A., Bielecki, S., Brown Jr., R.M., 2006. Microbial cellulose—the natural power to heal wounds. Biomaterials. 27 (2), 145–151.
Czaja, W., Romanovicz, D., Brown Jr., R.M., 2004. Structural investigation of microbial cellulose produced in stationary and agitated culture. Cellulose. 11, 403–411.
Daniele, M.A., Knight, A.J., Roberts, S.A., Radom, K., Erickson, J.S., 2015. Sweet substrate: a polysaccharide nanocomposite for conformal electronic decals. Adv. Mater. 27 (9), 1600–1606.
Das, R., Das, D., Ghosh, P., Dhara, S., Panda, A.B., Pal, S., 2015. Development and application of a nanocomposite derived from crosslinked HPMC and Au nanoparticles for colon targeted drug delivery. RSC Adv. 5 (35), 27481–27490.
Das, S.R., Nian, Q., Cargill, A.A., Hondred, J.A., Ding, S., Saei, M., et al., 2016. 3D nanostructured inkjet printed graphene via UV-pulsed laser irradiation enables paper-based electronics and electrochemical devices. Nanoscale. 8 (35), 15870–15879.
Davalos, R.V., Mir, L., Rubinsky, B., 2005. Tissue ablation with irreversible electroporation. Ann. Biomed. Eng. 33 (2), 223–231.
Derikvand, F., Yin, D.T., Barrett, R., Brumer, H., 2016. Cellulose-based biosensors for esterase detection. Anal. Chem. 88 (6), 2989–2993.
Dhar, P., Kumar, A., Katiyar, V., 2015. Fabrication of cellulose nanocrystal supported stable Fe(0) nanoparticles: a sustainable catalyst for dye reduction, organic conversion and chemo-magnetic propulsion. Cellulose. 22 (6), 3755–3771.
Diekmann, S., Siegmund, G., Roecker, A., Klemm, D.O., 2003. Regioselective nitrilotriacetic acid–cellulose–nickel-complexes for immobilisation of His6-tag proteins. Cellulose. 10 (1), 53–63.
Dong, S., Cho, H.J., Lee, Y.W., Roman, M., 2014. Synthesis and cellular uptake of folic acid-conjugated cellulose nanocrystals for cancer targeting. Biomacromolecules. 15 (5), 1560–1567.
Dubitsky, Y.A., Zhubanov, B.A., 1993. Polypyrrole–poly(vinyl chloride) and polypyrrole–cellulose acetate conducting composite films by opposite-diffusion polymerization. Synth. Met. 53 (3), 303–307.
Dufresne, A., 2010. Processing of polymer nanocomposites reinforced with polysaccharide nanocrystals. Molecules. 15 (6), 4111–4128.
Edwards, J.V., Fontenot, K.R., Haldane, D., Prevost, N.T., Condon, B.D., Grimm, C., 2016. Human neutrophil elastase peptide sensors conjugated to cellulosic and nanocellulosic materials: part I, synthesis and characterization of fluorescent analogs. Cellulose. 23 (2), 1283–1295.
Edwards, J.V., Prevost, N., Sethumadhavan, K., Ullah, A., Condon, B., 2013. Peptide conjugated cellulose nanocrystals with sensitive human neutrophil elastase sensor activity. Cellulose. 20 (3), 1223–1235.
Esmaeili, C., Abdi, M.M., Mathew, A.P., Jonoobi, M., Oksman, K., Rezayi, M., 2015. Synergy effect of nanocrystalline cellulose for the biosensing detection of glucose. Sensors. 15 (10), 24681–24697.
Fang, J.Y., Sung, K., Wang, J.J., Chu, C.C., Chen, K.T., 2002. The effects of iontophoresis and electroporation on transdermal delivery of buprenorphine from solutions and hydrogels. J. Pharm. Pharmacol. 54 (10), 1329–1337.
Fang, Z., Wu, W., Lu, X., Zeng, L., 2014. Lateral flow biosensor for DNA extraction-free detection of salmonella based on aptamer mediated strand displacement amplification. Biosens. Bioelectron. 56, 192–197.
Fink, H., Ganster, J., Fraatz, J., Nywlt, M., 1994. Akzo-Nobel viskose chemistry seminar-challenges in cellulosic man-made fibers. Stockholm. Suecia.

Fink, H.-P., Weigel, P., Purz, H., Ganster, J., 2001. Structure formation of regenerated cellulose materials from NMMO-solutions. Prog. Polym. Sci. 26 (9), 1473–1524.

Frenning, G.R., Razaq, A., Gelin, K., Nyholm, L., Mihranyan, A., 2009. Ionic motion in polypyrrole–cellulose composites: trap release mechanism during potentiostatic reduction. J. Phys. Chem. B. 113 (14), 4582–4589.

Gao, K., Shao, Z., Wu, X., Wang, X., Li, J., Zhang, Y., et al., 2013. Cellulose nanofibers/reduced graphene oxide flexible transparent conductive paper. Carbohydr. Polym. 97 (1), 243–251.

Gao, W., Sattayasamitsathit, S., Orozco, J., Wang, J., 2011. Highly efficient catalytic microengines: template electrosynthesis of polyaniline/platinum microtubes. J. Am. Chem. Soc. 133 (31), 11862–11864.

Gao, X., Sadasivuni, K.K., Kim, H.-C., Min, S.-K., Kim, J., 2015. Designing pH-responsive and dielectric hydrogels from cellulose nanocrystals. J. Chem. Sci. 127 (6), 1119–1125.

George, J., Sajeevkumar, V.A., Ramana, K.V., Sabapathy, S.N., 2012. Augmented properties of PVA hybrid nanocomposites containing cellulose nanocrystals and silver nanoparticles. J. Mater. Chem. 22 (42), 22433–22439.

Gibbs, J., Zhao, Y.-P., 2009. Autonomously motile catalytic nanomotors by bubble propulsion. Appl. Phys. Lett. 94 (16), .

Hassan, M.L., Moorefield, C.M., Elbatal, H.S., Newkome, G.R., Modarelli, D.A., Romano, N.C., 2012. Fluorescent cellulose nanocrystals via supramolecular assembly of terpyridine-modified cellulose nanocrystals and terpyridine-modified perylene. Mater. Sci. Eng., B. 177 (4), 350–358.

Hayashi, J., Sufoka, A., Ohkita, J., Watanabe, S., 1975. The confirmation of existences of cellulose IIII, IIIII, IVI, and IVII by the X-ray method. J. Polym. Sci. Polym. Lett. Ed. 13 (1), 23–27.

Hebeish, A., Hashem, M., El-Hady, M.A., Sharaf, S., 2013. Development of CMC hydrogels loaded with silver nano-particles for medical applications. Carbohydr. Polym. 92 (1), 407–413.

Hestrin, S., Schramm, M., 1954. Synthesis of cellulose by *Acetobacter xylinum* 2. Preparation of Freeze-dried cells capable of polymerizing glucose to cellulose. Biochem. J. 58 (2), 345–352.

Higashi, K., Miki, N., 2014. A self-swimming microbial robot using microfabricated nanofibrous hydrogel. Sens. Actuators, B: Chem. 202, 301–306.

Ho, T.T., Zimmermann, T., Ohr, S., Caseri, W.R., 2012. Composites of cationic nanofibrillated cellulose and layered silicates: water vapor barrier and mechanical properties. ACS Appl. Mater. Interfaces. 4 (9), 4832–4840.

Hoeng, F., Denneulin, A., Bras, J., 2016. Use of nanocellulose in printed electronics: a review. Nanoscale. 8 (27), 13131–13154.

Hornung, M., Ludwig, M., Schmauder, H.P., 2007. Optimizing the production of bacterial cellulose in surface culture: a novel aerosol bioreactor working on a fed batch principle (Part 3). Engineering in Life Sciences. 7 (1), 35–41.

Hu, H., Hou, X.-J., Wang, X.-C., Nie, J.-J., Cai, Q., Xu, F.-J., 2016. Gold nanoparticle-conjugated heterogeneous polymer brush-wrapped cellulose nanocrystals prepared by combining different controllable polymerization techniques for theranostic applications. Polym. Chem. 7 (18), 3107–3116.

Hu, H., Yuan, W., Liu, F.-S., Cheng, G., Xu, F.-J., Ma, J., 2015. Redox-responsive polycation-functionalized cotton cellulose nanocrystals for effective cancer treatment. ACS Appl. Mater. Inter. 7 (16), 8942–8951.

Hu, W., Chen, S., Yang, J., Li, Z., Wang, H., 2014. Functionalized bacterial cellulose derivatives and nanocomposites. Carbohydr. Polym. 101, 1043–1060.
Hutchison, D.N., Morrill, N.B., Aten, Q., Turner, B.W., Jensen, B.D., Howell, L.L., et al., 2010. Carbon nanotubes as a framework for high-aspect-ratio MEMS fabrication. Journal of Microelectromech. Sys. 19 (1), 75–82.
Idumah, C.I., Hassan, A., 2016. Emerging trends in eco-compliant, synergistic, and hybrid assembling of multifunctional polymeric bionanocomposites. Rev. Chem. Eng. 32 (3), 305–361.
Jaehne, E., Kowalik, T., Adler, H.J.P., Plagge, A., Stratmann, M., 2002. Ultra-thin layers of phosphorylated cellulose derivatives on metal surfaces. Macromol. Sym.. Wiley Online Library, pp. 97–110.
Jang, H.-S., Song, K.-G., Kim, S.-H., 2011. Fabrication and Characterization of Organic Thin-Film Transistors by Using Polymer Gate Electrode. Polym. Korea. 35 (4), 370–374.
Javadi, A., Zheng, Q., Payen, F., Javadi, A., Altin, Y., Cai, Z., et al., 2013. Polyvinyl alcohol-cellulose nanofibrils-graphene oxide hybrid organic aerogels. ACS Appl. Mater. Interfaces. 5 (13), 5969–5975.
Jones, P., Wegner, T.H., 2009. Wood and paper as materials for the 21st century. MRS Proceedings. Cambridge Univ Press, pp 1187-KK04-06.
Joo-Hyung, K., Seongcheol, M., Hyun, U.K., Gyu-Young, Y., Jaehwan, K., 2014. Disposable chemical sensors and biosensors made on cellulose paper. Nanotechnology. 25 (9), 092001.
Kaihovirta, N., Mäkelä, T., He, X., Wikman, C.-J., Wilén, C.-E., Österbacka, R., 2010. Printed all-polymer electrochemical transistors on patterned ion conducting membranes. Org. Electron. 11 (7), 1207–1211.
Ke, H., Ye, S., Carroll, R.L., Showalter, K., 2010. Motion analysis of self-propelled Pt^- silica particles in hydrogen peroxide solutions. J. Phys. Chem. A. 114 (17), 5462–5467.
Ke, H., Zhou, J., Zhang, L., 2006. Structure and physical properties of methylcellulose synthesized in NaOH/urea solution. Polym. Bull. 56 (4-5), 349–357.
Khan, A., Huq, T., Khan, R.A., Riedl, B., Lacroix, M., 2014. Nanocellulose-based composites and bioactive agents for food packaging. Crit. Rev. Food Sci. Nutr. 54 (2), 163–174.
Kievit, F.M., Veiseh, O., Bhattarai, N., Fang, C., Gunn, J.W., Lee, D., et al., 2009. PEI–PEG–chitosan-copolymer-coated iron oxide nanoparticles for safe gene delivery: synthesis, complexation, and transfection. Adv. Funct. Mater. 19 (14), 2244–2251.
Klemm, D., Heublein, B., Fink, H.P., Bohn, A., 2005. Cellulose: fascinating biopolymer and sustainable raw material. Angew. Chem. Int. Ed. Engl. 44 (22), 3358–3393.
Klemm, D., Schumann, Udhardt, U., Marsch, S., 2001. Bacterial Synthesized Cellulose-Artifical Blood Vessels for Microsurgery. Prog. Polym. Sci. 26, 1560–1603.
Kowalik, T., Adler, H.J.P., Plagge, A., Stratmann, M., 2000. Ultrathin layers of phosphorylated cellulose derivatives on aluminium surfaces. Macromol. Chem. Phys. 201 (15), 2064–2069.
Laaksonen, P., Walther, A., Malho, J.M., Kainlauri, M., Ikkala, O., Linder, M.B., 2011. Genetic engineering of biomimetic nanocomposites: diblock proteins, graphene, and nanofibrillated cellulose. Angew. Chem. Int. Ed. 50 (37), 8688–8691.
Lam, C.-W., James, J.T., McCluskey, R., Arepalli, S., Hunter, R.L., 2006. A review of carbon nanotube toxicity and assessment of potential occupational and environmental health risks. Crit. Rev. Toxicol. 36 (3), 189–217.
Laocharoensuk, R., Burdick, J., Wang, J., 2008. Carbon-nanotube-induced acceleration of catalytic nanomotors. ACS Nano. 2 (5), 1069–1075.

Lavoine, N., Desloges, I., Bras, J., 2014. Microfibrillated cellulose coatings as new release systems for active packaging. Carbohydr. Polym. 103, 528–537.

Lee, S., Pan, H., Hse, C.Y., Gunasekaran, A.R., Shupe, T.F., 2013. Characteristics of regenerated nanocellulosic fibers from cellulose dissolution in aqueous solutions for wood fiber/polypropylene composites. J. Therm. Comp. Mater. 27 (4), 558–570.

Li, F., Biagioni, P., Bollani, M., Maccagnan, A., Piergiovanni, L., 2013. Multi-functional coating of cellulose nanocrystals for flexible packaging applications. Cellulose. 20 (5), 2491–2504.

Li, L., Sun, Y., Bao, H., Li, X., Wang, G., 2011. Synthesis and characterization of a poly (aniline-based disulfide)/diisocyanate-modified graphite oxide hybrid by a grafting technique. Eur. Polym. J. 47 (8), 1630–1635.

Li, W., Fan, Q., Ji, Z., Qiu, X., Li, Z., 2011. The effects of irreversible electroporation (IRE) on nerves. PloS one. 6 (4), e18831.

Lin, N., Dufresne, A., 2014. Nanocellulose in biomedicine: current status and future prospect. Eur. Polym. J. 59, 302–325.

Lin, X., Zhao, N., Yan, P., Hu, H., Xu, F.-J., 2015. The shape and size effects of polycation functionalized silica nanoparticles on gene transfection. Acta. Biomaterialia. 11, 381–392.

Liu, A., Walther, A., Ikkala, O., Belova, L., Berglund, L.A., 2011. Clay nanopaper with tough cellulose nanofiber matrix for fire retardancy and gas barrier functions. Biomacromol. 12 (3), 633–641.

Liu, H., Wang, D., Song, Z., Shang, S., 2011. Preparation of silver nanoparticles on cellulose nanocrystals and the application in electrochemical detection of DNA hybridization. Cellulose. 18 (1), 67–74.

Mahadeva, S.K., Walus, K., Stoeber, B., 2015. Paper as a Platform for Sensing Applications and Other Devices: A Review. ACS Appl. Mater. Interfaces. 7 (16), 8345–8362.

Majumdar, R., Singh, N., Rathore, J.S., Sharma, N.N., 2013. In search of materials for artificial flagella of nanoswimmers. J. Mater. Sci. 48 (1), 240–250.

Mandal, B., Das, D., Rameshbabu, A.P., Dhara, S., Pal, S., 2016. A biodegradable, biocompatible transdermal device derived from carboxymethyl cellulose and multi-walled carbon nanotubes for sustained release of diclofenac sodium. RSC Adv. 6 (23), 19605–19611.

Manesh, K.M., Cardona, M., Yuan, R., Clark, M., Kagan, D., Balasubramanian, S., et al., 2010. Template-assisted fabrication of salt-independent catalytic tubular microengines. ACS Nano. 4 (4), 1799–1804.

Maor, E., Ivorra, A., Leor, J., Rubinsky, B., 2007. The effect of irreversible electroporation on blood vessels. Technol. Cancer Res. Treat. 6 (4), 307–312.

Marr, K.M., Chen, B., Mootz, E.J., Geder, J., Pruessner, M., Melde, B.J., et al., 2015. High aspect ratio carbon nanotube membranes decorated with Pt nanoparticle urchins for micro underwater vehicle propulsion via H2O2 decomposition. ACS Nano. 9 (8), 7791–7803.

Martins, N.C., Freire, C.S., Neto, C.P., Silvestre, A.J., Causio, J., Baldi, G., et al., 2013. Antibacterial paper based on composite coatings of nanofibrillated cellulose and ZnO. Colloids Surf. A. 417, 111–119.

Maruyama, S., Xiang, R., 2012. Chemical vapor deposition growth, optical, and thermal characterization of vertically aligned single-walled carbon nanotubes. J. Heat Transf. 134 (5), 051024.

Masaoka, S., Ohe, T., Sakota, N., 1993. Production of cellulose from glucose by *Acetobacter xylinum*. J. Ferment. Bioeng. 73 (1), 18–22.

Mei, Y., Huang, G., Solovev, A.A., Ureña, E.B., Mönch, I., Ding, F., et al., 2008. Versatile approach for integrative and functionalized tubes by strain engineering of nanomembranes on polymers. Adv. Mater. 20 (21), 4085–4090.

Meng, H., Xie, F., Chen, J., Sun, S., Shen, P.K., 2011. Morphology controllable growth of Pt nanoparticles/nanowires on carbon powders and its application as novel electro-catalyst for methanol oxidation. Nanoscale. 3 (12), 5041–5048.

Merino, S., Martín, C., Kostarelos, K., Prato, M., Vázquez, E., 2015. Nanocomposite hydrogels: 3D polymer–nanoparticle synergies for on-demand drug delivery. ACS Nano. 9 (5), 4686–4697.

Mihranyan, A., Nyholm, L., Bennett, A.E.G., Strømme, M., 2008. A novel high specific surface area conducting paper material composed of polypyrrole and Cladophora cellulose. J. Phys. Chem. B. 112 (39), 12249–12255.

Mirkovic, T., Zacharia, N.S., Scholes, G.D., Ozin, G.A., 2010. Fuel for thought: chemically powered nanomotors out-swim nature's flagellated bacteria. ACS Nano. 4 (4), 1782–1789.

Momtazi, L., Bagherifam, S., Singh, G., Hofgaard, A., Hakkarainen, M., Glomm, W.R., et al., 2014. Synthesis, characterization, and cellular uptake of magnetic nanocarriers for cancer drug delivery. J. Colloid Interface Sci. 433, 76–85.

Morales-Narváez, E., Golmohammadi, H., Naghdi, T., Yousefi, H., Kostiv, U., Horák, D., et al., 2015. Nanopaper as an optical sensing platform. ACS Nano. 9 (7), 7296–7305.

Nasrolahi Shirazi, A., Tiwari, R., Chhikara, B.S., Mandal, D., Parang, K., 2013. Design and biological evaluation of cell-penetrating peptide–doxorubicin conjugates as prodrugs. Molecular pharmaceutics. 10 (2), 488–499.

Ng, H.-M., Sin, L.T., Tee, T.-T., Bee, S.-T., Hui, D., Low, C.-Y., et al., 2015. Extraction of cellulose nanocrystals from plant sources for application as reinforcing agent in polymers. Composites, Part B. 75, 176–200.

Ning, Y., Li, W., Duan, Y., Yang, M., Deng, L., 2014. High Specific DNAzyme-Aptamer Sensor for Salmonella paratyphi A Using Single-Walled Nanotubes–Based Dual Fluorescence-Spectrophotometric Methods. J. Biomol. Screen.1087057114528538.

Nogi, M., Komoda, N., Otsuka, K., Suganuma, K., 2013. Foldable nanopaper antennas for origami electronics. Nanoscale. 5 (10), 4395–4399.

Nunes, A., Amsharov, N., Guo, C., Van den Bossche, J., Santhosh, P., Karachalios, T.K., et al., 2010. Hybrid Polymer-Grafted Multiwalled Carbon Nanotubes for In vitro Gene Delivery. Small. 6 (20), 2281–2291.

Olivier, C., Moreau, C., Bertoncini, P., Bizot, H., Chauvet, O., Cathala, B., 2012. Cellulose nanocrystal-assisted dispersion of luminescent single-walled carbon nanotubes for layer-by-layer assembled hybrid thin films. Langmuir. 28 (34), 12463–12471.

Olsson, R.T., Samir, M.A., Salazar-Alvarez, G., Belova, L., Ström, V., Berglund, L.A., et al., 2010. Making flexible magnetic aerogels and stiff magnetic nanopaper using cellulose nanofibrils as templates. Nat. Nanotechnol. 5 (8), 584–588.

Park, S., Baker, J.O., Himmel, M.E., Parilla, P.A., Johnson, D.K., 2010. Cellulose crystallinity index: measurement techniques and their impact on interpreting cellulase performance. Biotechnol. Biofuels. 3 (1), 1–10.

Pettersen, R.C., 1984. The chemical composition of wood. Chem. Solid Wood. 207, 57–126.

Pinto, R.J., Marques, P.A., Martins, M.A., Neto, C.P., Trindade, T., 2007. Electrostatic assembly and growth of gold nanoparticles in cellulosic fibres. J. Colloid Interface Sci. 312 (2), 506–512.

Pissuwan, D., Niidome, T., Cortie, M.B., 2011. The forthcoming applications of gold nanoparticles in drug and gene delivery systems. J. Controlled Release. 149 (1), 65–71.

Probst, C.E., Zrazhevskiy, P., Bagalkot, V., Gao, X., 2013. Quantum dots as a platform for nanoparticle drug delivery vehicle design. Adv. Drug Delivery Rev. 65 (5), 703–718.

Qi, H., Liebert, T., Meister, F., Heinze, T., 2009. Homogenous carboxymethylation of cellulose in the NaOH/urea aqueous solution. React. Funct. Polym. 69 (10), 779–784.

Ribeiro-Viana, R.M., Faria-Tischer, P.C., Tischer, C.A., 2016. Preparation of succinylated cellulose membranes for functionalization purposes. Carbohydr. Polym. 148, 21–28.

Rouabhia, M., Asselin, J., Tazi, N., Messaddeq, Y., Levinson, D., Zhang, Z., 2014. Production of biocompatible and antimicrobial bacterial cellulose polymers functionalized by RGDC grafting groups and gentamicin. ACS Appl. Mater. Interfaces. 6 (3), 1439–1446.

Ruan, D., Huang, Q., Zhang, L., 2005. Structure and properties of CdS/regenerated cellulose nanocomposites. Macromol. Mater. Eng. 290 (10), 1017–1024.

Ruiz-Palomero, C., Soriano, M.L., Valcárcel, M., 2016. Gels based on nanocellulose with photosensitive ruthenium bipyridine moieties as sensors for silver nanoparticles in real samples. Sens. Actuators, B: Chem. 229, 31–37.

Salas, C., Nypelo, T., Rodriguez-Abreu, C., Carrillo, C., Rojas, O.J., 2015. Nanocellulose properties and applications in colloids and interfaces. Curr. Opin. Colloid Interface Sci. 19, 383–396.

Sattayasamitsathit, S., Gu, Y., Kaufmann, K., Jia, W., Xiao, X., Rodriguez, M., et al., 2013. Highly ordered multilayered 3D graphene decorated with metal nanoparticles. J. Mater. Chem. A. 1 (5), 1639–1645.

Schyrr, B., Pasche, S.P., Voirin, G., Weder, C., Simon, Y.C., Foster, E.J., 2014. Biosensors based on porous cellulose nanocrystal–poly (vinyl alcohol) scaffolds. ACS Appl. Mater. Interfaces. 6 (15), 12674–12683.

Shi, Z., Phillips, G.O., Yang, G., 2013. Nanocellulose electroconductive composites. Nanoscale. 5 (8), 3194–3201.

Sillett, S.C., Van Pelt, R., Carroll, A.L., Kramer, R.D., Ambrose, A.R., Trask, D.A., 2015. How do tree structure and old age affect growth potential of California redwoods? Ecol. Monogr. 85 (2), 181–212.

Sirvio, J.A., Visanko, M., Liimatainen, H., 2016. Acidic Deep Eutectic Solvents As Hydrolytic Media for Cellulose Nanocrystal Production. Biomacromolecules.

Soler, L., Magdanz, V., Fomin, V.M., Sanchez, S., Schmidt, O.G., 2013. Self-propelled micromotors for cleaning polluted water. ACS Nano. 7 (11), 9611–9620.

Solovev, A.A., Mei, Y., Bermúdez Ureña, E., Huang, G., Schmidt, O.G., 2009. Catalytic microtubular jet engines self-propelled by accumulated gas bubbles. Small. 5 (14), 1688–1692.

Song, J., Jensen, D.S., Hutchison, D.N., Turner, B., Wood, T., Dadson, A., et al., 2011. Carbon-nanotube-templated microfabrication of porous silicon-carbon materials with application to chemical separations. Adv. Funct. Mater. 21 (6), 1132–1139.

Sun, S., Jaouen, F., Dodelet, J.P., 2008. Controlled growth of Pt nanowires on carbon nanospheres and their enhanced performance as electrocatalysts in PEM fuel cells. Adv. Mater. 20 (20), 3900–3904.

Sun, S., Zhang, G., Geng, D., Chen, Y., Banis, M.N., Li, R., et al., 2010. Direct growth of single-crystal Pt nanowires on Sn@ CNT nanocable: 3D electrodes for highly active electrocatalysts. Chemistry. 16 (3), 829–835.

Svensson, A., Nicklasson, E., Harrah, T., Panilaitis, B., Kaplan, D.L., Brittberg, M., et al., 2005. Bacterial cellulose as a potential scaffold for tissue engineering of cartilage. Biomaterials. 26 (4), 419–431.

Tanaka, M., Sackmann, E., 2006. Supported membranes as biofunctional interfaces and smart biosensor platforms. Physica status solidi A. 203 (14), 3452–3462.
Tasaltin, C., Basarir, F., 2014. Preparation of flexible VOC sensor based on carbon nanotubes and gold nanoparticles. Sens. Actuators, B: Chem. 194, 173–179.
Thomson, K.R., Cheung, W., Ellis, S.J., Federman, D., Kavnoudias, H., Loader-Oliver, D., et al., 2011. Investigation of the safety of irreversible electroporation in humans. J. Vasc. Interv. Radiol. 22 (5), 611–621.
Trakakis, G., Anagnostopoulos, G., Sygellou, L., Bakolas, A., Parthenios, J., Tasis, D., et al., 2015. Epoxidized multi-walled carbon nanotube buckypapers: A scaffold for polymer nanocomposites with enhanced mechanical properties. Chem. Eng. J. 281, 793–803.
Ullah, H., Santos, H.A., Khan, T., 2016. Applications of bacterial cellulose in food, cosmetics and drug delivery. Cellulose. 23 (4), 2291–2314.
USDFA Bacteriological Analytical Manual (BAM). http://www.fda.gov/food/foodscienceresearch/laboratorymethods/ucm2006949.htm.
Valentini, L., Bon, S.B., Cardinali, M., Fortunati, E., Kenny, J.M., 2014. Cellulose nanocrystals thin films as gate dielectric for flexible organic field-effect transistors. Mater. Lett. 126, 55–58.
Vanegas, D.C., Taguchi, M., Chaturvedi, P., Burrs, S., Tan, M., Yamaguchi, H., et al., 2014. A comparative study of carbon–platinum hybrid nanostructure architecture for amperometric biosensing. Analyst. 139 (3), 660–667.
Walther, A., Timonen, J.V., Díez, I., Laukkanen, A., Ikkala, O., 2011. Multifunctional high-performance biofibers based on wet-extrusion of renewable native cellulose nanofibrils. Adv. Mater. 23 (26), 2924–2928.
Wang, Y., Hernandez, R.M., Bartlett, D.J., Bingham, J.M., Kline, T.R., Sen, A., et al., 2006. Bipolar electrochemical mechanism for the propulsion of catalytic nanomotors in hydrogen peroxide solutions. Langmuir. 22 (25), 10451–10456.
Wang, Y., Zhao, Y., Deng, Y., 2008. Effect of enzymatic treatment on cotton fiber dissolution in NaOH/urea solution at cold temperature. Carbohydr. Polym. 72 (1), 178–184.
Watt, G.D., 2014. Kinetic evaluation of the viologen-catalyzed carbohydrate oxidation reaction for fuel cell application. Renewable Energy. 63, 370–375.
Weng, Z., Su, Y., Wang, D.W., Li, F., Du, J., Cheng, H.M., 2011. Graphene–cellulose paper flexible supercapacitors. Adv. Energy Mater. 1 (5), 917–922.
Wernimont, E.J. In *Monopropellant hydrogen peroxide rocket systems: optimum for small scale*, 42nd AIAA Joint Propulsion Conference and Exhibit, AIAA-2006-5235, Sacramento, CA, 2006.
Wuts, P.G., Greene, T.W., 2006. Greene's protective groups in organic synthesis. John Wiley & Sons.
Xiong, R., Hu, K., Grant, A.M., Ma, R., Xu, W., Lu, C., et al., 2016. Ultrarobust Transparent Cellulose Nanocrystal-Graphene Membranes with High Electrical Conductivity. Adv. Mater. 28 (7), 1501–1509.
Yoon, S.H., Jin, H.-J., Kook, M.-C., Pyun, Y.R., 2006. Electrically conductive bacterial cellulose by incorporation of carbon nanotubes. Biomacromolecules. 7 (4), 1280–1284.
Zeinali, E., Haddadi-Asl, V., Roghani-Mamaqani, H., 2014. Nanocrystalline cellulose grafted random copolymers of *N*-isopropylacrylamide and acrylic acid synthesized by RAFT polymerization: effect of different acrylic acid contents on LCST behavior. RSC Adv. 4 (59), 31428–31442.
Zhang, L., Mao, Y., Zhou, J., Cai, J., 2005. Effects of coagulation conditions on the properties of regenerated cellulose films prepared in NaOH/urea aqueous solution. Ind. Eng. Chem. Res. 44 (3), 522–529.

Zhang, L., Shi, T., Xi, S., Liu, D., Tang, Z., Li, X., et al., 2011. Carbon nanotube integrated 3-dimensional carbon microelectrode array by modified SU-8 photoresist photolithography and pyrolysis. Thin solid films. 520 (3), 1041–1047.

Zhang, S., Li, F.-X., Yu, J.-Y., Hsieh, Y.-L., 2010. Dissolution behaviour and solubility of cellulose in NaOH complex solution. Carbohydr. Polym. 81 (3), 668–674.

Zhao, X., Lu, X., Tze, W.T., Wang, P., 2010. A single carbon fiber microelectrode with branching carbon nanotubes for bioelectrochemical processes. Biosens. Bioelectron. 25 (10), 2343–2350.

Zheng, G., Cui, Y., Karabulut, E., Wågberg, L., Zhu, H., Hu, L., 2013. Nanostructured paper for flexible energy and electronic devices. MRS Bulletin. 38 (04), 320–325.

Zhou, L.L., Sun, D.P., Hu, L.Y., Li, Y.W., Yang, J.Z., 2007. Effect of addition of sodium alginate on bacterial cellulose production by *Acetobacter xylinum*. J. Ind. Microbiol. Biotechnol. 34 (7), 483–489.

Zhu, H., Fang, Z., Preston, C., Li, Y., Hu, L., 2014. Transparent paper: fabrications, properties, and device applications. Energy Environ. Sci. 7 (1), 269–287.

Zhu, H., Zhu, S., Jia, Z., Parvinian, S., Li, Y., Vaaland, O., et al., 2015. Anomalous scaling law of strength and toughness of cellulose nanopaper. Proc. Natl. Acad. Sci. 112 (29), 8971–8976.

Sonochemical preparation of hybrid polymer nanocomposites: properties and applications

11

Bharat A. Bhanvase[1] *and Shirish H. Sonawane*[2]
[1]Department of Chemical Engineering, Laxminarayan Institute of Technology, Rashtrasant Tukadoji Maharaj Nagpur University, Nagpur, Maharashtra, India, [2]Department of Chemical Engineering, National Institute of Technology, Warangal, Telangana, India

Chapter Outline

11.1 Introduction 321
11.2 Properties of polymer nanocomposite: role of ultrasound 322
11.3 Sonochemical preparation of polymer nanocomposite: case studies 327
11.3.1 PANI/ZM nanocomposite 329
11.3.2 PMMA/$CaCO_3$ nanocomposite 332
11.3.3 Poly(acrylamide)–kaolin composite hydrogel 332
11.3.4 Layer-by-layer assembled nanocontainer 335
11.4 Future perspectives 338
11.5 Conclusion 339
References 339

11.1 Introduction

Hybrid polymer nanocomposites have established enormous attention in various applications due to their noteworthy changes in the different properties such as mechanical, electrical, thermal, optical, anticorrosive, gas sensing, magnetic, and others compared to pristine polymer (Siegel, 1994; He et al., 2001; Huang et al., 2004; Lu et al., 2004; Chen et al., 2005; Li and Zhu, 2003; Avella et al., 2001b; Wu et al., 2006; Bhanvase et al., 2009, 2011, 2012, 2015; Bhanvase and Sonawane, 2010). The polymer nanocomposites have been prepared with the specific properties for the target application. The fillers are generally selected on the basis of the target properties, which are generally incorporated in the polymer matrix by various available methods such as emulsion polymerization (in situ and ex situ), solution casting method, hybrid latex polymerization, intercalative polymerization, and melt processing (Tiarks et al., 2001; Kong et al., 1999; Liu et al., 2004; Tissot et al., 2001;

Hybrid Polymer Composite Materials: Applications. DOI: http://dx.doi.org/10.1016/B978-0-08-100785-3.00011-5

Ryu et al., 2004; Garcia et al., 2004). The methods generally focus on the preparation of hybrid polymer nanocomposites with higher yield. In these methods, addition of nanofillers is generally accomplished by surface grafting or encapsulation. This surface grafting or encapsulation must be properly controlled in order to get finely dispersed or encapsulation polymer nanocomposite; otherwise, polymer and inorganic particle phases get separated, and discrete phases are formed (Bhanvase et al., 2011). Simultaneous addition of nanofillers in the polymer matrix during its preparation has problem of agglomeration as smaller nanofiller particles have tendency of agglomeration due to their high surface energy (Qi et al., 2006; Caris et al., 1989). Functionalization of nanofiller particles with various organic or surfactant molecules is an alternative approach to reduce the surface energy of nanofiller particles that will improve the encapsulation efficiency of nanofillers in the polymer nanocomposites (Bhanvase and Sonawane, 2010, 2014; Wang et al., 2006, 2007; Sh et al., 2006). Wang et al. (2006) have prepared hydrophobic $CaCO_3$ nanoparticles in aqueous mixture by carbonation technique. In this report, it has been reported that the use of prepared solution of $C_{17}H_{35}COONa$ acts as an organic compound for functionalization and for the growth medium for $CaCO_3$. Further, it has been reported that this prepared organic compound is responsible for the nucleation and growth of hydrophobic (active ratio equal to 99%) $CaCO_3$ nanoparticles.

Agglomeration of the dispersed phase (i.e., nanofillers) in continuous phase (i.e., polymer) can be substantially reduced by the addition of nanofillers in the polymer matrix in the presence of ultrasound (Bhanvase and Sonawane, 2010; Bhanvase et al., 2011, 2012, 2015). It has been reported that the use of ultrasound in the preparation of polymer nanocomposite improves the dispersion of the nanofillers in the polymer matrix that offers the properties of the final formed product. The cavitational effect of the ultrasound leads to generation of intense turbulence, free radical formation, and formation of liquid circulation currents. In the case of ultrasound-assisted emulsion polymerization, formation of very fine emulsion droplets takes place because of the collapse of the cavitation bubble at the interface of the immiscible liquids that leads disruption of the present phases because of formed microjets and related turbulence (Bhanvase et al., 2011). Therefore in the present chapter, preparation of the polymer nanocomposite by ultrasound-assisted method with its property profile is explained with some suitable case studies.

11.2 Properties of polymer nanocomposite: role of ultrasound

Encapsulation of inorganic nanoparticles into polymer matrix has great applications in the preparation of paints, inks, cosmetic formulations, and pharmaceuticals. For the preparation of polymer hybrid nanocomposite is generally being achieved by emulsion polymerization (Yanase et al., 1993), suspension polymerization (Daniel et al., 1982), and precipitation polymerization (Deng et al., 2003). In these methods, initially inorganic nanoparticles are suspended in dispersed phase, and then polymerization process is carried out in the presence of these nanoparticles

(Xu et al., 2004; Qiu et al., 2007). In the emulsion polymerization process, in the initial stage, inorganic nanoparticles are being dispersed in the monomer phase with the help of suitable surfactant. Further with the help of high-speed disperser, the monomer and aqueous phases are thoroughly mixed, which will result into formation of emulsion and is attributed to high shear force. Then the polymerization of monomer takes place on the surface of inorganic nanoparticles, and it results in the formation of encapsulated hybrid polymer nanocomposites (Zhang et al., 2005; Xie et al., 2003; Liu et al., 2004; Csetneki et al., 2004).

Further during the incorporation of inorganic nanoparticles in the polymer latex, the problem of agglomeration is crucial as easy agglomeration of inorganic nanoparticles takes place because of their higher surface energy. This agglomeration of inorganic nanoparticles in the formed polymer nanocomposite leads to its inferior property profile. This agglomeration of nanoparticles can be reduced with the help of surface modification of inorganic nanoparticles with the help of various coupling agents (Bhanvase et al., 2011; Monte and Sugerman, 1976; Nakai et al., 2004; Demjén et al., 1997). Further dispersion of inorganic nanoparticles is possible to improve with the use of ultrasound-assisted methods (Bhanvase and Sonawane, 2010; Bhanvase et al., 2011, 2012, 2015; Bhanvase and Sonawane, 2014). It has been found that ultrasound-assisted methods helps in the enhancement of inorganic nanoparticles dispersion in the polymer matrix and that controls the overall encapsulation process and surface properties of the prepared hybrid polymer nanocomposite as explained in earlier section. Further with the use of ultrasound-assisted process, fine dispersion of inorganic nanoparticles in the polymer matrix can be achieved which results in the enhanced property profile of the prepared polymer nanocomposite.

Qiu et al. (2007) have successfully used ultrasonically initiated miniemulsion polymerization for the polystyrene/Fe_3O_4 nanocomposite. It has been reported that with an increase in the quantity of Fe_3O_4 nanoparticles in the miniemulsion polymerization reaction medium, the rate of polymerization was substantially enhanced, which can be attributed to increase in the number of radicals in the presence of Fe_3O_4 nanoparticles due to ultrasonic irradiations. Further, fine dispersion of Fe_3O_4 nanoparticles was observed from their report, which is attributed to cavitational effects of the ultrasonic irradiations. Due to this fine and uniform dispersion of Fe_3O_4 nanoparticles in the polymer matrix, the thermal and magnetic properties of formed polymer nanocomposite were reported to be enhanced drastically. Lu et al. (2006) have used ultrasonic irradiations for the preparation of PANI/Fe_3O_4 nanocomposite. It has been reported in this article that the use of ultrasonication improves the dispersion of Fe_3O_4 nanoparticles in the PANI nanotubes and that enhances the conducting and magnetic properties of the polyaniline nanotubes containing Fe_3O_4 nanoparticles. Further, Zhang et al. (2007) have prepared core–shell structure of SiO_2/poly(3-aminophenylboronic acid) composite in the presence of ultrasound. It has been observed that with the use of ultrasonication aggregation of nanosilica gets reduced; deposition of poly(3-aminophenylboronic acid) units takes place on the nanosilica surface in the presence of ultrasound, and it results into formation of core–shell structure.

The crystalline and thermal properties of core–shell SiO_2/poly(3-aminophenylboronic acid) composite were reported to be better for ultrasonication method when it is being compared with nanocomposite preparation by conventional method. It is attributed to the fine and uniform dispersion of the silica nanoparticles in poly(3-aminophenylboronic acid) matrix. There are various benefits of the use of the ultrasonic irradiations for the preparation of nanomaterials as well as polymer nanocomposites that are higher reaction rate, better dispersion of the nanoparticles, higher yield, and others. It has been observed that the ultrasonic irradiation produces a large number of microbubbles that grow and collapse in milliseconds. This phenomenon generates extreme pressure and temperature conditions, and generation of localized hot spot takes place. These extreme conditions are responsible for the chemical and physical effects that are responsible for the enhancement in the reaction rate, improved dispersion of nanoparticles with higher loading, and higher yield (Bhanvase et al., 2011).

Xia and Wang (2002) have used ultrasound-assisted approach for the synthesis of PANI/TiO_2 composite nanoparticles. As per their report, ultrasound-assisted polymerization of aniline was carried out in the presence of TiO_2 nanoparticles that leads to the formation of finely dispersed PANI/TiO_2 composite. It is attributed to the reduction in the agglomeration and redispersion of the TiO_2 nanoparticles in the presence of ultrasonic irradiations that leads to the formation of core–shell structure of PANI/TiO_2 nanocomposite. Further, it has been stated that the ultrasonic irradiation helps in the enhancement in the doping level of TiO_2 nanoparticles in the polyaniline matrix, and also it helps in the enhancement of the conductivity of prepared nanocomposite compared to conventional method of synthesis.

Bhanvase et al. (2011) have explained the formation mechanism of polymer nanocomposite by in the presence of ultrasonic irradiations and initiator which is shown in Fig. 11.1. As per their report, the formation of the radicals takes place due to chemical transformation of cavitational effects and dissociation of initiator. Further, because of physical effects (Shearing and turbulence effect) of the ultrasonic irradiations, the droplet size of the monomer gets reduced substantially, and simultaneous entry of radicals and $CaCO_3$ nanoparticles takes place in the monomer droplet. Further as per their mechanism, the entered radicals in the monomer droplet initiates the polymerization, and polymerization takes place around the inorganic nanoparticles leading to formation of encapsulated structure of polymer nanocomposite which is depicted in Fig. 11.1. Further, Bhanvase et al. (2011) encapsulated the structure of the poly(methyl methacrylate) (PMMA)–$CaCO_3$ nanocomposite is also confirmed by TEM image (Fig. 11.2).

Further, Bhanvase et al. (2012) have also elaborated the ultrasound-assisted approach for the preparation of poly(styrene-*co*-methyl methacrylate)/montmorillonite nanocomposite. As per the schematic mechanism (Fig. 11.3) reported by Bhanvase et al. (2012), ultrasonic irradiations not only enhances the exfoliation of the MMT but also improves the encapsulation/dispersion process of MMT platelets in the copolymer. This is attributed to intense shear forces due to high turbulence in the presence of ultrasound that exfoliates the MMT into single platelets and therefore its encapsulation/dispersion in the polymer matrix becomes easier.

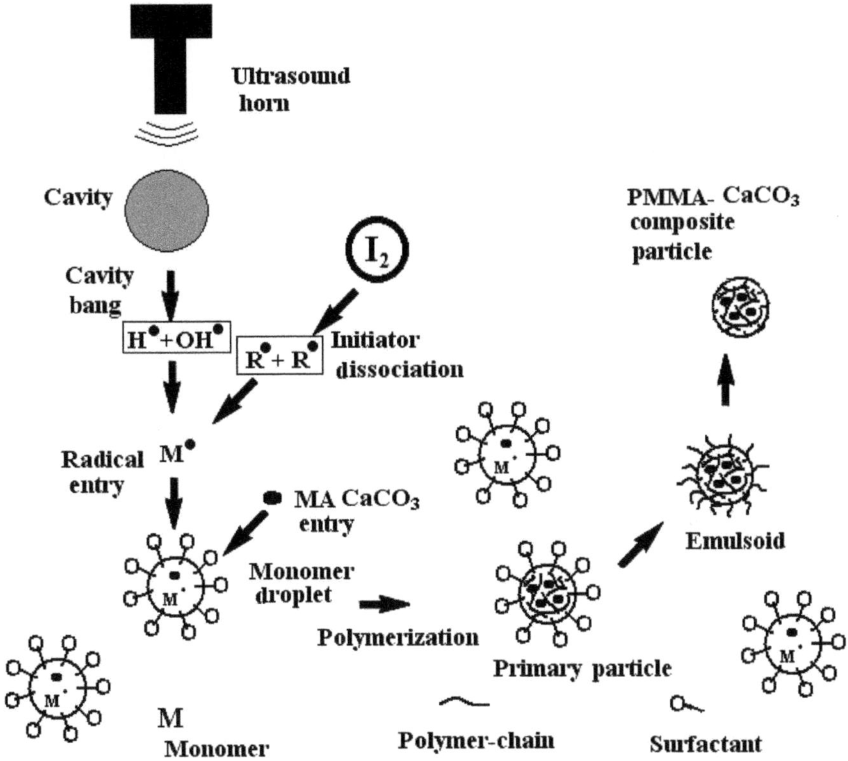

Figure 11.1 Schematic mechanism of the formation process of PMMA/$CaCO_3$ nanocomposite.
Reprinted with permission from Bhanvase et al., *Process intensification of encapsulation of functionalized $CaCO_3$ nanoparticles using ultrasound assisted emulsion polymerization*, (Elsevier: Chemical Engineering and Processing: Process Intensification, 2011), 50: 1160−1168.

After ultrasound-assisted emulsion polymerization, formation of polymer nanocomposite takes place as depicted in Fig. 11.3.

Bhanvase et al. (2012) have further reported that the better encapsulation/dispersion of MMT platelets in the polymer matrix enhances the thermal properties of the prepared polymer nanocomposite (Fig. 11.4). It is also further stated that the exfoliated structure of the clay in the polymer acts as a heat sink that toughens the nanocomposite structure and shows resistance to degradation process.

Further, the effect of encapsulation of MMT in the polymer matrix in the presence of ultrasonic irradiation gets significantly enhanced and due to which property profile gets enhanced (depicted in Table 11.1). It has been reported that the average particle size of the polymer nanocomposite depicted in Table 11.1 is considerably lesser, which is attributed to the decrease in the monomer droplet size due to continuous ultrasonic irradiation. Also the stability of the polymer nanocomposite

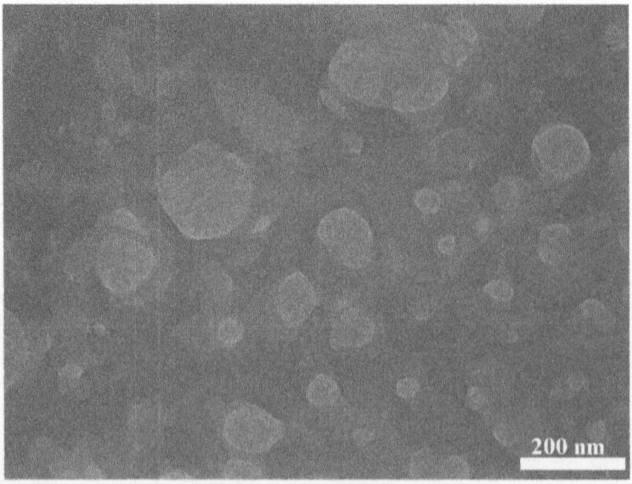

Figure 11.2 TEM images of PMMA/$CaCO_3$ nanocomposite.
Reprinted with permission from Bhanvase et al., Process intensification of encapsulation of functionalized $CaCO_3$ nanoparticles using ultrasound assisted emulsion polymerization, (Elsevier: Chemical Engineering and Processing: Process Intensification, 2011), 50: 1160–1168.

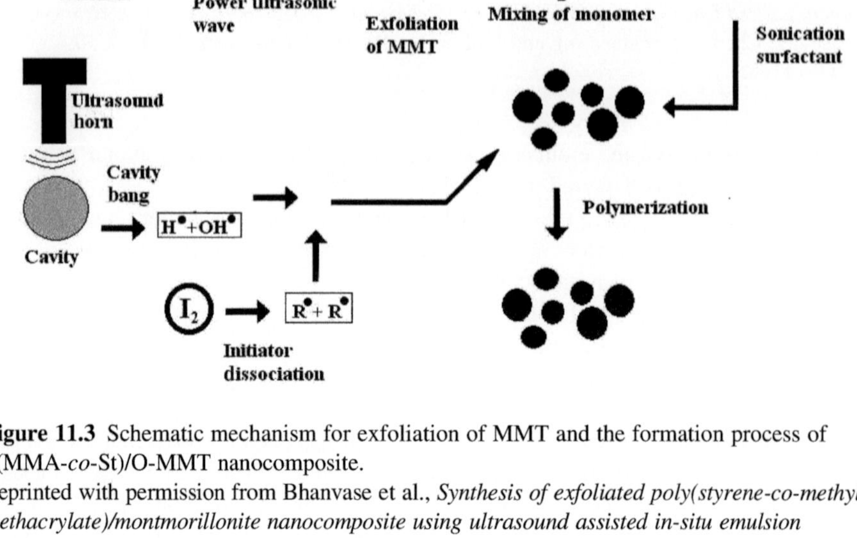

Figure 11.3 Schematic mechanism for exfoliation of MMT and the formation process of P(MMA-*co*-St)/O-MMT nanocomposite.
Reprinted with permission from Bhanvase et al., *Synthesis of exfoliated poly(styrene-co-methyl methacrylate)/montmorillonite nanocomposite using ultrasound assisted in-situ emulsion copolymerization*, (Elsevier: Chemical Engineering Journal, 2012), 181–182: 770–778.

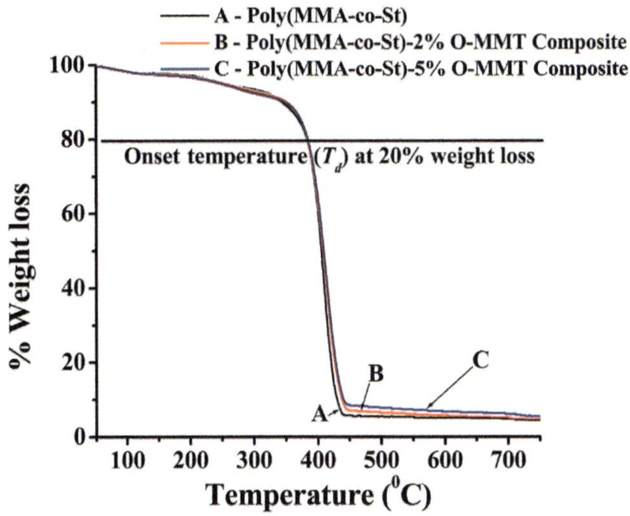

Figure 11.4 Thermogravimetric analysis (TGA) plots of (A) P(MMA-*co*-St) nanoparticles; (B) P(MMA-*co*-St)-2% O-MMT nanocomposite; and (C) P(MMA-*co*-St)-5% O-MMT nanocomposite.
Reprinted with permission from Bhanvase et al., *Synthesis of exfoliated poly(styrene-co-methyl methacrylate)/montmorillonite nanocomposite using ultrasound assisted in-situ emulsion copolymerization*, (Elsevier: Chemical Engineering Journal, 2012), 181–182: 770–778.

reported by Bhanvase et al. (2012). The reason behind this is an application of ultrasonic irradiation during in situ emulsion polymerization that causes fine and stable emulsion with lesser droplet size. This also leads to formation lesser size polymer nanocomposite.

All researchers are using ultrasound-assisted approach effectively for the preparation of the polymer nanocomposite with enhanced property profile as depicted in earlier section. Also they are able to produce finely and uniformly dispersed polymer nanocomposite with again lesser composite particle size that gives better property profile. The physical and chemical effects of the ultrasound have significant effect on the same.

11.3 Sonochemical preparation of polymer nanocomposite: case studies

As discussed in the previous section, it has been observed that the use of ultrasonic irradiation during the preparation of polymer nanocomposite shows better encapsulation/dispersion of inorganic nanoparticles in the polymer matrix that in turn enhances the property profile of the polymer nanocomposite. In this section, some case studies are put forth for better understanding.

Table 11.1 Effect of O-MMT loading on glass transition temperature, Z-average diameter, zeta potential values, pencil hardness, and heat of reaction of the P(MMA-*co*-St)/O-MMT nanocomposite latexes (temperature = 60°C, weight of monomer = 21.52 g, MMA:St ratio = 1:1, weight of KPS = 0.86 g, weight of SLS = 1.08 g, weight of water = 120 g)

Sr. No.	Sample	Wt. of O-MMT (g) (% of O-MMT loading)	Glass transition temperature (°C)	Heat of reaction (J/g)	Zeta potential (mV)	Z-average diameter (nm)	Pencil hardness
1	Neat P(MMA-*co*-St)	0.00 (0.0)	127.3	−437.5	−46.4	52.64	HB Fail
2	P(MMA-*co*-St)-0.5% O-MMT	0.11 (0.5)	129.1	−292.5	−46.7	156.58	HB Fail
3	P(MMA-*co*-St)-1% O-MMT	0.22 (1.0)	152.7	−265	−48.8	168.88	HB Fail
4	P(MMA-*co*-St)-2% O-MMT	0.43 (2.0)	131.8	−350	−42.8	170.74	HB Fail
5	P(MMA-*co*-St)-3% O-MMT	0.65 (3.0)	130.4	−458	−40.4	178.28	HB Pass
6	P(MMA-*co*-St)-4% O-MMT	0.86 (4.0)	129.8	−448	−37.1	185.72	HB Pass
7	P(MMA-*co*-St)-5% O-MMT	1.08 (5.0)	129.0	−382	−35.8	191.23	HB Pass

Reprinted with permission from Bhanvase et al., *Synthesis of exfoliated poly(styrene-co-methyl methacrylate)/montmorillonite nanocomposite using ultrasound assisted in-situ emulsion copolymerization*, (Elsevier: Chemical Engineering Journal, 2012), 181–182: 770–778.

11.3.1 PANI/ZM nanocomposite

PANI is being investigated for the various properties such as anticorrosion, electrical, conducting, and gas sensing properties. The conducting polymer PANI has better environmental stability and can be modified by oxidation state and protonation mechanism (Pud et al., 2003). The only limitation lying with polyaniline is the difficulty in the formation of films and immiscibility with other solvents. Further with the incorporation of the various compatible nanofillers in the PANI, it is possible to improve the properties simultaneously such as anticorrosion, thermal, gas sensing, and mechanical properties. Further application of ultrasonic irradiation during the preparation of polymer nanocomposite enhances the dispersion of inorganic nanoparticles in the polymer matrix.

Bhanvase et al. (2015) have used ultrasound-assisted in situ emulsion polymerization for the preparation of PANI/$ZnMoO_4$ nanocomposite for varying loading of $ZnMoO_4$ nanoparticles. It has been reported that the use of ultrasonic irradiations during the preparation of PANI/$ZnMoO_4$ nanocomposite by in situ emulsion polymerization shows better and uniform dispersion to of the $ZnMoO_4$ (ZM) nanoparticles in the PANI matrix that is depicted in Fig. 11.5 for better understanding. The surface modification of ZM nanoparticles with myristic acid and the use of ultrasonication during the preparation of PANI/ZM nanocomposite shows uniform dispersion of ZM nanoparticles with a size of 250 nm in the PANI matrix.

Due to this fine dispersion of the ZM nanoparticles in the PANI matrix, the property profile of formed PANI/ZM nanocomposite is superior, which is explained

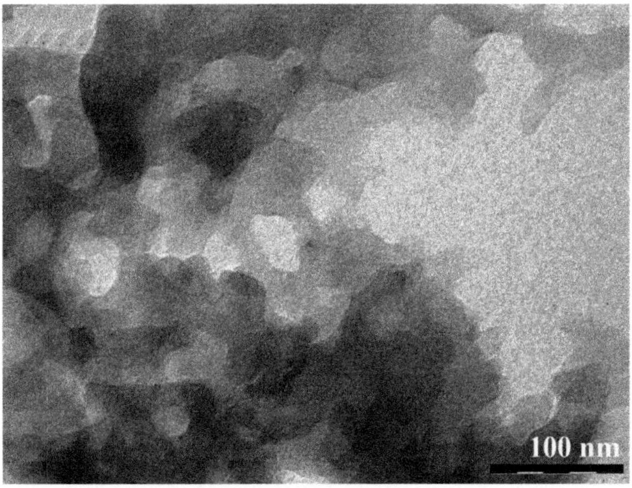

Figure 11.5 TEM Image of PANI/ZM nanocomposite.
Reprinted with permission from Bhanvase et al., Ultrasound assisted synthesis of PANI/$ZnMoO_4$ nanocomposite for simultaneous improvement in anticorrosion, physico-chemical properties and its application in gas sensing, (Elsevier: Ultrasonics Sonochemistry, 2015), 24: 87–97.

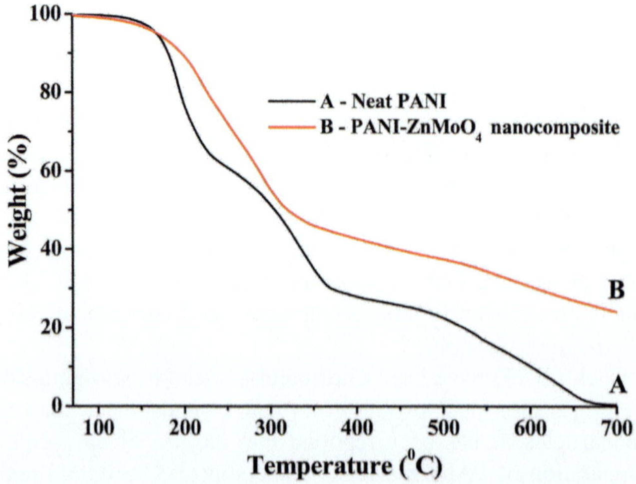

Figure 11.6 TGA plot of (A) neat PANI and (B) PANI/ZM nanocomposite.
Reprinted with permission from Bhanvase et al., Ultrasound assisted synthesis of PANI/ $ZnMoO_4$ nanocomposite for simultaneous improvement in anticorrosion, physico-chemical properties and its application in gas sensing, (Elsevier: Ultrasonics Sonochemistry, 2015), 24: 87–97.

by Bhanvase et al. (2015) in a better way. Fig. 11.6 depicted TGA curves of ultrasonically prepared pristine PANI and PANI/ZM nanocomposite. It has been clearly observed that the PANI/ZM nanocomposite shows better thermal stability than that of pristine PANI that is attributed to fine dispersion of the ZM nanoparticles in the PANI matrix in the presence of ultrasonic irradiations in this; the role of ZM nanoparticles is as a barrier that prevents the degradation of PANI in PANI/ZM nanocomposite.

Further Bhanvase et al. (2015) have reported the efficient use of the ultrasonically prepared PANI/ZM nanocomposite for the enhancement of the anticorrosion properties in alkyd resin. As reported by Bhanvase et al. (2015), Fig. 11.7 explains the corrosion mechanism for PANI/alkyd and PANI-ZM nanocomposite/alkyd coatings. It has been reported that the adhesion of the prepared coating on the substrate is a key factor for controlling the corrosion rate. An incorporation of finely dispersed PANI/ZM nanocomposite in alkyd resin gives better compactness of the coatings and that shows better adhesion towards the substrate. Further in the corrosion mechanism, it has been depicted that with the attack of acid, alkali, and salt on the coated samples, the formation of iron molybdate complex takes place effectively due to fine dispersion of ZM in PANI matrix. This formed iron molybdate complex passivates the substrate, and further attack of the corrosion elements is substantially reduced.

Bhanvase et al. (2015) have further reported the application of prepared PANI/ZM nanocomposite for the LPG sensing. In general, the conducting polymers, such as PANI, polypyrrole have problems like poor selectivity towards particular gas

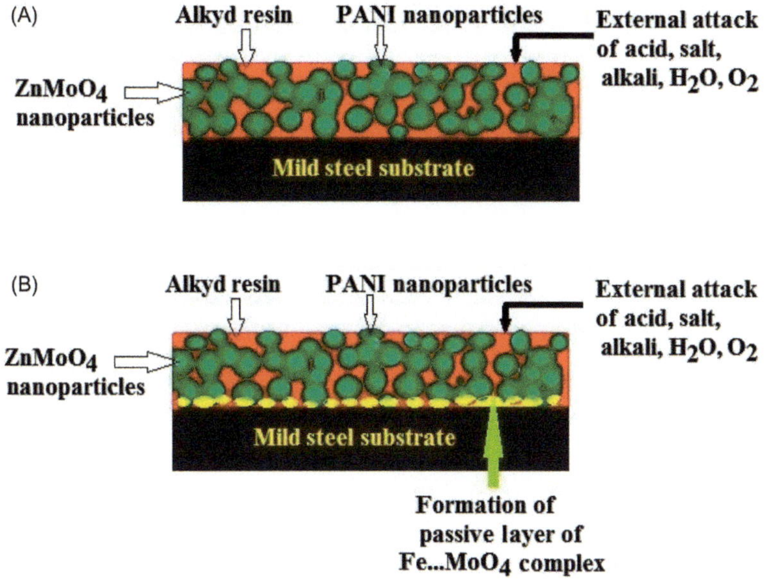

Figure 11.7 Corrosion mechanism in PANI/Alkyd and PANI-ZM anocomposite/alkyd coatings.
Reprinted with permission from Bhanvase et al., *Ultrasound assisted synthesis of PANI/ $ZnMoO_4$ nanocomposite for simultaneous improvement in anticorrosion, physico-chemical properties and its application in gas sensing*, (Elsevier: Ultrasonics Sonochemistry, 2015), 24: 87–97.

with lesser sensitivity (Geng et al., 2007). Therefore, Bhanvase et al. (2015) have suggested use of PANI (p-type)/ZM (n-type) heterojunction nanocomposite to overcome above-stated difficulties at room temperature. Bhanvase et al. (2015) have tested the ultrasonically prepared PANI/ZM nanocomposite for the LPG sensing at various concentrations as depicted in Fig. 11.8. It has been reported that the sensor prepared with the help of PANI/ZM nanocomposite shows considerably higher response compared to that of only PANI material. As per their report, the presence of ZM nanoparticles in the ultrasonically prepared PANI/ZM nanocomposite sensor helps the development of a range of p–n semiconductor sites that act as an LPG adsorption sites which result into an improvement in the response of sensor (Edwin Suresh Raj et al., 2002).

Overall, ultrasonically prepared PANI/ZM nanocomposite in the presence of myristic acid modified ZM nanoparticles shows better dispersion of ZM nanoparticles in the PANI matrix. The fine and uniform dispersion of ZM nanoparticles in PANI matrix is attributed to cavitational effect of the ultrasonic irradiations that enhances the distribution of ZM nanoparticles and controls the size of PANI/ZM nanocomposite. This in turn leads to enhancement in the various properties such as anticorrosion, thermal, and gas sensing.

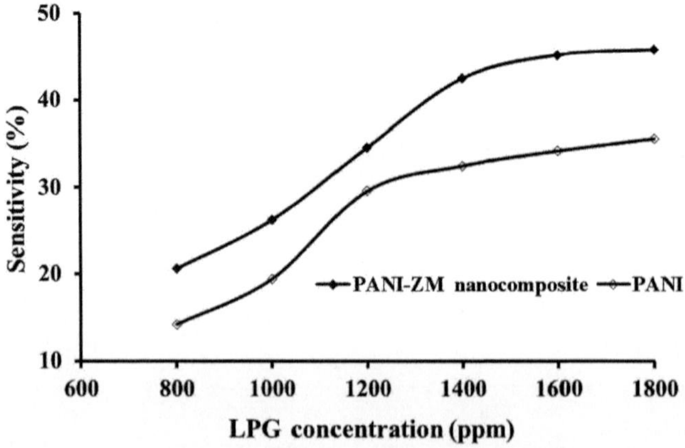

Figure 11.8 Sensitivity (response) of PANI-ZM nanocomposite film sensor at different LPG concentration.
Reprinted with permission from Bhanvase et al., *Ultrasound assisted synthesis of PANI/ZnMoO$_4$ nanocomposite for simultaneous improvement in anticorrosion, physico-chemical properties and its application in gas sensing*, (Elsevier: Ultrasonics Sonochemistry, 2015), 24: 87–97.

11.3.2 PMMA/$CaCO_3$ nanocomposite

Poly(methyl methacrylate) (PMMA) is a key polymer that has various uses in lighting, aircraft glazing, contact lenses, and others (Avella et al., 2001a). Encapsulation of $CaCO_3$ in PMMA by conventional polymerization process was attempted by Avella et al. (2001a). It has been reported that with the incorporation of 2% $CaCO_3$ in PMMA, this composite shows enhancement in the abrasion resistance.

Bhanvase et al. (2011) have attempted ultrasound-assisted in situ emulsion polymerization for the preparation of PMMA/$CaCO_3$ nanocomposite, and the formation mechanism of the said nanocomposite is presented in Fig. 11.1. The use of ultrasound shows positive effect during the formation of PMMA/$CaCO_3$ nanocomposite, which are (1) enhancement in the polymerization rate due to generation of radical because of ultrasonic irradiations and faster dissociation of initiator, (2) substantial reduction in the monomer droplet size due to shearing and turbulence effects of ultrasonic irradiations, and (3) better encapsulation/dispersion of the $CaCO_3$ nanoparticles in PMMA matrix. The formation mechanism of the PMMA/$CaCO_3$ nanoparticles is already discussed in earlier section in Fig. 11.1. The enhanced dispersion of $CaCO_3$ nanoparticles in the PMMA matrix is also depicted in Fig. 11.2. The fine dispersion/encapsulation of $CaCO_3$ nanoparticles in PMMA matrix leads to substantial enhancement in the glass transition temperature (from 119.5°C for neat PMMA to 210.6°C for the composite) of the PMMA/$CaCO_3$ nanocomposite.

11.3.3 Poly(acrylamide)–kaolin composite hydrogel

Recently, hydrogels have been used for the waste water treatment by several researchers. The main process in waste water treatment by hydrogel is adsorption

(Shirsath et al., 2011). Hydrogels have 3D crosslinked flexible chain network that adsorbs the pollutants effectively and retains it in its structure. This phenomenon is attributed to the presence of the hydrophilic group in the hydrogel network (Haraguchi, 2007; Fu and Soboyejo, 2010; Guilherme et al., 2010). In order to improve the adsorption capacity of the hydrogel, in recent years, several researchers have modified the properties of the polymeric hydrogel with the addition of inorganic nanomaterials/clays that enhances the properties such as elasticity and permeability of the hydrogel with higher adsorption capacity. Further ultrasound-assisted synthesis of the hydrogel can be carried out to have better control on the molecular weight of the hydrogel and is due to higher value of shearing effect because of the cavitational effects of the ultrasonic irradiations. Also use of this novel process enhances the dispersion/encapsulation of the clays/inorganic nanoparticles in the polymer matrix that, in turn, enhances the adsorption capacity and mechanical properties of the formed hydrogel nanocomposite.

Shirsath et al. (2015) have attempted ultrasound-assisted preparation of poly (acrylamide)–kaolin (PAAm-K) hydrogel nanocomposite by in situ emulsion polymerization. It has been reported in Fig. 11.9 (TEM of nanocomposite) that kaolin nanoparticles of size 50 nm are finely and uniformly dispersed/encapsulated in the hydrogel matrix. This is attributed to the intense environment generated by acoustic cavitation due to ultrasonic irradiations that enhance the dispersion/encapsulation of inorganic nanoparticles such as clay, $CaCO_3$ into hydrogel matrix (Bhanvase et al., 2012; Bhanvase and Sonawane, 2014). It has been further reported that in the presence of ultrasound, exfoliation of kaolin clay takes place due to shearing effects of the ultrasound.

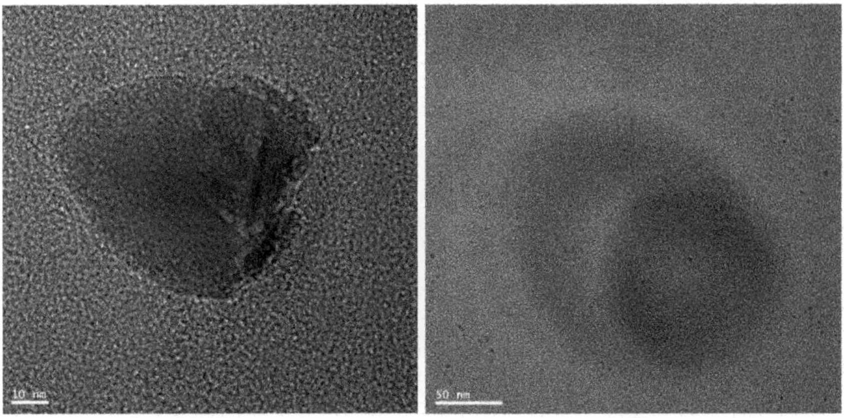

(A) (B)

Figure 11.9 Transmission electron microscopic images of PAAm-K hydrogel. Reprinted with permission from Shirsath et al., *Ultrasonically prepared poly(acrylamide)-kaolin composite hydrogel for removal of crystal violet dye from wastewater*, (Elsevier: Journal of Environmental Chemical Engineering, 2015), 3: 1152–1162.

Shirsath et al. (2015) have further studied the combined effect of ultrasound and hydrogel on the adsorption of Crystal Violet dye, and comparison was depicted with neat hydrogel (Fig. 11.10). It has been observed that the combined use of ultrasound and hydrogel shows better dye adsorption rate compared to neat hydrogel. It is attributed to the generation of OH· due to cavitational effect of the ultrasonic irradiation that is responsible for the degradation of dye resulting into more removal of dye by hydrogel in the presence of ultrasound. Also the drastic decrease in the time for the adsorption of dye was reported, which can be attributed to enhancement in the surface area of the hydrogel due to incorporation of kaolin in the presence of ultrasound that provides more adsorption sites. Also the presence of ultrasound facilitates the transport of the dye molecules to active site results in the drastic decrease in the dye adsorption time was observed.

Further, ultrasound-assisted process enhances the loading of kaolin clay nanoparticles in the PAAm matrix. It is again attributed to formation of very fine monomer droplet during in situ emulsion polymerization due to which a large number of kaolin nanoparticles get entered into the monomer droplet that in turn increases the loading of kaolin nanoparticles with fine dispersion of it in the presence of ultrasonic irradiations. Further, the loaded kaolin clay in hydrogel enhances the adsorption of crystal violet dye. The percentage removal of crystal violet dye at different loading of kaolin clay in hydrogel is depicted in Fig. 11.11. It has been reported that the % removal of crystal violet dye increases with an increase of the loading of kaolin clay in hydrogel composite. It is attributed to increase in the number

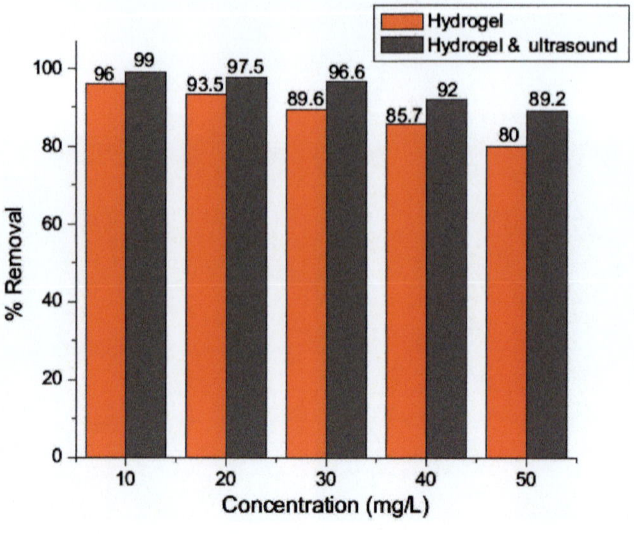

Figure 11.10 Comparison of hydrogel & hydrogel with ultrasound on adsorption of crystal violet dye at pH 10 with a temperature of 35°C.
Reprinted with permission from Shirsath et al., *Ultrasonically prepared poly(acrylamide)-kaolin composite hydrogel for removal of crystal violet dye from wastewater*, (Elsevier: Journal of Environmental Chemical Engineering, 2015), 3: 1152–1162.

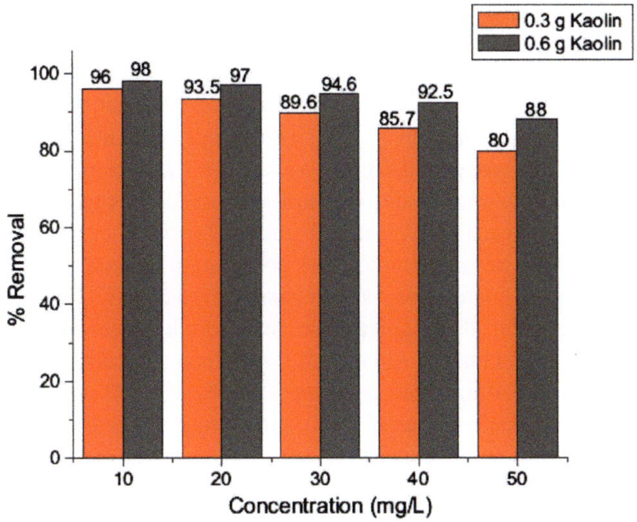

Figure 11.11 Effect of clay loading in PAAm-K hydrogel for the adsorption of CV dye at pH 10, temperature 35°C and equilibrium time 5 h.
Reprinted with permission from Shirsath et al., *Ultrasonically prepared poly(acrylamide)-kaolin composite hydrogel for removal of crystal violet dye from wastewater*, (Elsevier: Journal of Environmental Chemical Engineering, 2015), 3: 1152–1162.

of active sites for the adsorption in the hydrogel composite with an increase it the kaolin clay loading.

11.3.4 Layer-by-layer assembled nanocontainer

Now-a-days, anticorrosion properties are being enhanced with the use of active corrosion inhibitors in various coatings. For the same, nanocontainers prepared with layer-by-layer assembly are useful as it contains various species that contribute to enhancement in the anticorrosion properties (Sonawane et al., 2012; Bhanvase et al., 2013, 2014; Tyagi et al., 2014). Several researchers have used micron-sized nanocontainer for an enhancement in the anticorrosion properties (Abu and Aoki, 2005). However, size of the nanocontainer has an important role in the anticorrosion properties of the multilayer coatings. Further, after release of the corrosion inhibitor for the micron-sized nanocontainer leaves behind the cavities in the coating, and these cavities weaken the coatings applied on the surface of the substrate. This problem can be overcome by the use of nanometer-sized nanocontainers in the coatings that will enhance the compactness to the coating, and also release of the corrosion inhibitor can be enhanced. The distribution of the nanometer-sized nanocontainers takes place homogeneously in the coating and thereby homogeneous distribution of corrosion inhibitor in the coating that enhances anticorrosion performance to the substrate (Shchukin et al., 2006). The size of the nanocontainer can be controlled by adjusting the size of nanoparticles which acts as a template for the nanocontainer and thickness of the polyelectrolyte layers. The substantial reduction in the template (nanomaterials) size can be achieved by the preparation of these nanomaterials in the presence of ultrasonic irradiation (Patel et al., 2013; Deosarkar et al., 2013, 2014). The reduction in the size of the nanomaterials is attributed to the cavitational effects generated by the ultrasonic irradiations that enhance the nucleation events and solute transfer rate in the presence of the ultrasound leading to reduction in the nanoparticles size and thereby nanocontainer size.

Tyagi et al. (2014) have attempted the preparation of silica nanoparticles by Stober's process in the presence of ultrasonic irradiations. Further, they have attempted the preparation of the silica nanocontainers by layer-by-layer deposition of poly(diallyl dimethylammonium chloride), poly(styrene sulfonate) and a corrosion inhibitor (benzotriazole). It has been reported (Fig. 11.12) that there is substantial reduction in the silica nanoparticle (170 nm) size in the presence of ultrasound that can be attributed to faster nucleation and solute transfer rate in the presence of ultrasonic irradiations that result in particle size significantly. Also the growth rate of the silica nanoparticles is properly controlled in the presence of ultrasonic irradiations. Further, Fig. 11.12 shows no aggregation of spherical silica nanoparticles and nanocontainers (240 nm), and they are monodispersed. The spherical nature of the silica nanoparticles and nanocontainers was also confirmed by TEM analysis, which is depicted in Fig. 11.13. Further, Fig. 11.13 shows encapsulation of silica nanoparticles by polyelectrolyte layers and corrosion inhibitor layer.

Further, Bhanvase et al. (2014) have also attempted the preparation of cerium zinc molybdate nanoparticles by ultrasound-assisted approach and its nanocontainer for anticorrosion application. In this report also, they have achieved the reduction in the size of cerium zinc molybdate in the presence of ultrasound. The possible reasons for the same are reported in the earlier sections. The reported particle size of cerium zinc molybdate nanoparticles was around 26 nm. This reduction in the

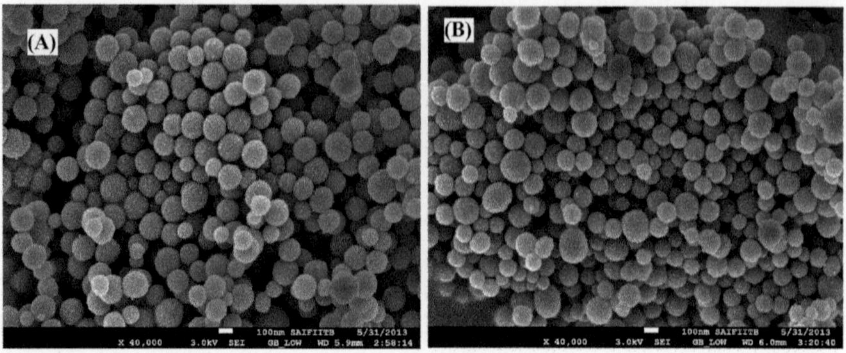

Figure 11.12 FEG-SEM images: (A) silica nanoparticles and (B) silica nanocontainers. Reprinted with permission from Tyagi et al., *Computational studies on release of corrosion inhibitor from layer-by-layer assembled silica nanocontainer*, (American Chemical Society: Industrial & Engineering Chemistry Research, 2014), 53: 9764–9771.

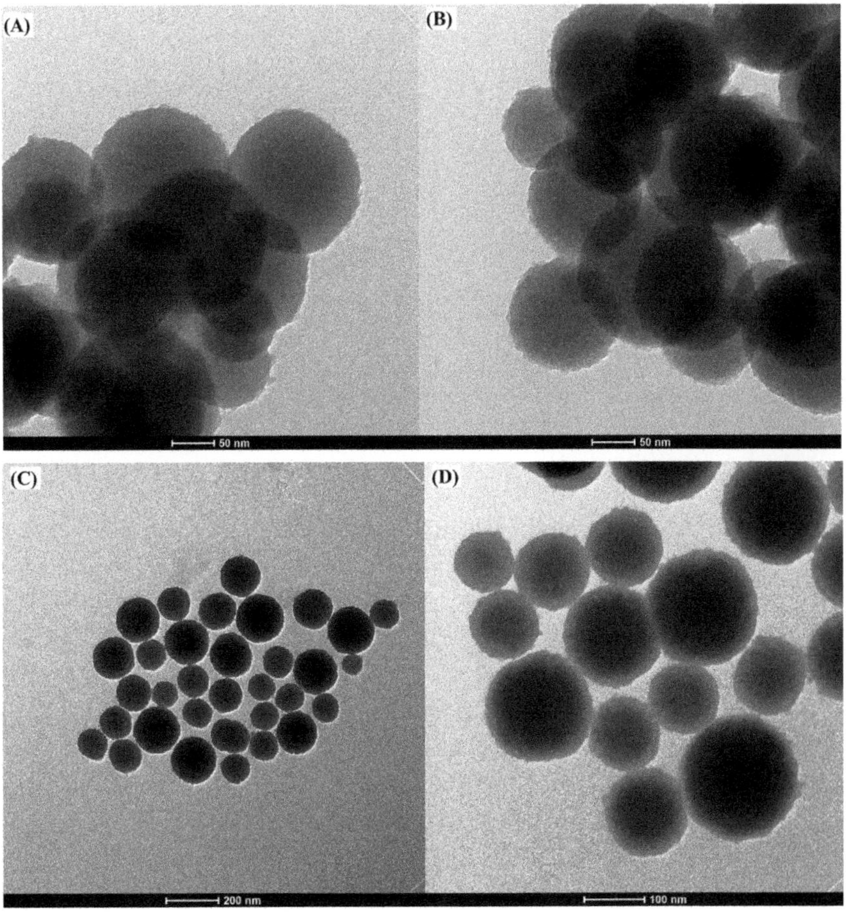

Figure 11.13 TEM images of (A and B): silica nanoparticles; and (C and D) silica nanocontainers at different magnifications.
Reprinted with permission from Tyagi et al., *Computational studies on release of corrosion inhibitor from layer-by-layer assembled silica nanocontainer*, (American Chemical Society: Industrial & Engineering Chemistry Research, 2014), 53: 9764–9771.

particle size of cerium zinc molybdate nanoparticles is attributed to faster nucleation rate, solute transfer rate, and controlled growth rate in the presence of the ultrasonication. Bhanvase et al. (2014) have used cerium zinc molybdate nanoparticles prepared by ultrasound-assisted approach for the preparation of cerium zinc molybdate nanocontainers by deposition of polyaniline, imidazole, and polyacrylic acid layers on cerium zinc molybdate template. The morphology of the formed nanocontainer is depicted in Fig. 11.14. The reported size of cerium zinc molybdate nanocontainer size was in the range of 110 to 150 nm. Due to this lesser size, these nanocontainers show some agglomeration due to their higher surface energy. Further, the lesser cerium zinc molybdate nanocontainer size is attributed to the

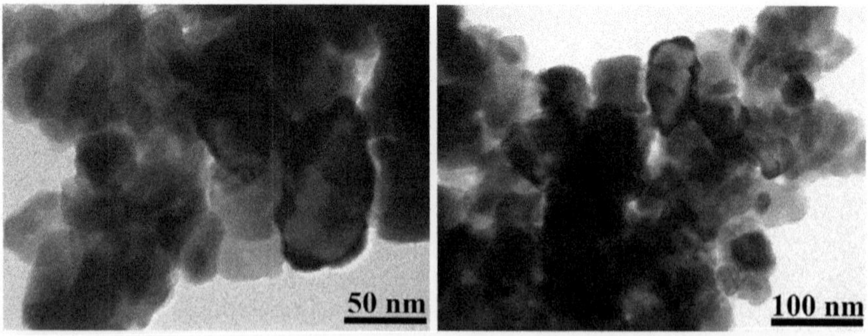

Figure 11.14 TEM images of CZM nanocontainer at different magnifications.
Reprinted with permission from Bhanvase et al., *Kinetic properties of layer-by-layer assembled cerium zinc molybdate nanocontainers during corrosion inhibition*, (Elsevier: Corrosion Science, 2014), 88: 170–177.

cavitational effects of the ultrasonic irradiations and is also attributed to use of smaller sized cerium zinc molybdate nanoparticles as a template, that is, core material for the preparation of cerium zinc molybdate nanocontainer. Also the aggregation effect of cerium zinc molybdate nanocontainers is reduced to some extent due to ultrasonic irradiations.

Overall, the nanocontainers prepared with the use of ultrasonically prepared template shows drastic reduction in the size with a better shape control; therefore, it can be easily incorporated/dispersed in the various coating where active corrosion inhibitor gets released in the presence of corrosion elements and forms the complex that protects the substrate in a better way.

11.4 Future perspectives

Many researchers have used ultrasound-assisted method for the preparation of hybrid polymer nanocomposite in the presence of the initiator and surfactant. However, the separation used of initiator and surfactant from the formed hybrid polymer nanocomposite is very difficult and that adds the impurities in the resulted polymer nanocomposite. Further, during the preparation of the polymer nanocomposite, generally higher temperature was maintained. With the use of ultrasound-assisted emulsion polymerization, it is possible to carry out the polymerization in the presence of inorganic nanoparticles that can result in the formation of finely dispersed/encapsulated polymer nanocomposite. Further, it can be possible to perform the polymerization in the absence of the initiator and surfactant (or for lesser amount of surfactant) that result into pure polymer nanocomposite. This will be a green process and is ecofriendly for the environment. Also new aspect of this work is to study the properties of the polymer nanocomposite prepared by ultrasound-assisted in situ semibatch emulsion polymerization process at lower temperature and in the absence of initiator and surfactant.

11.5 Conclusion

Ultrasound-assisted polymerization has been successfully employed by the several researchers for the preparation of the hybrid polymer nanocomposite. It has been found that with the use of ultrasound-assisted process, it is possible to reduce the size of the polymer nanocomposite with fine dispersion of the inorganic nanoparticles in the polymer matrix compared with conventional method of preparation. Further, it has been observed that the use of ultrasonic irradiations during the preparation of polymer nanocomposite enhances the loading of inorganic nanoparticles in the polymer nanocomposite that substantially enhances the various properties such as mechanical, anticorrosive, thermal, rheological, gas sensing properties of the hybrid polymer nanocomposite. The nanocontainers prepared with the use of ultrasonically prepared inorganic nanoparticles as a template show drastic decrement in the size. This decrease in the nanocontainer size shows significant enhancement in the anticorrosion properties. With the use of ultrasound-assisted process for the preparation of the inorganic nanoparticles, the reduction in time and growth rate was observed with enhanced nucleation rate and solute transfer rate due to which the reaction conditions gets milder compared to conventional method of synthesis.

References

Abu, Y.M., Aoki, K., 2005. Corrosion protection by polyaniline-coated latex microspheres. J. Electroanal. Chem. 583, 133–139.

Avella, M., Errico, M., Martuscelli, E., 2001a. Novel PMMA/$CaCO_3$ nanocomposites abrasion resistant prepared by an in situ polymerization process. Nano Lett. 1, 213–217.

Avella, M., Errico, M.E., Martelli, S., Martuscelli, E., 2001b. Preparation methodologies of polymer matrix nanocomposites. Appl. Organomet. Chem. 15, 435–439.

Bhanvase, B.A., Sonawane, S.H., 2010. New approach for simultaneous enhancement of anticorrosive and mechanical properties of coatings: application of water repellent nano $CaCO_3$-PANI emulsion nanocomposite in alkyd resin. Chem. Eng. J. 156, 177–183.

Bhanvase, B.A., Sonawane, S.H., 2014. Ultrasound assisted in-situ emulsion polymerization for polymer nanocomposite: a review. Chem. Eng. Process. 85, 86–107.

Bhanvase, B.A., Gumfekar, S.P., Sonawane, S.H., 2009. Water based PMMA-nano $CaCO_3$ nanocomposite by in-situ polymerization technique: synthesis, characterization and mechanical properties. Polym. Plast. Technol. Eng. 48, 939–944.

Bhanvase, B.A., Pinjari, D.V., Gogate, P.R., Sonawane, S.H., Pandit, A.B., 2011. Process intensification of encapsulation of functionalized $CaCO_3$ nanoparticles using ultrasound assisted emulsion polymerization. Chem. Eng. Process. 50, 1160–1168.

Bhanvase, B.A., Pinjari, D.V., Gogate, P.R., Sonawane, S.H., Pandit, A.B., 2012. Synthesis of exfoliated poly(styrene-*co*-methyl methacrylate)/montmorillonite nanocomposite using ultrasound assisted in-situ emulsion copolymerization. Chem. Eng. J. 181–182, 770–778.

Bhanvase, B.A., Kutbuddin, Y., Borse, R.N., Selokar, N., Pinjari, D.V., Sonawane, S.H., et al., 2013. Ultrasound assisted synthesis of calcium zinc phosphate pigment and its application in nanocontainer for active anticorrosion coatings. Chem. Eng. J. 231, 345–354.

Bhanvase, B.A., Patel, M.A., Sonawane, S.H., 2014. Kinetic properties of layer-by-layer assembled cerium zinc molybdate nanocontainers during corrosion inhibition. Corros. Sci. 88, 170–177.

Bhanvase, B.A., Darda, N.S., Veerkar, N.C., Shende, A.S., Satpute, S.R., Sonawane, S.H., 2015. Ultrasound assisted synthesis of PANI/$ZnMoO_4$ nanocomposite for simultaneous improvement in anticorrosion, physico-chemical properties and its application in gas sensing. Ultrason. Sonochem. 24, 87–97.

Caris, C.H.M., Louisa, P.M., Herk, A.M., 1989. Polymerization of MMA at the surface of inorganic submicron particles. Br. Polym. J. 21, 133–140.

Chen, M., Zhou, S., You, B., Wu, L., 2005. A novel preparation method of raspberrylike PMMA/SiO_2 hybrid microspheres. Macromolecules. 38, 6411–6417.

Csetneki, I., Faix, M.K., Szilagyi, A., Kovacs, A.L., Nemeth, Z., Zrinyi, M., 2004. Preparation of magnetic polystyrene latex via the miniemulsion polymerization technique. J. Polym. Sci. Part A: Polym. Chem. 42, 4802–4808.

Daniel J.C., J.L. Schuppiser, M. Tricot deceased, Magnetic polymer latex and preparation process, US Patent, 4358388, 1982.

Demjén, Z., Pukánszky, B., FÖldes, E., Nagy, J., 1997. Interaction of silane coupling agents with $CaCO_3$. J. Colloid Interface Sci. 190, 427–436.

Deng, Y.H., Yang, W.L., Wang, C.C., Fu, S.K., 2003. A novel approach for preparation of thermoresponsive polymer magnetic microspheres with core-shell structure. Adv. Mater. 15, 1729–1732.

Deosarkar, M.P., Pawar, S.M., Sonawane, S.H., Bhanvase, B.A., 2013. Process intensification of uniform loading of SnO_2 nanoparticles on graphene oxide nanosheets using a novel ultrasound assisted in situ chemical precipitation method. Chem. Eng. Process. 70, 48–54.

Deosarkar, M.P., Pawar, S.M., Bhanvase, B.A., 2014. In-situ sonochemical synthesis of Fe_3O_4-graphene nanocomposite for lithium rechargeable batteries. Chem. Eng. Process. 83, 49–55.

Edwin Suresh Raj, A.M., Mallika, C., Swaminathan, K., Sreedharan, O.M., Nagaraja, K.S., 2002. Zinc(II) oxide-zinc(II) molybdate composite humidity sensor. Sensor. Actuat. B-Chem. 81, 229–236.

Fu, G., Soboyejo, W.O., 2010. Swelling and diffusion characteristics of modified poly (N-isopropylacrylamide) hydrogels. Mater. Sci. Eng. C. 30 (1), 8–13.

Garcia, M., Vliet, G.V., Cate, M.G.J.T., Chávez, F., Norder, B., Kooi, B., et al., 2004. Large-scale extrusion processing and characterization of hybrid nylon-6/SiO_2 nanocomposites. Polym. Adv. Technol. 15, 164–172.

Geng, L., Zhao, Y., Huang, X., Wang, S., Zhang, S., Wu, S., 2007. Characterization and gas sensitivity study of polyaniline/SnO_2 hybrid material prepared by hydrothermal route. Sensor. Actuat. B—Chem. 120, 568–572.

Guilherme, M.R., Fajardo, A.R., Moia, T.A., Kunita, M.H., Gonçalves, M.C., Rubira, A.F., et al., 2010. Porous nanocomposite hydrogel of vinyled montmorillonite-crosslinked maltodextrin-*co*-dimethylacrylamide as a highly stable polymer carrier for controlled release systems. Eur. Polym. J. 46, 1465–1474.

Haraguchi, K., 2007. Nanocomposite hydrogels. Curr. Opin. Solid State Mater. Sci. 11 (3–4), 47–54.

He, W., Pan, C., Lu, T., 2001. Soapless emulsion polymerization of butyl methacrylate through microwave heating. J. Appl. Polym. Sci. 80, 2455–2459.

Huang, Z., Lin, Z., Cai, Z., Mai, K., 2004. Physical and mechanical properties of nano-CaCO3/PP composites modified with acrylic acid. Plast. Rubber Compos. 33, 343–346.

Kong, Y.X., Kan, C., Sun, C., 1999. Encapsulation of calcium carbonate by styrene polymerization. Polym. Adv. Technol. 10, 54–59.
Li, Z., Zhu, Y., 2003. Surface-modification of SiO_2 nanoparticles with oleic acid. Appl. Surf. Sci. 211, 315–320.
Liu, P., Liu, W.M., Xue, Q.J., 2004. In situ radical transfer addition polymerization of styrene from silica nanoparticles. Eur. Polym. J. 40, 267–271.
Liu, X., Guan, Y., Ma, Z., Liu, H., 2004. Surface modification and characterization of magnetic polymer nanospheres prepared by miniemulsion polymerization. Langmuir. 20, 10278–10282.
Lu, X., Mao, H., Chao, D., Zhang, W., Wei, Y., 2006. Ultrasonic synthesis of polyaniline nanotubes containing Fe_3O_4 nanoparticles. J. Solid State Chem. 179, 2609–2615.
Lu, Y., McLellan, J., Xia, Y., 2004. Synthesis and crystallization of hybrid spherical colloids composed of polystyrene cores and silica shells. Langmuir. 20, 3464–3470.
Monte, S.J., Sugerman, G., 1976. Organo-titanate coupling agents for filled polymer systems. Am. Chem. Soc. Div. Org. Coat. Plast. Chem. Prepr. 36, 207.
Nakai G., T. Fukuda, K. Hosoi, Surface-coated calcium carbonate particles, method for manufacturing same, and adhesive, U.S. Patent 6,686,044, 2004.
Patel, M.A., Bhanvase, B.A., Sonawane, S.H., 2013. Production of cerium zinc molybdate nano pigment by innovative ultrasound-assisted approach. Ultrason. Sonochem. 20, 906–913.
Pud, A., Ogurtsov, N., Korzhenko, A., Shapoval, G., 2003. Some aspects of preparation methods and properties of polyaniline blends and composites with organic polymers. Prog. Polym. Sci. 28, 1701–1753.
Qi, D.M., Bao, Y.Z., Huang, Z.M., Weng, Z.X., 2006. Synthesis and characterization of poly (butyl acrylate)/silica poly(butyl acrylate)/silica/poly(methyl methacrylate) composite particles. J. Appl. Polym. Sci. 99, 3425–3432.
Qiu, G., Wang, Q., Wang, C., Lau, W., Guo, Y., 2007. Polystyrene/Fe_3O_4 magnetic emulsion and nanocomposite prepared by ultrasonically initiated miniemulsion polymerization. Ultrason. Sonochem. 14, 55–61.
Ryu, J.G., Kim, H., Lee, J.W., 2004. Characteristics of polystyrene/polyethylene/clay nanocomposites prepared by ultrasound-assisted mixing process. Polym. Eng. Sci. 44, 1198–1204.
Sh, Y., Zhou, B., Wang, C., Zhao, X., Deng, Y., Wang, Z., 2006. In situ preparation of hydrophobic $CaCO_3$ in the presence of sodium oleate. Appl. Surf. Sci. 253, 1983–1987.
Shchukin, D.G., Zheludkevich, M.L., Yasakau, K.A., Lamaka, S.V., Ferreira, M.G.S., Möhwald, H., 2006. LbL Nanocontainers for self-healing corrosion protection. Adv. Mater. 18, 1672–1678.
Shirsath, S.R., Hage, A.P., Zhou, M., Sonawane, S.H., Ashokkumar, M., 2011. Ultrasound assisted preparation of nanoclay bentonite-FeCo nanocomposite hybrid hydrogel: a potential responsive sorbent for removal of organic pollutant from water. Desalination. 281, 429–437.
Shirsath, S.R., Patil, A.P., Bhanvase, B.A., Sonawane, S.H., 2015. Ultrasonically prepared poly(acrylamide)-kaolin composite hydrogel for removal of crystal violet dye from wastewater. J. Environ. Chem. Eng. 3, 1152–1162.
Siegel, R.W., 1994. Nanostructured materials-mind over matter. Nanostr. Mater. 4, 121–138.
Sonawane, S.H., Bhanvase, B.A., Jamali, A.A., Dubey, S.K., Kale, S.S., Pinjari, D.V., et al., 2012. Improved active anticorrosion coatings using layer-by-layer assembled ZnO nanocontainers with benzotriazole. Chem. Eng. J. 189–190, 464–472.

Tiarks, F., Landfester, K., Antonietti, M., 2001. Silica nanoparticles as surfactants and fillers for latexes made by miniemulsion polymerization. Langmuir. 17, 5775–5780.

Tissot, I., Novat, C., Lefebvre, F., Bourgeat-Lami, E., 2001. Hybrid latex particles coated with silica. Macromolecule. 34, 5737–5739.

Tyagi, M., Bhanvase, B.A., Pandharipande, S.L., 2014. Computational studies on release of corrosion inhibitor from layer-by-layer assembled silica nanocontainer. Ind. Eng. Chem. Res. 53, 9764–9771.

Wang, C., Sheng, Y., Zhao, X., Pan, Y., Bala, H., Wang, Z., 2006. Synthesis of hydrophobic $CaCO_3$ nanoparticles. Mater. Lett. 60, 854–857.

Wang, C., Sheng, Y., Bala, H., Zhao, X., Zhao, J., Ma, X., et al., 2007. A novel aqueousphase route to synthesize hydrophobic $CaCO_3$ particles in situ. Mater. Sci. Eng. C. 27, 42–45.

Wu, W., He, T., Chen, J., Zhang, X., Chen, Y., 2006. Study on in situ preparation of nano calcium carbonate/PMMA composite particles. Mater. Lett. 60, 2410–2415.

Xia, H., Wang, Q., 2002. Ultrasonic irradiation: a novel approach to prepare conductive polyaniline/nanocrystalline titanium oxide composites. Chem. Mater. 14, 2158–2165.

Xie, G., Zhang, H.P., Zhang, Q.Y., Li, T.H., 2003. Preparation of magnetic composite microspheres by miniemulsion polymerization and their characterization. Acta Polym. Sin. 1, 626–630.

Xu, Z.Z., Wang, C.C., Yang, W.L., Deng, Y.H., Fu, S.K., 2004. Encapsulation of nanosized magnetic iron oxide by polyacrylamide via inverse miniemulsion polymerization. J. Magn. Magn. Mater. 277, 136–143.

Yanase, N., Noguchi, H., Asakura, H., Suzuta, T., 1993. Preparation of magnetic latex particles by emulsion polymerization of styrene in the presence of a ferrofluid. J. Appl. Polym. Sci. 50, 765–776.

Zhang, S.W., Zhou, S.X., Weng, Y.M., Wu, L.M., 2005. Synthesis of SiO_2/polystyrene nanocomposite particles via miniemulsion polymerization. Langmuir. 21, 2124–2128.

Zhang, Y., Lee, S., Reddy, K.R., Gopalan, A., Lee, K., 2007. Synthesis and characterization of core–shell SiO_2 nanoparticles/poly(3-aminophenylboronic acid) composites. J. Appl. Polym. Sci. 104, 2743–2750.

Biomedical applications of hybrid polymer composite materials

12

Burhan Ates[1], *Suleyman Koytepe*[1], *Sevgi Balcioglu*[1], *Ahmet Ulu*[1] *and Canbolat Gurses*[2]
[1]Department of Chemistry, Inonu University, Malatya, Turkey, [2]Department of Molecular Biology and Genetics, Inonu University, Malatya, Turkey

Chapter Outline

12.1 Introduction 343
12.2 Tissue engineering applications 347
12.2.1 Scaffolds 347
12.2.2 Bone scaffolds 348
12.2.3 Biomimetic scaffolds 356
12.2.4 Injectable scaffolds 357
12.2.5 Wound-healing materials 361
12.2.6 Other scaffold applications 362
12.2.7 Implant materials 367
12.3 Drug delivery 373
12.3.1 Drug delivery with hybrid composites prepared from natural polymers 374
12.3.2 Drug delivery with hybrid composites prepared from synthetic polymer 378
12.4 Biosensor applications 384
12.4.1 Glucose biosensors 385
12.4.2 Urea biosensors 388
12.4.3 Hydrogen peroxide biosensors 391
12.4.4 Cholesterol biosensors 394
12.4.5 Acetylcholinesterase biosensors 395
12.5 Other biomedical applications 399
12.6 Conclusion 400
References 401

12.1 Introduction

Today, new products appeared with the development in material technology affect all technologies from health to industrial areas. Especially in the last 30 years, the increasingly important field of advanced materials is to make daily life easier and to extend the average human life span. The improvements in this area continuously

Hybrid Polymer Composite Materials: Applications. DOI: http://dx.doi.org/10.1016/B978-0-08-100785-3.00012-7

support the medical field to provide new generation of drug combinations, medical diagnosis, treatment facilities, and implants for biomedical field. Thus, works in this field continue developing the cutting-edge technology. The need for materials which show superior performance, improved efficiency and reliability, having superior physical, chemical, and mechanical properties, occupying minimum space, lighter, cheaper, and more suitable to the new functions increases especially in the field of medicine and biomedical fields. During this new material search process, multifunctional materials that combine many features have come to the forefront. For this purpose, numerous composite and hybrid material synthesis are carried out. In particular, organic–inorganic hybrid materials, which is a class of combining both organic and inorganic materials, are today's one of the most important major work areas (Sanchez et al., 2005).

In the preparation of hybrid materials, in order to reduce cost or improve the matrix characteristics the methods of covalent or a strong secondary binding interactions of a polymeric material to an inorganic structured compound are generally preferred. Hybrid material is a general term for coordinated polymers having high crystallinity, amorphous sol–gel components, and different materials that have organic or inorganic components interacting with each other or not with a wide range of applications (Hood et al., 2014). Commonly, a hybrid material is the material which two materials nest at the molecular level. Mostly, one of the components is a synthetic or natural organic ingredient. A more detailed classification could be distinguished by the interactions between inorganic and organic species. Expressed as the weak interactions between hybrid materials such as Van der Waals, hydrogen bonds, or weak electrostatic forces of attraction are formed as the 1st class. In the 2nd class hybrid materials, strong chemical interactions have been seen between components. These characteristics caused by the gradual changes in the chemical interaction force reveal the basic performance of the hybrid material. In addition, looking at the binding characteristics of structure helps the separation of different hybrid materials or to define the material. The functional group comprising an organic compound allows us to connect on an inorganic network structure. For example, trialkoxysilane group can act as a network modifying compound because the final product, inorganic network structure, has been modified only by the organic group. An example of such components is the structure shown in Fig. 12.1 (Koytepe et al., 2015). In this structure, the organic structure is provided by a polyimide chain, whereas the inorganic structure is silica nanospheres. Both structures are connected to each other by covalent bonds. In this way, a new compound type having high thermal stability and a network structure is revealed.

If there are no strong chemical interactions between the inorganic and organic component blocks, different mixtures are concerned. These are not strong interactions between inorganic clusters or particles and organic polymers. A material in this condition is either the inorganic component separated from the organic polymer network may be stuck in the polymer during polymerization or being together caused by the weak secondary interactions and functional groups in the organic polymer chains. If structures having an organic and inorganic network are formed without a strong chemical interactions in each other, which is called interpenetrating polymer networks. If a sol–gel material has a network in this case, it is

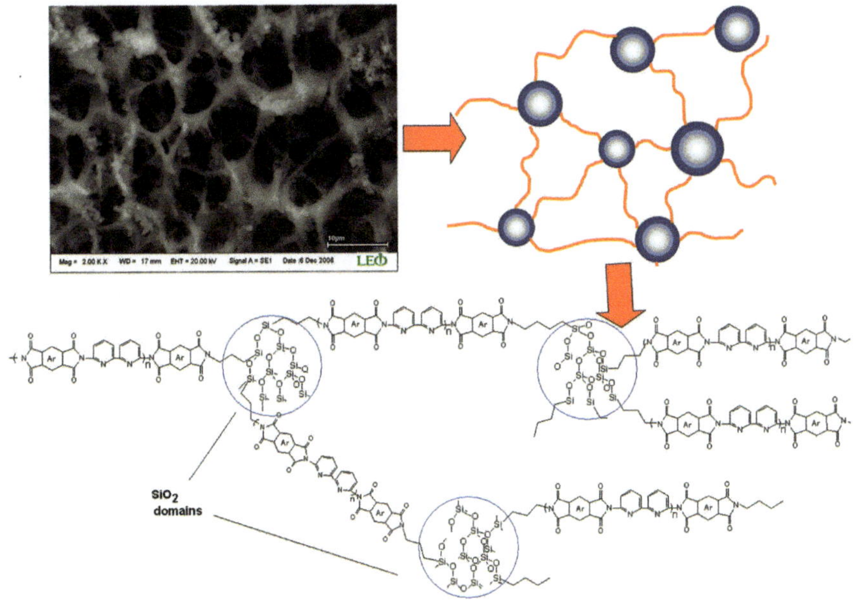

Figure 12.1 Silica-based hybrid polyimide material.
Reprinted with permission from Koytepe, S., Kucuk, I., Seckin, T., & Adiguzel, H.I. (2015). Preparation, characterization, and properties of novel polyimide–SiO2 Hybrid composites based on bipyridine for low dielectric applications. Polym. Plast. Technol. Eng., 54 (12), 1251–1262. Copyright © 2015, Taylor & Francis.

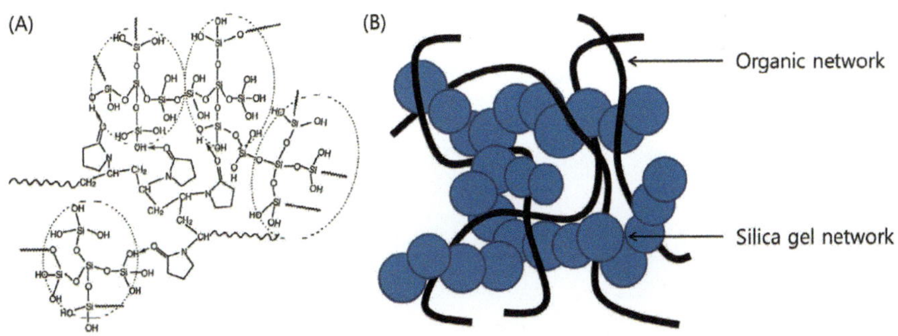

Figure 12.2 Schematic diagram of (A) molecular structure of PVP-silica hybrid and (B) hybrid with interpenetrating SiO_2 and organic gel networks structure (Owens et al., 2016).
Reprinted with permission from Owens, G.J., Singh, R.K., Foroutan, F., Alqaysi, M., Han, C. M., Mahapatra, C., et al. (2016). Sol–gel based materials for biomedical applications. Prog. Mater. Sci., 77, 1–79. Copyright © 2016, Elsevier.

composed due to the presence of organic polymer or the inorganic structure. As a result, both materials are indicated as 1st class hybrid materials. Polyvinylpyrrolidone (PVP)-silica hybrid structure in Fig. 12.2 is an example of this type of material. Second class hybrids are formed with covalent bonds between different inorganic building blocks or clusters and organic polymers or each other covalently bindings of organic and inorganic polymers (Owens et al., 2016).

In general, in biomedical applications of hybrid materials well known as biodegradable or natural polymers such as polyvinyl alcohol (PVA), poly(caprolactam), polyurethane (PU), poly(2-hydroxyethyl methacrylate) (pHEMA) gelatin, chitosan (Chi), polyethylene glycol (PEG), alginate, chitin, cyclodextrin, and starch are preferred as organic structures. Such materials are confined in different inorganic network structures in situ with sol−gel reactions. In addition, while creating from inorganic structures they are simultaneously synthesized. These structures having high polarity are attached to the free-surface hydroxyl groups on the inorganic structures with strong secondary interactions. Clay, C allotropes, $CaCO_3$, TiO_2, SiO_2, zinc oxide (ZnO), Al_2O_3, and other nanometal and metal oxides may be used as inorganic structure.

Polymer hybrid materials with their structurally multifunctional and versatile features are widely used in the fields of biomedical implants, drug delivery systems, sensors, bone, and tissue scaffolds applications (Fig. 12.3). For the construction of implant or medical prosthesis having high mechanical strength, the capabilities of resistant against infections itself and self-repair, and some of the specific properties of many materials are necessary to bring together. Similarly, for a smart drug delivery system or the synthesis of scaffolds that leave its location to natural tissue in time, a blend material structure is desired. Therefore, the designs of polymer hybrid materials are inevitable to achieve more effective properties in these areas. In this chapter, polymeric hybrid material structures and their use in the biomedical field are specifically detailed.

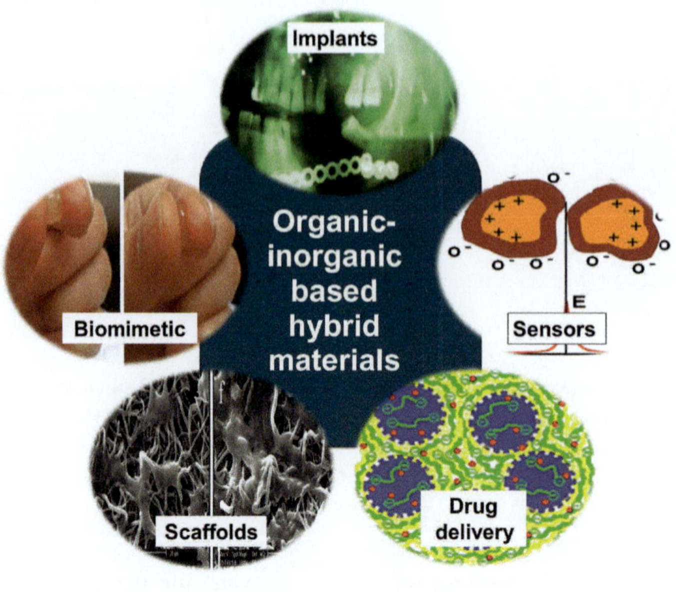

Figure 12.3 Biomedical usage areas for polymer hybrid composites.

12.2 Tissue engineering applications

Over the last 30 years, tissue engineering is one of the most researched biomedicine fields which scientists study. The main objective of tissue engineering is to create organs or tissues in laboratory conditions if tissue and/or organ damage or loss happens, and it consists of three main components. These are human body biocompatible scaffolds that can be integrated without any damage using various stem cells or primary cells that heal the damaged area efficiently and biosignal molecules that can stimulate cellular behaviors. With these studies, it is intended to provide imitating tissues in biological systems as the best way by utilizing of these components together or separately. Cells are surrounded by a three-dimensional (3D) support material also known as the extracellular matrix (ECM) in their natural environment. ECM is a more protein made of biological material being in each tissue and organs, and it is designed and produced by the cells depending on their needs (Geiger et al., 2001). ECM is formed by the 3D organization of proteins and polysaccharides such as collagen (Col), hyaluronic acid (HA), proteoglycans, glycosaminoglycans, and elastin, which supports cells physically and organizes the cell–cell interactions and gives a variety of biochemical and biophysical signals for cells adhesion, migration, proliferation, differentiation, and matrix accumulation (Streuli, 1999). Furthermore, it is also known that Col in ECM provides tissue elasticity (Aumailley and Gayraud, 1998). ECM structure is basically similar in each tissue but the ratio of the molecules that comprise the content and type varies among different tissues. Also, depending on the tissue function, molecules that are specific to any tissue are included in ECM structure. Thus, tissue scaffolds used in tissue engineering are designed to mimic the ECM, and they are made of metal, ceramic, natural or synthetic polymers, or the combination of several these materials.

As its natural texture is similar to a hybrid like structure, to use hybrid approaches in the preparation of ECM's mimicking is very important. In this sense, polymer hybrid composites have come to the fore.

12.2.1 Scaffolds

Tissue defects caused by injuries, diseases, or natural ways are the completely unsolved problems that people are trying to cope with them over the years. In contrast, for many years, scientists have been developing a variety of treatment methods to overcome these problems and this trend still has not been ended. One of the most important fundamentals of these therapies is scaffolds. Scaffolds are natural, synthetic, or semisynthetic materials that do not disrupt the natural flow in the system as if they act as natural parts of the applied area and provide mechanical force and architecture and also support cell proliferation. A lot of scaffold studies are done to be used especially in bone fractures, osteoporosis disease occurred by many reasons, unhealed wounds in diseases such as diabetes, cartilage defects by providing of the osteoinduction, connective tissue formation, and tissue remodeling.

According to the so far results produced in the tissue engineering studies, tissue scaffolds need to be hosting many features at the same time in accordance with the applied region.

- The most important parameters for the choice of tissue implant are that produced scaffold has to need two requirements as being completely biocompatible and not showing toxic effects. Thus, very unhealthy tissue has not already developed an immune response against the material that will be added (Levenberg and Langer, 2004).
- Another important parameter is the requirement of being adjustable biodegradability property of scaffold according to its applied area. Because of material intervening as proliferating of the cells after closing the damaged tissue, scaffold must degrade. As recovery time in all parts of the body are not the same, the degradability of material must be adjusted according to this case (Antoniou et al., 2000).
- Newly manufactured scaffold should stick firmly to the target area and ensure tissue regeneration. Thus, migration, proliferation, and differentiation of the target cell in the material will be provided (Griffith, 2002).
- Scaffold must have 3D networks and appropriate porosity. So, both cell growth in material and the transportation of oxygen, nutrients, and metabolic waste produced by cells in the region will be provided (Hennink and Nostrum, 2012; Zhu and Marchant, 2011; Karageorgiou and Kaplan, 2005).
- The mechanical properties of the material must be very similar to host tissue. During the patient's normal activities, the required elasticity and tensile/compressive strength should be provided (Lee and Mooney, 2001; DiMarco and Heilshorn, 2012).
- Finally, it must meet the requirements in terms of clinical use. It should be easily manufacturable and commercially producible, have long storage life, easily sterilizable, and must be infection resistant (Langer and Vacanti, 1993; Stevens and George, 2005; Kai et al., 2012; Yang et al., 2013; Gaharwar et al., 2011; Lau and Kiick, 2015).

Materials providing all of the properties mentioned above at the same time and at the desired level are difficult to produce. Therefore, materials such as pure polymers or fibrins or bioactive glasses (BGs) can meet only some of these requirements. At this point, hybrid structures were proposed for using the unique properties of each material and synthesizing materials having better properties in results. For example, inorganic structures like BGs promote a high level of cell growth and proliferation but do not provide the necessary conditions for the elasticity of the body because of the fragile structure. On the other hand, organic groups like polymeric structures are elastic and biocompatible, but they cannot promote 3D cell proliferation sufficiently. When these types of materials are hybridized, synthesized materials will become having the properties of both groups, and scaffold will completely meet the body needs with a new superior feature (Fattahi et al., 2014; Zhou and Lee, 2011).

12.2.2 Bone scaffolds

Bone is a dynamic structure with superior features such as healing and remodeling itself. Bone-tissue scaffolds are the easiest ECM mimicking, and therefore, they are the most preferred type of scaffolds. In the structure, hydroxyapatite (HAP) nanocrystalline inorganic molecules are available in the Col nanofiber organic matrix. Therefore, produced scaffolds are much preferred as the hybrid material for the naturally structured hybrid bone#tissue. Studies involving hybrid polymer composites made in recent years regarding to the bone scaffold are available below.

In a study by Chen et al. (2015), the different rates of polydimethylsiloxane–BG–poly(caprolactone) (PDMS–BG–PCL) hybrid scaffold materials were produced via sol–gel method. PDMS–BG was used as a strengthening for PCL in this study, and it has been observed that the biomineralization activity, the mechanical properties of the material, and the osteoblast cell compatibility were increased. Thanks to the bone-like apatite mineral phase BG is chemically linked when it contacts with living tissues and significantly helps to cell proliferation (Lei et al., 2012; Hench and Polak, 2002; Zhou et al., 2013). However, due to the fragility and lack of mechanical strength, it does not provide enough flexibility in body conditions. On the other hand, polymeric structures are biocompatible, but they cannot provide sufficient interaction with body fluids. Therefore, in recent years the use of these two main structures is more preferred as the hybrid material. In the study, by mixing with BG sol and PGL solution with PDMS, the hybridization reaction was performed as in Fig. 12.4. Si–O–Si groups in PDMS create strong interaction with BG sol and alkyl side chains allow the hybridization with

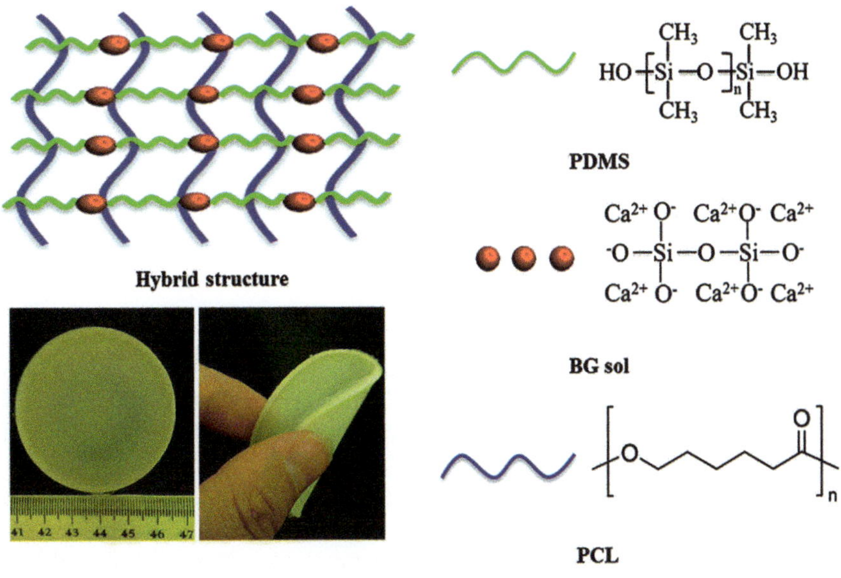

Figure 12.4 Chemical structure and optical images of crack-free PDMS–BG–PCL hybrid monoliths fabricated using representative sol–gel route.
Reprinted with permission from Chen, J., Du, Y., Que, W., Xinga, Y., & Lei, B. (2015). Content-dependent biomineralization activity and mechanical properties based on polydimethylsiloxane–bioactive glass–poly(caprolactone) hybrids monoliths for bone tissue regeneration. RSC Adv., 5, 61309–61317. Copyright © 2015, The Royal Society of Chemistry.

PCL. Thus, it was obtained a hybridized image without cracks. Then, the mechanical forces of materials doped with various ratios (0%, 30%, 40%, 50%, and 60%) of PDMS–BG were examined, and it was observed that when PDMS–BG contribution rate increases, tensile strength decreases from 26 to 8 MPa. On the other hand, it was also observed that the highest Young's modulus value among the materials is 329 MPa in the material containing of 30% PDMS–BG. Based on these results, the fragile structure of PDMS–BG has turned into a rigid structure with PCL. The study team also carried out a particularly important biomineralization test for biomaterials used in bone tissues. In the test, the produced polymer hybrid materials were incubated for 7 days in simulated body fluid, and then scanning electron microscope (SEM) images were taken. According to the images, it is seen that PCL alone did not provide mineralization and when the PDMS–BG ratio increased, needle-like apatite layers also increased. The latest study in the research is in-vitro osteoblast compatibility test. In this test, MC3T3-E1 osteoblast cell line was incubated 1, 3, and 5 days over the material and analyzed proliferation values. It was observed that an increase in PDMS–BG ratio increases the cell viability, and the relative intensity in the sample containing 60% PDMS–BG was 20 in 3 days of incubation. As a result, it was reported that the produced material might be used as scaffold thanks to both the mechanical force, and the compatibility demonstrated to bone tissue (Chen et al., 2015).

In the study of John et al. (2016), strontium (Sr)-doped organic–inorganic 3D hybrid porous bone scaffolds were produced by sol–gel process. Calcium, Sr, and phosphate ions were doped into the scaffolds made using triethoxyvinylsilane and 2-HEMA monomers (Fig. 12.5). After being characterized of the materials with Fourier transform infrared spectroscopy (FTIR), SEM, X-ray, scratch test, and mechanical tests, the efficacy of hybrids was investigated with biomineralization studies. Biomineralization is the formation of apatite-like natural structures while incubating the material in a biological fluid and to change the material's morphology. This new structure allows easier proliferation of bone tissue in material. In the study, in order to observe the natural mineral formations on the material and the surface crystallization of inorganic salts, static experimental procedure was followed and materials were incubated in Dulbecco's Modified Eagle Medium solution for specific periods, and the mineralization was observed with X-ray diffraction (XRD). As shown in Fig. 12.6, the material surface was covered with synthetic hydroxyapatite, and the mineralization was occurred. However, this situation did not close macro pores as shown in the figure, and thus the cells communication between the network's was not blocked. Mineralization, porosity, and roughness of the material surface are important for the proliferation and growth of bone tissue. For the porosity, the tissue mineralization was observed at the pores larger than 50 μm, at the pores larger than 100 μm favorable environment was provided for the rebuilding of hard tissues (Karageorgiou and Kaplan, 2005; Hatano et al., 1999). Pore sizes created during the research were around 150–350 μm that means that the suitable environment for cell growth was occurred. The cell viability of the material was measured by sulforhodamine B test over NHOst cells and according to the results, up to 96% viability of the cell was observed. Synthesized

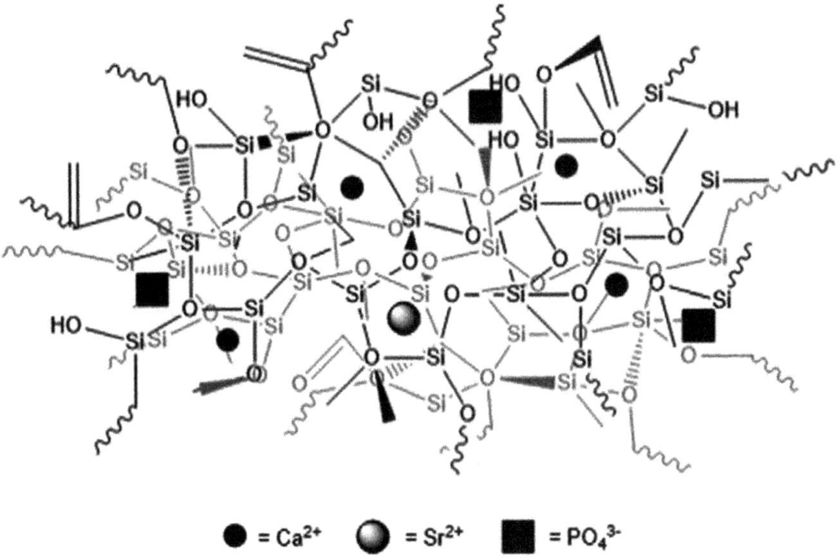

Figure 12.5 Ramified structure of the organic–inorganic hybrid network.
Reprinted with permission from John, Ł., Podgórska, M., Nedelec, J., Cwynar-Zając, Ł., & Dzięgiel, P. (2016). Strontium-doped organic-inorganic hybrids towards three-dimensional scaffolds for osteogenic cells. Mater. Sci. Eng., C, 68, 117–127. Copyright © 2016, Elsevier.

materials during the work can be used as a possible bone scaffold (John et al., 2016).

Col is a main component of ECM in natural bones. It plays a crucial role in bone regeneration. The biological activity of Col is supported as advantageously biomaterial for bone-tissue engineering. However, mechanical properties of these scaffolds alone are inadequate. More importantly, porous structure is not stable in wet states. For the effective solution of these problems, hybrid structures comprising biological originated and synthetic materials are needed for accomplishing the bioactivity and mechanical strength for the synthesis of bone. In the study of Long et al. (2015), Col fiber and bioglass-based 3D macroporous bone scaffold was produced via slurry-dipping technique. Then, its mechanical and biological properties were evaluated. The scaffold consisting of Col fiber and bioglass is linked each other with an approximately pore size of 40–200 μm. The water absorption of the hybrid scaffold consisting of Col fibers and bioglass compared to the scaffold made from pure Col decreased dramatically from 889% to 52%. This leads to significantly prevention in the swelling of Col and enhanced the stability of scaffold. The synthesized scaffold has the compression strength of 5.8 ± 1.6 MPa and an elastic modulus of 0.35 ± 0.01 GPa, which are the similar values with trabecular bones. Moreover, in-vitro cell experiments show that the scaffold presents good biocompatibility for the spread and proliferation of human bone marrow stromal cells. Therefore, Long et al. (2015) think that the scaffold comprising bioglass and Col fibers has a great potential for bone-tissue engineering applications.

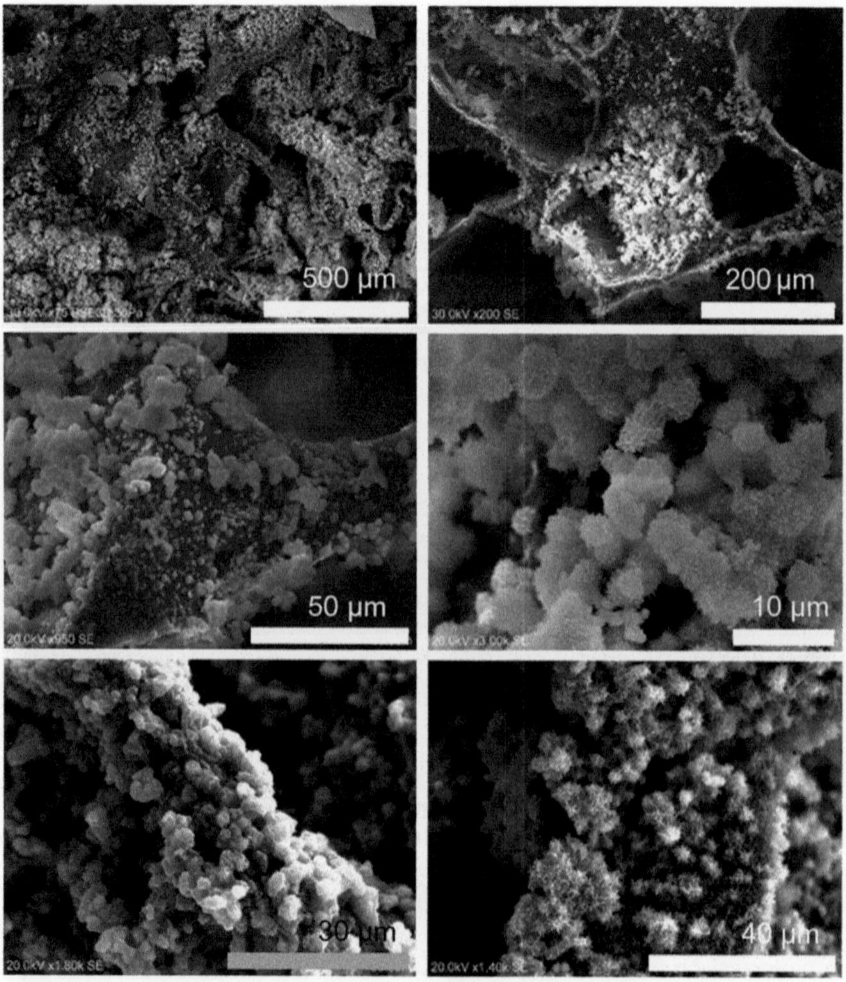

Figure 12.6 Scaffolds "decorated" by synthetic hydroxyapatite.
Reprinted with the permission from John, Ł., Podgórska, M., Nedelec, J., Cwynar-Zając, Ł., & Dzięgiel, P. (2016). Strontium-doped organic-inorganic hybrids towards three-dimensional scaffolds for osteogenic cells. *Mater. Sci. Eng., C*, 68, 117–127. Copyright © 2016, Elsevier.

Polylactic acid (PLA) has been extremely investigated in especially biomedical engineering applications due to its superior mechanical strength and in-vivo biocompatibility. However, the inherent fragility, slow solubility, and extreme hydrophilicity greatly impede its successful implementations. In the study of Shi et al. (2015), biodegredable, crosslinked elastomer polyglycerol sebacate was used to modify PLA scaffolds for bone-tissue engineering. A 3D, large porous PLA-based scaffold was prepared. In the study, large porous, 3D and a highly interconnected PLA-based scaffold was produced with a NaCl particulate-leaching method. The poly(glycerol sebacate) prepolymer (pre-PGS) was prepared either by premolding binary blend or by surface coating of a homogeneous PGS onto PLA-based scaffolds (Fig. 12.7). The resulted PLA/PGS scaffolds showed well-interconnected open-cell structures after curing at 130°C. The addition of PGS to PLA both by binary blend and surface coating developed the hydrophilicity, degradation, toughness, and ductility. The best efficacy was observed in the surface coating with the oxygen plasma pretreatment (Fig. 12.8). At the ratios of PLA/PGS 9:1 and 7:3, the fracture strain of the PLA/PGS scaffolds improved from 8% (pure PLA) to 13% and 24%, respectively. Further studies demonstrated that increased hydrophilicity and surface roughness were the main parameters for the positive effect of oxygen-based plasma treatment. Moreover, these hybrid PLA/PGS scaffolds presented good mineralization, high cell biocompatibility, enhanced cell adhesion, and osteogenic differentiation for bone mesenchymal stem cells. The research related to the surface coating of PGS with oxygen-based plasma pretreatment as an effective strategy in order to modulate the properties of PLA and the hybrid PLA/PGS scaffold was an excited candidate for the applications in bone-tissue regeneration (Shi et al., 2015).

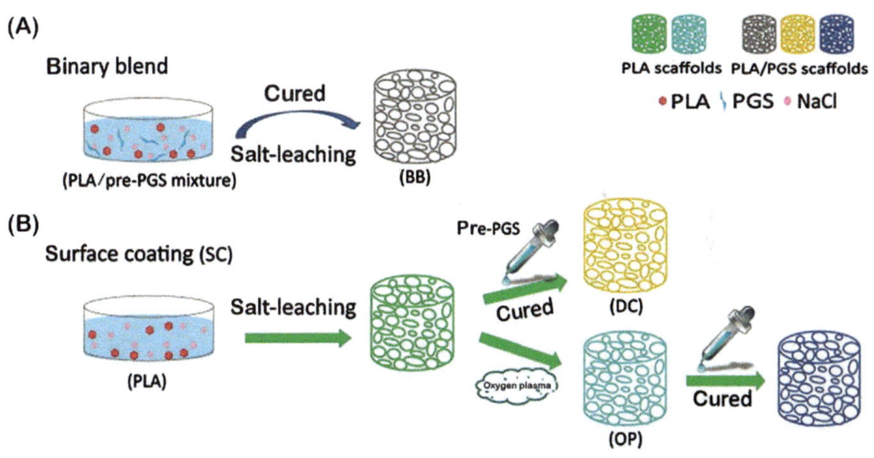

Figure 12.7 Schematic diagram of three different approaches for PLA/PGS scaffolds. (A) Premolding binary blend (BB): PLA and pre-PGS were mixed directly and PLA/PGS porous scaffolds prepared by a NaCl particulate-leaching technique. (B) Surface coating (SC): PLA-based porous scaffolds were prepared first, and then coated with specific proportion of PGS directly (DC) and PLA-based porous scaffolds surface treated with oxygen plasma, finally coated with PGS (OP). All resulting PLA/pre-PGS scaffolds were cured at 130°C in a vacuum oven for 48 h.
Reprinted with the permission from Shi, H., Gan, Q., Liu, X., Ma, Y., Hu, J., Yuan, Y., et al. (2015). Poly(glycerol sebacate)-modified polylactic acid scaffolds with improved hydrophilicity, mechanical strength and bioactivity for bone tissue regeneration. RSC Adv.,5, 79703–79714. Copyright © 2015, The Royal Society of Chemistry.

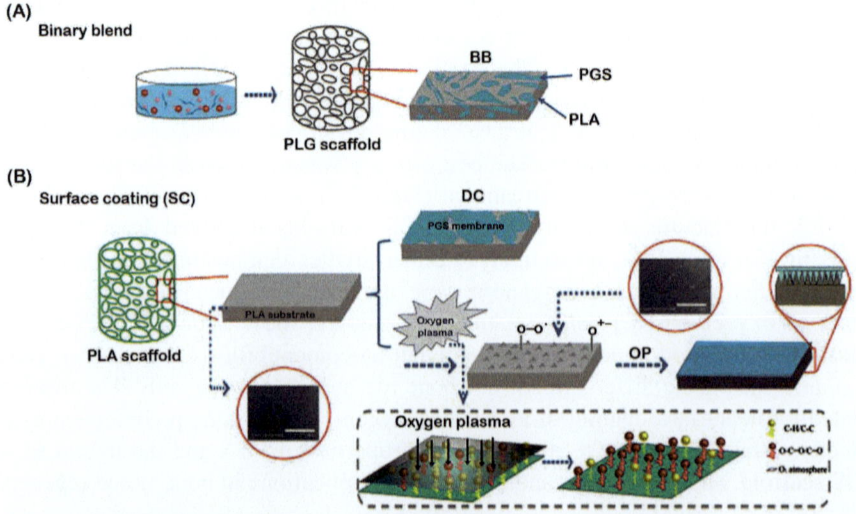

Figure 12.8 Schematic illustration of PLG scaffolds fabricated by BB (A) and SC (B). Strengthening mechanism of oxygen-based plasma treatment modify the microtopography of PLA substrate and allow for a homogeneous coating of PGS tightly covered on the PLA surface (scale bar: 500 mm).
Reprinted with the permission from Shi, H., Gan, Q., Liu, X., Ma, Y., Hu, J., Yuan, Y., et al. (2015). Poly(glycerol sebacate)-modified polylactic acid scaffolds with improved hydrophilicity, mechanical strength and bioactivity for bone tissue regeneration. *RSC Adv.*,5, 79703–79714. Copyright © 2015, The Royal Society of Chemistry.

For fabricating micro and nanofiber matrices, electrospinning method has been using for many years because it is simple and effective. As electrospun fiber matrices have numerous advantages to be used as tissue engineering scaffolds such as high surface area-to-volume ratio, mass production capability, and structural similarity to the natural ECM, electrospun matrices comprising biocompatible polymers and various biomaterials have been developed as biomimetic scaffolds for the tissue engineering applications. Specifically, graphene oxide (GO) has recently been considered as a new biomaterial for skeletal muscle regeneration due to the fact that it can promote the growth and differentiation of myoblasts. In the study of Shin et al. (2015), the hybrid fiber matrices which induce myoblasts differentiation for skeletal muscle regeneration were prepared. Hybrid fiber matrices consisting of poly(lactic-*co*-glycolic acid, PLGA) and Col with GO (GO–PLGA–Col) were produced via an electrospinning process. The results showed that the GO–PLGA–Col hybrid matrices having randomly oriented continuous fibers with a 3D porous structure dispersed uniformly throughout the GO–PLGA–Col matrices. Furthermore, the hydrophilicity of produced matrices was significantly increased by mixing with a small amount of Col and GO. The attachment and proliferation of the C2C12 skeletal myoblast cells increased on the GO–PLGA–Col hybrid matrices. Moreover, GO–PLGA–Col matrices induced the myogenic differentiation of C2C12 skeletal myoblasts. Lastly, based on the results of the study, the researchers concluded that this new GO–PLGA–Col hybrid fiber matrices can be used as potential biomimetic scaffolds for skeletal tissue engineering and regeneration due to the induction of spontaneous myogenesis with exhibiting superior bioactivity and biocompatibility (Shin et al., 2015).

Three-dimensional human made artificial scaffolds that increase the cell adhesion, and proliferation have gained much interest especially in bone-tissue engineering. As preparing of 3D scaffolds is so complex related to its micro-and nanofeatures, a couple of materials that need mechanical and biological characterizations have been produced in recent years. In the study of Terzaki et al. (2013), three materials with various chemical compositions were researched for bone-tissue regeneration. A hybrid material consisting of methacryloxypropyl trimethoxysilane and zirconium propoxide, a hybrid organic–inorganic material of the above containing 50% 2-(dimethylamino)ethyl methacrylate (DMAEMA) and a pure organic material based on poly(DMAEMA) were synthesized, and their mechanical properties as well as the biological response over preosteoblastic cells were also characterized. By using 3D scaffold technology and direct femtosecond laser writing (DLW), the complex structure was fabricated. The results demonstrated that all three materials could be used in tissue engineering, and more specifically 50% DMAEMA composite presented the best mechanical properties for the fabrication of 3D structure with DLW and strong biological response among other materials (Terzaki et al., 2013).

Numerous materials have been proposed for bone-tissue engineering day-by-day. In the study of Yang et al. (2014), a novel designed polymeric hybrid composite scaffold comprising PLGA and nanonized pearl powder, which also means a naturally bioceramic hybrid material, was produced. The biological activities and physical properties of the scaffold were assessed for bone-tissue engineering. Natural pearl is a composite consisting of calcium carbonate crystal in an aragonite structure with an organic matrix. Due to a promising osteoinductive and biocompatible material, nanosized pearl powder was used in the experiment. In the study, the prepared biohybrid of nanonized pearl powder/PLGA biocomposite scaffolds were used as seeding MC3T3-E1 cells. Moreover, the seeded tissues were characterized by using SEM and biochemical assays. Finally, the study showed that the efficiency of nanosized pearl powders in bone cell differentiation is so different from proteins. The future studies in the same research group will be focused on the new generation 3D bone substitute scaffolds (Yang et al., 2014).

In the another study of Dorati et al. (2013), the aim was to develop multiphase composite 3D scaffolds and to examine in-vitro degradation performance, cell seeding capacity, and cells proliferation of the new composite scaffolds. For multiphase composite scaffolds, via solvent casting and particulate-leaching techniques, hydroxyapatite was used to develop the mechanical properties of PLGA polymer scaffold. On the other hand, Chi was contributed to the scaffold structure for its bioactivity, osteoconductivity, and bioadhesive properties. The in-vitro degradation results showed that the composite scaffold had a similar degradation rate with their composition and structural features. Fibroblast cells were seeded on the surface of the composite scaffolds. Then, cell viability and long-term proliferation on composite scaffolds were evaluated. Consequently, the results exhibited that viable cells attached on the surface of scaffolds migrated into the porous scaffolds. These results indicated that the two-phase scaffold, namely PLGA/HAP and PLGA/Chi composite scaffolds, is a promising graft for bone-tissue engineering. The further studies focus on designing and producing an improved three-phase composite structure consisting of PLGA, HAP, and Chi in a special type of scaffold (Dorati et al., 2013).

The bioactivity of sol–gel synthesized poly(e-caprolactone) (PCL)/TiO_2 or poly(e-caprolactone)/ZrO_2 particles has already known. In order to make new two--dimensional composite substrates for hard-tissue engineering, by using the possibility to embed PCL/TiO_2 or PCL/ZrO_2 hybrid fillers into a PCL matrix, morphologically controlled scaffolds consisting of PCL reinforced with PCL/TiO_2 or PCL/ZrO_2 hybrid fillers were synthesized in the study of Russo et al. (2013). The results from the small punch test and the Young's modulus of the materials with mechanical and biological tests were obtained. Thus, the performance of prototyped scaffolds was tested to understand the effects of the inclusion of the hybrid fillers in 3D porous structures. The inclusion of the hybrid fillers increased the compressive modulus about 90 MPa and also the cell viability/proliferation (Russo et al., 2013).

Chi, hydroxyapatite (HAP), and magnetite (Fe_3O_4) have been extensively researched for bone treatment applications. In the research of Heidari et al. (2016), pure Chi, Chi/HAP, Chi/HAP/magnetite, and Chi/magnetite were produced. The mechanical properties of all were evaluated by the measurements of bending strength, elastic modulus, compressive strength, and hardness values. Furthermore, the morphology of the bending fracture surfaces were characterized using a SEM and an image analyzer. In the studies, they also examined the biological response of the human mesenchymal stem cells on different composites. The research team found that although all of these composites show in vitro biocompatibility, adding hydroxyapatite and magnetite into the Chi matrix significantly improves the mechanical properties of the pure Chi (Heidari et al., 2016).

12.2.3 Biomimetic scaffolds

Scaffolds are made by imitating ECM in order to ensure compliance of tissue and balance the mechanical strength and cell proliferation in body conditions. Besides, the approaches which have been exactly imitated the structure of bone in terms of micro/nano architecture are available. The principle of biomimetic architecture in creating scaffold includes such approaches. In a very specific example for this, Jiang et al. (2013a) prepared Chi/cellulose/nanohydroxyapatite (Chi/CMC/n-HAP)-based biomimetic anisotropic spiral-cylindrical scaffold membranes and tested them in the bone-tissue regeneration. In the study, samples containing 60% n-HAP were used in scaffold fabrication. The microstructure of natural bone is shown in Fig. 12.9. In order to mimic this system Chi/CMC/n-HAP hybrid membrane having 300-μm thickness was made and for increasing the integration of bone cells holes having 300-μm diameter were mechanically opened on the surface. The distance

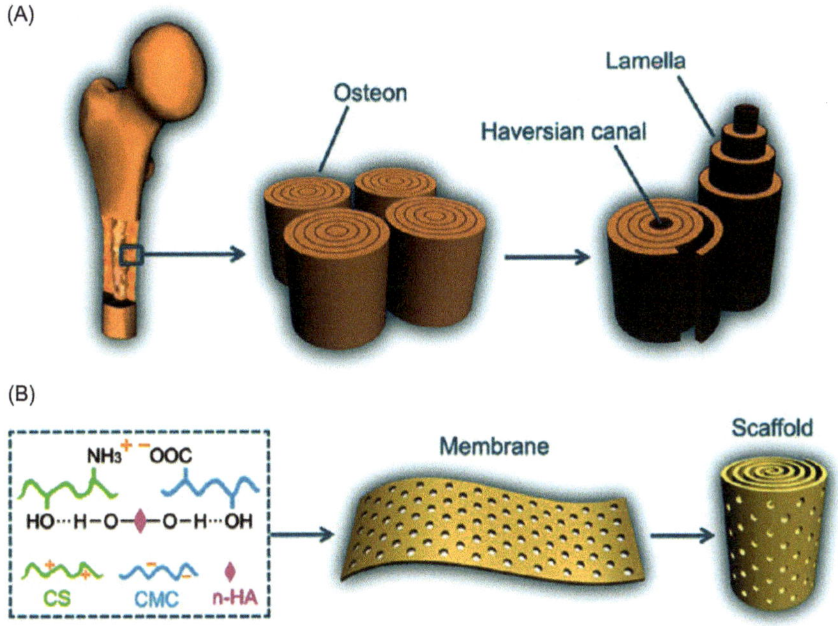

Figure 12.9 (A) Microstructure of natural bone and (B) fabrication process of biomimetic spiral-cylindrical scaffold based on the hybrid Chi/CMC/n-HAP membrane.
Reprinted with the permission from Jiang, H., Zuo, Y., Zou, Q., Wang, H., Du, J., Li, Y., et al. (2013a). Biomimetic spiral-cylindrical scaffold based on hybrid chitosan/cellulose/ nano-hydroxyapatite membrane for bone regeneration. ACS Appl. Mater. Interfaces, 5(22), 12036–12044. Copyright © 2013, American Chemical Society.

between these holes was set to 1 mm. Then, by tightly wrapping of the porous membrane spiral-cylindrical structure formation was achieved. In rabbit model applied in vivo osteogenesis study, by creating concave defects at the left radius as 10 mm in length and 3 mm in depth of animals the same sized spiral-cylindrical membranes were placed in this area as in Fig. 12.10. In-vivo biodegradability of the materials was measured to be about 16% in the material containing 60% n-HAP after 12 weeks. The compressive strength was measured as 4.91 ± 0.54 MPa. This corresponds to a comparable strength with cancellous bone (1–12 MPa). According to in-vivo osteogenetic experiments, it was found to be highly complementary of the material in bone tissue after 12 weeks (Jiang et al., 2013a).

12.2.4 Injectable scaffolds

Unlike other approaches, injectable scaffold technology is the direct injection of the gel to be applied into the shape of the tissue in terms of adapting and gelling the structures in there. This mannered hydrogels can be applied in many different areas from the hard tissues to the soft tissues.

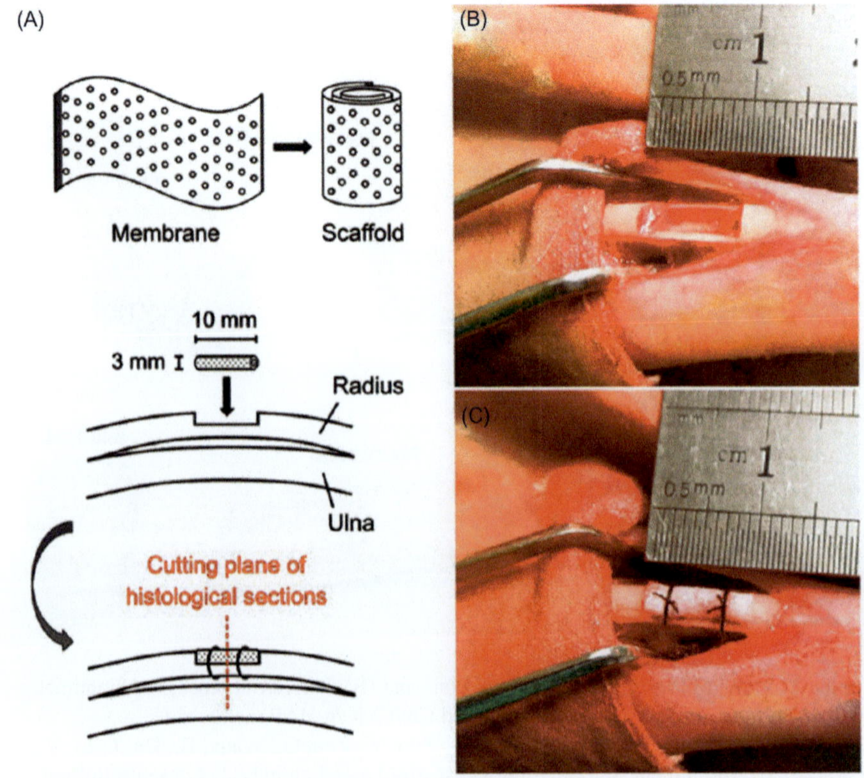

Figure 12.10 (A) Schematic diagram of surgical procedures, (B) concave defect of rabbit radius, and (C) implantation and fixation of the scaffold.
Reprinted with the permission from Jiang, H., Zuo, Y., Zou, Q., Wang, H., Du, J., Li, Y., et al. (2013a). Biomimetic spiral-cylindrical scaffold based on hybrid chitosan/cellulose/nano-hydroxyapatite membrane for bone regeneration. *ACS Appl. Mater. Interfaces*, 5(22), 12036–12044. Copyright © 2013, American Chemical Society.

Su et al. (2016) produced the hydrogels having injectable, promoting cell growth, and slow degradability characteristics in order to treat bone damage. With regenerated silk fibroin (RSF) and laponite (LAP), a clay type, nanoplate they designed a hybrid hydrogel scaffold for bone tissue. The main purpose in their study was to prepare hybrid composite scaffolds using biocompatible, biodegradable, and natural fibrin material for the relief of bone damage properties of RSF and increasing the osteoinduction property of LAP (like other clay structures) in bone tissue. As RSF does not contain integrin and growth factors that have an important role in osteoinduction, it does not act a scaffold role alone. At this point, LAP molecule increases the osseointegration by increasing the hydrophobic interactions between RSF molecules and accomplishing the release of having its own ions such as Mg^{2+} without the need for extra growth factors (Hoppe et al., 2011; Gaharwar et al., 2013). In this way, the insufficient points of two different

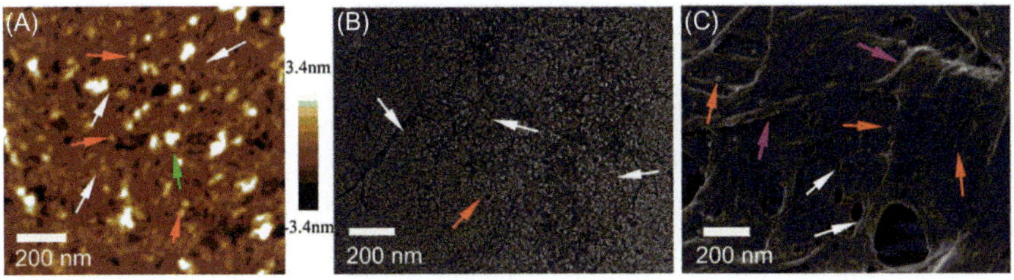

Figure 12.11 (A) AFM height image of the dilute RSF/LAP blends. (B) RSF nanofibrils in RSF/LAP blend observed by TEM. (C) FESEM image of RSF/LAP blends on the glass substrate.
Reprinted with the permission from Su, D., Jiang, L., Chen, X., Dong, J., & Shao, Z. (2016). Enhancing the gelation and bioactivity of injectable silk fibroin hydrogel with laponite nanoplatelets. ACS Appl. Mater. Interfaces, 8, 9619 − 9628. Copyright © 2016, American Chemical Society.

molecules would be eliminated. In the study, RSF and LAP structures were formed by homogeneous mixing of the two molecules together in an ultrasonic bath, and sol−gel formation was observed qualitatively. After prepared hybrid material was characterized by transmission electron microscopy (TEM), atomic-force microscopy (AFM), and field emission SEM (FESEM), LAP nanoplates dispersed in an orderly manner in the structure shown as the red arrows in Fig. 12.11A, but it was still seemed to be aggregation tendency in some places (the green arrow). In order to test the viability and proliferation of osteoblastic cells, cells were placed into the different RSF/LAP hydrogels and incubated for 7 days. As clearly seen in Fig. 12.11B, the 3D reproduction of cells in the RSF gel alone is less compared to the RSF/1% LAP vs RSF/5% LAP samples. In the same manner of SEM images, hybrid gels enabled more proliferation compared to the pure RSF (Fig. 12.11C). From these data, we can say that the LAP molecule increases the adhesion and proliferation of the cells. Consequently, RSF/LAP hydrogels have a potential to be used in the treatment of irregular bone damages (Su et al., 2016).

Neves et al. (2016) designed injectable scaffolds from alginate and Sr-rich ceramic microspheres. The reason of using Sr in scaffolds is that it is known to have an important function as bone remodeling in the treatment of osteopenic disorders and osteoporosis (Marie et al., 2011; Grynpas et al., 1996; Bonnelye et al., 2008; Yang et al., 2011a; Baron and Tsouderos, 2002; Peng et al., 2011). In the study, hydroxyapatite microspheres having a 555-μm diameter were prepared and were added in 3.5% sodium alginate solution in situ, and then the gelation was provided with calcium/Sr carbonate and glucone-δ-lactone agents. Mechanical strength tests and injectability tests were done to the scaffolds prepared in the study, and the optimum strength for injectable sample between these tests was found to be the system comprising 35% microsphere. The gaps between the microspheres were calculated as 220 μm. Also, at 37°C the gelation time was 7 min for Sr-based gelling, and 25 min for Ca-based gelling. In the study, in terms of founding the data scaffolds having suitable porosity and biocompatibility were produced for the bone-tissue regeneration and cell migration (Neves et al., 2016).

Nowadays, stimuli-responsive nanocomposite-derived hydrogels have gained much interest in tissue engineering as they can be applied specifically as injectable scaffolds in especially bone and cartilage repair. In the study of Moreira et al. (2016), they tried to synthesize and characterize novel thermosensitive Chi-based composites. These Chi-based composites were also chemically modified with Col and reinforced with BG nanoparticles for the improvement of injectable nanohybrids in regenerative medicine applications. Then, the structural, morphological, rheological, and biological properties of the novel prepared composite hydrogels were characterized with SEM, FTIR, and 3-(4,5-Dimethylthiazol-2-yl)-2,5-Diphenyltetrazolium Bromide(MTT), LIVE/DEAD assays. The composites showed thermosensitive response with the gelation temperature approximately at 37°C, which is proper with the human body temperature. Furthermore, from SEM analysis 3D porous structures were seen in the Chi hydrogels. Col in the system increased the average pore size. Moreover, rheological measurements were carried out to get information about the viscoelastic behavior of the hydrogels as a function of the temperature. The results exhibited that after the gelation process, the addition of Col and BG increased the mechanical properties. The addition of 2 wt% of BG nanoparticles caused an increase of approximately 39% for the stiffness value compared to pure Chi. Moreover, the addition of 30 wt% Col also increased the stiffness by 95%. The results of MTT and LIVE/DEAD assays showed no toxic effect of the composites on the human osteosarcoma cell culture and kidney cell line of human embryo (HEK 293T) (Fig. 12.12). Thus, the authors said that the novel produced

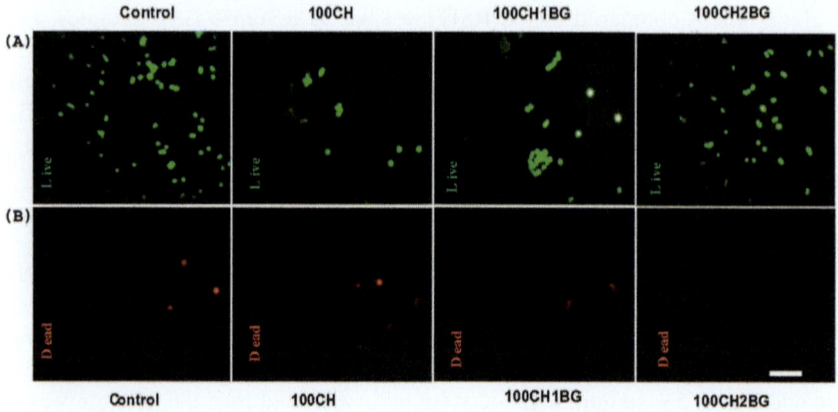

Figure 12.12 Live/dead assay HEK cells after 24 h of contact with the gel. (A) Live cells (green). (B) Dead cells (red) (bar = 200 μm, 10 ×).
Reprinted with the permission from Moreira, C.D.F., Carvalho, S.M., Mansur, H.S., & Pereira, M.M. (2016). Thermogelling chitosan–collagen–bioactive glass nanoparticle hybrids as potential injectable systems for tissue engineering. Mater. Sci. Eng.: C, 58, 1207–1216. Copyright © 2016, Elsevier.

composites might be successfully used as having promised potential of the next generation thermoresponsive biomaterials in bone-tissue bioapplications (Moreira et al., 2016).

12.2.5 Wound-healing materials

Wound dressing materials are one of the main areas of tissue engineering applications. They are used in burns, wounds which are difficult to heal such as diabetes and even the inner organ defects as transplantation agents. They are indispensable for tissue regeneration due to containing growth factors or cells and micro/nanoarchitectures.

Shahverdi et al. (2014) made PLGA/SF-based hybrid scaffolds with electrospinning method and investigate the effects of that hybrid scaffold as a wound dressing material. After performing the characterization of synthesized PLGA and PLGA/SF materials in the study, the mechanical properties of the materials were examined. According to the results, pure SF exhibited stiff structure, whereas the contribution of PLGA made the material more flexible and significantly increased the tensile and elongation values [elongation; SF:11.08 and PLGA/SF(2:1):31.66]. In another test, the hydrophilic character, which is important in terms of cell adhesion to the material, was measured by the liquid contact angle test and it was reported that 110° angle of PLGA decreased up to 88° when there was 2:1 ratio of PLGA/SF. In vitro cell culture and cell proliferation tests were conducted with L-929 fibroblast cells and it was reported that in MTT assay hybrid material was statistically higher than the control group and pure PLGA. Moreover, in order to prove that the interaction occurred between cells adhered on material, they incubated cells with scaffold for 24, 48, and 72 h and SEM images were taken (Fig. 12.13). As clearly seen in the figure, PLGA/SF material provided higher adhesion and proliferation for each incubation period compared to pure PLGA. Finally, the wound healing process induced by in vivo tests are also in agreement with these images. In the light of these studies, the produced material is considered to create a scaffold specifically for diabetic wounds (Shahverdi et al., 2014).

In the study of Maharjan et al. (2017), PU-zein fibrin-based wound dressing scaffold was prepared using the Ag-metal in situ with electrospinning. Ag metal in here is known with its antimicrobial character for many years and used in biomaterials (Maillard and Hartemann, 2013). For this purpose, Ag-polyurethane-zein (AgNP) scaffolds were prepared and characterized by FTIR, FESEM, TGA, and XRD. Furthermore, antimicrobial properties were examined against g + *Staphylococcus aureus* and g − *Escherichia coli* bacteria. As seen in Fig. 12.14, pure PU and PU-zein samples show no antibacterial property, whereas the inhibition zone is significantly observed for Ag-doped samples. Then, the biocompatibility of samples over fibroblast cells was analyzed, and it was seen that it enhanced the proliferation of the cells. In light of the work, it was proposed that produced materials could be used as wound dressing scaffolds (Maharjan et al., 2017).

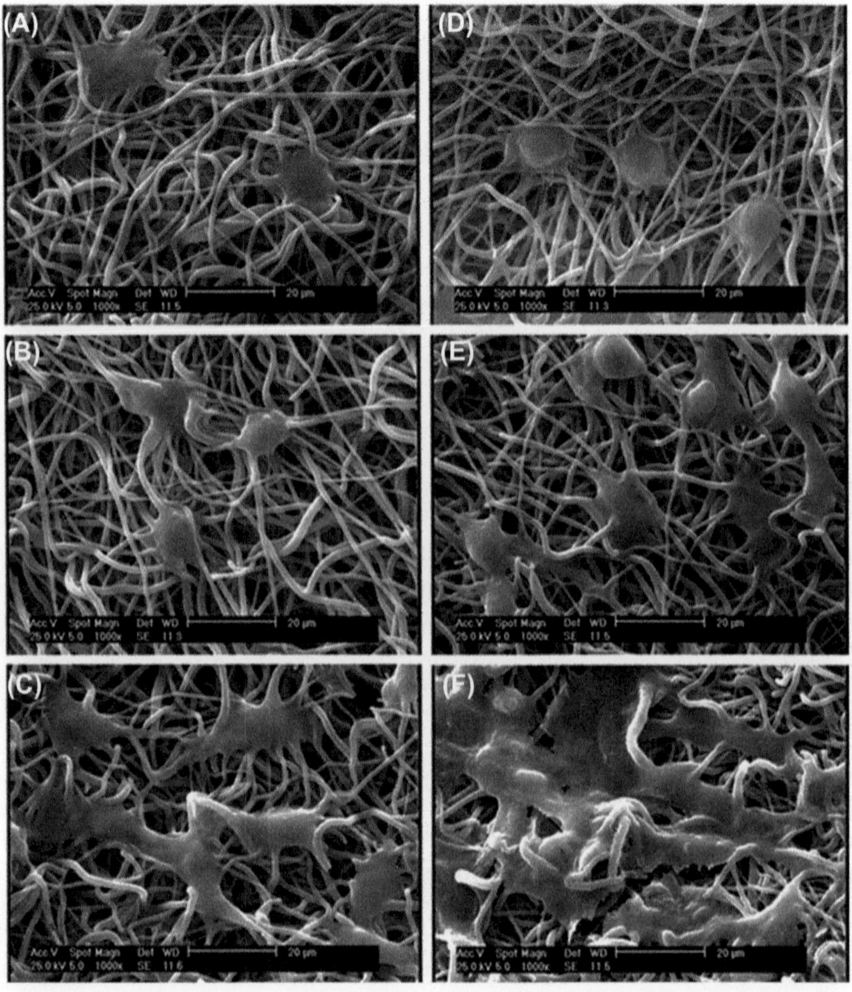

Figure 12.13 Morphological evaluation of L-929 on PLGA (A–C) or PLGA/SF (2:1) (D–F) membranes by SEM. Cells were cultured for 24 h (A and D), 48 h (B and E) and 72 h (C and F).
Reprinted with the permission from Shahverdi, S., Hajimiri, M., Esfandiari, M.A., Larijani, B., Atyabi, F., Rajabiani, A., et al. (2014). Fabrication and structure analysis of poly(lactide-*co*-glycolic acid)/silk fibroin hybrid scaffold for wound dressing applications. Int. J. Pharm., 473, 345–355. Copyright © 2014, Elsevier.

12.2.6 Other scaffold applications

For the tissue engineering scaffold approach, systems produced from many different tissues are available. Besides much studied tissues, less common but clinically significant approaches are also available. As the first one of these, in the research of

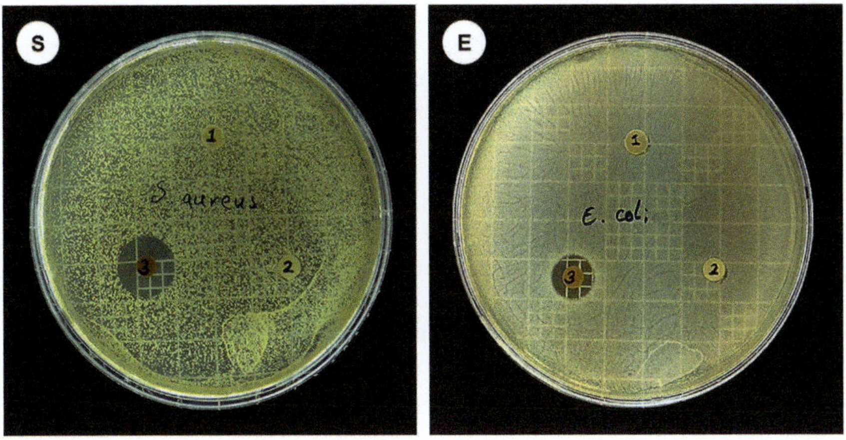

Figure 12.14 (S) and (E) represent the antibacterial activity of different scaffolds against *S. aureus* and *E. coli*, respectively. PU, PU-Zein, Ag/PU-Zein samples are represented as 1, 2, 3 in each plate, respectively.
Reprinted with the permission from Maharjan, B., Joshi, M.K., Tiwari, A.P., Park, C.H., & Kim, C.S. (2017). In-situ synthesis of AgNPs in the natural/synthetic hybrid nanofibrous scaffolds: Fabrication, characterization and antimicrobial activities. J. Mech. Behav. Biomed. Mater., 65, 66–76. Copyright © 2017, Elsevier.

Mauretti et al. (2016) for being used in the renewal of the heart muscle the mechanical and biological properties of PEG-fibrinogen hydrogels were improved. Then, the proliferation and differentiation of human Sca-1 cardiac progenitor cells were supported. PEG-fibrinogen hydrogels were modified with embedded air or perfluorohexane fulfilled bovine serum albumin (BSA) microbubbles and characterized. The changes in cell morphology were observed in the microbubbles with PEG-fibrinogen hydrogel. It was found that microbubbles increase the formation of cell and affect the growth through axis direction. As active molecular carriers, the properties of microbubbles also increased too much. For the first time, microbubbles coated with an enzyme were used as a system for the manufacture of hydrogen sulfide releasing scaffolds. The new produced PEG-fibrinogen hydrogel as releasing hydrogen sulfide improved the growth of cardiac progenitor cells. This new 3D cell-scaffold system will be used to evaluate the effects of hydrogen sulfide as a tissue repair application for the heart muscle regeneration (Mauretti et al., 2016).

Feng et al. (2015) examined the effect of nonhomogeneous dispersion of gelatin/polycaprolactone hybrid fibrous scaffolds in cell proliferation for the treatment of skin tissue. The different compositional and morphological homogeneity of two different groups of fibrous scaffolds prepared from the gelatin/polycaprolactone mixture as 50:50 wt% dissolved in trifluoroethanol or acetic acid doped trifluoroethanol solvent system via electrospining technique were investigated by SEM vs FTIR. Then, cell adhesion, morphology, and proliferation using green fluorescent protein labeled mouse fibroblasts and human keratinocyte cell line cells as model cells were evaluated via laser scanning confocal microscope, SEM and cell counting kit-8 test. Morphological and compositional uniformity of the scaffold in the first group significantly affected the cell adhesion and proliferation. In contrast, in the second group a significant difference was not observed because of good fiber morphology and composition. The phase separation originated from the differences in gelatin content of the scaffolds in group 1 was shown as one of the main reasons. This study focuses on the importance of the production of morphologically uniform and compositionally homogeneous composite nanofibers produced from natural and synthetic polymer blends using electrospining technique (Feng et al., 2015).

In the another study of Liao et al. (2015), hybrid scaffolds were produced with biodegradable methacrylated chondroitin sulfate (CSMA), poly(ethylene glycol) methyl ether-ε-caprolactone-acryloyl chloride (PECA), and GO in order to be used in cartilage defects. Then, they determined the efficacy of the material by porosity, swelling rate, conductivity, and biocompatibility tests. Chondroitin sulfate is a natural polymer found in the structure of cartilage tissue. It acts in the communication between cell surfaces and ECM and mutual recognition of cells. Therefore, it is thought to contribute to cell proliferation. Due to the adjustable mechanical properties, PCL is a Food and Drug Administration-approved molecule (Pitt et al., 1981). However, due to its partial hydrophobic character, it is insufficient for the recognition of the cells and has a low biodegradation property (Zhu et al., 2002). The certain amount of GO contribution (up to 4.5%) provides the material more than twice the compression modulus value than the average and increases the conductivity of the material. Therefore, the acquisition of superior characteristics is provided with the combination of these materials by forming hybrid structures. In the study, CSMA/PECA/GO hybrid materials were characterized and the average pore sizes of CSMA:PECA ratio for 5:1, 4:2, and 3:3 were measured as 193.5, 175.2, and 152.8 μm, respectively. The reason of this was reported that due to the partial hydrophilic character of CSMA, water absorption of the material increased and depending on this the pore diameters also increased. In order to test the cell biocompatibility of the scaffolds indirect cytotoxicity tests were perform in 3T3 cell line for 1, 3, and 5 days. It was reported that samples containing 3% GO showed in the average of 80% vitality and during the in vivo biocompatibility test performed in rats inflammatory parameters were no presence (Fig. 12.15). They also reported that in in vitro cell adhesion test cells successfully adhered to the scaffold and ensured the proliferation. Finally, it was stated that in full-thickness cartilage defect repair test scaffolds stuck to the tissue as desired and after 12 weeks the wound was closed (Liao et al., 2015).

With freeze drying and the sol–gel methods. Wang et al. (2016) prepared the porous silica/Chi hybrid scaffolds which can be used for many applications in general. Then, the inorganic/organic ratios over the effects of material properties were investigated. After preparing the materials, the morphology was studied with FESEM and mechanical tests, water adsorption, in vitro biocompatibility tests were carried out. In the study, 3-glycidoxypropyl trimethoxysilane (GPTMS) was used as the coupling agent. Thus, the hybrid structure between silica network and Chi was formed (Fig. 12.16). The resulted hybrid structure formed by covalent bonds was

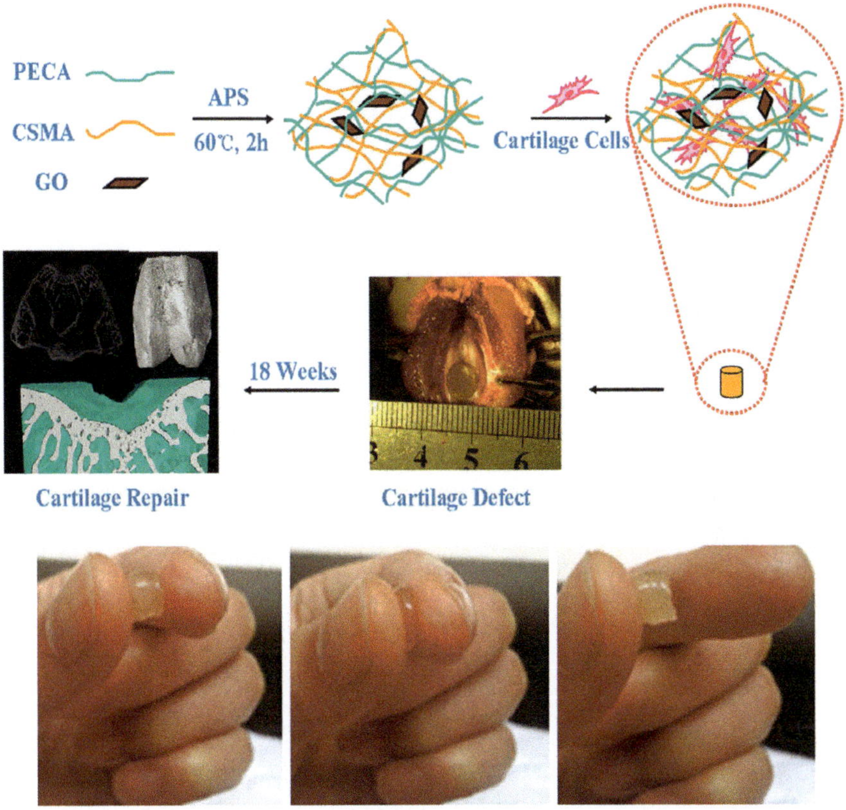

Figure 12.15 CSMA/PECA/GO hybrid scaffold for cartilage tissue engineering.
Reproduced with the permission from Liao, J.F., Qu, Y., Chu, B.Y., Zhang, X.N., & Qian, Z.Y. (2015). Biodegradable CSMA/PECA/Graphene porous hybrid scaffold for cartilage tissue engineering. Sci. Rep., 5(9879), 1–16. Copyright © 2015, Nature.

confirmed by FTIR. It was reported that thanks to the epoxide ring of GPTMS agent with primary amines of Chi molecules, crosslinkings occurred between Chi and silica networks. In the work, it was shown that when GPTMS rate increased, networks of silica and thus the porosity of the polymer also increased from 150 to 400 μm. Based on the SEM images of the samples, a bulk and homogeneous structure was found and increasing in GPTMS rate improved mechanical properties. Lastly, the cell proliferation assay was performed on CAL-27 cell line. When GPTMS of cells and thus silica ratio increased, more proliferation was observed (Wang et al., 2016).

In the study of Sun et al. (2015), polymer hybrid hydrogel platform was developed in terms of the improvement in cellular interactions and the demonstration in the effect of gelatin hydrogels over protein releasing. The biodegradable, biocompatible hybrid hydrogel platform was made from the appropriate properties of gelatin and arginine via Ultraviolet (UV) photo-crossing (Fig. 12.17). The hydrogels were characterized based on their internal morphologies, mechanical properties,

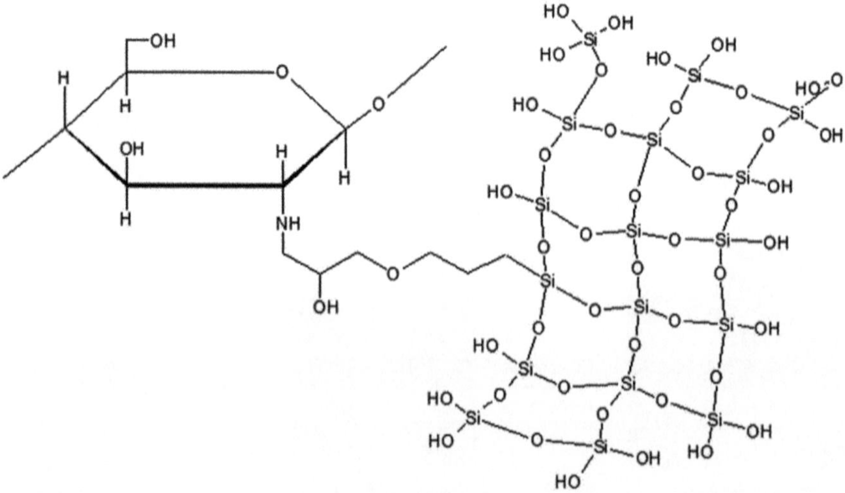

Figure 12.16 Schematic for the chemical structure of silica/chitosan hybrid scaffolds. Reprinted with the permission from Wang, D., Liu, W., Feng, Q., Dong, C., Liu, Q., Duan, L., et al. (2017). Effect of inorganic/organic ratio and chemical coupling on the performance of porous silica/chitosan hybrid scaffolds. Mater. Sci. Eng.: C.70, 2017, 969-975. Copyright © 2017, Elsevier.

swelling mechanics, and biodegredability capacities. In vitro biocompatibility studies showed that hybrid hydrogels had better performance than gelatin methacrylamide (Gel-MA) hydrogels in terms of cell proliferation and adhesion. Furthermore, the therapeutic proteins were loaded into the hydrogels and their releasing behavior were examined. The loading and releasing profiles of the new cationic gelatin hydrogels demonstrate that protein loading capacity and in vitro protein releasing significantly increased. Finally, structure and function studies indicate that the composition of a material has a large influence on the properties of the hydrogels (Sun et al., 2015).

Abalone nacre is a natural ceramic-based composite. It consists of 95 wt% stacked $CaCO_3$ tiles and 5 wt% organic layers coming into a multilayer structure. A couple of toughening mechanisms such as crack deflection at the organic/inorganic interfaces, viscoelastic organic glue, nanoasperities, and interconnected mineral bridges between tiles prevent deformation and failure of abalone nacre. Observed from abalone nacre, zirconia, and polyimide multilayers were synthesized by a hybrid system via sputtering and pulsed laser deposition methods. By adding thin polyimide interlayers between zirconia layers, the fracture toughness of multilayer coatings (5.2 MPa · m1/2) was significantly increased and approached a six times higher value than zirconia monolayer (1.0 MPa m1/2). In order to investigate the effect of organic/inorganic interfaces on the mechanical properties of the coatings, the thickness ratio of zirconia and polyimide was kept as 10:1, while the thickness and number of interfaces were changed. The results showed that multilayer structure could increase the

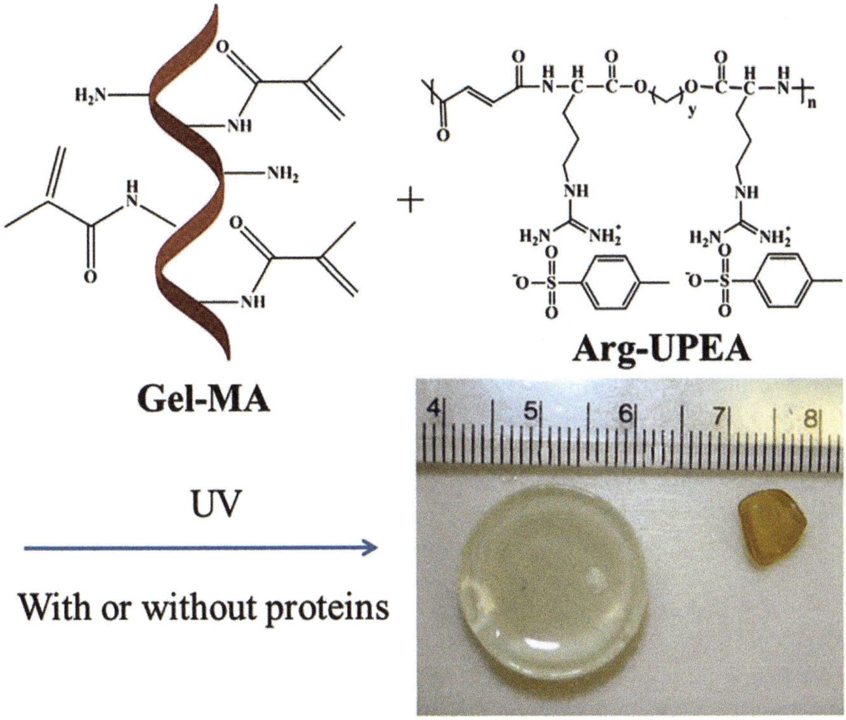

Figure 12.17 Illustration of the chemical structure of hydrogel precursors [Gel-MA (left) and Arg-UPEA (right)] and the photo image of Gel-MA/2-UArg-2-S (4: 1, w:w) hydrogel [swollen (left) and dry (right)].
Reprinted with the permission from Sun, X., Zhao, X., Zhao, L., Li, Q., D'Ortenzio, M., Nguyen, B., et al. (2015). Development of a hybrid gelatin hydrogel platform for tissue engineering and protein delivery applications. *J. Mater. Chem. B*, 3, 6368–6376. Copyright © 201, Royal Society of Chemistry.

fracture toughness of coatings. The fracture toughness enhanced with increasing number of interfaces but the hardness slightly decreased. SEM images proved that the major toughening mechanism of the multilayer coating was crack deflection at organic/inorganic interfaces. With certain critical interfacial roughness, fracture toughness of multilayer can be developed. As a result, bioinspired organic/inorganic multilayers could enhance the intrinsically toughness of brittle ceramic or glass coatings. Thus, such multilayers extend their applications in protection, corrosion resistance, optical, and biomedical fields (Yang et al., 2015).

12.2.7 Implant materials

Today, if we search the latest developments in implant technology compared to implants made from a single type of material such as stainless steel, titanium, gold, platinum, different ceramic, or polymeric structures, we can see that implant composites produced by the combination of more than one substance spread fairly (Dorozhkin, 2009). The reason for this is that in biomedical implants, the combination of many desirable features such as mechanical strength, biocompatibility, biostability, blood compatibility, low toxicity, and nonallergic was wanted (Valliant and Jones, 2011). Even though these multi featured materials are attempted to obtain with composite structures during the history, it replaces its place by other material pursuits due to the fact that ensuring the perfect alignment between matrix and additive material is not possible in composite structures. In biomedical and inside the body applications, in order to meet the forces resulting from mechanical influences, a very good interaction between the matrix and additive should have. Any micro gap may occur the destruction or may result in the breakage of composite. Therefore, in order to increase the interactions between the matrix and the additive two components with secondary interactions or covalent binding are concerned. Such bindings are possible with the use of hybrid materials (Sierra-Martin and Fernandez-Barbero, 2015). Hybrid structures made from the combination of natural or synthetic polymers with various inorganic other materials have many advantages such as increasing the cellular interactions, promoting the full integration of host tissue, the adjustment of material properties, and degradation kinetics (Mazaheri et al., 2015; Sanchez et al., 2005; Zhao et al., 2015; Hood et al., 2014). This relentless development in the hybrid composite materials increases the success rate in tissue engineering and regenerative medicine day-by-day. In this section, current hybrid polymer composite materials used in implants and the related studies are given.

Typical implanted biomaterials to support bone fractures are stainless steel, cobalt, titanium, magnesium alloy, HA, alumina, zirconia, poly(methyl methacrylate) (PMMA), PLA, carbonfiber-polyetherketone, and carbon fiber-ultrahigh molecular massed polyethylene structures. With the recent advances in nanotechnology, it has been a significant improvement in orthopedic and dental implants. In particular, the mechanical strength, biocompatibility and bone-cell compatibility of implants have been significantly increased (Sanchez et al., 2005). Nanostructured implants compared to common materials stimulate the bone-cell growth and the binding of bone-cell (osseointegration). For example, a metallic implant surface is not suitable for the adhesion and growth of bone cells, whereas it would be a suitable surface to hold the bone cells with a porous, rough, and functional groups when coated with a nanostructured hybrid polymer.

Today, although bulk Ti implants are one of the most preferred implant types, in the case of complex stress as a bulk metal or a powder material, it may present the high plastic resistance which may cause breakage and allergic effects caused by the elements such as V and Al at the form of alloy in its structure. To eliminate these types of disadvantages synthetic and natural polymers are an excellent candidate. Natural polymers such as alginate, Col, fibrin, Chi, and hyaluronic acid form a bioactive surface on pure metal surfaces. As this resulted biocompatible surface provides an appropriate condition for the adhesion and growth of cells, it stimulates the healing for bone tissue. While applying related polymers on metallic implants, they are administered to include inorganic structures such as CA, P, CAP, or $CaCO_3$ to improve the bone-tissue compatibility or antibacterial nanoparticles such as Ag and Au not as pure polymer films in general.

Although, metallic prostheses made from stainless steel or titanium alloys usually are used in the correction of bone fractures, these materials have various disadvantages such as showing the magnetic effect, corrosion, metal-body mismatch, the anode-cathode reactions together with the reduction of bone mass, increasing of the porosity and ultimately delay of the fracture healing (Fig. 12.18) (Huang and Fujihara, 2005; Acero et al., 1999). Furthermore, as such metallic implants have the heating and cooling features as distinct from body tissues, they may cause pain in different atmospheric conditions. Even, occurring of refractures due to insufficient bone growth after the removal of the prosthesis is reported a lot. Moreover, the use of permanent implants is prohibited for young patients because such implants prevent the development of bone and they also put an extra burden related to their heavy weight on the body.

Despite of their light weight and good biocompatibility properties, ceramic composites can be brittle, devastating for and cutting the tissues which they are attached. As polymer composite materials are selected as a much better alternative instead of ceramic composites and metallic bone plates can be varied easily, they have different properties, performances, and shapes (Bagheri et al., 2013). We quite often find the use of high mechanical strength polyimide, polyamide and high-density polyethylene or some of the special polymeric composites as implants in orthopedic applications. Moreover, in these types of polymeric materials disadvantages such as stretch and monomer release are frequently observed. Therefore, the search for new materials is maintained in the field of orthopedic implants.

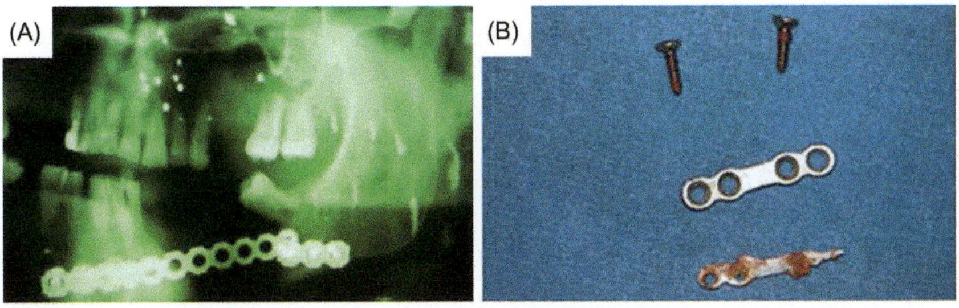

Figure 12.18 Titanium plate fracture (A) and removal of a miniplate with bone neoformation over it (B)
Reprinted with the permission from Cifuentes S.C. (2015). Copyright © 2015 (Cifuentes, S.C. (2015). Processing and characterization of novel biodegradable and bioresorbable PLA/Mg composites for osteosynthesis. PhD Thesis, Universidad Carlos III de Madrid, Madrid, Spain; Acero, J., Calderon, J., Salmeron, J., Verdaguer, J., Concejo, C., et al.(1999). The behaviour of titanium as a biomaterial: microscopy study of plates and surrounding tissues in facial osteosynthesis. J. Cranio-Maxillofac Surg., 27, 117–123).

For the aim of Sandra Cifuentes et al. (2012) to overcome the disadvantages of the permanent or intrabody implants, poly(L,D lactic acid) (PLA)/Mg hybrid polymer composites were prepared from PLA as the organic and Mg and Mg_5Zn particles as inorganic dopant hybrid material. So, the obtained hybrid composite material was examined in the eligibility for osteosynthesis and industrial applications by utilizing the advantages of both materials. In-vitro degradation, physicochemical, and mechanical characterizations of PLA/Mg composites for medical applications were studied. Moreover, during the shaping of these materials, high-temperature processing techniques were used such as injection molding and extrusion/compression. The thermal stability and thermal resistance of the composites were examined. It was shown that the increase of Mg volume in hybrid dramatically reduced the thermal degradation of composite. The temperatures above 200°C allowing for the PLA/Mg composites used in industrial production processes is the upper limit for the processing. Extradur process allowed a reduction in the thermal degradation of the material and the participation of Mg particles in both materials in polymeric materials was provided up to 15%. The microindentation device test, which Young's modulus and hardness informations are obtained, was performed and mechanical characterization was achieved by applying compression/tensile tests. The tests showed that the addition of Mg particles increased the plastic fluidity and the polymer resistance against hardness (Cifuentes et al., 2012). Especially, this type of hybrid composites thought to be used in intrabody implants was carried out in-vivo and in-vitro biocompatibility tests. In vitro studies of these materials show the degradation rates largely depending on the crystallinity degree, the component ratio of Mg and Mg particle size. When all composites were immersed in PBS, proton was released as tolerating the human body and did not exceed the buffering capacity of the solution. In-vitro studies performed on composites having 10% particle demonstrated that the material consisting of spherical particles of Mg doped amorphous PLA matrix showed the best degradation behavior. The same material lost its 4% of the strength after 7 days, and 40% after 28 days. Consequently, due to the controllable degradation rates of new PLA/Mg composites for osteosynthesis and sufficient mechanical properties, it was determined that they have great potential as the resorbable and biocompatible implant material.

In another study of Wan et al. (2014), for orthopedic implants the synthesis and the structural characterization of polymers-like poly-L-lactide (PLLA)/magnesium and PLLA/magnesium Fluoride hybrid composites were performed. PLLAs are used as the first commercial bioabsorbent and biodegradable implant materials. However, some minimized restrictions as a result of inflammation are desired such as exhibiting poor mechanical properties and the acid degradation due to forming as byproducts. PLLA hybrid composites doped with 3% and 7% of magnesium and magnesium fluoride particles were produced to overcome these limitations. The morphological change of hybrid composites was observed using SEM in Fig. 12.19. Mg particles were dispersive embedded on the surface of PLLA matrix. Moreover, no obvious crack was found on the surface, which demonstrated a good adhesion between the matrix and the incorporated Mg particle. It was seen that the surface of hybrid PLLA became rougher with some bulges due to the incorporation of

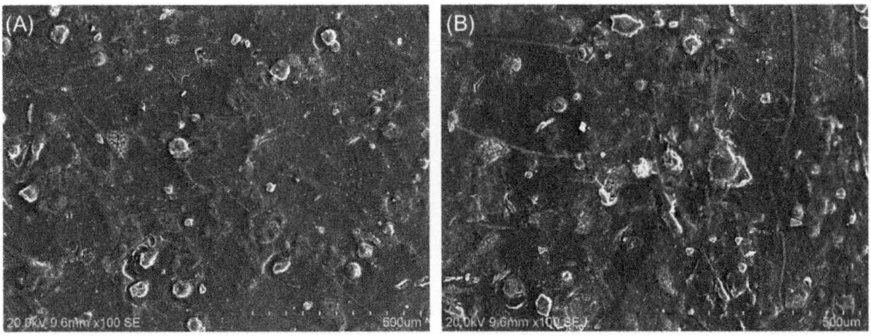

Figure 12.19 SEM images of PLLA/Mg composites (Wan et al., 2014).
Reprinted with the permission from Wan, P., Yuan, C., Tan, L.L., Li, Q., & Yang, K. (2014). Fabrication and evaluation of bioresorbable PLLA/magnesium and PLLA/magnesium fluoride hybrid composites for orthopedic implants. Compos. Sci. Technol., 98, 36–43. Copyright © 2014, Elsevier.

magnesium fluoride compared to PLLA/Mg. The morphology, mechanical, and thermal properties belonging to PLLA/Mg composites were determined by the SEM, tensile test device, and differential scanning calorimetry (DSC). In addition, in-vitro cytotoxicity and degradation assays were performed using MTT method. It was found that the addition of Mg particles partially decreased the tensile stress. The stress fracture morphologies show that the interval adhesion between Mg particles and PLLA matrix could join the effect on mechanism. The participations of Mg and $MgFl_2$ into PLLA are effective in neutralizing the acid environment due to induced degradation intermediates. Furthermore, cell viability tests showed that with the release of Mg ions composite had better cyto compatibility. As a result, the research team think that PLLA/Magesium and PLLA/Magnesium Fluoride hybrid composites have a promising potential in orthopedic implant applications. In particular, it was observed that the bone-tissue compatibility of these implants was higher than other implants (Wan et al., 2014).

In the development of implant success rate, following the implantation the improvement in response against healing conditions is important. The uncontrolled movement of inflammatory cells and fibroblasts is one of the in-vivo problems. Implant's porosity characteristics have a very strong impact on it. In another study by Vrana et al. (2011), a hybrid system was obtained from macropore titanium structure filled with the microporous biodegradable polymer (Vrana et al., 2011). This polymer matrix allows the different cell types such as fibroblasts and epithelial cells to settle with having a different porosity gradient. The main clinical application of this system would protect the restenosis depending on the excessive fibroblast migration and proliferation seen in tracheal implants. Using the freeze-extraction method, a microbead-based titanium template filled with porous PLLA having different porosity values of inner and outer surfaces of the implants were determined with photo analysis and mercury porosimetry (9.862.2 vs 36.7611.4 mm, $P \leq 0.05$). A thin film on the surface was added to optimize the growth of epithelial cells. For controlling the movement of fibroblasts, PKH26-labeled fibroblasts were seeded onto the titanium and titanium/PLLA implants. With a confocal microscope by observing the weekly cell movement, it was found that weekly samples moved deeper in Ti samples compared to Ti/PLLA samples. Results of in-vitro experiments presented that this new implant was shown to be effective in directing different cell types contacting with implant. This system allows spatial and temporal controls over cell migration in tracheal implants having macropores and nanopores. Furthermore, the mechanical strength properties will often be based on the titanium frame.

The another application in implant technology relating to the use of hybrid materials is metallic implants coated with biocompatible, high blood, and bone-tissue compatible surface films. In these coatings, thin ceramic coatings on metal implants and polymer ceramic composites are the examples of this type of hybrid materials. In such coatings, particularly structures containing of hydroxyapatite and Calcium phospate (CaP) ceramics come to the fore (LeGeros, 2002). The bone compatibility of CaP ceramics and hydroxyapatite structures is high; however, the surface adhesion properties are low. In order to overcome the poor mechanical properties of most CaP ceramics, it is aimed to use polymeric or sol–gel coatings in combination with CaP ceramics. CaP doped into the different types of adhesive polymeric films are extremely used as synthetic bone graft substitutes. The properties of used CaP and method in contribution to the hybrid structure are decisive for the bioactivity of final structure.

In order to compare the abilities of bone marrow-derived human mesenchymal stromal cells to support the osteogenic differentiation and proliferation, Birgani et al. (2016) were investigated new monolithic composites comprising nano dimensional CaP, PLA, and CaP coated PLA. PLA/CaP composites produced by using the physical mixture and extrusion were coated with different thicknesses on the implant surfaces (Birgani et al., 2016). CaP-doped coating that includes biomimetic coating in the environment close to physiological conditions supported the proliferation of human mesenchymal stromal cells in higher rates.

In hybrid-based implants used for orthopedic purpose, implants having nanoceramic and nanopolymer structures exhibit a more appropriate structure compared to other implants. In these structures, ceramic nanostructure tasks a matrix, whereas nanopolymer structure serves as an additive that provides flexibility, durability, porosity, and lightness. Gain et al. (2015) with a similar approach investigated the availability of porous HA/ZrO_2-PMMA structures as bone implants. With this approach, it was found that the hybrid structures containing the porous HA provide very high mechanical strength compared to the conventional implant structures. These structures also can be applied as more lighter and flexible. In another study, PLLA-based hybrid nanocomposite structures were synthesized by Hickey et al. (2015) as orthopedic implants. In these structures, there is a nanoceramic inorganic phase containing HA/MgO. This structure improves the function of the mechanical properties as well as significantly increases bone cell growth. An important new alternative additive for bone implants is the carbon nanotube structures. Liao et al. (2013) prepared HA obtained polypropylene structures containing a combination of carbon nanotube additives. This application proves us that carbon nanotube structures can be used to improve the mechanical properties of the bone-implants. Then, other carbon allotropes such as graphene, graphite, and carbon fiber were tested as an additive in orthopedic implants (Yang et al., 2011b; Wu et al., 2013; Ahmed et al., 2013).

In another presented paper, for biomedical applications, a new hybrid approach that consists of a complex ceramic coating on Ti plates comprising TiO_2 nanotubes (TNTs) was obtained by anodizing, multiwalled carbon nanotubes (MWCNTs) functionalized with −COOH groups, and hydroxyapatite. The research team functionalized MWCNTs with −COOH groups in order to obtain TNTs. Then, the electro-deposition method was applied on the novel hybrid material. By using FTIR and SEM, morphological features, and the structure of the hybrid material were determined. The contact angle measurements and micro hardness tests were occurred to identify the wettability and mechanical resistance of the new hybrid material. Moreover, in-vitro cell response of the hybrid material was investigated. For the osteoblast adhesion, viability, and proliferation this new hybrid material showed a proper surface with an optimal microstructure and morphology. The ceramic coating on Ti plates has induced an increase of micro hardness and the hemolytic index in the same time with a decrease of contact angle; however, the values have remained in the domains of hydrophilic character and nonhemolytic materials. Thus, according to the tests results, researchers think that this new ceramic coating may promise in the field of implant biomaterials (Prodana et al., 2015).

12.3 Drug delivery

In recent years, controlled drug-delivery systems are being widely investigated in the pharmaceutical and medical fields. Compared to conventional formulations, controlled drug-delivery systems have tremendous advantages such as controlled release rate, reducing toxic side effects, and drug efficacy. An ideal drug carrier system has to be inert, biocompatible, mechanically stable, comfortable for the patient, simple and effective drug loading, ease of manufacture, removed when not desired, and economic (Contessotto et al., 2009). Most of the current researches are directed to the preparation of pH and temperature-sensitive smart drug-delivery matrixes. Thanks to the pH and temperature-sensitive matrixes, drug delivery can be increased or reduced based on the targeted site. For this purpose, thermoresponsive biopolymers (such as Chi and alginate) are widely used. As biopolymers have chemical structure diversity, stereo functionality, a certain and broad molecular weight, biological characteristics, are nontoxic and green chemistry products, they are commonly used by researchers as a drug delivery matrix (Bosio et al., 2014). Biocompatible and biodegradable synthetic polymers are also used as drug carrier matrix. Recently, a new polymeric hybrid composite in combination with inorganic porous materials is being prepared in order to increase the efficiency of drugs loaded into natural and synthetic polymers. Due to their biocompatible, biomimetic and pH-sensitive properties, organic–inorganic hybrid materials have recently gained great interest in controlled drug delivery systems (Shi et al., 2014). In the studies, because of the advantages of hybrid polymer composite materials, it has been proven that they are more effective than pure polymers and composites as drug carrier matrix. In the studies conducted with both natural and synthetic polymer hybrid composites, carrier systems for doxorubicin (DOX), ibufen, dopamine (DOP), paraastemol, indomethacin, and BSA drugs have been prepared.

12.3.1 Drug delivery with hybrid composites prepared from natural polymers

Natural polymers in preparing polymer hybrid composites as drug-delivery systems have been used quite widely in the researches. The most fundamental point of preferring the natural polymers is to eliminate the biocompatibility concerns in drug-delivery systems. In this context, studies including hybrid composites that use natural polymers with drug delivery systems are illustrated below.

DOX is a hydrophilic drug that is widely used in the treatment of breast, lung, ovary, and other cancers. This drug is highly toxic and can cause serious side effects in cancerous cells as well as nontarget tissues (e.g., heart, liver, kidneys, stomach, and blood-cell lineages). Therefore, new methods of encapsulation and carrier matrixes are designed in order to control releasing of the drug. In the literature, for the transport of the drug hybrid polymer composites are prepared with natural or synthetic polymers doped in various inorganic materials (mostly $CaCO_3$). For the controlled transportation of DOX drug, $CaCO_3$/carboxymethyl Chi (CMC) hybrid composite micro and nanospheres were prepared by Wang et al. (2010) (Fig. 12.20). CMC is a biocompatible and biodegradable derivative of Chi with better water solubility and an ideal polymeric material as a drug delivery matrix. With the $CaCO_3$ content of hybrid polymer composites, the drug loading efficiency was increased by absorbing of more DOX drug. In this study, DOX was encapsulated into the porous of hybrid polymer composites and the release of drug was determined spectrophotometrically. The study team reported that the encapsulation efficiency for both microsphere and nanosphere hybrid polymer composites is more than 60%. According to the releasing results, there has been a release initially due to the hydrophilic character of CMC chain. However, after 24 h, this ratio was decreased. It was mentioned that the cumulative DOX releasing in ~260 h is 60% from hybrid microspheres, while it is 90%—150 h for hybrid nanospheres. The authors emphasized that the amount of $CaCO_3$ and the surface area affected drug release and also hybrid polymer composites prepared according to the in-vitro drug release rates were an ideal matrix for the transport of water-soluble drugs.

In another study, carrageenan (Car)–$CaCO_3$ hybrid polymer composites containing Car biopolymer as DOX carrier was prepared by Bosio et al. (2014). Car extracted from red marine algae is made of sulfated linear polygalacturonase. The aim of that study was to design nanostructured polymer-inorganic hydrogel-based hybrid polymer composite in order to transport DOX drug into target cancer cells.

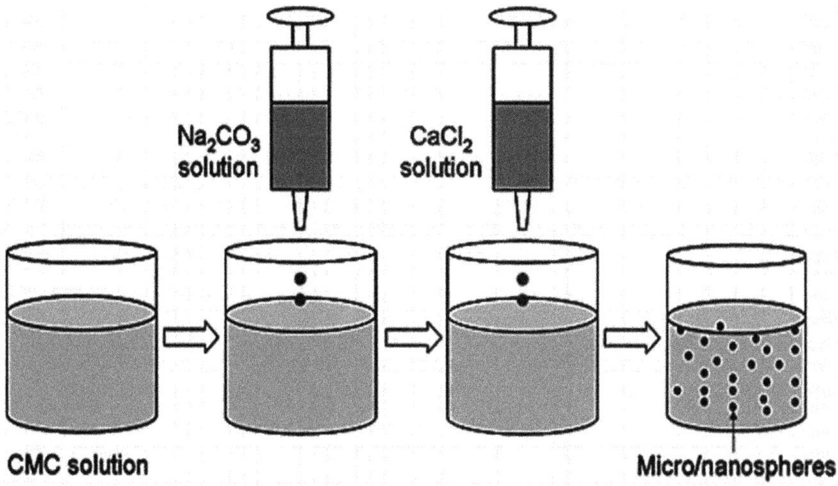

Figure 12.20 Schematic diagram of preparation procedure of hybrid microspheres and nanospheres.
Reprinted with the permission from Wang, J., Chen, J.S., Zong, J.Y., Zhao, D., Li, F. Zhuo, R.X., et al. (2010). Calcium carbonate/carboxymethyl chitosan hybrid microspheres and nanospheres for drug delivery. J. Phys. Chem. C, 114, 18940–18945. Copyright © 2010, American Chemical Society.

At the same time, prepared hybrid material was derivatized with folic acid (FA), which is a polar molecule, having functionalized with carboxylic groups. The morphological, structural, and in vitro drug release properties of prepared hybrid materials were investigated (Fig. 12.21). DOX drug loading results have been reported as 29% for Car–$CaCO_3$ and 83% for Car–$CaCO_3$, respectively. FA drug loading efficiency increased approximately threefold. The authors commented on the cause of this increase that the polar character increased the interaction between FA and DOX because of the functionalization of carrier matrix with FA. The hydrophilic character of DOX drug was also effected. As a result of 25 days of incubation, the release rate of DOX from DOX–Car–$CaCO_3$ was 73%, while the rate of drug released from the hybrid FA–Car–$CaCO_3$ material derivatized with FA was only 26%. At the end of this study, for controlled drug delivery and targeting cancer cells hybrid biopolymer–inorganic materials are considered to be used as the new model system and therapeutic strategies in chemotherapy treatment.

Kamari and Ghiaci (2016) prepared modified Chi (MC)/TiO_2 hybrid composites to control the transport of ibuprofen (IBU), which has serious side effects because of the short half-life (Fig. 12.22). Chi is a natural polysaccharide containing numerous hydroxyl and amine groups. Chi is widely used in drug delivery systems as a drug carrier because of the features such as biodegradability, biocompatibility, high mechanical strength, nontoxicity, mucoadhesivity, and gel-forming capacity. The study team first prepared in a series of IBU/MC composites and release studies were performed for drug loading in the composites with UV–Vis

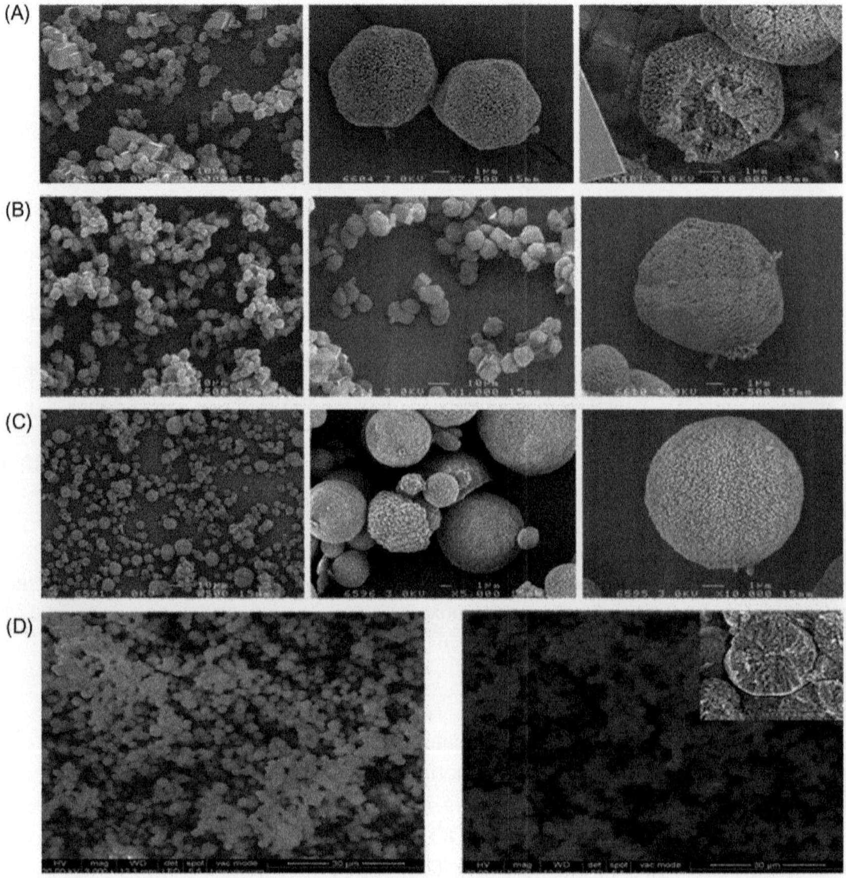

Figure 12.21 SEM images of the $CaCO_3$ microparticles synthesized. (A) $CaCO_3$ MP without Gly buffer and without biopolymers. (B) $CaCO_3$ MP without biopolymers. (C) hNPs with biopolymer and (D) optimization of the synthesis for Car and FA−Car (left and right, respectively); longitudinal cut particle of FA−Car on the right inset figure.
Reprinted with the permission from Bosio, V.E., Cacicedo, M.L., Calvignac, B., León, I., Beuvier, T., Boury, F., et al. (2014). Synthesis and characterization of CaCO3−biopolymer hybrid nanoporous microparticles for controlled release of doxorubicin. Colloids Surf., B: Biointerfaces, 123, 158−169. Copyright © 2014, Elsevier.

spectrophotometer. It was reported that the structure of the composite containing 40% IBU exhibited the most regular release graph and during the first 10 min the drug release rate was higher than 32%, after 1 h this ratio was higher than 52%. In the study, composites prepared to reduce the drug release rate were coated with different amounts of TiO_2 (30−60 wt%) to obtain hybrid composites. Being coated with TiO_2 improved the retention of drugs and drug release was reduced. In addition, IBU/MC/TiO_2 hybrid composites showed that drug release was pH sensitive. In simulated gastric fluid (pH 1.2), IBU release rate was significantly slower than

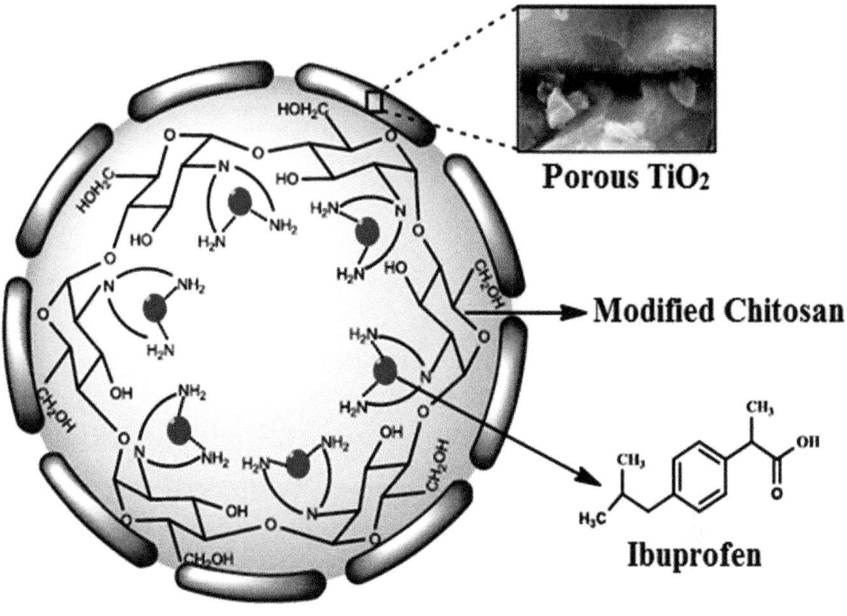

Figure 12.22 Schematic presentation of IBU/MC/TiO_2 nanohybrid composite and the interactive effect of the ibuprofen with chemically modified chitosan.
Reprinted with the permission from Kamari, Y., & Ghiaci, M. (2016). Preparation and characterization of ibuprofen/modified chitosan/TiO2 hybrid composite as a controlled drug-delivery system. Microporous Mesoporous Mater., 234, 361–369. Copyright © 2016, Elsevier.

pH 7.4. As a result of the study, an inclusion of TiO_2 as porous inorganic coating into the composite did not only improve the stability of the compound but also slowed the drug release keeping in the pores at the same time.

The hybrid composites obtained from the coating of Chi with TiO_2 in different pH values were also used as the DOP (3,4-dihydroxyphenylethylamine) carrier (Fig. 12.23) (Safari et al., 2015). DOP is an important drug using as an intravenous injection in the treatment of nerve diseases such as Parkinson disease and schizophrenia. In the study, DOP/Chi (10–40 wt%) was primarily prepared and nano hybrid organic–inorganic Chi/DOP/TiO_2 composites coated with different amounts of TiO_2 (10–50 wt%) were prepared. It was determined that the TiO_2-uncoated composites released completely DOP drug in 10 min, whereas the drug release rate of TiO_2 coated nanohybrid composites was 70% in 5 h and was 87% in 12 h. After 16 h, all of the DOP was released. These results showed that hybrid composites may be used in the extension of DOP releasing time.

Biocompatible and biodegradable polyelectrolytes are also used as a drug delivery matrix in the literature. Polyelectrolytes are the blends of anionic and cationic charged polymers at the molecular level, and it has been reported that polyelectrolytes are stimulated with the inorganic minerals such as hydroxyapatite, $CaCO_3$ and

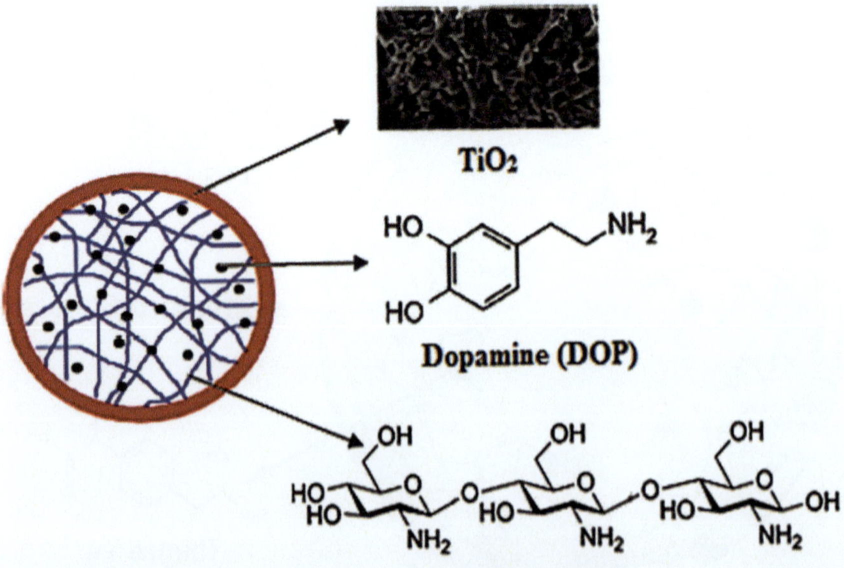

Figure 12.23 Schematic presentation of TiO_2-coated dopamine/chitosan nanohybrid composite as a new drug delivery system.
Reproduced with the permission from Safari, M., Ghiaci, M., Jafari-Asl, M., & Ensafi, A.A. (2015). Nanohybrid organic–inorganic chitosan/dopamine/TiO2 composites with controlled drug-delivery properties. Appl. Surf. Sci., 342, 26–33. Copyright © 2015, Elsevier.

$CaSO_4$ in the literature. By taking advantage of carboxymethyl cellulose and Chi natural polymers, CMC/Chi/calcium phosphate hybrid polymer composites were prepared with biomimetic mineralization method (Fig. 12.24) and were used as carrier matrix for BSA (Salama and El-Sakhawy, 2014). Due to the structural similarity, there is a strong interaction between CMC and Chi. According to the drug release data, the drug release rate of CMC/Chi polyelectrolyte after 10 h was 92%. However, the inorganic content increased, BSA release was decreased up to 70%. These results indicate that polyelectrolytes doped with inorganic materials block the permeability of encapsulated drug and may reduce the effective drug release.

12.3.2 Drug delivery with hybrid composites prepared from synthetic polymer

Hybrid composites prepared from synthetic polymers are widely used in drug-delivery systems. Polymer hybrid composites and synthetic polymers selected from studies comprising drug-delivery systems are polymers known for their biocompatibility. In this context, PU, polyamine, PEG, polystyrene sulfonate (PSS), pHEMA,

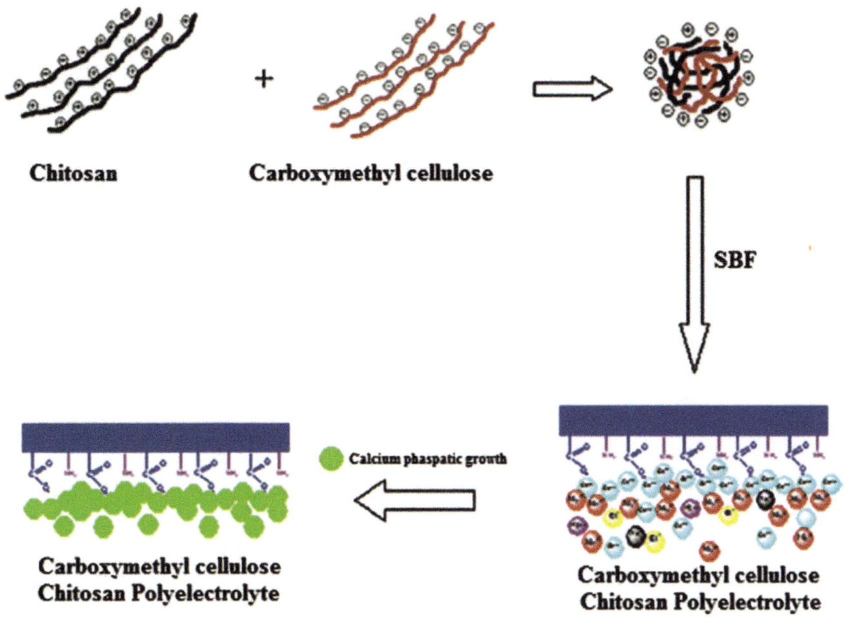

Figure 12.24 The preparation of CMC/Chi polyelectrolyte hydrogel and biomimetic mineralization process.
Reproduced with the permission from Salama, A., & El-Sakhawy, M. (2014). Preparation of polyelectrolyte/calcium phosphate hybrids for drug delivery application. Carbohydr. Polym., 113, 500–506. Copyright © 2014, Elsevier.

and *N*-isopropylacrylamid (NIPAAM) stand out as polymers, and they are available in the following examples.

In drug-delivery studies including hybrid composites used synthetic polymers, polymer hybrid composite structures are forefront for the transport of DOX drug. For example, Shi et al. (2014) prepared $CaCO_3$/aliphatic poly(urethane-amine) (PUA)/sodium PSS hybrid polymer composites as a controlled DOX carrier matrix using the electrostatic interactions between molecules (Fig. 12.25). $CaCO_3$ is widely used in drug delivery systems, because it has a biocompatible, biodegradable, pH-sensitive, and porous structure. Aliphatic PUA in composite structure is a biodegradable, biocompatible, and temperature sensitive polymer. On the other hand, PSS is used as crystal growth additive to control the morphology of $CaCO_3$ microparticles. Due to the fact that cancer cells have more acidic and higher heat values compared to normal cells, the temperature sensitive (PUA provides) as well as the pH-sensitive ($CaCO_3$) hybrid polymer composites were chosen to be loaded DOX. Thus, DOX release will be sensitive to pH and temperature and will be started rapidly releasing in the area where cancer cells are. It was reported that DOX loading percentage into the samples were reported as 82.42%–92.52%. The authors stated that this high yield rate of DOX loading was related to nanoporous structure of

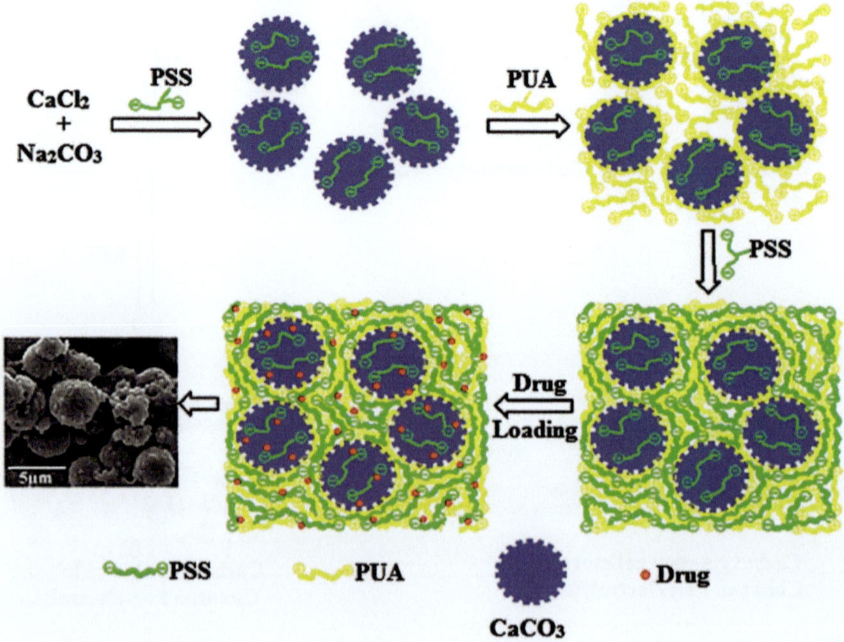

Figure 12.25 The preparation of $CaCO_3$/aliphatic PUA/sodium PSS hybrid composites and DOX loading.
Reproduced with the permission from Shi, J., Shi, J., Feng, D., Yue, P., & Cao, S. (2014). Stimuli-responsive hybrid composites based on CaCO3 microparticles and smart polyelectrolytes for controllable drug delivery. Polym. Bull., 71, 1857–1873. Copyright © 2014, Springer-Verlag Berlin Heidelberg.

$CaCO_3$ and the electrostatic interaction between PSS (negatively charged) and DOX (positively charged). Moreover, the study team-made DOX releasing experiments at two different pH and temperature values. DOX release at pH 7.4 (intestinal pH) was 40% in 36 h and at pH 2.1 (gastric pH), it was recorded as 90% under the same conditions. Also, in the same time, the release of DOX at 55°C was 51%, and at 37°C it was 40%. This study showed that pH and heat-sensitive hybrid polymer composites increased DOX drug release at low pH and high temperature. This increases the amount of DOX loaded into cancerous cells. This study shows that hybrid polymer composites prepared from the addition of inorganic $CaCO_3$ microparticles into synthetic polymer structure with the electrostatic interactions may have an advantageous effect in controlled drug loading efficiency and release.

Kaamyabi et al. (2016) prepared a new pH and heat-sensitive hybrid polymer composite with the radical polymerization of NIPAAM and methacrylate-functional Fe_3O_4 (Fig. 12.26) and used as DOX carrier. DOX release researches were studied in blood mimicking, different pH, and temperature conditions. As cancer cells are acidic (pH: 5.8), pH values were chosen 5.8 and 7.4. According to the results in the pH 5.8 and 7.4, DOX was released as 94% and 83%, respectively. The authors

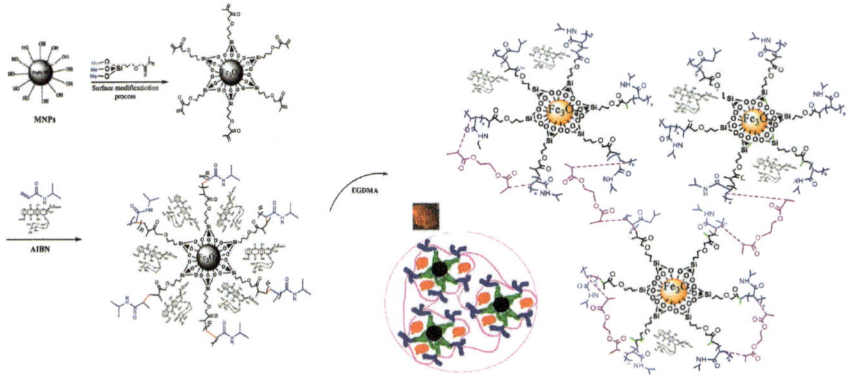

Figure 12.26 A schematic diagram for the synthesis of poly(NIPAAM@Fe_3O_4 MNPs/TMSPMC/DOX.
Reproduced with the permission from Kaamyabi, S., Habibi, D., & Amini, M.M. (2016). Preparation and characterization of the pH and thermosensitive magnetic molecular imprinted nanoparticle polymer for the cancer drug delivery. Bioorg. Med. Chem. Lett., 26, 2349–2354. Copyright © 2016, Elsevier.

explained the cause of increased DOX drug release in acidic conditions that as hydrogen bonds between DOX drug and hybrid composite weakened, this led to an increase in the total DOX release. In order to investigate the effect of the temperature over total DOX release, studies were conducted at 37 °C for normal cell and at 40 °C for cancer cell temperature imitation. When temperature increased, the total DOX release also increased in a high rate. The authors commented these temperatures as the lower critical solution temperature of drug loaded hybrid composites. The effect of time over DOX release was also investigated. With this study, hybrid polymer composites containing magnetic nanoparticles that are sensitive to a new pH, and temperature were synthesized and used as drug carriers for DOX.

In the drug-delivery studies done with hybrid composites prepared from synthetic polymers, ibofurene drug also stands out in the foreground. For instance, hybrid composite membranes were prepared with poly(vinylidene fluoride-trifluoroethylene)/NaY zeolite as the carrier matrix for IBU (Salazar et al., 2015). As zeolite has a porous structure, it was used in order to reduce the absorption and emission of the drug. Poly(vinylidene fluoride-trifluoroethylene) is a biocompatible and biostable polymer. Moreover, thanks to the polymer composite membrane thickness porosity can be adjusted as desired. IBU was encapsulated into polymer hybrid composite membranes prepared with these two materials with physical vapor deposition method. The composite membrane containing 32% zeolite showed two times more IBU release than the composite containing 16% zeolite (Fig. 12.27).

In order to improve the efficacy of IBU, Popat et al. (2012) researched a pH responsive organic–inorganic system. The system containing of covalent binding of positively charged polymer Chi onto phosphonate functionalized mesoporous silica nanoparticles (MSN) was prepared and pure MSN (without Chi) was also used

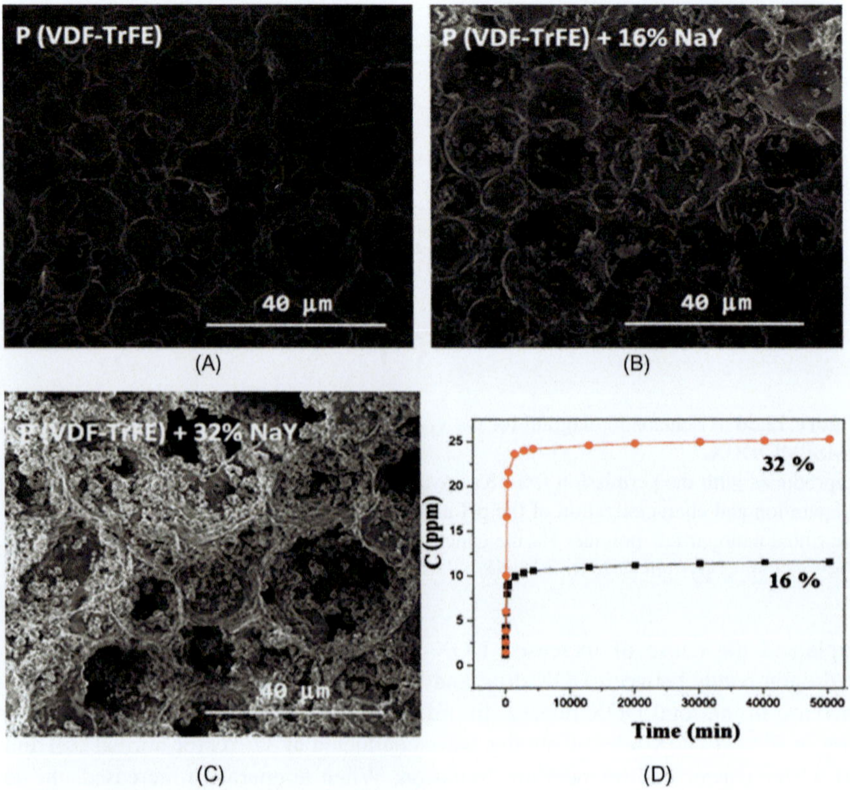

Figure 12.27 SEM images for the copolymer and corresponding composite membranes with different zeolite content (3000 ×) (A–C). IBU release from the P(VDF–TrFE)/NaY membranes, quantified by UV–visspectroscopy (D).
Reproduced with the permission from Salazar, H., Lima, A.C. Lopes, A.C., Botelho, G., & Lanceros-Mendez, S. (2015). Poly(vinylidene fluoride-trifluoroethylene)/NAY zeolite hybrid membranes as a drug release platform applied to ibuprofen release. Colloids Surf., A: Physicochem. Eng. Aspects, 469, 93–99. Copyright © 2015, Elsevier.

as the control group. As illustrated in Fig. 12.28, phosphonate functionalized MCM-41 nanoparticles were activated by 1-ethyl-3-(3-dimethylaminopropyl) carbodiimide. In the study, two different pH media, pH 5 acetate buffer (simulating endosome pH), and pH 7.4 (simulating normal tissue pH), were used. The authors obtained the data that pure MSN system is able to reach the saturation of the drug release in both pH after 4 h. MSN coated Chi showed different release rates. At pH 7.4, only 20% of IBU was released and the releasing rate reached 90% at pH 5.0 after 8 h. The authors commented on these behaviors by the fact that Chi has low degradability and solubility properties at pH 7.4. This system offers an important advantage as pH-responsive nanocarrier and shows a drug modified release behavior with increasing the effectiveness of treatment and reducing side effects.

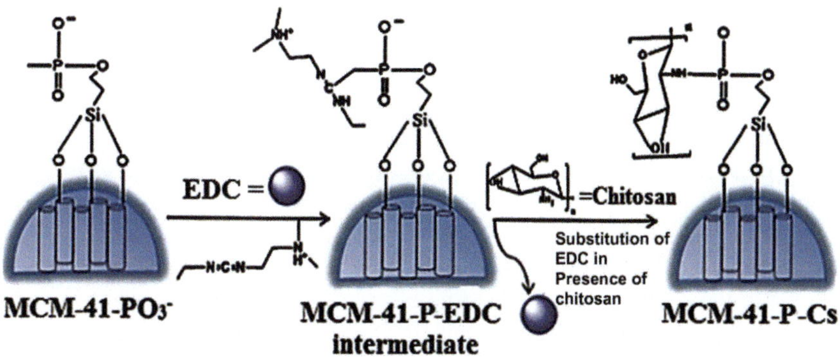

Figure 12.28 The schematic procedure for the preparation of chitosan coated mesoporous silica nanoparticles.
Reprinted with the permission from Popat, A., Liu, J., Lu, G.Q., & Qiao, S.Z. (2012). A pH-responsive drug delivery system based on chitosan coated mesoporous silica nanoparticles. J. Mater. Chem., 22, 11173–11178. Copyright © 2012, The Royal Society of Chemistry.

Poly(NIPAAM)(PNIPAAm)/$CaCO_3$ micro/nano hybrid composites were used as the vitamin B_2 (VB_2) carrier by Shi et al. (2013). For pure PNIPAAm nanogels, the drug release rate was around 91% after 12 h, and the value for the composites prepared with 2.37 and 3.55 mM of Ca^{2+} was only 82%. For the hybrid composites prepared with 1.18 mM of Ca^{2+}, the drug release rate was exactly the same speed with pure PNIPAAm nanogels. This suggested that low content of $CaCO_3$ microparticles could not achieve decreasing both the permeability of VB_2 and the drug release. Eventually, the authors concluded that VB_2 release behaviors of the hybrid composites showed that vaterite microparticles could block the permeation of the encapsulated drug and gain sustained release properties to the hybrid composites and maintain the thermo- and pH-sensitivity of the hybrid composites in the meantime.

The silica/PEG (SiO_2/PEG) polymeric hybrid composite material was prepared via sol–gel method as an indomethacin (nonsteroidal antiinflammatory drug) carrier (Catauro et al., 2014). Due to some important advantages of silica such as chemical resistance and thermal and electrical stability as well as biocompatible and environment-friendly material, it was previously used in drug delivery systems for controlled drug release. In this study, PEG was used due to its biocompatible and immune system-regulating properties. Indomethacin, nonsteroidal antiinflammatory drug, was used as the drug model. In the composite structure, tetraethoxysilane, (inorganic precursor), PEG, and PEG 400 (organic precursor) at different rates (0, 6, 12, 24, and 50 wt%) were used. Indomethacin was loaded as 5, 10, and 15 wt% rates into these prepared composites. According to the drug release results, drug release was decreased with increasing rate of PEG. The authors suggested that the reason of this is the weak interactions between PEG and drug. The composite containing 6 wt% PEG increased the drug release rate independently of indomethacin concentration. It was reported that after 48 h, the drug release rates of the composites comprising 5% and 15% of the drug were 96.6% and 95.5%, respectively.

In another study, pHEMA/clay nanocomposites were prepared by in-situ free radical polymerization and the potential use of paracetamol (reducing mild-to-moderate pain and fever) in biomedical applications was examined as drug-delivery matrix. Novel pHEMA/montmorillonite (MMT) hybrid composites prepared by the addition of different amounts of MMT were characterized by various structural and thermal methods. Paracetemol drug was loaded into these prepared hybrid composites. DSC study revealed that in drug-loaded hydrogels paracetamol became amorph structure. This indicates that the drug is homogeneously distributed in the polymer matrix. In-vitro drug release studies were carried out in both gastric and intestinal pHs. According to the obtained results, the drug completely released within 80 h, composite with the highest clay rate was around 70% cumulative in the same period. The results showed that the drug release rate can be changed by appropriately adjusting of the clay content in the hydrogel. The reason of this was described that the crosslinking between the clay and the polymer was the reduction of drug release rate. Consequently, the authors reported that composites are more advantageous and effective than pure polymers in order to use as drug carrier matrix (Bounabi et al., 2016).

As a result, hybrid polymer composites containing organic–inorganic structures are presented as an alternative method instead of conventional methods in drug release studies. The use of such materials in drug release studies offers many advantages such as low toxicity, high drug loading and improving the drug release efficiency. These devices may offer a new approach for the treatment of various diseases. However, more studies are needed to develop these materials.

12.4 Biosensor applications

Biosensors are analytical devices incorporating with biological materials such as enzymes, nucleic acids (DNA and RNA), microorganisms, cells, antibodies and cell receptors, or biologically derived materials with a physicochemical detector (Dhawan et al., 2009). Electrochemical biosensors are commonly used to provide an alternative to the analysis of certain molecules because of their simplicity, low cost, fast response, and high sensitivity. In recent years, hybrid composite materials have been improved instead of single component materials for developing biosensor technology. Thereby, a new material having superior properties of the two components is fabricated. Organic part of the hybrid polymer composite materials is based on conducting natural or synthetic polymers [Chi, polyaniline (PANI), etc.]. The metal oxide nanoparticles as inorganic part are preferred because of their large surface area. It was reported that hybrid polymer composites incorporating with metal oxide nanoparticles showed superior catalytic property and enhanced sensing performance compared to unsupported polymer (Jia et al., 2011). Because of these advantages, hybrid composite materials have been used more than pure materials in biosensor studies. There are many biosensor that are based on hybrid polymer composites in literature. In this part, electrochemical biosensors will be illustrated for the detection of biological samples which are frequently used in biology and medicine.

12.4.1 Glucose biosensors

Diabetes is caused by a lack of insulin, and this disease is determined by the concentration of glucose in the blood. High glucose levels can cause serious disorders in body (eyes, heart, nerves, feet, and kidneys) over time (Wang and Lee, 2015). Therefore, early and accurate detection of blood glucose level is very important to reduce the above risks. To determine the glucose concentration, optical, capacitive detection, electro-chemiluminescence, colorimetric, and electrochemical techniques are present in literature. Yet, electrochemical-based biosensor received attention due to their rapid response, high sensitive, low cost, and simplicity (Choudhary et al., 2014). Glucose biosensors are usually enzyme-based that glucose oxidase (GOx) is most widely used as the enzyme. The GOx (EC. 1.1.3.4) is an oxidoreductase that catalyses the oxidation of glucose to hydrogen peroxide (H_2O_2) and gluconolactone.

To improve a third generation glucose biosensor, graphene–polyethyleneimine–gold nanoparticles (GNS–PEI–AuNPs) hybrid composites were prepared using microwave-irradiation method (Rafighi et al., 2016). Au electrode was modified with the hybrid composite mixtures for proving amino groups at the electrode surface (Fig. 12.29). GOx was then immobilized on gold electrodes modified with GNS–PEI–AuNPs hybrid material and electrochemical performance of the glucose biosensor was determined. GNS–PEI–AuNPs provides both a conductive platform with high charge transfer property and binding sites for enzyme immobilization. The electrochemical behavior of the GOx modified electrode was investigated with glucose and without oxygen. The glucose biosensor exhibited a detection range of 10–100 μM with the sensitivity about 93 μA mM^{-1} cm^{-2} and detection limit of 0.32 μM ($S/N = 3$). It was reported that the amino groups of hybrid composite lead binding more GOx on the surface. The obtained results demonstrated that

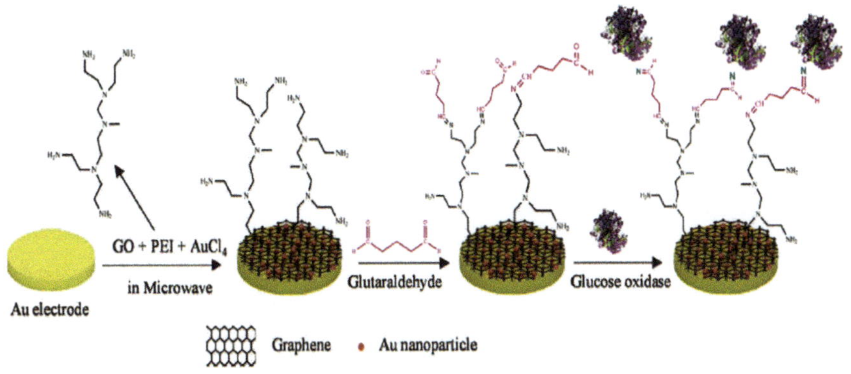

Figure 12.29 Schematic presentation of the route of biosensor, Au/GNS–PEI–AuNPs/Glu–GOx.
Reprinted with the permission from Rafighi, P., Tavahodi, M., & Haghighi, B. (2016). Fabrication of a third-generation glucose biosensor using graphene-polyethyleneimine-gold nanoparticles hybrid. Sens. Actuators, B: Chem., 232, 454–461. Copyright © 2016, Elsevier.

GNS–PEI–AuNPs hybrid material can be a good candidate for immobilization of biomolecules and production of glucose biosensors.

In another study, it was reported cheap, flexible, and disposable glucose biosensor using cellulose/tin oxide (SnO_2) hybrid material as carrier matrix (Mahadeva and Kim, 2011). The authors used cellulose membrane due to biodegradable, biocompatible, low cost, and flexible properties of it. SnO_2 was selected because of optically transparent, and well-known for electrical conductivity. Cellulose/SnO_2 hybrid composite was prepared via liquid phase deposition technique. The prepared hybrid composite exhibited electrical properties of SnO_2 and the natural properties of cellulose. GOx was then immobilized on cellulose/SnO_2 hybrid nanocomposites via physical adsorption method (Fig. 12.30). GOx immobilization was confirmed by the X-ray photoelectron spectroscopy analysis. The biosensor demonstrated wide linear response to glucose in the concentration ranges of 5–12 mM and coefficient value of 0.96. Based on these results, the authors commented that cellulose/SnO_2 hybrid nanocomposite can be used to fabricated cheap, flexible, and disposable glucose biosensors.

PANI–TNT hybrid composites were reported for glucose biosensor having biocompatible, high electrical conductivity, low electrochemical interference, and high signal/noise ratio (Zhu et al., 2015). The PANI–TNT hybrid composite was prepared by polymerizing aniline on TNTs using chemical oxidation polymerization method (Fig. 12.31). Then, the hybrid composite film was used to immobilize GOx on the surface modified glass carbon electrode. Electrochemical studies revealed

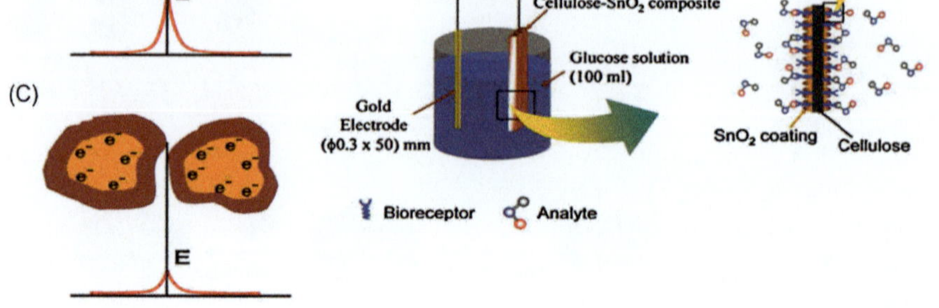

Figure 12.30 Schematic representation of the detection mechanism of cellulose–SnO_2 hybrid nanocomposite glucose biosensor. Increase in potential barrier at the grain boundaries (A), enzymatic reaction occurs between GOx and glucose (B), increase in potential barrier at the grain boundaries (C).

Reprinted with the permission from Mahadeva, S.K., Kim, J., 2011. Conductometric glucose biosensor made with cellulose and tin oxide hybrid nanocomposite. Sens. Actuators, B: Chem. 157, 177–182. Copyright © 2011, Elsevier.

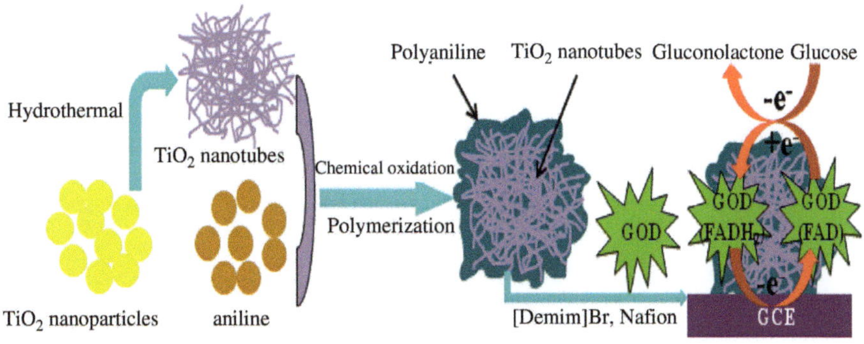

Figure 12.31 Fabrication process of the PANI–TNTs modified biosensor and the reaction scheme for catalytic oxidation of glucose under nitrogen atmosphere on it.
Reprinted with the permission from Zhu, J., Liu, X., Wang, X., Huo, X., & Yan, R. (2015). Preparation of polyaniline–TiO2 nanotube composite for the development of electrochemical biosensors. Sens. Actuators, B: Chem., 221, 450–457. Copyright © 2015, Elsevier.

that the GOx/PANI–TNT bioelectrode exhibited linearity from 10 to 2500 μM of glucose with a detection limit as 0.5 μM and sensitivity as 11.4 μA mM^{-1}. According to the results, research team reported that the designed biosensor was superior than other TNT-based sensor and other TiO_2 related composite materials in terms of catalytic capability and analytical performance.

In above study, the electrochemical properties of biosensor were examined after the GOx enzyme was immobilized on electrode surface. In the current study, nonenzymatic biosensors were designed without GOx enzyme. Choudhary et al. (2014) prepared polyaniline-supported copper(I) composite by using in situ, one-step chemical synthesis route (Fig. 12.32). The designed nonenzymatic glucose biosensor showed high sensitivity, fast response, low detection limit, and good stability. As a result, nonenzymatic biosensors showed a high electrocatalytic activity toward glucose just as enzymatic sensor. Besides, the authors interpreted that prepared inorganic–organic composite may be used a wide range of applications such as bioelectronics and next-generation energy storage electrode.

In addition to the aforementioned hybrid polymer composite, a lot of electrochemical glucose biosensors have been developed on the basis of hybrid polymer composites such as poly(pyrrole)-latex hybrit composite (Kros et al., 2001), Chi /silica hybrid composite film (Tan et al., 2005), polyaniline/carbon nanotubes composites (Li et al., 2008; Pilan and Raicopol, 2014), titanium dioxide–cellulose hybrid nanocomposite (Maniruzzaman et al., 2012), poly(*m*-phenylenediamine)–Prussian blue hybrid film (Tao et al. 2013), protein (BSA) and polysaccharide (Chi)-composite sol–gel silicate film (Matsuhisa et al., 2013), copper nanoparticles/poly(*o*-phenylenediamine) hybrid nanocomposites (Liu et al., 2015), polyaniline-ZrO_2 hybrid composite (Pahurkar et al., 2015), reduced GO doped poly(3,4-ethylenedioxythiophene) hybrid polymer composite (Hui et al. 2015), so on.

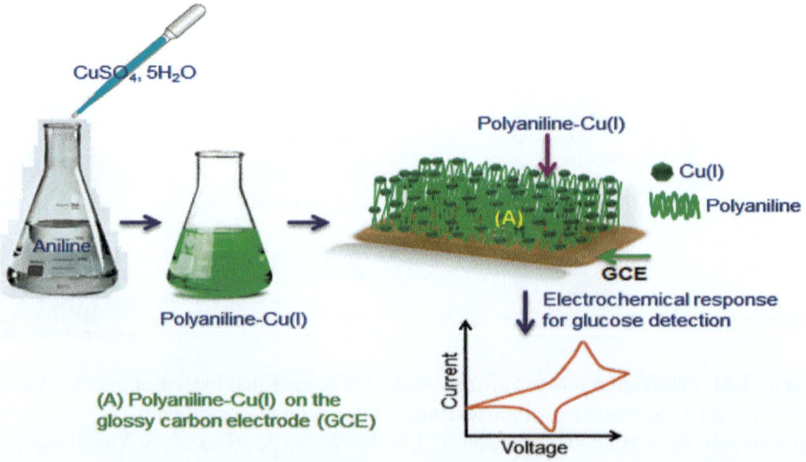

Figure 12.32 Schematic illustration of fabrication process of the PANI–Cu(I) modified biosensor.
Reprinted with the permission from Choudhary, M., Shukla, S.K., Taher, A., Siwal, S., & Mallick, K. (2014). Organic – inorganic hybrid supramolecular assembly: an efficient platform for nonenzymatic glucose sensor. Sustainable Chem. Eng., 2, 2852–2858. Copyright © 2014, American Chemical Society.

12.4.2 Urea biosensors

Urea is the end product of protein metabolism in the human body. The normal level of urea in the blood serum is between 15 and 40 mg dL^{-1} (Rahmanian et al., 2015). High concentration of urea can cause renal failure (acute or chronic), urinary tract obstruction, dehydration, shock, burns, and gastrointestinal bleeding. On the other hand, low concentration of urea can cause liver failure, nephrotic syndrome, cachexia (low-protein and high carbohydrate diets) (Kaushik et al., 2009). Therefore, the accurate and rapid detection of urea has recently been considered very important. For detection of urea, various chromatographic, calorimetry analysis, and colorimetric methods have been reported. In comparison, electrochemical sensors can allow simple, rapid, sensitive, and cost-effective urea detection. For urea detection, urease (Ur, E.C 3.5.1.5) is immobilized on carrier matrix. It catalyzes the hydrolysis of urea into carbon dioxide and ammonia. There are many polymers and composites as carrier matrix. However, due to their low surface area, electron transfer, it is not efficient between analytes and electrode. Therefore, for enhancing surface area, hybrid polymer composites have been prepared using nanometal or metal oxides as additives in the functionalization of polymers. The prepared materials have been used frequently to increase the biosensor stability and sensitivity because of their high surface/volume ratio, high-surface reaction activity, high catalytic efficiency, and strong adsorption capability, recently.

For instance, Kaushik et al. (2009) reported Chi/superparamagnetic iron oxide (Fe_3O_4) nanoparticles hybrid composite to develop a new type of amperometric urea biosensor (Fig. 12.33). The hybrid nanobiocomposite film was used to coimmobilize

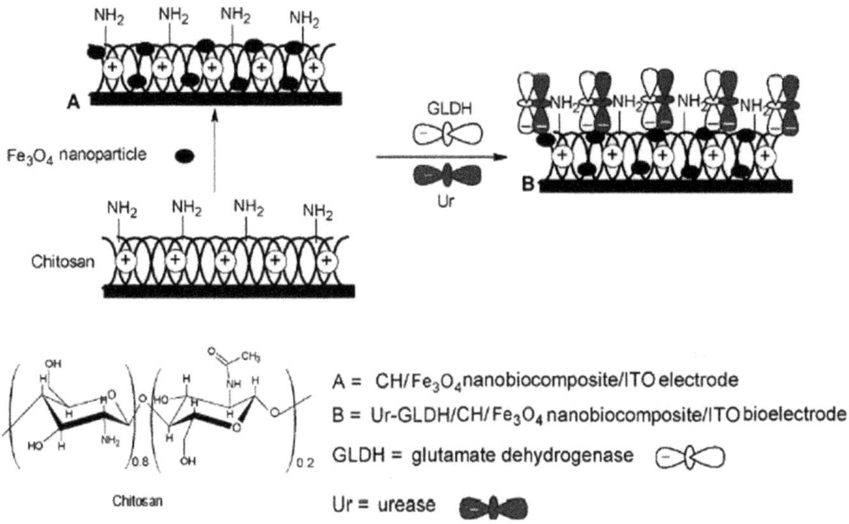

Figure 12.33 Proposed mechanism for the preparation of CH−Fe_3O_4 nanobiocomposite and immobilization of Ur−GLDH onto CH−Fe_3O_4 nanobiocomposite film.
Reprinted with the permission from Kaushik, A., Solanki, P.R., Ansari, A.A., Sumana, G., Ahmad, S., & Malhotra, B.D. (2009). Iron oxide-chitosan nanobiocomposite for urea sensor. Sens. Actuators, B: Chem., 138, 572−580. Copyright © 2009, Elsevier.

Ur and glutamate dehydrogenase on the surface of indium-tin oxide-modified glass carbon electrode. The bioelectrode was linear with the concentration of urea in the range of 5−100 mg dL^{-1} with the sensitivity 12.5 μA mM^{-1} cm^{-2}. The detection limit was 0.5 mg dL^{-1} and the fast response achieved within 10 s. The electrochemical results reveal that the biosensor showed wide linearity, fast response, good selectivity, and antiinterference performance. In addition, Fe_3O_4 nanoparticles enhanced electron transfer and increased shelf-life of the biosensor due to the increase in active surface area of the hybrid nanobiocomposite. The authors underlined that the designed biosensor can be used for measurement of real serum sample.

In another paper, inexpensive, flexible, and disposable urea biosensor based on cellulose-SnO_2 hybrid nanocomposite was reported (Mahadeva and Kim, 2013). Initially, the research team fabricated cellulose−SnO_2 hybrid nanocomposite film by liquid-phase deposition technique. Then, urease was immobilized on the electrode based on the hybrid composite film using physical absorption technique. Fig. 12.34 demonstrates the schematic representation of detection mechanism of urea biosensor. The urea concentration in blood serum is less than 10 mM. The fabricated biosensor exhibited a linearity in the range from 0 to 42 mM and 0.5 mM as detection limit. It was also found that the sensitivity of sensor did not change for 7 days. Therefore, the fabricated urea biosensor can be tested in real samples with qualitative promising responses.

A novel impedimetric urea biosensor based on PVA/ZnO nanostructured hybrid film was developed by Rahmanian and Mozaffari (2015). Fabrication of biosensor

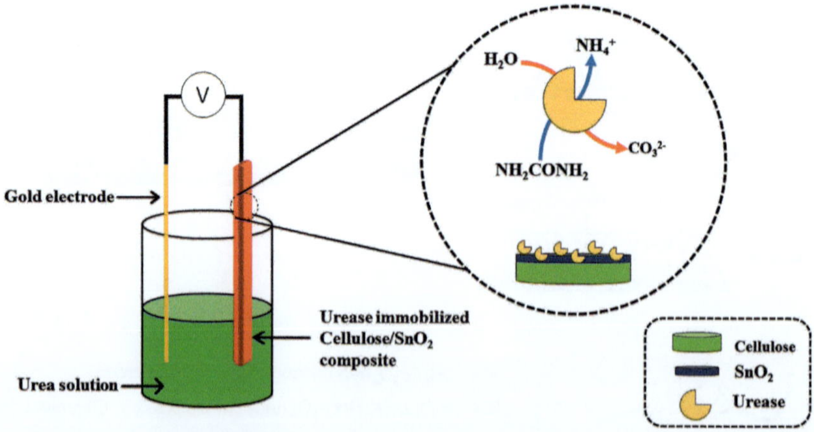

Figure 12.34 Schematic representation of detection mechanism of cellulose–SnO_2 hybrid nanocomposite urea biosensor.
Reprinted with the permission from Mahadeva, S.K., Kim, J., 2013. Porous tin-oxide-coated regenerated cellulose as disposable and low-cost alternative transducer for urea detection. IEEE. Sens. J. 13, 2223–2228. Copyright © 2013, IEEE Sensors Council.

was schematized in Fig. 12.35. PVA, as polymeric organic additive, is a nontoxic and biocompatible synthetic polymer. It has good electron communication and it is an ideal material for making an enzyme electrode (Pundir et al., 2010). In this study, because of their hydroxyl functional groups, PVA provides excellent surface to enhanced urease immobilization via covalent linking. The performance of proposed biosensor was carried out using cyclic voltametry and electrochemical impedance spectroscopy techniques. The linear response of the biosensor was in the range from 5.0 to 125.0 mg dL^{-1}. The response time and detection limit of biosensor were determined as 3 s and 3.0 mg dL^{-1}, respectively. Moreover, this biosensor has a number of advantages such as high selectivity, reducing in signal of negatively charged interferents (such as ascorbic and uric acids), high stability, and reusability. It was concluded that the rapid and sensitive urea biosensor may be used for the detection of urea in blood serum.

PANI is one of the most intensively polymers investigated by researchers, due to its excellent environmental stability, good electronic properties, and strong biomolecular interactions. The Pt nanoflower/PANI composite nanofibers were fabricated by Jia et al. (2011) as urea biosensor. First, PANI nanofibers were prepared by the in-situ polymerization of aniline. Then, urease was immobilized in Pt nanoflower/PANI composite nanofibers matrix, and this sensor was used for urea detection in a flow-injection-analysis system. Linear range of the biosensor was up to 20 mM, and a limit of detection of 10 μM ($S/N = 3$). Also, it showed an outstanding antiinterference property against chloride ion. The strong interactions both of Pt nanoflowers-amine groups in urea and PANI-ammonia (formed by enzymatic hydrolysis of urea) led to selective detection of urea. The author concluded that the hybrid composite material consists of conductive polymer and metal nanoparticles can be used in a wide range of applications.

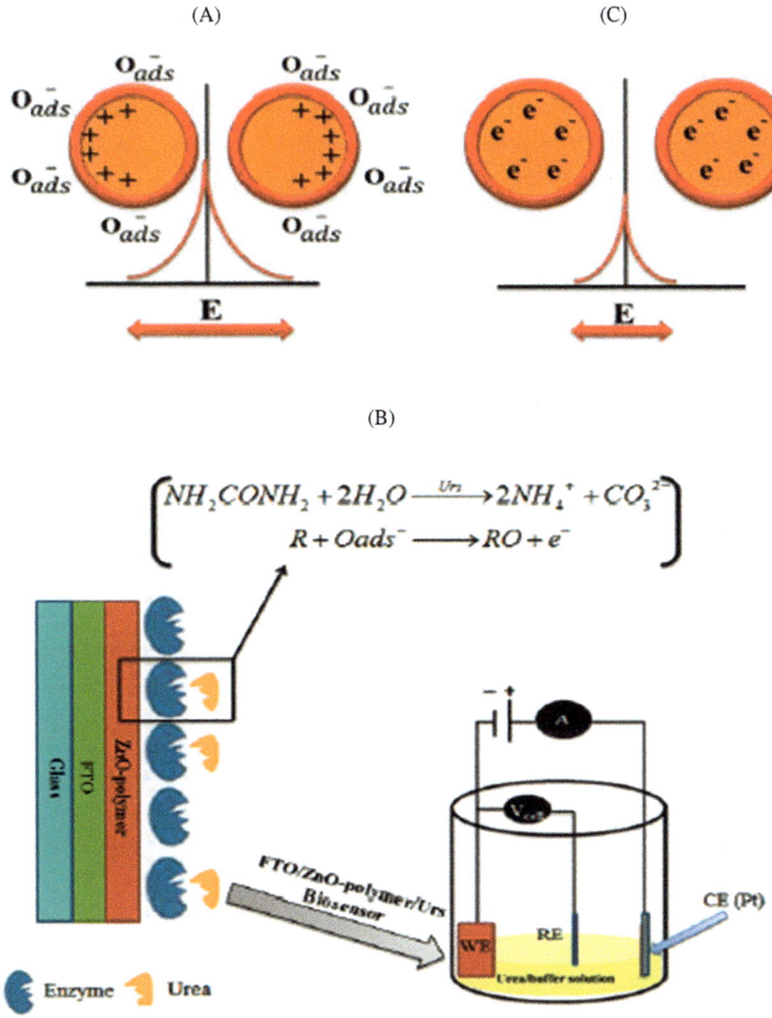

Figure 12.35 Schematic representation of the detection mechanism of FTO/ZnO–Polymer/Urs urea biosensor, increase in potential barrier at the grain boundaries (A), enzymatic reaction occurs between urease and urea (B), decrease in potential barrier at the grain boundaries (C).
Reprinted with the permission from Rahmanian, R., Mozaffari, S.A., & Abedi, M. (2015). Disposable urea biosensor based on nanoporous ZnO film fabricated from omissible polymeric substrate. Mater. Sci. Eng. C, 57, 387–396. Copyright © 2015, Elsevier.

12.4.3 Hydrogen peroxide biosensors

The fast and precise detection of H_2O_2 is of great importance in terms of clinical, biological, food, medicine, and the environment analysis (Chirizzi et al., 2016). Until now, many strategies have been developed such as titrimetric, spectrophotometric, chemiluminescence, and electrochemical methods for the detection of H_2O_2. Among these strategies, electrochemical technique is one of the most commonly used methods due to its relatively high sensitivity, inexpensive, and good selectivity. Generally, researches have been done on the amperometric, potentiometric, and impedimetric or conductometric H_2O_2 biosensors based on the horseradish peroxidase (HRP). For more information, the recent articles for H_2O_2 biosensors were highlighted and summarized in below.

A novel Au/Cerium dioxide (CeO_2)−Chi composite biosensor was designed, fabricated, and characterized for H_2O_2 detection. Furthermore, the hybrid composite was used for HRP immobilization. The components of hybrid polymer composite have been used due to electron transfer capability of Au nanoparticles, enhancing the biocompatibility of CeO_2 nanoparticles and film-forming ability of Chi, respectively. The resulting Au/CeO_2−Chi composite showed high conductivity and biocompatibility. The electrochemical performance of the prepared sensor was investigated by cyclic voltametry, electrochemical impedance spectroscopy, and typical amperometric methods. The results demonstrated that the sensor exhibits linearity in a wide range of 0.05−2.5 mM ($r = 0.998$) with a detection limit of 7 μM ($S/N = 3$). In addition, the biosensor showed satisfactory affinity ($K_m^{app} = 1.93$ mM), storage stability and good reproducibility. The present biosensor promises potential applications in electrochemical biosensing (Zhang et al., 2012).

The biosensors consist of enzyme have some problems such as half-life and stability of enzyme. Furthermore, the enzymes are expensive and enzyme immobilization is a time-consuming procedure. Therefore, hybrid polymer composite-based nonenzyme biosensors have become an increasing trend, nowadays.

A polypyrrole (PPy)/platinum nanoparticles (Pt NPs) hybrid nanocomposite was prepared by a simple, ultrafast, and microwave-assisted polyol process (Xing et al., 2015). The structure of this PPy/Pt nanocomposite biosensor is shown in Fig. 12.36.

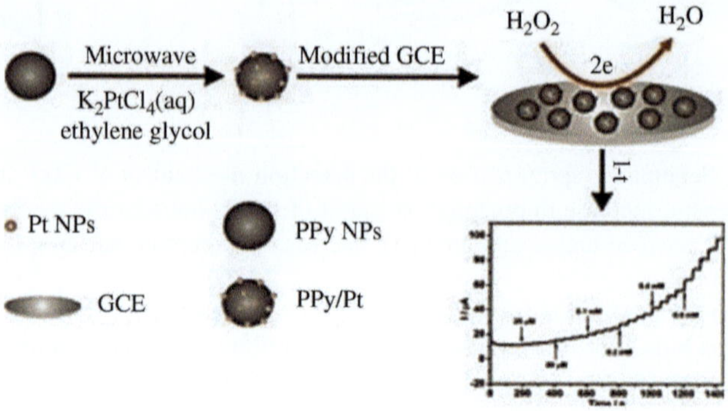

Figure 12.36 Schematic illustration of a nonenzyme electrochemical sensor of H_2O_2 based on PPy/Pt nanocomposite.
Reprinted with the permission from Xing, L., Rong, Q., & Ma, Z. (2015). Non-enzymatic electrochemical sensing of hydrogen peroxide based on polypyrrole/platinum nanocomposites. Sens. Actuators, B: Chem., 221, 242−247. Copyright © 2015, Elsevier.

PPy is widely used due to its high electronic conductivity and environmental stability for the detection of various molecules. Pt NPs were also selected as the inorganic molecules due to their superior electrocatalytic activities. The glass carbon electrode was modified with hybrid composite. Under the optimal conditions, nonenzymatic electrochemical H_2O_2 sensor exhibited the sensitivity of 164.94 μA mM^{-1} cm^{-2} with the detection limit of 0.6 μM and a linear H_2O_2 concentration range from 500 μM to 6.3 mM. The nonenzymatic biosensor also showed high sensitivity, good repeatability, good reproducibility, and long-term stability.

With the same strategy, Fang et al. (2012) reported poly(vinyl alcohol)–MWCNTs. Pt NP hybrids were chemically fabricated and were used for modification of glassy carbon electrode for development of a nonenzymatic H_2O_2 biosensors. The biosensor showed a rapid response (5 s) to H_2O_2 in the linear range of 0.002–3.8 mM and had a high sensitivity of 122.63 μA mM^{-1} cm^{-2}.

In the recent, the enzyme-mimetic catalyst works are trend for biosensor technology. For this purpose, researchers have given intense efforts to develop the artificial enzyme mimetics. In the work, a reversible cellulose-based colorimetric H_2O_2 biosensor was fabricated by immobilization of peroxidase-like Fe^{3+} onto mesoporous silica film precoated natural cellulose substance for enzyme-mimetic H_2O_2 detection. First, mesoporous silica–cellulose hybrid composites were successfully fabricated by surface sol – gel coating method. After, hexadecyl trimethyl ammonium bromide template in the silica films was removed and thus, specific surface area (80.7 m^2 g^{-1}) was obtained higher than that of natural cellulose. The resultant hybrid composite material demonstrated excellent sensitivity for the H_2O_2 just as the peroxidase. Fig. 12.37 showed schematic illustration of the fabrication process for cellulose-based peroxidase-like catalyst. The detection limit was determined about 1 μmol L^{-1}. The proposed H_2O_2 biosensor can be recycled for 10 times

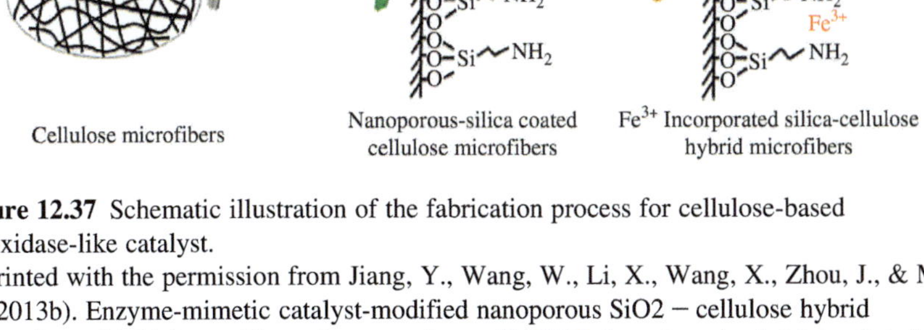

Figure 12.37 Schematic illustration of the fabrication process for cellulose-based peroxidase-like catalyst.
Reprinted with the permission from Jiang, Y., Wang, W., Li, X., Wang, X., Zhou, J., & Mu, X. (2013b). Enzyme-mimetic catalyst-modified nanoporous SiO2 – cellulose hybrid composites with high specific surface area for rapid H2O2 detection. Appl. Mater. Interfaces, 5, 1913 – 1916. Copyright © 2013, American Chemical Society.

without losing the peroxidase-like activity. In addition, it has various advantages such as stable hybrid structure, extraordinary sensitivity, and reusability. Thus, this enzyme-mimetic catalyst method can guide to future applications (Jiang et al., 2013b).

12.4.4 Cholesterol biosensors

Cholesterol is an essential molecule in the assessment of the risk of thrombosis and myocardial infarction disease as well as diagnosis of coronary heart disease, atherosclerosis, and other clinical (lipids) disease (Gomathi et al., 2011). Therefore, it is necessary to detect as accurate/quickly of cholesterol in the blood of human and so it is important to develop high sensitive cholesterol biosensors. Amperometric biosensors have been used due to low limit of detection and enzyme stabilization. Enzyme-based cholesterol sensors are a leader among biosensor systems because of advantages such as high sensitivity and wide linear range. Cholesterol oxidase (ChOx) has widely been used for the production of cholesterol-biosensor. It catalyzes the oxidation of cholesterol to H_2O_2 and cholest-4-en-3-one in the presence of oxygen. Organic–inorganic (hybrid) composites have been used as a cholesterol sensor by researchers.

The PANI-hybrid inorganic/organic composites have a wide range of applications (particularly electronics design and electrochemical sensors) due to advantages such as easily preparation and high conductivity. Zhang et al. (2013) developed polyaniline/Au nanocomposite-based hybrid cholesterol biosensor via seed-mediated strategy (Fig. 12.38). The glass carbon electrode was modified with prepared hybrid composite and ChOx was then immobilized on the electrode. The analytic performance of the sensor was tested by cyclic voltametry and electrochemical impedance spectroscopy methods. The resulted biosensor exhibited a high sensitivity linear response to cholesterol.

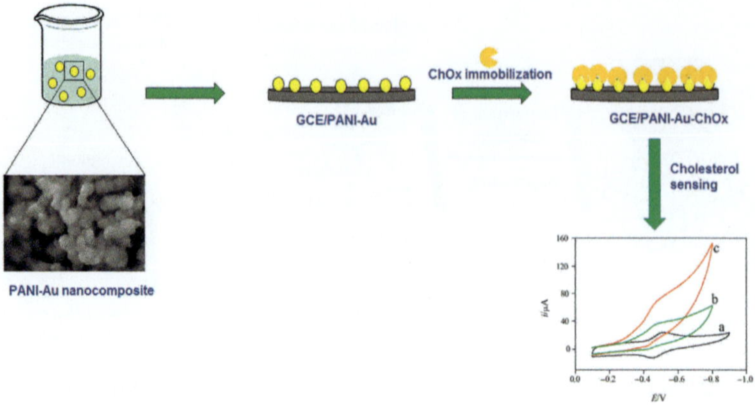

Figure 12.38 Schematic illustration of the fabrication process for cholesterol sensor. Reproduced with the permission from Zhang, H., Liu, R., & Zheng, J., (2013). Seed-mediated synthesis of polyaniline/Au nanocomposite and its application for a cholesterol biosensor. Synth. Met., 167, 5–9. Copyright © 2013, Elsevier.

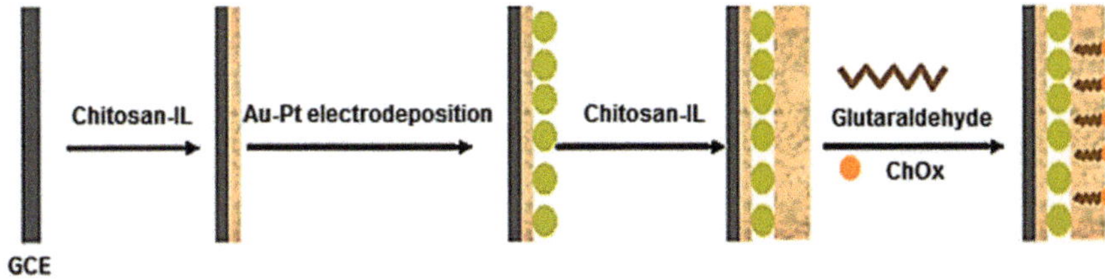

Figure 12.39 Schematic illustration of the preparation procedures of ChOx/AuPt–Ch–IL/GCE. Reprinted with the permission from Safavi, A., Farjami, F., 2011. Electrodeposition of gold–platinum alloy nanoparticles on ionic liquid–chitosan composite film and its application in fabricating an amperometric cholesterol biosensor. Biosens. Bioelectron. 26, 2547–2552. Copyright © 2013, Elsevier.

In another amperometric cholesterol sensor work, the gold–platinum alloy nanoparticles/ Chi /ionic liquid hybrid composites were prepared for modifying of glass carbon electrode (Fig. 12.39). Alloys have been used as inorganic matrix due to its high sensitivity, fast response time, wide linear range, high selectivity, and reproducibility. Chi may accumulate metal ions through various chelating mechanisms depending on metal ions and pH. Due to high ionic conductivity and good thermal stability, the ionic liquids have been used in different applications such as lithium battery, electrochemical capacitors, and solar cells.

The amperometric reduction peak linearly increased depending on cholesterol concentration (0.05–6.2 and 6.2–11.2 mM). The proposed bioelectrode was selective to cholesterol with high sensitivity (90.7 $\mu A\ mM^{-1}\ cm^{-2}$), lower detection limit (10 μM) and response time of less than 7 s. The sensor was applied to the determination of cholesterol in blood serum samples and the results showed that the recoveries were in the range of 92%–106% (Safavi and Farjami, 2011).

Apart from the aforementioned work, free enzyme-cholesterol sensors have also been used for cholesterol detection. In the study, a novel nonenzymatic cholesterol biosensor based on β-cyclodextrin (β-CD) functionalized graphene was produced by Agnihotri et al. (2015). Graphene-β-CD hybrid composite is hydrophilic, electroactive and has a high surface area. Graphene layers accelerate electrochemical signal and cholesterol molecules can enter into the hydrophobic inner cavity of β-CD due to the host–guest interaction (Fig. 12.40). The electrochemical results demonstrated that the detection limit for cholesterol was 1 mM. This value was relatively low concentration compared with the other data in the literature. Consequently, β-CD functionalized graphene material is an advantage hybrid material to detection of cholesterol in low concentration.

12.4.5 Acetylcholinesterase biosensors

Because of its high toxicity, extensive use of organophosphate pesticides (OPs) causes serious problems in human life (Marinov et al., 2010). Therefore, rapid and reliable determination of OPs has become increasingly important for the protection

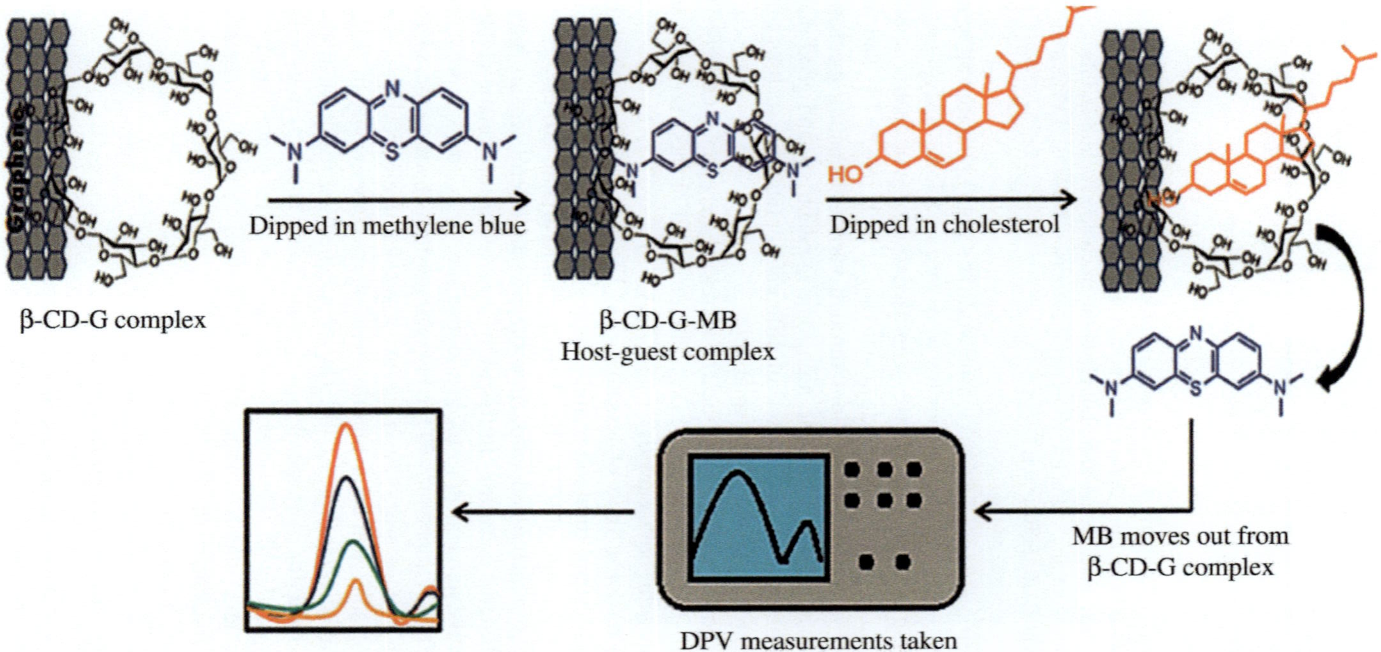

Figure 12.40 Demonstrating the mechanism of cholesterol sensing using Grp-β-CD as the working matrix.
Reprinted with the permission from Agnihotri, N., Chowdhury, A.D., & De, A. (2015). Non-enzymatic electrochemical detection of cholesterol using β-cyclodextrin functionalized graphene. Biosens. Bioelectron., 63, 212–217. Copyright © 2015, Elsevier.

of public health and safety. Although there are many methods, enzymatic methods have frequently used because of some advantages such as fast response, low cost, and high sensitivity. Among them, amperometric acetylcholinesterase (AChE) biosensors have shown satisfactory results for the analysis of pesticides. For the AChE immobilization, hybrid composite systems have been gained much attention by researchers due to their advantages.

A new nanosized calcium carbonate Chi hybrid composites (nano$CaCO_3$–chi) were prepared through one-step coelectrodeposition method and were used as AChE immobilization matrix. The $CaCO_3$–Chi hybrid composite facilitated equal distribution of the substrate, electron transfer and determination of pesticide. As Fig. 12.41 shows, the hybrid composite had biocompatible and 3D porous structure. The solid electrode surface was modified with AChE/$CaCO_3$–chi and biosensor was used to determine methyl parathion as a model OPs (Gong et al., 2009). The biosensor exhibited linearity in the range from 0.005–0.2 to 0.75–3.75 $\mu g\ mL^{-1}$

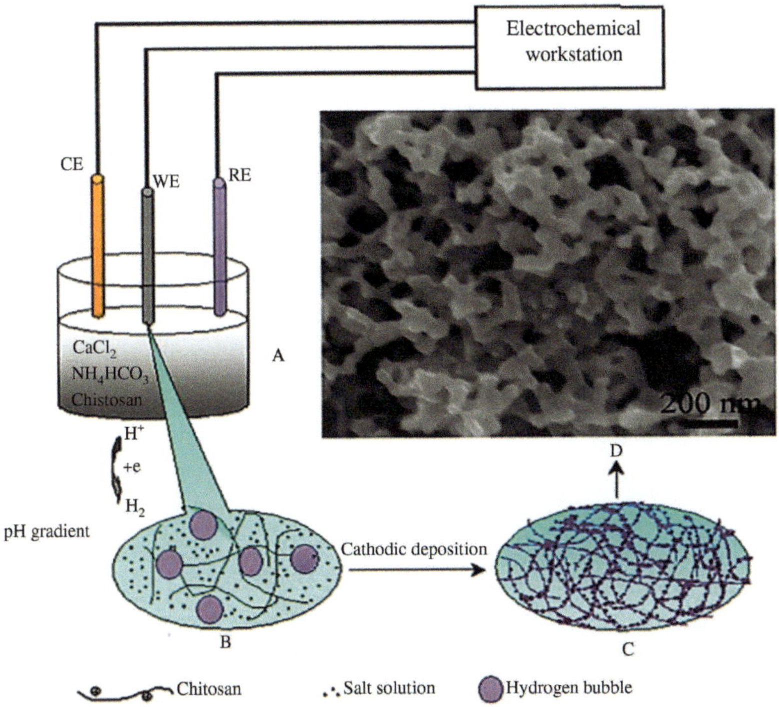

Figure 12.41 Illustration for the formation of porous nano $CaCO_3$–chi composite by coelectrodeposition (A–C) and SEM image of nano $CaCO_3$–chi composite film onto GCE (D)
Reprinted with the permission from Gong, J., Liu, T., Song, D., Zhang, X., & Zhang, L. (2009). One-step fabrication of three-dimensional porous calcium carbonate–chitosan composite film as the immobilization matrix of acetylcholinesterase and its biosensing on pesticide. Electrochem. Commun., 11, 1873–1876. Copyright © 2009, Elsevier.

and a detection limit of 1 ng mL^{-1} for inhibition methyl parathion. Besides, the sensor showed good reproducibility and acceptable stability. The work team interpreted that $CaCO_3$–Chi hybrid composite would be useful in bioelectrochemistry application.

Zhai et al. (2013) studied Chi/prussian blue/multiwall carbon nanotubes/hollow gold nanospheres nanocomposite film and evaluated its potential application for acetylcholinesterase sensor (Fig. 12.42). Multiwalled carbon nanotubes are used for chemical stability, excellent electrical conductivity, and high mechanical strength. Prussian blue is a hexacyanoferrate material increasing interest for applications such as electrochemical sensors. The gold electrode surface was modified by one-step electrodeposition method using hybrid composite film. Then, AChE and nafion were immobilized on hybrid composite film. In this study, malathion, chlorpyrifos, monocrotophos, and carbofuran were used as pesticides models. According to pesticide measurement results of prepared sensor, biosensor exhibited a wide concentration range, low detection limit, high stability, and reusability. Moreover, the biosensor can also be used for direct analysis of practical samples, which would be a new promising tool for pesticide analysis.

In a similar study, Yang et al. (2005) studied determination of pesticides in vegetable samples using an acetylcholinesterase biosensor based on nanoparticles ZrO_2/Chi composite film (Fig. 12.43). To avoid interference from ascorbic acid, the

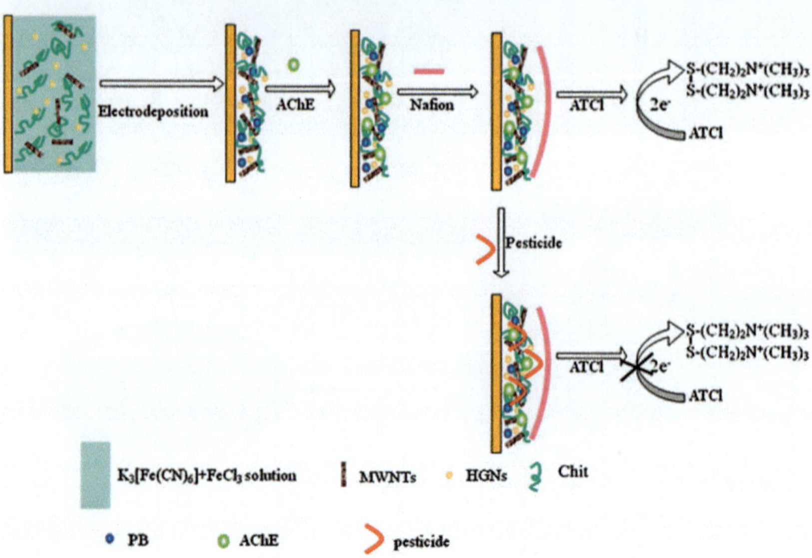

Figure 12.42 Schematic illustration of the stepwise AChE biosensor fabrication process and immobilized AChE inhibition in pesticide solution.
Reprinted with the permission from Zhai, C., Sun, X., Zhao, W., Gong, Z., & Wang, X. (2013). Acetylcholinesterase biosensor based on chitosan/prussian blue/multiwall carbon nanotubes/hollow gold nanospheres nanocomposite film by one-step electrodeposition. Biosens. Bioelectron., 42, 124–130. Copyright © 2013, Elsevier.

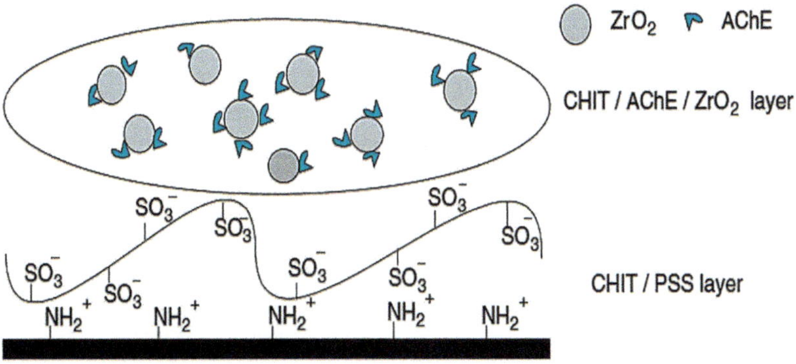

Figure 12.43 Schematic diagram of the preparation of the AChE electrode.
Reprinted with the permission from Yang, Y., Guo, M., Yang, M., Wang, Z., Shen, G., & Yu, R. (2005). Determination of pesticides in vegetable samples using an acetylcholinesterase biosensor based on nanoparticles ZrO2/chitosan composite film. Int. J. Environ. Anal. Chem., 85, 163–175. Copyright © 2005, Taylor & Francis Group Ltd.

electrode was coated with Chi–PSS. The electrode had a linear response to acetylthiocholine within 9.90×10^{-6} to 2.03×10^{-3} M, and detection limit is 1.3×10^{-6} M. In addition, ascorbic acid and other possible interferents (glycine, salicin, citric acid, and oxalic acid) had no significant interference owing to the specificity of the enzyme. As a result, this biosensor can be used to determine pesticides in a real vegetable sample.

12.5 Other biomedical applications

The use and spread of hybrid materials in the biomedical field particularly grow in parallel with the development of polymer chemistry. Functional structures and molecular designs improved in polymer chemistry are quickly transferred into the hybrid material field. Apart from the aforementioned common practices within this process, the gas-barrier applications, smart packaging, optic selective packaging, indicator featuring hybrid structures for different biomolecules also show itself quickly in some current practices in biomedical field. In these applications, gas barrier property is a very important application. Gas permeability is used in making tested packings. Clay doped polymeric hybrid structures show such properties. Clay is a laminated structure. During the adjustment of hybrid structure, if these layers are regularly and evenly distributed, it stops the progression of molecules such as oxygen, carbondioxide, and moisture into the package. With this way, the growth of bacteria is prevented in package. In biomedical field, the bacterial contamination is not desirable in medical and medical supplies. The shelf life of biomedical materials sterilized with such hybrid packages is extended. When there is a bacterial contamination in the package, color changes happen with antibodies or enzymes entrapped in sol–gel structures and thus user is warned. Gas passes during the prolonged waiting of bandages, medicated bandages, bumpers, bone cements, tissue adhesives, sterile instruments, and packaging materials in time. During this gas transition, bacterial activities begin and distress the use of medical supplies. This type of smart packaging alerts the user for either the bacterial activity or a reducing in sterilization by changing its color.

A new potential application that will be used of hybrid structures is biomedical materials which can be used in the detection of chemicals in the body from sweat and body fluids. For example, in order to immediately see the inflammation, the color change in a sticker which will affix forehead of the sick child is very important when body temperature rises. Thus, with this immediate intervention the child can be prevented from the seizures. Such structures are formed by doping of the thermochromic materials in polymers such as poly-*N*-isopropylacyrylamide. In another similar application, for diabetic contact lenses which show glucose levels are produced by Badugu et al. (2003). Lens structure linked boronic acid groups was obtained and the color changes were examined at different glucose concentrations. As a result, a significantly color change with successfully increasing glucose concentration in the body was identified (Badugu et al. 2003).

Another application area where hybrid materials are widely used is body imaging systems for diseases and cancer. Especially in this area, hybrid structures are obtained in binding of structures such as polymers, small organic biomarker molecules, proteins, peptides, nucleic acids, and antibodies on the surface of different size and shape of nanoparticles. These obtained hybrid compounds can be detected by a display system provided nanoparticles which are carried at the center. For different imaging systems in this area, different structures have been studied quite extensively (Lee et al., 2012; Owens et al., 2016). Apart from these applications, there are commercial and academic many hybrid structured systems such as antibacterial surfaces, the protective surfaces to avoid the exposure of biomedical devices from body fluids, new dental fillings, moisture absorbent systems in biomedical packaging, membranes used in air cleansing systems, and some private dialysis membranes in biomedical field.

12.6 Conclusion

The curiosity for recognition the world and the structures in nature of human beings shows that there are mixed and composite materials more than pure ones. In particular, we see mixed or binary structures connected by covalent bonds or secondary interactions in the shells of crustaceans, teeth and bones, and tree trunk. The most important reason for this is that the features of such dual structures do not present the average functionality of both components but superior than their total characteristics. By knowing these structures with the development in monitoring and structure enlightment systems, humans want to produce similar materials and use them in different areas. The hybrid materials area born with these approaches has begun to be used widely in chemistry, materials science, and medical field in recent years. Hybrid materials have formed an important class of materials very quickly in medicine and medical areas due to their high tissue compatibilities, superior performances, easily changeable features and structural diversities. Because of such advantages, they have replaced many traditional materials in biomedical product market. For instance, conventional dental fillings, traditional drug delivery matrix, and metallic implants left their places to hybrid materials today. In this section, the importance of organic–inorganic hybrid materials in biomedical field and their applications is tried to explain in detail. In particular, widespread use of available drug delivery systems, biomedical implants, biomedical sensors, scaffolds, and tissue engineering applications with current examples were described in detail. This introduction has presented the significance of the interface between inorganic and organic materials which has to be changed to overcome critical problems in the preparation of hybrid materials. Different building blocks and approaches can be produced by hybrid materials which show a bridge between inorganic and organic materials. In addition, the nano/micro-structure formation, processing and analysis of hybrid materials are also important. Lots of techniques can be used for these issues and several of them have been introduced to the reader.

References

Acero, J., Calderon, J., Salmeron, J., Verdaguer, J., Concejo, C., Somacarrera, M., 1999. The behaviour of titanium as a biomaterial: microscopy study of plates and surrounding tissues in facial osteosynthesis. J. Cranio-Maxillofac Surg. 27, 117–123.

Agnihotri, N., Chowdhury, A.D., De, A., 2015. Non-enzymatic electrochemical detection of cholesterol using β-cyclodextrin functionalized graphene. Biosens. Bioelectron. 63, 212–217.

Ahmed, R.A., Fekry, A.M., Farghali, R.A., 2013. A study of calcium carbonate/multiwalled-carbon nanotubes/chitosan composite coatings on Ti–6Al–4V alloy for orthopedic implants. Appl. Surf. Sci. 285 (B), 309–316.

Antoniou, G., Mikos, A.G., Temenoff, J.S., 2000. Formation of highly porous biodegradable scaffolds for tissue engineering. Electron. J. Biotechnol. 3 (2), 114–119.

Aumailley, M., Gayraud, B., 1998. Structure and biological activity of the extracellular matrix. J. Mol. Med. 76, 253–265.

Badugu, R., Lakowicz, J.R., Geddes, C.D., 2003. A glucose sensing contact lens: a non-invasive technique for continuous physiological glucose monitoring. J. Fluoresc. 13 (5), 371–374.

Bagheri, Z.S., Sawi, I.E., Schemitsch, E.H., Deroa, R.Z., Bougherar, H., 2013. Biomechanical properties of an advanced new carbon/flax/ epoxy composite material for bone plate applications. J. Mech. Behav. Biomed. Mater.,. 20, 398–406.

Baron, R., Tsouderos, Y., 2002. *In vitro* effects of S12911-2 on osteoclast function and bone marrow macrophage differentiation. Eur. J. Pharmacol. 450, 11–17.

Birgani, Z.T., Blitterswijk, C.A., Habibovic, P., 2016. Monolithic calcium phosphate/poly (lactic acid) composite versus calcium phosphate-coated poly(lactic acid) for support of osteogenic differentiation of human mesenchymal stromal cells. J. Mater. Sci.: Mater. Med.,. 54, 1–12.

Bonnelye, E., Chabadel, A., Saltel, F., Jurdic, P., 2008. Dual effect of strontium ranelate: stimulation of osteoblast differentiation and inhibition of osteoclast formation and resorption *in vitro*. Bone. 42, 129–138.

Bosio, V.E., Cacicedo, M.L., Calvignac, B., León, I., Beuvier, T., Boury, F., et al., 2014. Synthesis and characterization of $CaCO_3$–biopolymer hybrid nanoporous microparticles for controlled release of doxorubicin. Colloids Surf., B: Biointerfaces. 123, 158–169.

Bounabi, L., Mokhnachi, N.B., Haddadine, N., Ouazib, F., Barille, R., 2016. Development of poly(2-hydroxyethyl methacrylate)/clay composites as drug delivery systems of paracetamol. J. Drug Deliv. Sci. Technol. 33, 58–65.

Catauro, M., Bollino, F., Papale, F., Gallicchio, M., Pacifico, S., 2014. Synthesis and chemical characterization of new silica polyethylene glycol hybrid nanocomposite materials for controlled drug delivery. J. Drug Deliv. Sci. Technol. 24, 320–325.

Chen, J., Du, Y., Que, W., Xinga, Y., Lei, B., 2015. Content-dependent biomineralization activity and mechanical properties based on polydimethylsiloxane–bioactive glass–poly(caprolactone) hybrids monoliths for bone tissue regeneration. RSC Adv. 5, 61309–61317.

Chirizzi, D., Guascito, M.R., Filippo, E., Malitesta, C., Tepore, A., 2016. A novel nonenzymatic amperometric hydrogen peroxide sensor base on CuO@Cu_2O nanowires embedded into poly(vinylalcohol). Talanta. 147, 124–131.

Choudhary, M., Shukla, S.K., Taher, A., Siwal, S., Mallick, K., 2014. Organic – inorganic hybrid supramolecular assembly: an efficient platform for nonenzymatic glucose sensor. Sustainable Chem. Eng. 2, 2852–2858.

Cifuentes, S.C., 2015. Processing and characterization of novel biodegradable and bioresorbable PLA/Mg composites for osteosynthesis. *PhD Thesis*. Universidad Carlos III de Madrid, Madrid, Spain.

Cifuentes, S.C., Frutos, E., González-Carrasco, J.L., Muñoz, M., Multigner, M., Chao, J., et al., 2012. Novel PLLA/magnesium composite for orthopedic applications: a proof of concept. Mater. Lett. 74, 239–242.

Contessotto, L., Ghedini, E., Pinna, F., Signoretto, M., Cerrato, G., Crocell, V., 2009. Hybrid organic–inorganic silica gel carriers with controlled drug-delivery properties. Chemistry. 15, 12043–12049.

Dhawan, G., Sumana, G., Malhotra, B.D., 2009. Recent developments in urea biosensors. Biochem. Eng. J. 44, 42–52.

DiMarco, R.L., Heilshorn, S.C., 2012. Multifunctional materials through modular protein engineering. Adv. Mater. 24, 3923–3940.

Dorati, R., Colonna, C., Tomasi, C., Genta, I., Modena, T., Conti, B., 2013. Design of 3D hybrid composite scaffolds: effect of composition on scaffold structure and cell proliferation. Macromol. Symp. 334, 106–116.

Dorozhkin, S.V., 2009. Calcium orthophosphate-based biocomposites and hybrid biomaterials. J. Mater. Sci. 44, 2343–2387.

Fang, Y., Zhang, D., Qin, X., Miao, Z., Takahashi, S., Anzai, J., et al., 2012. A nonenzymatic hydrogen peroxide sensor based on poly(vinyl alcohol)–multiwalled carbon nanotubes–platinum nanoparticles hybrids modified glassy carbon electrode. Electrochim. Acta. 70, 266–271.

Fattahi, P., Yang, G., Kim, G., Abidian, M.R., 2014. A review of organic and inorganic biomaterials for neural interfaces. Adv. Mater. 26, 1846–1885.

Feng, B., Duan, H., Fu, W., Cao, Y., Zhang, W.J., Zhang, Y., 2015. Effect of inhomogeneity of the electrospun fibrous scaffolds of gelatin/polycaprolactone hybrid on cell proliferation. J. Biomed. Mater. Res. A. 103A, 431–438.

Gaharwar, A.K., Dammu, S.A., Canter, J.M., Wu, C.J., Schmidt, G., 2011. Highly extensible, tough, and elastomeric nanocomposite hydrogels from poly(ethylene glycol) and hydroxyapatite nanoparticles. Biomacromolecules. 12 (5), 1641–1650.

Gaharwar, A.K., Mihaila, S.M., Swami, A., Patel, A., Sant, S., Reis, R.L., et al., 2013. Bioactive silicate nanoplatelets for osteogenic differentiation of human mesenchymal stem cells. Adv. Mater. 25, 3329–3336.

Gain, A.K., Zhang, L., Liu, W., 2015. Microstructure and material properties of porous hydroxyapatite-zirconia nanocomposites using polymethyl methacrylate powders. Mater. Des. 67, 136–144.

Geiger, B., Bershadsky, A., Pankov, R., Yamada, K.M., 2001. Transmembrane crosstalk between the extracellular matrix and the cytoskeleton. Nat. Rev. Mol. Cell. Biol. 2, 793–805.

Gomathi, P., Ragupathy, D., Choi, J.H., Yeum, J.H., Lee, S.C., Kim, J.C., et al., 2011. Fabrication of novel chitosan nanofiber/gold nanoparticles composite towards improved performance for a cholesterol sensor. Sens. Actuators, B: Chem. 153, 44–49.

Gong, J., Liu, T., Song, D., Zhang, X., Zhang, L., 2009. One-step fabrication of three-dimensional porous calcium carbonate–chitosan composite film as the immobilization matrix of acetylcholinesterase and its biosensing on pesticide. Electrochem. Commun. 11, 1873–1876.

Griffith, L.G., 2002. Emerging design principles in biomaterials and scaffolds for tissue engineering. Ann. N. Y. Acad. Sci. 961, 83–95.

Grynpas, M.D., Hamilton, E., Cheung, R., Tsouderos, Y., Deloffre, P., Hott, M., et al., 1996. Strontium increases vertebral bone volume in rats at a low dose that does not induce detectable mineralization defect. Bone. 18, 253–259.

Hatano, K., Inoue, H., Kojo, T., Matsunaga, T., Tsujisawa, T., Uchiyama, C., et al., 1999. Effect of surface roughness on proliferation and alkaline phosphatase expression of rat calvarial cells cultured on polystyrene. Bone. 25 (4), 439–445.

Heidari, F., Razavi, M., Bahrololoom, M.E., Bazargan-Lari, R., Vashaee, D., Kotturi, H., et al., 2016. Mechanical properties of natural chitosan/hydroxyapatite/magnetite nanocomposites for tissue engineering applications. Mater. Sci. Eng.: C. 65, 338–344.

Hench, L.L., Polak, J.M., 2002. Third-generation biomedical materials. Science. 295 (5557), 1014–1017.

Hennink, W.E., Nostrum, C.F., 2012. Novel crosslinking methods to design hydrogels. Adv. Drug. Deliv. Rev. 64, 223–236.

Hickey, D.J., Ercan, B., Sun, L., Webster, T.J., 2015. Adding MgO nanoparticles to hydroxyapatite–PLLA nanocomposites for improved bone tissue engineering applications. Acta Biomater. 14, 175–184.

Hood, M.A., Mari, M., Muñoz-Espí, R., 2014. Synthetic strategies in the preparation of polymer/ınorganic hybrid nanoparticles. Materials. 7, 4057–4087.

Hoppe, A., Güldal, N.S., Boccaccini, A.R., 2011. A review of the biological response to ionic dissolution products from bioactive glasses and glass-ceramics. Biomaterials. 32, 2757–2774.

Huang, Z.M., Fujihara, K., 2005. Stiffness and strength design of composite bone plates. Compos. Sci. Technol. 65, 73–85.

Hui, N., Wang, S., Xie, H., Xu, S., Niu, S., Luo, X., 2015. Nickel nanoparticles modified conducting polymer composite of reduced graphene oxide doped poly(3,4-ethylenedioxythiophene) for enhanced nonenzymatic glucose sensing. Sens. Actuators, B: Chem. 221, 606–613.

Jia, W., Su, L., Lei, Y., 2011. Pt nanoflower/polyaniline composite nanofibers based urea biosensor. Biosens. Bioelectron. 30, 158–164.

Jiang, H., Zuo, Y., Zou, Q., Wang, H., Du, J., Li, Y., et al., 2013a. Biomimetic spiral-cylindrical scaffold based on hybrid chitosan/cellulose/nano-hydroxyapatite membrane for bone regeneration. ACS Appl. Mater. Interfaces. 5 (22), 12036–12044.

Jiang, Y., Wang, W., Li, X., Wang, X., Zhou, J., Mu, X., 2013b. Enzyme-mimetic catalyst-modified nanoporous SiO_2 – cellulose hybrid composites with high specific surface area for rapid H_2O_2 detection. Appl. Mater. Interfaces. 5, 1913–1916.

John, Ł., Podgórska, M., Nedelec, J., Cwynar-Zając, Ł., Dzięgiel, P., 2016. Strontium-doped organic-inorganic hybrids towards three-dimensional scaffolds for osteogenic cells. Mater. Sci. Eng., C. 68, 117–127.

Kaamyabi, S., Habibi, D., Amini, M.M., 2016. Preparation and characterization of the pH and thermosensitive magnetic molecular imprinted nanoparticle polymer for the cancer drug delivery. Bioorg. Med. Chem. Lett. 26, 2349–2354.

Kai, D., Prabhakaran, M.P., Stahl, B., Eblenkamp, M., Wintermantel, E., Ramakrishna, S., 2012. Mechanical properties and *in vitro* behavior of nanofiber–hydrogel composites for tissue engineering applications. Nanotechnology. 23 (9), 2012.

Kamari, Y., Ghiaci, M., 2016. Preparation and characterization of ibuprofen/modified chitosan/TiO_2 hybrid composite as a controlled drug-delivery system. Microporous Mesoporous Mater. 234, 361–369.

Karageorgiou, V., Kaplan, D., 2005. Porosity of 3D biomaterial scaffolds and osteogenesis. Biomaterials. 26, 5474–5491.

Kaushik, A., Solanki, P.R., Ansari, A.A., Sumana, G., Ahmad, S., Malhotra, B.D., 2009. Iron oxide-chitosan nanobiocomposite for urea sensor. Sens. Actuators, B: Chem. 138, 572–580.

Koytepe, S., Kucuk, I., Seckin, T., Adiguzel, H.I., 2015. Preparation, characterization, and properties of novel polyimide–SiO_2 Hybrid composites based on bipyridine for low dielectric applications. Polym. Plast. Technol. Eng. 54 (12), 1251–1262.

Kros, A., van Hövel, S.W.F.M., Nolte, R.J.M., Sommerdijk, N.A.J.M., 2001. Sens. Actuators, B: Chem. 80, 229–233.

Langer, R., Vacanti, J.P., 1993. Tissue engineering. Science. 260 (5110), 920–926.

Lau, H.K., Kiick, K.L., 2015. Opportunities for multicomponent hybrid hydrogels in biomedical applications. Biomacromolecules. 16 (1), 28–42.

Lee, K.Y., Mooney, D.J., 2001. Hydrogels for tissue engineering. Chem. Rev. 101 (7), 1869–1880.

Lee, D.E., Koo, H., Sun, I.C., Ryu, J.H., Kim, K., Kwon, I.C., 2012. Multifunctional nanoparticles for multimodal imaging and theragnosis. Chem. Soc. Rev. 41, 2656–2672.

LeGeros, R.Z., 2002. Properties of osteoconductive biomaterials: calcium phosphates. Clin. Orthop. Relat. Res.,. 395, 81–98.

Lei, B., Chen, X., Hanab, X., Zhou, J., 2012. Versatile fabrication of nanoscale sol–gel bioactive glass particles for efficient bone tissue regeneration. J. Mater. Chem. 22, 16906–16913.

Levenberg, S., Langer, R., 2004. Advances in tissue engineering. Curr. Top. Dev. Biol. 61, 113–134.

Li, L., Sheng, Q., Zheng, J., Zhang, H., 2008. Facile and controllable preparation of glucose biosensor based on Prussian blue nanoparticles hybrid composites. Bioelectrochemistry. 74, 170–175.

Liao, C.Z., Li, K., Wong, H.M., Tong, W.Y., Yeung, K.W.K., Tjong, S.C., 2013. Novel polypropylene biocomposites reinforced with carbon nanotubes and hydroxyapatite nanorods for bone replacements. Mater. Sci. Eng.: C. 33 (3), 1380–1388.

Liao, J.F., Qu, Y., Chu, B.Y., Zhang, X.N., Qian, Z.Y., 2015. Biodegradable CSMA/PECA/Graphene porous hybrid scaffold for cartilage tissue engineering. Sci. Rep. 5 (9879), 1–16.
Liu, L., Chen, Y., Lv, H., Wang, G., Hu, X., Wang, C., 2015. Construction of a non-enzymatic glucose sensor based on copper nanoparticles/poly(o-phenylenediamine) nanocomposites. J. Solid State Electrochem.,. 19, 731–738.
Long, T., Yang, J., Shi, S.S., Guo, Y.P., Ke, Q.F., Zhu, Z.A., 2015. Fabrication of three-dimensional porous scaffold based on collagen fiber and bioglass for bone tissue engineering. J. Biomed. Mater. Res. B Appl. Biomater. 103 (7), 1455–1464.
Mahadeva, S.K., Kim, J., 2011. Conductometric glucose biosensor made with cellulose and tin oxide hybrid nanocomposite. Sens. Actuators, B: Chem. 157, 177–182.
Mahadeva, S.K., Kim, J., 2013. Porous tin-oxide-coated regenerated cellulose as disposable and low-cost alternative transducer for urea detection. IEEE. Sens. J. 13, 2223–2228.
Maharjan, B., Joshi, M.K., Tiwari, A.P., Park, C.H., Kim, C.S., 2017. In-situ synthesis of AgNPs in the natural/synthetic hybrid nanofibrous scaffolds: Fabrication, characterization and antimicrobial activities. J. Mech. Behav. Biomed. Mater. 65, 66–76.
Maillard, J.Y., Hartemann, P., 2013. Silver as an antimicrobial: facts and gaps in knowledge. Crit. Rev. Microbiol. 39 (4), 373–383.
Maniruzzaman, M., Jang, S.D., Kim, J., 2012. Titanium dioxide–cellulose hybrid nanocomposite and its glucose biosensor application. Mater. Sci. Eng. B. 177 (2012), 844–848.
Marie, P.J., Felsenberg, D., Brandi, M.L., 2011. How strontium ranelate, via opposite effects on bone resorption and formation, prevents osteoporosis. Osteoporos. Int. 22, 1659–1667.
Marinov, I., Ivanov, Y., Gabrovska, K., Godjevargova, T., 2010. Amperometric acetylthiocholine sensor based on acetylcholinesterase immobilized on nanostructured polymer membrane containing gold nanoparticles. J. Mol. Catal. B: Enzym. 62, 67–75.
Matsuhisa, H., Tsuchiya, M., Yasushi, H., 2013. Protein and polysaccharide-composite sol–gel silicate film for an interference-free amperometric glucose biosensor. Colloids Surf., B: Biointerfaces. 111, 523–529.
Mauretti, A., Neri, A., Kossover, O., Seliktar, D., Nardo, P.D., Melino, S., 2016. Design of a novel composite H_2S-releasing hydrogel for cardiac tissue repair. Macromol. Biosci. 16, 847–858.
Mazaheri, M., Eslahi, N., Ordikhani, F., Tamjid, E., Simchi, A., 2015. Nanomedicine applications in orthopedic medicine: state of the art. Int. J. Nanomed. 10 (1), 6039–6054.
Moreira, C.D.F., Carvalho, S.M., Mansur, H.S., Pereira, M.M., 2016. Thermogelling chitosan–collagen–bioactive glass nanoparticle hybrids as potential injectable systems for tissue engineering. Mater. Sci. Eng.: C. 58, 1207–1216.
Neves, N., Campos, B.B., Almeida, I.F., Costa, P.C., Cabral, A.T., Barbosa, M.A., et al., 2016. Strontium-rich injectable hybrid system for bone regeneration. Mater. Sci. Eng.: C. 59, 818–827.
Owens, G.J., Singh, R.K., Foroutan, F., Alqaysi, M., Han, C.M., Mahapatra, C., et al., 2016. Sol–gel based materials for biomedical applications. Prog. Mater. Sci. 77, 1–79.
Pahurkar, V., Tamgadge, Y., Nagale, V., Gambhire, A., Muley, G., 2015. Glucose detection using polyaniline-ZrO_2 composite cladding modified fiber optic intrinsic biosensor. Bionano Front. 8, 261–264.
Peng, S., Liu, X.S., Huang, S., Li, Z., Pan, H., Zhen, W., et al., 2011. The cross-talk between osteoclasts and osteoblasts in response to strontium treatment: involvement of osteoprotegerin. Bone. 49, 1290–1298.
Pilan, L., Raicopol, M., 2014. Highly selective and stable glucose biosensors based on polyaniline/carbon nanotubes composites. UPB Sci. Bull., Ser. B: Chem. Mater. Sci.,. 76, 155–166.

Pitt, C.G., Gratzl, M.M., Kimmel, G.L., Surles, J., Schindler, A., 1981. Aliphatic polyesters II. The degradation of poly(DL-lactide), poly(epsilon-caprolactone), and their copolymers *in vivo*. Biomaterials. 2 (4), 215–220.

Popat, A., Liu, J., Lu, G.Q., Qiao, S.Z., 2012. A pH-responsive drug delivery system based on chitosan coated mesoporous silica nanoparticles. J. Mater. Chem. 22, 11173–11178.

Prodana, M., Duta, M., Lonita, D., Bojin, D., Stan, M.S., Dinischiotu, A., et al., 2015. A new complex ceramic coating with carbon nanotubes, hydroxyapatite and TiO_2 nanotubes on Ti surface for biomedical applications. Ceram. Int.,. 41 (5-A), 6318–6325.

Pundir, C.S., Singh, B.S., Narang, J., 2010. Construction of an amperometric triglyceride biosensor using PVA membrane bound enzymes. Clin. Biochem. 43, 467–472.

Rafighi, P., Tavahodi, M., Haghighi, B., 2016. Fabrication of a third-generation glucose biosensor using graphene-polyethyleneimine-gold nanoparticles hybrid. Sens. Actuators, B: Chem. 232, 454–461.

Rahmanian, R., Mozaffari, S.A., 2015. Electrochemical fabrication of ZnO-polyvinyl alcohol nanostructured hybrid film for application to urea biosensor. Sens. Actuators, B: Chem.: Chemical. 207, 772–781.

Rahmanian, R., Mozaffari, S.A., Abedi, M., 2015. Disposable urea biosensor based on nanoporous ZnO film fabricated from omissible polymeric substrate. Mater. Sci. Eng. C. 57, 387–396.

Russo, T., D'Amora, U., Gloria, A., Tunesi, M., Sandri, M., Rodilossi, S., et al., 2013. Advanced composites for hard-tissue engineering based on PCL/organic–inorganic hybrid fillers: from the design of 2D substrates to 3D rapid prototyped scaffolds. Proc. Eng. 59, 233–239.

Safari, M., Ghiaci, M., Jafari-Asl, M., Ensafi, A.A., 2015. Nanohybrid organic–inorganic chitosan/dopamine/TiO_2 composites with controlled drug-delivery properties. Appl. Surf. Sci. 342, 26–33.

Safavi, A., Farjami, F., 2011. Electrodeposition of gold–platinum alloy nanoparticles on ionic liquid–chitosan composite film and its application in fabricating an amperometric cholesterol biosensor. Biosens. Bioelectron. 26, 2547–2552.

Salama, A., El-Sakhawy, M., 2014. Preparation of polyelectrolyte/calcium phosphate hybrids for drug delivery application. Carbohydr. Polym. 113, 500–506.

Salazar, H., Lima, A.C., Lopes, A.C., Botelho, G., Lanceros-Mendez, S., 2015. Poly(vinylidene fluoride-trifluoroethylene)/NAY zeolite hybrid membranes as a drug release platform applied to ibuprofen release. Colloids Surf., A: Physicochem. Eng. Aspects. 469, 93–99.

Sanchez, C., Julián, B., Belleville, P., Popall, M., 2005. Applications of hybrid organic–inorganic nanocomposites. J. Mater. Chem. 15, 3559–3592.

Shahverdi, S., Hajimiri, M., Esfandiari, M.A., Larijani, B., Atyabi, F., Rajabiani, A., et al., 2014. Fabrication and structure analysis of poly(lactide-*co*-glycolic acid)/silk fibroin hybrid scaffold for wound dressing applications. Int. J. Pharm. 473, 345–355.

Shi, J., Qi, W., Du, C., Shi, J., Cao, S., 2013. Micro/nanohybrid hierarchical poly(*N*-isopropylacrylamide)/calcium carbonate composites for smart drug delivery. J. Appl. Polym. Sci. 129, 577–584.

Shi, J., Shi, J., Feng, D., Yue, P., Cao, S., 2014. Stimuli-responsive hybrid composites based on $CaCO_3$ microparticles and smart polyelectrolytes for controllable drug delivery. Polym. Bull. 71, 1857–1873.

Shi, H., Gan, Q., Liu, X., Ma, Y., Hu, J., Yuan, Y., et al., 2015. Poly(glycerol sebacate)-modified polylactic acid scaffolds with improved hydrophilicity, mechanical strength and bioactivity for bone tissue regeneration. RSC Adv. 5, 79703–79714.

Shin, Y.C., Lee, J.H., Jin, L., Kim, M.J., Kim, Y.J., Hyun, J.K., et al., 2015. Stimulated myoblast differentiation on grapheme oxide-impregnated PLGA-collagen hybrid fibre matrices. J. Nanobiotechnol. 13 (21), 1–11.

Sierra-Martin, B., Fernandez-Barbero, A., 2015. Multifunctional hybrid nanogels for theranostic applications. Soft Matter. 11, 8205–8216.

Stevens, M.M., George, J.H., 2005. Exploring and engineering the cell surface interface. Science. 310 (5751), 1135–1138.

Streuli, C., 1999. Extracellular matrix remodelling and cellular differentiation. Curr. Opin. Cell. Biol. 11, 634–640.

Su, D., Jiang, L., Chen, X., Dong, J., Shao, Z., 2016. Enhancing the gelation and bioactivity of ınjectable silk fibroin hydrogel with laponite nanoplatelets. ACS Appl. Mater. Interfaces. 8, 9619–9628.

Sun, X., Zhao, X., Zhao, L., Li, Q., D'Ortenzio, M., Nguyen, B., et al., 2015. Development of a hybrid gelatin hydrogel platform for tissue engineering and protein delivery applications. J. Mater. Chem. B. 3, 6368–6376.

Tan, X.C., Tian, Y.X., Cai, P.X., Zou, X.Y., 2005. Glucose biosensor based on glucose oxidase immobilized in sol–gel chitosan/silica hybrid composite film on Prussian blue modified glass carbon electrode. Anal. Bioanal. Chem. 381, 500–507.

Tao, J.Z., Xu, G.R., Hao, H.L., Yang, F.X., Ahn, K.S., Lee, W.Y., 2013. Poly(m-phenylenediamine)–Prussian blue hybrid film formed by one-step electrochemical deposition for glucose biosensor. J. Electroanal. Chem.,. 689, 96–102.

Terzaki, K., Kissamitaki, M., Skarmoutsou, A., Fotakis, C., Charitidis, C.A., Farsari, M., et al., 2013. Pre-osteoblastic cell response on three-dimensional, organic-inorganic hybrid material scaffolds for bone tissue engineering. J. Biomed. Mater. Res. A. 101A (8), 2283–2294.

Valliant, E.M., Jones, J.R., 2011. Softening bioactive glass for bone regeneration: sol–gel hybrid materials. Soft Matter. 7, 5083–5095.

Vrana, N.E., Dupret, A., Coraux, C., Vautier, D., Debry, C., Lavalle, P., 2011. Hybrid titanium/biodegradable polymer implants with an hierarchical pore structure as a means to control selective cell movement. PLoS. ONE. 6 (5), 1–11.

Wan, P., Yuan, C., Tan, L.L., Li, Q., Yang, K., 2014. Fabrication and evaluation of bioresorbable PLLA/magnesium and PLLA/magnesium fluoride hybrid composites for orthopedic implants. Compos. Sci. Technol. 98, 36–43.

Wang, H.C., Lee, A.R., 2015. Recent developments in blood glucose sensors. J. Food Drug Anal.,. 23, 191–200.

Wang, J., Chen, J.S., Zong, J.Y., Zhao, D., Li, F., Zhuo, R.X., et al., 2010. Calcium carbonate/carboxymethyl chitosan hybrid microspheres and nanospheres for drug delivery. J. Phys. Chem. C. 114, 18940–18945.

Wang, D., Liu, W., Feng, Q., Dong, C., Liu, Q., Duan, L., et al., 2017. Effect of inorganic/organic ratio and chemical coupling on the performance of porous silica/chitosan hybrid scaffolds. Mater. Sci. Eng.: C. 70, 969–975.

Wu, M., Wang, Q., Liu, X., Liu, H., 2013. Biomimetic synthesis and characterization of carbon nanofiber/hydroxyapatite composite scaffolds. Carbon. N. Y. 51, 335–345.

Xing, L., Rong, Q., Ma, Z., 2015. Non-enzymatic electrochemical sensing of hydrogen peroxide based on polypyrrole/platinum nanocomposites. Sens. Actuators, B: Chem. 221, 242–247.

Yang, Y., Guo, M., Yang, M., Wang, Z., Shen, G., Yu, R., 2005. Determination of pesticides in vegetable samples using an acetylcholinesterase biosensor based on nanoparticles ZrO_2/chitosan composite film. Int. J. Environ. Anal. Chem.,. 85, 163–175.

Yang, F., Yang, D., Tu, J., Zheng, Q., Cai, L., Wang, L., 2011a. Strontium enhances osteogenic differentiation of mesenchymal stem cells and *in vivo* bone formation by activating Wnt/catenin signaling. Stem Cells. 29, 981–991.

Yang, L., Zhang, L., Webster, T.J., 2011b. Carbon nanostructures for orthopedic medical applications. Nanomedicine. 6 (7), 1231–1244.

Yang, X., Bakaic, E., Hoare, T., Cranston, E.D., 2013. Injectable polysaccharide hydrogels reinforced with cellulose nanocrystals: morphology, rheology, degradation, and cytotoxicity. Biomacromolecules. 14 (12), 4447–4455.

Yang, Y.L., Chang, C.H., Huang, C.C., Kao, W.M.W., Liu, W.C., Liu, H.W., 2014. Osteogenic activity of nanonized pearl powder/poly(lactide-*co*-glycolide) composite scaffolds for bone tissue engineering. Biomed. Mater. Eng. 24 (1), 979–985.

Yang, H.M., Chan, Y.C., Hsu, T.H., Chen, H.W., Lee, J.W., Duh, J.G., et al., 2015. Synthesis and characterization of nacre-inspired zirconia/polyimide multilayer coatings by a hybrid sputtering and pulsed laser deposition technique. Surf. Coat. Technol. 284, 118–128.

Zhai, C., Sun, X., Zhao, W., Gong, Z., Wang, X., 2013. Acetylcholinesterase biosensor based on chitosan/prussian blue/multiwall carbon nanotubes/hollow gold nanospheres nanocomposite film by one-step electrodeposition. Biosens. Bioelectron. 42, 124–130.

Zhang, W., Xie, G., Li, S., Lu, L., Liu, B., 2012. Au/CeO_2–chitosan composite film for hydrogen peroxide sensing. Appl. Surf. Sci. 258, 8222–8227.

Zhang, H., Liu, R., Zheng, J., 2013. Seed-mediated synthesis of polyaniline/Au nanocomposite and its application for a cholesterol biosensor. Synth. Met. 167, 5–9.

Zhao, F., Yao, D., Guo, R., Deng, L., Dong, A., Zhang, J., 2015. Composites of polymer hydrogels and nanoparticulate systems for biomedical and pharmaceutical applications. Nanomaterials. 5, 2054–2130.

Zhou, H., Lee, J., 2011. Nanoscale hydroxyapatite particles for bone tissue engineering. Acta Biomater. 7 (7), 2769–2781.

Zhou, Z., Ou, B., Huang, T., Zeng, W., Liu, L., Liu, Q., et al., 2013. Biocompatibility *in-vitro* of Gel/HA composite scaffolds containing nano-bioactive glass for tissue engineering. J. Macromol. Sci., A. 50 (10), 1048–1053.

Zhu, J., Marchant, R.E., 2011. Design properties of hydrogel tissue-engineering scaffolds. Expert. Rev. Med. Devices. 8 (5), 607–626.

Zhu, Y., Gao, C., Liu, X., Shen, J., 2002. Surface modification of polycaprolactone membrane via aminolysis and biomacromolecule ımmobilization for promoting cytocompatibility of human endothelial cells. Biomacromolecules. 3, 1312–1319.

Zhu, J., Liu, X., Wang, X., Huo, X., Yan, R., 2015. Preparation of polyaniline–TiO_2 nanotube composite for the development of electrochemical biosensors. Sens. Actuators, B: Chem. 221, 450–457.

Liquid crystalline DNA: a smart polymer with a variety of applications ranging from photonics to plasmonics

13

Luciano De Sio[1,2], Ferdinanda Annesi[2], Tiziana Placido[3,4], Roberto Comparelli[4], Alfredo Pane[2], Maria L. Curri[4], Cesare Umeton[2,5] and Roberto Bartolino[2,5,6]
[1]Beam Engineering for Advanced Measurements Company, Orlando, FL, United States, [2]CNR-Lab. Licryl, Institute NANOTEC, Arcavacata di Rende, Italy, [3]University of Bari, Chemistry Department, Bari, Italy, [4]CNR-IPCF, Institute for Physical and Chemical Processes, Bari, Italy, [5]Department of Physics, Centre of Excellence for the Study of Innovative Functional Materials, University of Calabria, Arcavacata di Rende, Italy, [6]Interdisciplinary Institute B, Segre of the National Academy of Lincei, Rome, Italy

Chapter Outline

13.1 Deoxyribonucleic acid 409
- 13.1.1 DNA and its liquid crystalline phases 409
- 13.1.2 DNA extraction 410
- 13.1.3 Agarose gel electrophoresis analysis 412
- 13.1.4 DNA labeling 412

13.2 Plasmonic nanoparticles 413
13.3 Liquid crystalline DNA organization in polymeric template 413
13.4 Gold nanorods organized in liquid crystalline DNA 417
13.5 Conclusion 419
Acknowledgments 420
References 420

13.1 Deoxyribonucleic acid

13.1.1 DNA and its liquid crystalline phases

DNA is a semiflexible polymer made of two polynucleotide chains, held together by weak thermodynamic forces (Lander and Weinberg, 2000). The monomer units

Hybrid Polymer Composite Materials: Applications. DOI: http://dx.doi.org/10.1016/B978-0-08-100785-3.00013-9

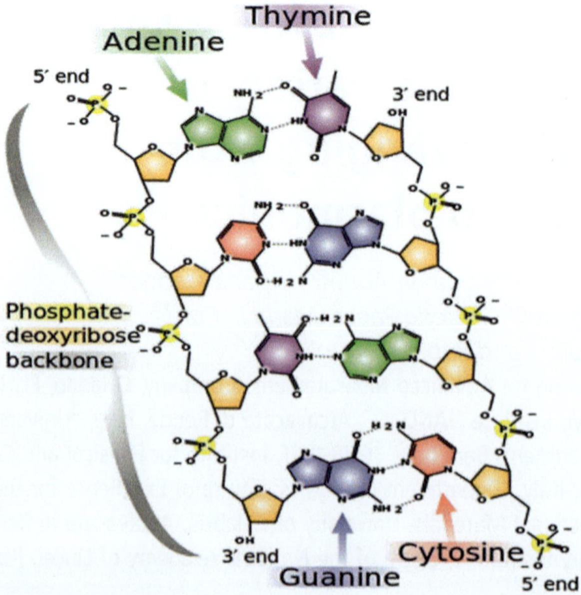

Figure 13.1 DNA chemical structure.

of DNA are the nucleotides. Each nucleotide consists of a 5-carbon sugar (deoxyribose), a nitrogen containing base attached to the sugar, and a phosphate group (Travers and Muskhelishvili, 2015). There are four different types of nucleotides found in DNA: adenine (A), thymine (T), cytosine (C), and guanine (G). A and T are connected by two hydrogen bonds while G and C are connected by three hydrogen bonds (Fig. 13.1).

An interesting and comprehensively studied phenomenon is represented by the tendency of flexible polymers in concentrated aqueous solutions to form liquid crystalline phases (Onsager, 1949; Nakata et al., 2007). Indeed, the ability of both long and ultrashort, hydrated, double-stranded DNA molecules to form liquid crystal (LC) phases has been known for more than 50 years and played a key role in the initial deciphering of the molecular DNA. Linear DNA fragments in aqueous solution form multiple LC phases whose nature depends on the polymer concentration: when increasing concentration, the isotropic solution transforms into either blue phase or precholesteric stage and then into a cholesteric phase which turns itself into columnar hexagonal (Strzelecka et al., 1988).

13.1.2 DNA extraction

The isolated DNA can be obtained by using the Wizard Genomic DNA Purification Kit, designed to extract DNA from different sources, such as white blood cells, tissue culture cells, animal tissue, and others. In our specific case, we have chosen to follow the DNA extraction method starting from fresh whole human blood. The sample of whole blood was collected in an EDTA anticoagulant tube to prevent the clotting and kept at −20°C. For the utilization, the frozen blood sample was thawed out and left at room temperature (3−4 h) before starting the DNA isolation method. The DNA extraction method is illustrated in Fig. 13.2:

1. A volume of 300 μL of blood was put in a sterile 1.5 mL microcentrifuge tube which contained 900 μL of cell lysis solution. The mixture was mixed by inverting the tube 5−6 times and was incubated for 10 min at room temperature to lyse the red blood cells. The mixture was then centrifuged for 50 s (14,000 × *g*) at room temperature. Supernatant was removed and discarded as much as possible without disturbing the visible white pellet. To obtain it, an additional aliquot of Cell Lysis Solution was added and the incubation and centrifugation steps were repeated. The pellet was resuspended by a brief vortexing.
2. The white blood cells and their nuclei (where the DNA is accumulated) underwent lysing. Nuclei lysis solution (300 μL for 300 μL sample volume) was added to the tube containing the resuspended cells and the mixture was again incubated for 1 h at 37°C. At this point, a RNase solution was added (1.5 μL for 300 μL sample volume) to the nuclear

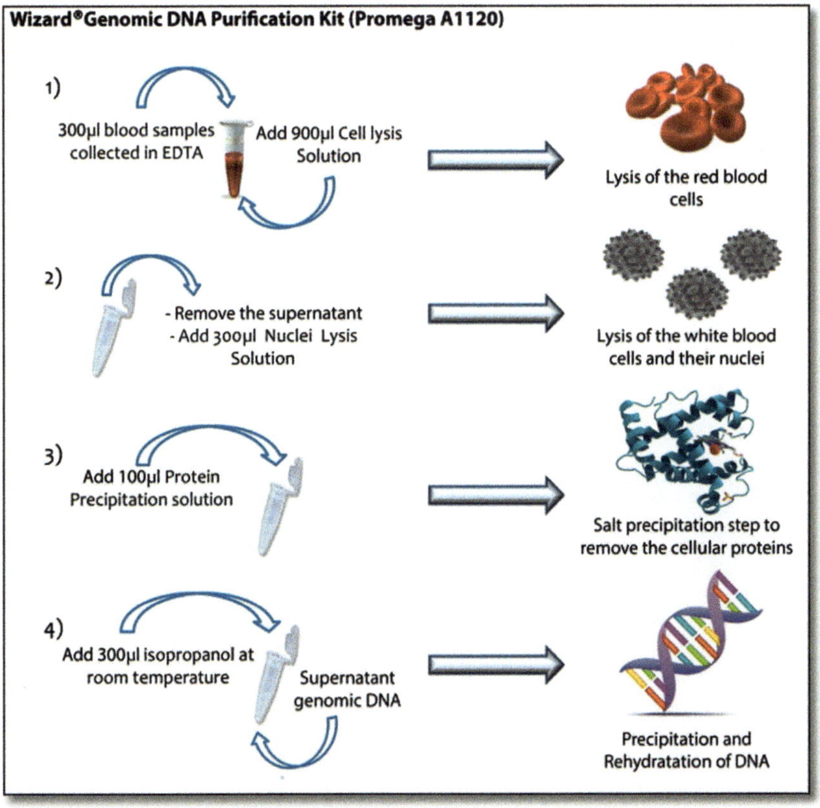

Figure 13.2 Sketch of the DNA extraction method.

lysate; the sample was mixed by inverting the tube 2–5 times. The mixture was incubated at 37°C for 15 min and then cooled down to room temperature.

3. The cellular proteins were removed by means of a salt precipitation. Protein precipitation solution was added (100 μL for 300 μL sample volume) to the nuclear lysate, after the mixture was centrifugated for 3 min (14,000 × *g*) at room temperature. In this way, the cellular proteins were removed while the high molecular weight genomic DNA was left in the solution.
4. The supernatant was transferred to a clean tube containing 300 μL of isopropanol. Therefore, the solution was gently mixed by inversion until the macromolecules of DNA formed a *visible mass*. After centrifugation for 1 min (14,000 × *g*) at room temperature, the DNA was visible as a small white pellet. The pellet was washed with 70% ethanol. After centrifugation for 1 min (14,000 × *g*) at room temperature, the DNA pellet and the sides of the tube were gently washed. By using a drawn Pasteur pipette, the ethanol was carefully aspired. The tube was inverted on a clean absorbent paper and the pellet was air-dried for 1–2 h. Finally, the DNA was resuspended in 100 μL of rehydration solution and incubated at 65°C for 1 h.

13.1.3 Agarose gel electrophoresis analysis

It is a technique particularly suited for accomplishing both separation and analysis of macromolecules (e.g., DNA, RNA, proteins) and their fragments, based on their length and charge (Lee et al., 2012). Charged molecules of the solution to be tested are forced to move through an agarose matrix under the influence of an electric field that is created by applying an external DC voltage. The DNA molecules, being negatively charged along their length, migrate toward the positive electrode. In this way, during their electrophoretic run, they are separated according to their length, since pores in the agarose matrix allow short DNA molecules to snake through more easily than long ones.

13.1.4 DNA labeling

Labeling is a basic technique of molecular biology that allows to determine the position of a particular nucleic acid molecule on a membrane or gel, on a chromosome or within a tissue or a cell; the emitted signal can be detected appropriately by fluorescence microscopy (Bellizzi et al., 2007). DNA molecules can be easily and optimally labeled with radioactive phosphates, biotin, fluorophores, and enzymes during their synthesis. Nonisotopic DNA labeling procedures are essential for integration of DNA diagnostics into the clinical laboratory. Labeling DNA can be performed at either the 5′ or 3′ ends or all along the molecule, depending on the application. For applications such as in situ hybridization, northern and southern blot hybridization, it is usually advantageous to generate probes with label distributed throughout the nucleic acid; for other applications like protein interactions, end-labeled probes are preferentially generated. The main procedures for labeling a double-stranded DNA fragment are 3′ end-labeling with DNA polymerase (enzymes that create deoxyribonucleic acid polymers) or at the 5′ end using T4 polynucleotide kinase (enzymes that transfer the phosphate group of [γ32-P]ATP to the 5′

hydroxyl terminus of a DNA molecule); nick translation, using DNA polymerase in the presence of at least one radioactive precursor, to "translate" a nick (single-strand cut in the double-stranded DNA molecule) along the molecule in the 5′ to 3′ direction; random primer labeling, where the DNA molecule is used as a probe and it is denaturated by heating and mixed with random hexanucleotides which prime the polymerization reactions; PCR (polymerase chain reaction) labeling. It is recommended to purify labeled probes by many methods: spin column chromatography, membrane filtration, adsorption to silica gel membranes, ethanol precipitation.

13.2 Plasmonic nanoparticles

Plasmonic nanoparticles (NPs) play a very important role in miniaturized modern photonics due to their capability to confine the light at the nanoscale beyond the diffraction limit of light by exploiting the so-called localized plasmonic resonance (LPR) (Pendry, 2000; Pendry et al., 2006). There are many ways to control the LPR frequency such as varying the NP shape, size, composition, and roughness (Liz-Marzán, 2004; Huang et al., 2009; De Sio et al., 2012). These parameters enable a "static" design of the LPR frequency throughout the visible and into the near-infrared spectra when keeping the particle size below 100 nm. The characteristic color of colloidal solutions of NPs therefore originates from the strong absorption occurring when the frequency of the electromagnetic field becomes resonant with the coherent electron motion. Since copper is easily oxidized, Au and Ag nanostructures are most attractive for optical and sensing applications. The spectral position, shape, and intensity of the LPR strongly depend on various factors: the dielectric constant of the surrounding medium, the electronic interactions between the stabilizing ligands and the NP, which alter the electron density inside the NP, and the size, shape, and monodispersity of the NPs. A particular class of NPs is gold nanorods (GNRs) (Pérez-Juste et al., 2005); they exhibit two plasmon resonances, one due to the transverse oscillation of the electrons, positioned around 520 nm for Au, and the other due to the longitudinal plasmon resonance at longer wavelengths, as shown for various aspect ratios in Fig. 13.3.

The transverse LPR does not depend on the aspect ratio and is at the same wavelength as the plasmon resonance of spherical NPs. The longitudinal LPR, on the contrary, increases with larger aspect ratios. Thus, the longitudinal LPR can be tuned across the visible region by changing the aspect ratio.

13.3 Liquid crystalline DNA organization in polymeric template

Liquid crystalline DNA can be stretched and immobilized in a holographic grating by using the self-organization properties of those filaments (De Sio et al., 2011). The structure is obtained by starting from a periodic liquid crystalline composite

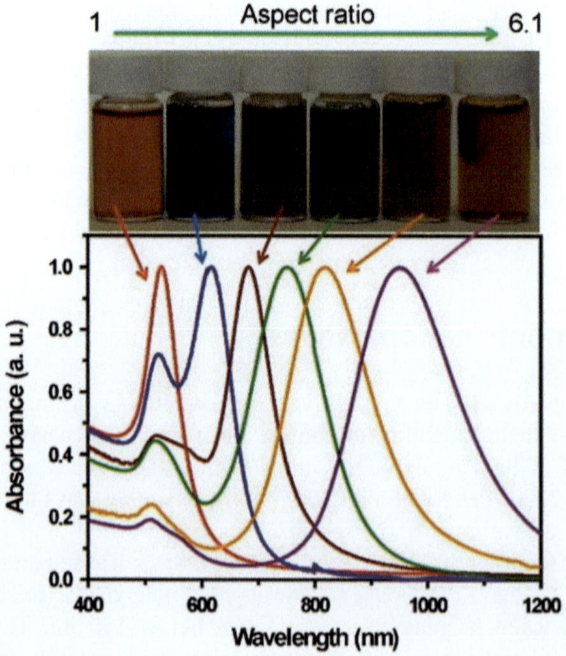

Figure 13.3 Shape and spectral position of surface plasmon resonance of gold NRs with different aspect ratio.

material, called POLICRYPS (acronym of polymer LC polymer slices), which is a nano/microcomposite holographic grating made of slices of almost pure polymer alternated to films of well-aligned nematic liquid crystals (NLCs) (De Sio et al., 2013b; Caputo et al., 2004). The NLC is removed by using a selective etching process, and the microchannels of the empty polymeric template are filled with liquid crystalline DNA. With this technique, we have immobilized and permanently aligned both human whole-genome and ultrashort (27 bp) DNA molecules in a geometry that depends on their length (De Sio et al., 2013a). DNA helices have been investigated by means of polarized optical microscopy (POM) and confocal microscopy (CM). In a first attempt, we have used a human whole-genome DNA; attempts performed on several, different sample cells, with varying cross-sectional dimensions, have shown that the best performances are exhibited by the POLICRYPS structure of $d = 10$ μm thickness, with a channel width $L = 2.5$ μm.

Fig. 13.4A represents a POM view of a droplet of water solution of DNA deposited on a glass substrate: the DNA is randomly aligned exhibiting no macroscopic birefringence. By capillary flow, we infiltrated the empty polymeric template (Fig. 13.4B) with the same DNA solution. During the process, the sample is kept at a high temperature (80°C); in this condition, the DNA strands are unpaired, meanwhile the aqueous solution induces a hydrodynamic flow due to the surface tension between the solution and the microchannel surfaces. Consequently, an extension

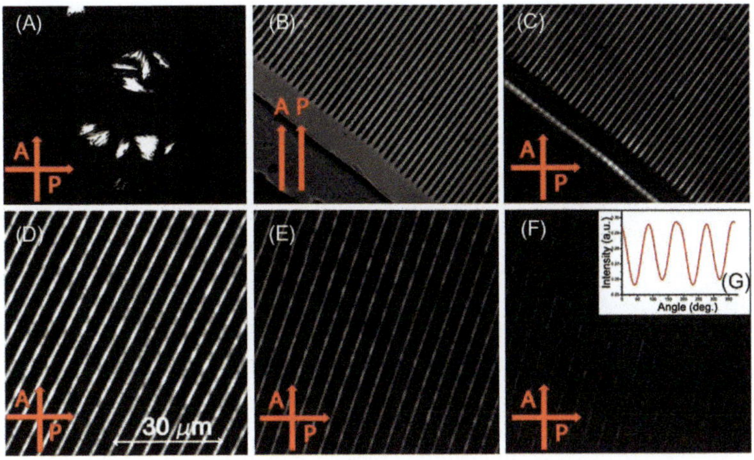

Figure 13.4 POM view of droplet of DNA water solution on a glass substrate (A); "empty POLICRYPS" template between parallel polarizers before being infiltrated with the DNA water solution (B) (between crossed polarizers, it appears as a uniform dark square—not shown); the same, between crossed polarizers, after being infiltrated with the DNA water solution (C); transmitted intensity (bright fringes) for different angles between the analyzer and the optical axis of the sample (D–F); dependence of the diffracted intensity of the structure on the polarization of the impinging probe light (G).
Reproduced from De Sio et al. 2013a. Directed organization of DNA filaments in a soft matter template, Langmuir, 29, 3398–3403.

force is induced and the single strands are elongated along the channel direction. Then, by keeping the filled sample at high temperature for a long time interval (20 min), evaporation of most of the aqueous solvent takes place. Subsequently, by slowly (0.5 deg min^{-1}) cooling down the sample to room temperature, a covalent bonding between the thiol group (S–H) present in the photo-polymer (NOA-61) and the DNA ends is created (Jin et al., 2004). Due to the optical contrast between the isotropic polymeric slices and the birefringent dehydrated DNA confined in the microchannels, the *POM micrograph of* Fig. 13.4C *highlights bright strips of stretched DNA microbundles.* Once realized, the sample behaves as a uniaxial retardation plate (Hecht et al., 1998) with its optical axis parallelly aligned to the microchannel direction. For an impinging white (noncollimeted) light, the optical contrast between bright (Fig. 13.4D) and dark (Fig. 13.4F) states, obtained by rotating the sample between the crossed polarizers of the POM, is quite high (better than 20:1), thanks to the high quality of the dark image, which reflects a good DNA alignment. Since a way to investigate well-aligned birefringent materials inside a periodic structure is to observe its diffractive pattern, we monitored the intensity diffracted by our sample while varying the polarization of a probe impinging light. The sinusoidal behavior reported in Fig. 13.4G is a well-known characteristics of periodic structures that contain an aligned birefringent material and is of relevance to us, since we believe that it can represent the first step of an approach for a totally holographic DNA analysis and examination. In order to characterize the DNA geometry inside the single polymeric microchannel, we carried out new experiments by using ultrashort DNA segments. In particular, we used a linear double-stranded 27-bp long (about 9 nm) oligonucleotide DNA. The sample was labeled at the 3′-terminal by using terminal deoxynucleotidyl transferase and fluorescein-12-dUTP; this label gives a fluorescent signal that enables CM analysis. A water solution of short molecules (27 bp) was injected in the empty polymeric template by using the already described procedure.

In the POM view of Fig. 13.5A, the presence of DNA microbridges between the polymeric slices of the channels is evident. This geometry can be explained by considering the ratio between the length of the ultrashort DNA filament and the width of the microchannels. We believe that, in the actual case, when the sample is kept at high temperature, the denatured DNA short molecules interact each other due to the presence of a slow nucleation process, which requires a nucleation cluster (usually 5–6 monomers); then, a rapid filament extension in long continuous

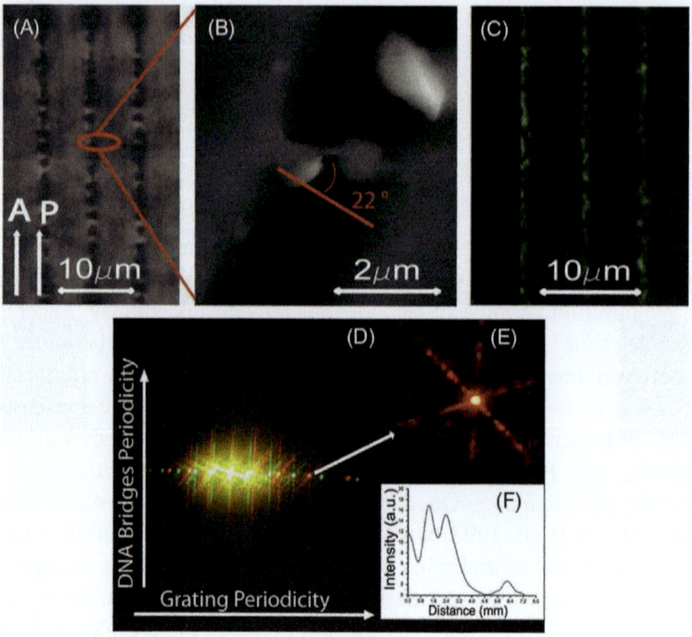

Figure 13.5 POM view (A) and ESEM picture of DNA bridge arrays (B), along with the confocal analysis image (C). Far field 2D diffraction pattern obtained by colaunching two laser beams of different color (D). High magnification of the "star-like" diffraction pattern (E). Intensity profile along one arm of the diffraction pattern (F).
Reproduced from De Sio et al. 2013a. Directed organization of DNA filaments in a soft matter template, Langmuir, 29, 3398–3403.

filaments follows. In order to investigate the geometry, we have analyzed the sample with an environmental scanning electronic microscope (*ESEM*). The *ESEM* picture of Fig. 13.5B confirms the elongated shape of the DNA bridges and also a certain distribution of their tilt angle with respect to the normal to the polymeric slices. By using the fluorescence signal of the labeled filament head, we have also analyzed the sample under a CM: the result is reported in Fig. 13.5C. Each microchannel appears quasi periodically illuminated, due to the strong fluorescence signal of DNA bridges, which possess several light responsive markers. We have probed the sample with two, colaunched laser beams: the far field view of the obtained 2D diffraction pattern is reported in Fig. 13.5D, which indicates the existence of a 2D quasi crystal structure. The pattern is a result of the polymeric grating periodicity in one direction (Fig. 13.4B) and the quasi-periodic "zig-zag" distribution of the DNA bridges, which generates the X-shaped diffraction spot reported in Fig. 13.5E.

13.4 Gold nanorods organized in liquid crystalline DNA

Homogeneous mixtures of DNA and GNRs have been prepared by adding an aliquot of the GNRs solution (2×10^{-9} M) in the dispersion of DNA, ranging the GNRs concentration from 2% to 50% in weight. A preliminary investigation of the interaction between GNRs and DNA has been carried out by performing a gel electrophoresis analysis (De Sio et al., 2015). The mixture of DNA and GNRs has been added of a small amount (1% in weight) of a positively charged blue dye (ethidium bromide, EtBr), which is an intercalating agent commonly used to act as a fluorescent tag; when EtBr is exposed to UV light, its orange fluorescence increases of about 20 folds upon binding to DNA.

The obtained solution has been pipetted in the *agarose matrix* molded with wells (Fig. 13.6A) containing also DNA filaments of known size (in the first well on the left of Fig. 13.6A), which are used as markers. The *agarose matrix* is then placed on a UV lamp and covered with a UV glass filter. When the system power supply is turned on, the DC voltage (120 V, 0.78 mA) induces a migration of the pure DNA (Fig. 13.6A, second well on the right) toward the positive electrode. Observation of a single fluorescent band indicates that DNA molecules have almost the same length, which includes approximately 10,000 base pairs. By increasing the GNRs concentration up to 5%, the solution shows a remarkable fluorescence very close to the well, pointing out an inhibited migration toward the positive electrode. Such an evidence can be accounted for by taking into account the presence of the CTAB bilayer, which makes GNRs positively charged, and thus able to electrostatically interact with the DNA molecules, without affecting their average length, but varying the effective charge of the whole system. By increasing the GNRs concentration up to 20%, the fluorescence light appears as split and coming from two main sites: a first one, closer to the well, reasonably ascribable to the electrostatic conjugation between GNRs and DNA, and a second one, at about a 10,000 base

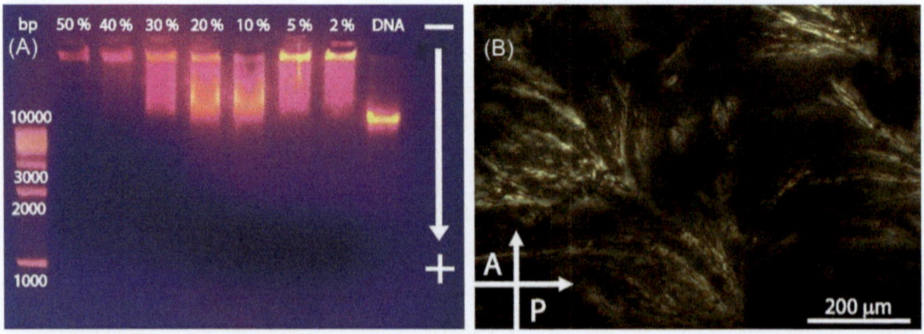

Figure 13.6 Gel electrophoresis analysis for different GNRs concentrations (A), along with the POM view of the DNA–GNR mixture deposited on a glass substrate by drop casting (B). Reproduced from De Sio et al. 2015. Templating gold nanorods with liquid crystalline DNA. J. Opt., 17, 025001.

pairs distance, which is due to the excess of DNA, not electrostatically conjugated to GNRs, and hence more mobile. For concentrations of GNRs exceeding 20%, the heavy GNRs damage the DNA strands (when the value is about 30%–40%) affecting also their mobility (for values $\approx$ 50%). POM analysis has been also performed. Preliminary observations confirm that a liquid crystalline phase of DNA molecules exists and is maintained up to a 20% of GNRs concentration; by exceeding this value, the large quantity of GNRs strongly affects the liquid crystalline structure of the DNA and the whole solution becomes isotropic. Therefore, in the following experiments, a solution with 20% of GNRs has been used in order to both maximize the presence of GNRs and still keep the DNA in a liquid crystalline phase, avoiding DNA damage. A droplet of this solution has been drop-cast onto an untreated glass substrate and allowed to dry under controlled temperature (45°C); meanwhile, a stretching of the droplet on the surface, along a preferred direction, has been induced by means of an air-flow, thus obtaining a quite thin and uniform film of large area ($\approx 1\ mm^2$). Fig. 13.6B shows a POM view of this film, acquired at room temperature, after evaporating all the solvent; at this level of magnification, details on the local organization of the GNRs around the filaments of DNA cannot be seen and they appear oriented, in average, along the shearing direction. The bright strips, clearly visible in Fig. 13.6B, confirm that the DNA exhibits a LC phase, with a local birefringence of about 0.010 (measured by the optical compensator method). Furthermore, we have performed a far field, polarized, spectral analysis of the same sample area (Fig. 13.7B) by utilizing white light. Fig. 13.7A shows the absorption spectra of the sample for light linearly polarized (1) along the shearing direction, (2) perpendicularly to it, and (3) at 45° (indicated in Fig. 13.7B with red, blue, and green arrows, respectively).

Since GNRs that embed the DNA filaments are, in average, oriented along the shearing direction, light polarized in this direction experiences only the longitudinal

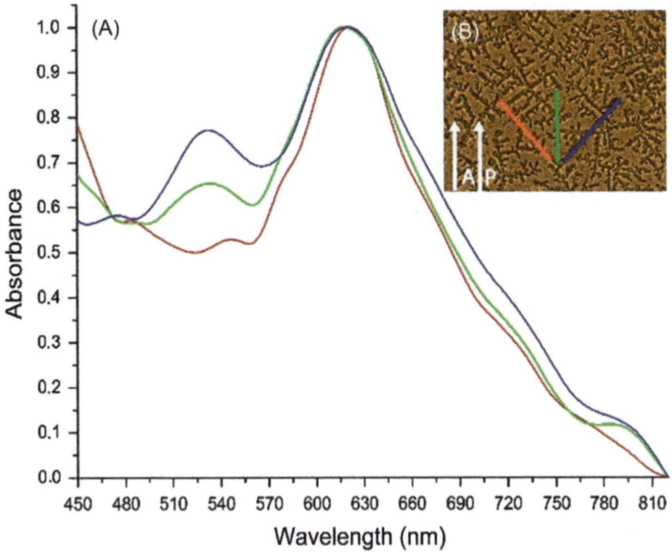

Figure 13.7 Polarized spectral response of the sample (A), along with an optical micrograph (B) of the analyzed area.
Reproduced from De Sio et al. 2015. Templating gold nanorods with liquid crystalline DNA. J. Opt., 17, 025001.

plasmon response, as it is clear from Fig. 13.7A (red curve), while the transversal band is, in this case, completely missing. In addition, this spectroscopic response shows a small bump at $\lambda \approx 800$ nm, which is ascribable to interparticle interaction phenomena between close packed GNRs (Funston et al., 2009); this represents a further evidence of the macroscopic organization of GNRs along the shearing direction, induced by the liquid crystalline phase of the DNA. Above considerations are supported by the spectral features exhibited by the same sample when probed with light polarized perpendicularly to the shearing direction (Fig. 13.7A, blue curve): in this case, the transversal band is well evident, while the amplitude of the longitudinal band is reduced of about 2 times. Finally, in the spectral response to light polarized at 45° with respect to the shearing direction (Fig. 13.7A, green curve), both plasmonic bands are observed, with amplitudes that are in between the ones of the two previously described cases, as expected on the basis of above considerations.

13.5 Conclusion

To summarize, we have studied the capabilities of a liquid crystalline DNA in terms of template-assisted and self-organized properties. An empty polymeric template realized through a selective etching process has been used to realize a surfactant free method for permanently aligning both human whole-genome and ultrashort DNA molecules (top-down approach). It turns out that the human whole-genome DNA is stretched along the polymeric channels while the ultrashort DNA produces a quasi-periodic "zig-zag" distribution of DNA bridges. The self-organization properties of a whole-genome DNA (bottom-up approach) are used to induce both short and long-range order of GNRs. The assembly mechanism at a molecular level, as investigated by means of a gel electrophoresis analysis, results to be driven by electrostatic interactions between negatively charged DNA molecules and positively charged GNRs particles. The system exhibits the existence of a strong correlation between the amplitude of the plasmon bands and the polarization direction of the impinging radiation. Both approaches highlight the possibility to successfully employ the liquid crystalline properties of DNA for realizing several applications ranging from photonics to plasmonics such as holographic biosensors (Marshall et al., 2004) and bio-inspired metamaterials (Meza et al., 2015).

Acknowledgments

We acknowledge the cooperation of all coauthors of the papers we published on the argument: P. D'Aquila, E. Brunelli, G. Strangi, D. Bellizzi, G. Passarino, V. Bruno, and G. Palermo from the University of Calabria.

References

Bellizzi, D., Dato, S., Cavalcante, P., Covello, G., Di Cianni, F., Passarino, G., et al., 2007. Characterization of a bidirectional promoter shared between two human genes related to aging: SIRT3 and PSMD13. Genomics. 89, 143–150.

Caputo, R., De Sio, L., Veltri, A., Umeton, C., Sukhov, A.V., 2004. Development of a new kind of switchable holographic grating made of liquid-crystal films separated by slices of polymeric material. Opt. Lett. 29, 1261–1263.

De Sio, L., Ferjani, S., Strangi, G., Umeton, C., Bartolino, R., 2011. Universal soft matter template for photonic applications. Soft Matter. 7, 3739–3743.

De Sio, L., Cunningham, A., Verrina, V., Tone, C.M., Caputo, R., Burgi, T., et al., 2012. Double active control of the plasmonic resonance of a gold nanoparticle array. Nanoscale. 4, 7619–7623.

De Sio, L., D'aquila, P., Brunelli, E., Strangi, G., Bellizzi, D., Passarino, G., et al., 2013a. Directed organization of DNA filaments in a soft matter template. Langmuir. 29, 3398–3403.

De Sio, L., Annesi, F., Placido, T., Comparelli, R., Bruno, V., Pane, A., Palermo, G., Curri, M.L., Umeton, C., Bartolino, R., 2015. Templating gold nanorods with liquid crystalline DNA. J. Opt. 17, 025001.

De Sio, L., Tabiryan, N., Bunning, T., Kimball, B.R., Umeton, C., 2013b. Chapter 1—Dynamic photonic materials based on liquid crystals. In: Emil, W. (Ed.), Progress in Optics. Elsevier, pp. 1–64.

Funston, A.M., Novo, C., Davis, T.J., Mulvaney, P., 2009. Plasmon coupling of gold nanorods at short distances and in different geometries. Nano Lett. 9, 1651–1658.

Hecht, J.D., Eifler, A., Riede, V., Schubert, M., Krauß, G., Krämer, V., 1998. Birefringence and reflectivity of single-crystal ${\mathrm{CdAl}}_{2}{\mathrm{Se}}_{4}$ by generalized ellipsometry. Phys. Rev. B. 57, 7037–7042.

Huang, X., Neretina, S., El-Sayed, M.A., 2009. Gold nanorods: from synthesis and properties to biological and biomedical applications. Adv. Mater. 21, 4880–4910.

Jin, B., Ji, X., Nakamura, T., 2004. Voltammetric study of interaction of Co(phen)33 + with DNA at gold nanoparticle self-assembly electrode. Electrochim. Acta. 50, 1049–1055.

Lander, E.S., Weinberg, R.A., 2000. Journey to the center of biology. Science. 287, 1777.

Lee, P.Y., Costumbrado, J., Hsu, C.-Y., Kim, Y.H., 2012. Agarose gel electrophoresis for the separation of DNA fragments. J. Vis. Exp.e3923, pii: 3923. http://dx.doi.org/10.3791/3923.

Liz-Marzán, L.M., 2004. Nanometals: formation and color. Mater. Today. 7, 26–31.

Marshall, A.J., Young, D.S., Blyth, J., Kabilan, S., Lowe, C.R., 2004. Metabolite-sensitive holographic biosensors. Anal. Chem. 76, 1518–1523.

Meza, L.R., Zelhofer, A.J., Clarke, N., Mateos, A.J., Kochmann, D.M., Greer, J.R., 2015. Resilient 3D hierarchical architected metamaterials. Proc. Natl. Acad. Sci. 112, 11502–11507.

Nakata, M., Zanchetta, G., Chapman, B.D., Jones, C.D., Cross, J.O., Pindak, R., et al., 2007. End-to-end stacking and liquid crystal condensation of 6- to 20-base pair DNA duplexes. Science. 318, 1276.

Onsager, L., 1949. The effects of shape on the interaction of colloidal particles. Ann. N.Y. Acad. Sci. 51, 627–659.

Pendry, J.B., 2000. Negative refraction makes a perfect lens. Phys. Rev. Lett. 85, 3966–3969.

Pendry, J.B., Schurig, D., Smith, D.R., 2006. Controlling electromagnetic fields. Science. 312, 1780–1782.

Pérez-Juste, J., Pastoriza-Santos, I., Liz-Marzán, L.M., Mulvaney, P., 2005. Gold nanorods: synthesis, characterization and applications. Coord. Chem. Rev. 249, 1870–1901.

Strzelecka, T.E., Davidson, M.W., Rill, R.L., 1988. Multiple liquid crystal phases of DNA at high concentrations. Nature. 331, 457–460.

Travers, A., Muskhelishvili, G., 2015. DNA structure and function. FEBS J. 282, 2279–2295.

Index

Note: Page numbers followed by "*f*" and "*t*" refer to figures and tables, respectively.

A

Abalone nacre, 366–367
Acetylcholinesterase biosensors, 395–399
Actuators, 114–117
 biopolymer-carbonaceous particles hybrid actuators, 116–117
 biopolymer–polymer hybrid actuators, 115–116
Aerospace structural applications, 37
Agarose gel electrophoresis analysis, 412
Agarose matrix, 417–418
AGMA1–hydrogel composites, synthesis of, 247*f*
AGMA1–MMT composites, 246
Aircraft structures, 37
Alginate, 119, 346
Amorphous calcium carbonate (ACC), 132–134
Amperometric acetylcholinesterase (AChE) biosensors, 395–397
Amperometric biosensors, 394
Angular performance, 83–84
Anochitin/chitosan-based scaffolds, 128–129
Anticorrosion coatings, 25–26
Antimicrobial poly(e-caprolactone) (PCL)/clay nanocomposites, 235
Aramid-reinforced aluminum laminate (ARAL), 37
Artificial muscles, 114
Atomic-force microscopy (AFM), 358–359
Attenuation coefficient, 93
Au/Cerium dioxide (CeO_2)–Chi composite biosensor, 392
Automotive components, 37–38

B

Bacterial cellulose membranes, 299–300
Bandwidth-to-thickness ratio, 78–83
$BaTiO_3$ nanoparticles, 91–92
BC, 116, 127
Bicycle featuring carbon-based composite components, 43, 44*f*
Binary conducting composites, electrical conductivity variation in, 4–5
Bioactive glasses (BGs), 348
Biocompatible and biodegradable polyelectrolytes, 377–378
Biomedical applications of hybrid polymer composite materials, 343, 399–400
 biosensor applications, 384–399
  acetylcholinesterase biosensors, 395–399
  cholesterol biosensors, 394–395
  glucose biosensors, 385–387
  hydrogen peroxide biosensors, 391–394
  urea biosensors, 388–390
 drug delivery, 373–384
  with hybrid composites prepared from natural polymers, 374–378
  with hybrid composites prepared from synthetic polymer, 378–384
 tissue engineering applications, 347–373
  biomimetic scaffolds, 356–357
  bone scaffolds, 348–356
  implant materials, 367–373
  injectable scaffolds, 357–361
  scaffold applications, 362–367
  scaffolds, 347–348
  wound-healing materials, 361
Biomimetic scaffolds, 356–357
 for tissue engineering, 129–134
  liquid-crystalline-based scaffolds for tissue engineering, 132–134
Biomineralization, 350–351
Biopolymer-carbonaceous particles hybrid actuators, 116–117
Biosensing, hybrid NCC materials in, 302–307

Biosensor applications, 384–399
acetylcholinesterase biosensors, 395–399
cholesterol biosensors, 394–395
glucose biosensors, 385–387
hydrogen peroxide biosensors, 391–394
urea biosensors, 388–390
Biosensors, 110–113
for cholesterol detection, 111–112
for detection of biomolecules, 113
for glucose detection, 110–111
Bitumen, 216–217
Blood glucose monitoring, 110–111
Bone and cartilage (BC), 123–126
Bone scaffolds, 348–356
Boron nitride (BN), 153–154
Bouligand structure, 132
Bovine serum albumin (BSA) drugs, 362–363, 373–374
Brodie's method, 16–17

C

$CaCO_3$/carboxymethyl Chi (CMC) hybrid composite micro and nanospheres, 374
Carbon black (CB), 85–86
Carbon FRP (CFRP), 38–41
Carbon nanofibers (CNFs), 11, 16–17, 157*f*
Carbon nanotubes (CNTs), 3, 11, 14, 23–24, 152–153, 156, 174–175
-based composites, 44
and polymer composites TE materials, 179–183
5-Carbon sugar (deoxyribose), 409–410
Carbon-based composite components, bicycle featuring, 43, 44*f*
Carbon-based fillers, 16–17
Carbon-based nanostructures/polymer nanocomposites TE materials, 179–186
CNTs and polymer composites TE materials, 179–183
conducting composites of polymer/graphite, 186
conducting polymer/graphene composites, 183–185
Carbon-reinforced aluminum laminate (CARAL), 37
Carboxymethyl cellulose (CMC), 309
Carrageenan (Car)–$CaCO_3$ hybrid polymer composites, 374–375
Cartilage, 125–126
Cellulose, 108–109, 289–291
and cellulose-based materials, 125
chemical structure of, 108*f*
Cellulose nanofibril, 307
Cellulose nanofibrilated, 127
Cellulose nanohybrid materials, synthesis and applications of, 289
applications, 295–310
propulsion, 296–307
applications of hybrid NCC materials in biosensing, 302–307
bacteria detection, 305–306
in vivo detection and other sensing applications, 306–307
small molecules detection, 304–305
cellulose-based electronics, 298–302
functionalization, 295
nanocrystalline cellulose, 293–295
packaging materials and drug delivery systems, 308–310
regenerated cellulose, 291–292
Cellulose/SnO_2 hybrid composite, 386
Cellulose–heparin fibers, 126
Cellulose-SnO_2 hybrid nanocomposite, 389
Cement-based materials, 202–203
Cerium zinc molybdate nanoparticles, 336–338
Chi/cellulose/nanohydroxyapatite (Chi/CMC/n-HA)-based biomimetic anisotropic spiral-cylindrical scaffold membranes, 356–357
Chi/prussian blue/multiwall carbon nanotubes/hollow gold nanospheres nanocomposite film, 398
Chi/silica hybrid composite film, 387
Chi/superparamagnetic iron oxide (Fe_3O_4) nanoparticles hybrid composite, 388–389
Chi-based composites, 360–361
Chitin, 109, 346
chemical structure of, 108*f*
Chitosan (Chi/CS), 109, 118–119, 346, 375–377
biomedical application of, 124*t*
chemical structure of, 108*f*

Chitosan–palygorskite composite, chemical structure of, 251*f*
Cholesteric chitin gel, 133*f*
Cholesterol biosensors, 394–395
Cholesterol detection, polysaccharide-based biosensors for, 111–112
Cholesterol oxidase (ChOx), 394
Chondroitin sulfate, 364
Civil infrastructures, 38–40
Clay, 227–228
  organic modification of, 231*f*
  structure of, 228–229
Clay minerals, classification of, 229–230
Clay surface treatments, 230–232
  activation, 230–231
  grafting, 232
  ion exchange reactions, 231–232
  purification, 230
  reactions with biomolecules, 232
CNCs, 120, 127
CNF-montmorillonite, 308
CNT actuators, 116–117
Collagen (Col), 129–130, 347, 351, 354–355, 360–361
Computer simulation, 270
Concrete-filled FRP circular tubes (CFFTs), 44
Conducting polymer composites (CPCs), 3–4, 14, 17–18, 22–24, 28
Conducting polymer-based thermoelectric composites, 169
  carbon-based nanostructures/polymer nanocomposites TE materials, 179–186
    CNTs and polymer composites TE materials, 179–183
    conducting composites of polymer/graphite, 186
    conducting polymer/graphene composites, 183–185
  conductive polymers for TE materials, 175
  polymer–inorganic hybrid TE materials, 175–178
    PANi–inorganic TE nanocomposites, 175–176
    PEDOT:PSS–inorganic TE nanocomposites, 177–178
    PTH-inorganic TE nano composites, 176–177
  thermoelectric composites, preparation and processing of, 172–175
Confocal microscope (CM), 413–414
Copper nanoparticles/poly(*o*-phenylenediamine) hybrid nanocomposites, 387
Corrosion, 25
Creatinine biosensor, 113
CSMA/PECA/GO hybrid materials, 364
Custom-shaped noise-suppressing absorbers, 67*f*
Cyclodextrin, 346

D

Defective graphene, 268
Density functional theory (DFT), 268–269
Deoxyribonucleic acid (DNA), 409–413
  agarose gel electrophoresis analysis, 412
  chemical structure, 410*f*
  extraction, 410–412, 411*f*
  and its liquid crystalline phases, 409–410
  labeling, 412–413
Dextran, biomedical application of, 124*t*
Diabetes, 385
Differential scanning calorimetry (DSC), 370–371
2-(Dimethylamino)ethyl methacrylate (DMAEMA), 355
Direct femtosecond laser writing (DLW), 355
Dopamine (DA), 113
Dopamine (DOP), 373–374, 377
Doxorubicin (DOX), 120, 122, 373–374, 379–381
Drug delivery systems, 373–384
  packaging materials and, 308–310
  with hybrid composites prepared from natural polymers, 374–378
  with hybrid composites prepared from synthetic polymer, 378–384
Dye-sensitized solar cells (DSSC), 24–25

E

Elastin, 347
Electrical conductivity variation in binary conducting composites, 4–5
Electrical double layer capacitors (EDLCs), 26–27

Electrical percolation threshold, factors affecting, 6–7
Electrically conducting polymers, chemical structures of, 172*f*
Electroactive paper (EAPap), 115, 134
Electroactive polymers (EAPs) actuators, 114
Electrochemical biosensors, 384
Electrochemical DA sensors, development of, 113
Electromagnetic (EM) waves, 22–23
Electromagnetic compatibility (EMC), 54–55
Electromagnetic emissions (EME), 54
Electromagnetic immunity, 54
Electromagnetic interference (EMI), 2, 22–24, 22*f*, 54, 65*f*
Electromagnetic shielding, 54, 57
Electromagnetic wave absorbers (EMWA), 65
Electrospun carbon nanofibers (ECNFs), 160
Energy sector components, 40–41
Energy technology, alternate, 19–22
Enhanced permeability and retention effect, 118
Environmental scanning electronic microscope (ESEM), 416–417
Enzymatic amperometric glucose biosensors, 110–111
Enzyme-based cholesterol sensors, 394
1-Ethyl-3-(3-dimethylaminopropyl) carbodiimide, 381–382
1-Ethyl-3-methylimidazolium tetrafluoroborate (EMI-BF4), 116
Exchange–coupling interaction, 89–90
Exfoliated graphite nanoplatelets (xGNPs), 13
Exfoliated nanocomposites, 233
Expanded graphite (ExGr), 7–9, 14, 16–17, 186, 187*f*
  formation of, 9*f*
  SEM picture of, 9*f*
  TEM picture of sonicated expanded graphite, 10*f*
Extracellular matrix (ECM), 347, 356–357

F

Faraday's law, 56
Fe_3O_4 nanoparticles, 388–389
Fed-Batch cultivation, 294
Fermi–Dirac distribution, 170
Ferrimagnetic inclusions, composite materials with, 88
Ferrites with hexagonal crystal structure (hexaferrites), 89
Fiber–metal laminates (FMLs), 37
Fiber-reinforced polymer (FRP) composites, 35–37, 41–42
  components of, 152*f*
  FRP–UHPFRC girder system, 44–47, 46*f*
Fibroblast cells, 355–356
Field emission SEM (FESEM), 358–359
Flagellated microbial-robot, 299*f*
Flax/polypropylene underbody components of the Mercedes Benz A-Class, 39*f*
Fluorine doped tin oxides, 24
Food packaging, 235–240
Formaldehyde, 277
Fractional bandwidth, 81–83
Frequency-selective surfaces (FSS), 82–83
Functionalized graphene and chitosan, interaction between, 269–271

G

Gas sensing, metal-doped graphene for, 272–279
Gelatin methacrylamide (Gel-MA) hydrogels, 365–366
Gene theraphy delivery process using a cellulose substrate, 307*f*
Geometrical models, 6
Glass fibers, 35–37
Glass FRP (GFRP), 38–40
Glass-reinforced aluminum laminate (GLARE), 37, 38*f*
Gluconacetobacter xylinus, 293–294
Glucose biosensors, 385–387
Glucose detection, polysaccharide-based biosensors for, 110–111
Glucose oxidase (GOx), 110–111, 243–246, 385–386
3-Glycidoxypropyl trimethoxysilane (GPTMS), 364–365
3-Glycidoxypropyltrimethoxysilane, 16–17
Glycosaminoglycans, 347
Gold nanorods (GNRs), 413, 417–419
Gold nanorods organized in liquid crystalline DNA, 417–419

Gold–platinum alloy nanoparticles/Chi/ionic liquid hybrid composites, 395
Grafted silane, 163
Grafting, 232
Graphene, 156, 299
Graphene nanoplatelets (GNPs), 156
Graphene oxide (GO), 10–11, 268–270, 354–355
Graphene oxide poly(p-phenylene benzobisoxazole) (PBO) fibers, 163
Graphene/PANI composites, 26–27
Graphene-based materials, 267
  applications, 269–281
    interaction between functionalized graphene and chitosan for tissue engineering, 269–271
    interaction between MgH_2 and graphene for hydrogen storage, 280–281
    metal-doped graphene for gas sensing, 272–279
  properties of different grapheme, 268
  theoretical study methodologies, 268–269
Graphene–polyethyleneimine–gold nanoparticles (GNS–PEI–AuNPs) hybrid composites, 385–386
Graphene-β-CD hybrid composite, 395
Graphite, 7–8, 186
  crystal structure of, 8*f*
  expanded, 8–9
Graphite intercalation compounds (GIC), 7–9
Graphite nanoplatelets (GNPs), 14, 16–17
Graphite nanosheets (GNS), 7–8
Graphite oxide (GO)/ordered PANi composites, 186
Graphite oxide to graphene, 10–11
Ground tire rubber (GTR), 216–217

H

Halloysite nanotubes (HNTs), 211–212
Hartree–Fock equation, 268–269
Hexadecylethyldimethylammonium (HDEDMA), 253
Hexagonal crystal structure (hexaferrites), ferrites with, 89
High density polyethylene (HDPE), 200–203
Holographic grating, 413–414
Horseradish peroxidase (HRP), 391–392
Human bone marrow (hBMSC), 123–125
Hyaluronic acid (HA), 347
Hyaluronic acid/hyaluronate (HA), 119
  biomedical application of, 124*t*
Hybrid composite materials, 35–37, 60, 89
Hybrid FRP composites for structural applications, 37–44
  aircraft structures, 37
  automotive components, 37–38
  civil infrastructures, 38–40
  energy sector components, 40–41
  marine transportation, 41–42
  smart structures, 42–43
  sporting goods, 43
  telecommunication components, 43–44
Hybrid FRP–concrete girder/deck systems, 44–47
Hybrid FRP–concrete tubular columns/piles, 47–49
Hybrid NCC materials, in biosensing, 302–307
  bacteria detection, 305–306
  in vivo detection and other sensing applications, 306–307
  small molecules detection, 304–305
Hybrid polymer composites for electromagnetic absorption in electronic industry, 53
  applications, 64–75
    measurements of material parameters of noise-suppressing thin sheet materials, 68–69
    noise-suppressing thin sheet materials, comparison of, 69–75
    of thin noise-suppressing sheet materials, 64–68
  electromagnetic shielding, 64–68
  material's design, 84–95
    reflected wave absorbers, 85–92
    transmitted wave absorbers, 92–95
  shielding effectiveness definitions, 60–64
  specular absorber for far field operation, 76–84
    angular performance, 83–84
    bandwidth-to-thickness ratio, 78–81
    fractional bandwidth, 81–83

Hybrid polymer/clay nanocomposites, synthesis of, 227
classification of clay minerals, 229–230
clay surface treatments, 230–232
activation, 230–231
grafting, 232
ion exchange reactions, 231–232
purification, 230
reactions with biomolecules, 232
polymer/clay nanocomposites (PCNs), 233–234
biomedical applications, 240–246
food packaging, 235–240
preparation and applications of, 234–259
wastewater pretreatment, 246–253
structure of clay, 228–229
Hybrid polysaccharide-based systems for biomedical applications, 107
actuators, 114–117
biopolymer-carbonaceous particles hybrid actuators, 116–117
biopolymer–polymer hybrid actuators, 115–116
biomimetic scaffolds for tissue engineering, 129–134
liquid-crystalline-based scaffolds for tissue engineering, 132–134
biosensors, 110–113
for cholesterol detection, 111–112
for detection of biomolecules, 113
for glucose detection, 110–111
future perspectives, 134–135
nanocellulose and nanochitin, tissue-engineering applications of, 127–129
anochitin/chitosan-based scaffolds, 128–129
nanocellulose-based scaffolds, 128
theranostic systems, 117–123
polysaccharides and gold hybrid systems, 120–121
polysaccharides and quantum dots hybrid systems, 121
polysaccharides and SPIONs hybrid systems, 121–123
polysaccharides in, 118–120
tissue-engineering applications of native cellulose and chitin/chitosan, 123–127
bone and cartilage, 123–126
skin, 126
Hydrogel nanocomposite, 332–333
Hydrogels, 332–333
Hydrogen peroxide biosensors, 391–394
Hydrophobic $CaCO_3$ nanoparticles, 321–322
Hydroxyapatite (HA), 355–356
Hydroxyapatite nanocrystalline inorganic molecules, 348–349
Hydroxypropylcellulose (HPC), 117

I

Ibufen, 373–374
Ibuprofen (IBU), 375–377, 381
Implant materials, 367–373
Indium tin oxide (ITO), 24
Indomethacin, 373–374, 383
Inherently conducting polymers (ICPs), 2–3
Injectable scaffolds, 357–361
Inorganic nanoparticles, encapsulation of, 322–323
Inorganic–polymer composites, 172–173
In-situ polymerization route, 13
Integrated circuits (ICs), 54
Intercalant, 7–8
Intercalated nanocomposites, 233
Iodine doped polyacetylene, 2
Ion exchange reactions, 231–232
Irreversible electroporation (IRE), 310
I-shaped CFRP/GFRP girders, 39*f*
Islamic Republic of Iran, 199

K

Kaolinite-doped polyacrylonitrile (PAN) NCs, 253
Ketorolac tromethamine, 240–241
Kramers–Krönig relations, 78–79

L

Laponite (LAP), 358–359
Layer-by-layer assembled nanocontainer, 335–338
LCR meters and impedance analyzers, 68–69
Li-ion batteries (LIBs), 44
Liquid crystal (LC) phases, 410
Liquid crystalline DNA, 409

deoxyribonucleic acid, 409–413
agarose gel electrophoresis analysis, 412
DNA and its liquid crystalline phases, 409–410
DNA extraction, 410–412
DNA labeling, 412–413
gold nanorods organized in, 417–419
gold nanorods organized in, 417–419
organization, in polymeric template, 413–417
plasmonic nanoparticles, 413
Liquid-crystalline-based scaffolds, 132–134
Localized plasmonic resonance (LPR), 413
Low-density polyethylene (LDPE), 204, 235–236

M

Magnetic-filler materials, 88–90
Magnetite, 356
Magnevist, 122
Maleic anhydride grafted PP (MAPP), 205–207
Marine transportation, 41–42
Material's design, 84–95
reflected wave absorbers, 85–92
magnetic-filler materials, 88–90
mixed-type filler materials, 90–92
nonmagnetic filler materials, 85–87
transmitted wave absorbers, 92–95
mixed-type filler materials, 94–95
nonmagnetic filler materials, 93–94
Materials absorbing electromagnetic energy, 58
MCM-41 nanoparticles, 381–382
Melt compounding, 13
Melt crystallized PP–ExGr/PPS–ExGr composites, 19
Mercedes-Benz, 37–38, 39*f*
Mesoporous silica nanoparticles (MSN), 381–382
Metal oxide fillers, 156–157
Metal-doped graphene, 268
for gas sensing, 272–279
Metallic nanoparticles, 306–307
Metamaterials, 82–83
3-Methacryloxypropyl-trimethoxysilane (3MPS), 201–202
MgH_2 and graphene for hydrogen storage, interaction between, 280–281
Microcrystalline cellulose (MCC) fibers, 205–207
Microfibrillated cellulose films (MFC), 296–297
Mixed-type filler materials, 90–92, 94–95
Modified Eagle Medium solution, 350–351
Molecular dynamics (MD) method, 268–269
Monolithic rectorite/starch composites (PRSs), modification of, 251
Montmorillonite (MMT), 202–203, 211, 326*f*, 384
MMT-polypyrrole (PPy) hybrids, 243
Multifunctional composites, 154–160
applications of, 161–163, 162*f*
multifunctional hybrid composites, 157
multifunctional hybrid nanocomposites, 158–160
multifunctional nanocomposites, 156–157
multifunctional natural fibers composites, 155
multifunctional synthetic fibers composites, 155
Multifunctional materials, 153–154
applications of, 161–163
Multiwalled carbon nanotube (MWCNT), 16–17, 86–87, 91, 93–94, 207–209

N

Nanocellulose and nanochitin, tissue-engineering applications of, 127–129
anochitin/chitosan-based scaffolds, 128–129
nanocellulose-based scaffolds, 128
Nanocellulose-based scaffolds, 128
Nanochitin, 127–128
Nanocomposites, 153
Nanocrystalline cellulose (NCC), 293–295
Nanofiller particles, functionalization of, 321–322
Nanopapers, 308
Natural fibers, different applications of, 36*f*
Natural polymers
drug delivery with hybrid composites prepared from, 374–378
Natural polysaccharides, 108–109

NCs, 199–217
obtained from other recycled polymers, 215–217
obtained from polyethylene terephthalate (PET), 207–212
obtained from polypropylene (PP), 204–207
obtained from polystyrene (PS), 212–214
obtained from recycled polyethylene (PE), 200–204
Nematic liquid crystals (NLCs), 413–414
Nicolson–Ross–Weir (NRW) method, 68–69
Nielsen model, 6
N-isopropylacrylamid (NIPAAM), 378–381
Noise suppressor sheet (NSS), 66–68, 66*f*, 68*f*, 71*f*
Noise-suppressing thin sheet materials
comparison of, 69–75
measurements of material parameters of, 68–69
Nonisotopic DNA labeling procedures, 412–413
Nonmagnetic filler materials, 85–87, 93–94

O

O(6)-benzylguanine, 122
Oil palm empty fruit bunch fibers (OPEFB), 153–154
Organic conducting polymers, 175
Organic–inorganic (hybrid) composites, 394
Organic–inorganic hybrid materials, 343–344
Organic–inorganic hybrid network, ramified structure of, 351*f*
Organometallic polymer conductors, 3
Organophosphate pesticides (OPs), 395–397
Organosilane, 16–17

P

Packaging materials and drug delivery systems, 308–310
PANI (p-type)/ZM (n-type) heterojunction nanocomposite, 330–331
PANi nanocomposites, 174–175
PANi/PbTe composite powders, 174–175
PANI/TiO_2 composite nanoparticles, 324
PANI/ZM nanocomposite, 329–331, 329*f*
PANi–inorganic TE nanocomposites, 175–176
Paper-based biosensors, 112
Paraastemol, 373–374
PDOS (partial density of state) analysis, 270, 274*f*
PEDOT:PSS–inorganic TE nanocomposites, 177–178
Peltier coefficient, 170
Peltier effect in TE refrigeration, 170
Peltier–Seebeck effect, 170
PEM fuel cell, structure of, 20*f*
Percolation theory, 5–6
Perdew–Burke–Ernzerhof-generalized gradient approximation, 269
PE-terephthalate (PET)-fiber-filled PP composites, 18–19
PHEMA/clay nanocomposites, 384
Physical mixing preparation methods, 172–173
Piezoelectricity, 115
Plasmonic nanoparticles, 413
Plastic recycling, 198, 198*f*
Platinum metal catalyst, 24–25
Platinum nanourchins (PNUs), 296–297
Platinum-coated fluorine doped tin oxide, 24–25
PLGA/SF-based hybrid scaffolds, 361
Polarizing optical microscopy (POM), 132, 413–414
POLICRYPS (acronym of polymer LC polymer slices), 413–414
Poly(2-(dimethylamino)ethyl methacrylate) (PDMAEMA), 120–121, 309–310
Poly(2-hydroxyethyl methacrylate) (pHEMA) gelatin, 346, 378–379
Poly(3,4-ethylenedioxythiophene) (PEDOT), 175
Poly(3,4-ethylenedioxythiophene) (PEDOT): PS sulfonate (PSS) composite electrodes, 24
Poly(3-aminophenylboronic acid) units, 323–324
Poly(3-hexylthiophene) (P3HT), 177
Poly(3-hydroxybutyrate) foams, 163
Poly(3-octylthiophene)/silver nanocomposites (P3OT/Ag), 176–177

Poly(acrylamide)–kaolin (PAAm-K) hydrogel nanocomposite, 333
Poly(acrylamide)–kaolin composite hydrogel, 332–335
Poly(acrylic acid) (PAA), 132–134
Poly(caprolactam), 346
Poly(D,L-lactide) (PLA)/MMT NC, 241–243
Poly(e-caprolactone) (PCL)/TiO_2 particle, 356
Poly(e-caprolactone)/ZrO_2 particle, 356
Poly(glycerol sebacate) prepolymer (pre-PGS), 352–353
Poly(L,D lactic acid) (PLA)/Mg hybrid polymer composites, 370
Poly(lacticco-glycolic acid) (PLGA), 355
Poly(methyl methacrylate) (PMMA)/$CaCO_3$ nanocomposite, 324, 325*f*, 326*f*, 332
Poly(methylmethacrylate) (PMMA)–clay NC, 243–246
Poly(*m*-phenylenediamine)–Prussian blue hybrid film, 387
Poly(NIPAAM)(PNIPAAm)/$CaCO_3$ micro/nano hybrid composites, 383
Poly(poly(ethylene glycol)ethyl ether methacrylate) (PPEGEEMA), 120–121
Poly(pyrrole)-latex hybrit composite, 387
Poly(styrene-co-methyl methacrylate)/montmorillonite nanocomposite, 324–325
Poly(vinyl) alcohol (PVA) films, 308
Poly(vinylidene fluoride) (PVDF), 255
Poly(vinylidene fluoride-trifluoroethylene), 381
Poly(ε-caprolactone) (PCL)/poly(lactic acid) (PLA) blend, 17–18
Polyamine, 378–379
Polyaniline (PANI), 23–24, 185, 390
  PANI-hybrid inorganic/organic composites, 394
  PANI–TNT hybrid composites, 386–387
Polyaniline/carbon nanotubes composites, 387
Polyaniline-ZrO_2 hybrid composite, 387
Polydimethylsiloxane–BG–poly (caprolactone) (PDMS–BG–PCL) hybrid scaffold materials, 349–350
Polyelectrolytes, 377–378
Polyethylene glycol (PEG), 346, 378–379, 383
Polyethylene terephthalate (PET)
  NCs obtained from, 207–212
  recycling, 208*f*
Polylactic acid (PLA), 215–216, 352–353
Polymer blending, 216–217
Polymer charge transfer complexes, 3
Polymer composites, 151–153
Polymer hybrid composites, biomedical usage areas for, 346*f*
Polymer nanocomposite, properties of, 322–327
Polymer recycling, 197–198
Polymer silicate nanocomposites, biomedical applications of, 242*f*
Polymer/clay nanocomposites (PCNs), 228–229, 233–234
  preparation and applications of, 234–259
    biomedical applications, 240–246
    food packaging, 235–240
    wastewater pretreatment, 246–253
Polymer–graphite binary and hybrid composites, 1
  applications, 19–28
    alternate energy technology, 19–22
    anticorrosion coatings, 25–26
    electromagnetic interference shielding devices, 22–24
    solar cells, 24–25
    supercapacitors, 26–27
  carbon nanofiber, 11
  conducting polymers and classification, 2–3
    conducting polymer composites (CPCs), 3
    inherently conducting polymers (ICPs), 2–3
    organometallic polymer conductors, 3
    polymer charge transfer complexes, 3
  electrical conductivity variation in binary conducting composites, 4–5
  electrical percolation threshold, factors affecting, 6–7
  future perspectives, 28
  geometrical models, 6
  graphite and graphite derived high aspect ratio fillers, 7–11

Polymer–graphite binary and hybrid composites (*Continued*)
expanded graphite, 8–9
graphite, 7–8
graphite oxide to graphene, 10–11
percolation theory, 5–6
processing methods, 11–13
in-situ polymerization route, 13
melt compounding, 13
powder mixing route, 12
solution blending route, 12–13
structure development and properties enhancement in conducting polymer composites, 13–19
electrical properties, 14–15
mechanical properties, 16–17
thermal properties, 17–18
thermoplastic–graphite composites, structure development in, 18–19
structure oriented model, 6
thermodynamic model, 6
Polymeric template, liquid crystalline DNA organization in, 413–417
Polymer–inorganic hybrid TE materials, 175–178
PANi–inorganic TE nanocomposites, 175–176
PEDOT:PSS–inorganic TE nanocomposites, 177–178
PTH-inorganic TE nano composites, 176–177
Polymers-like poly-L-lactide (PLLA)/ magnesium and PLLA/magnesium Fluoride hybrid composites, 370–371
Polyolefin plastics, 200
Polyphenylene sulfide (PPS)–graphite–ExGr hybrid composites, 14
Polypropylene (PP), NCs obtained from, 204–207
Polypyrrole (PPy), 25, 27
Polypyrrole (PPy)/platinum nanoparticles (Pt NPs) hybrid nanocomposite, 392–393
Polysaccharides, 107–108
and gold hybrid systems, 120–121
and quantum dots hybrid systems, 121
and SPIONs hybrid systems, 121–123
Polystyrene (PS), NCs obtained from, 212–214
Polystyrene sulfonate (PSS), 378–380
Polyurethane (PU), 93–94, 346, 378–379
Polyvinyl alcohol (PVA/PVOH), 346, 389–390
Polyvinylpyrrolidone (PVP)-silica hybrid structure, 344–346
Porous rectorite/starch xanthate composites (PRSX), 251
Potassium monopersulfate (KMPS), 256
Powder mixing route, 12
Power integrity (PI), 55
Poynting theorem, 64
PP/rubberwood flour (RWF) composites, 205–207
Printable electronics, 300
Printed circuit wearable biosensors, 300
Pristine graphene, 268
Proteoglycans, 347
Proton exchange membrane fuel cells (PEMFCs), 19
PSBA–montmorillonite clay (PSBA–MMT) films, 238
Pseudocapacitors, 26–27
PTH-inorganic TE nano composites, 176–177

Q

Quantum mechanics (QM), 268–269

R

Radar absorber, operating waveband of, 80–81
Recycled PET and MWCNTs-GI, possible interactions between, 208*f*
Recycled polyethylene (PE), 216–217
NCs obtained from, 200–204
Recycled polymer NCs, properties and applications of, 218*t*
Recycled polymers, NCs obtained from, 215–217
Recycled PP/PSP, 204–207
Reduced graphene oxide (rGO), 87
Reflected wave absorbers, 85–92
magnetic-filler materials, 88–90
mixed-type filler materials, 90–92
nonmagnetic filler materials, 85–87
Reflection bandwidth, 90
Reflection coefficient, 76
Reflection loss, 56–57, 62–63

Reflection over a wide frequency range, 56–57
Regenerated cellulose, 291–292
Regenerated silk fibroin (RSF), 358–359

S
Scaffold applications, 362–367
Scaffolds, 347–348
Schelkunoff decomposition, 62–63
Schrodinger equation, 268–269
Seebeck coefficient, 170, 175–176
Semisynthetic polysaccharides, 108
Sepiolite clay, 255
Sepiolite/PVI composite, 248–250
Shape memory alloys (SMA), 42–43
Shape memory polymers (SMPs), 42–43
Shielding effectiveness (SE), 54–55, 57, 60–61, 63
Shielding effectiveness definitions, 60–64
Shielding efficiency, 23–24
Signal integrity (SI), 55
Silane grafting, 232
Silica/chitosan hybrid scaffolds, chemical structure of, 366*f*
Silica/PEG (SiO_2/PEG) polymeric hybrid composite material, 383
Silica-based hybrid polyimide material, 345*f*
Silicate clays, layered, 230
Silicon, 298–299
Silicon-based solar cells, 28
Single-walled carbon nanotubes, structure of, 45*f*
Single-walled nanotubes (SWNTs), 174–175
Skin, 126
Small molecules detection, 304–305
Smart materials, 153–154
Smart structures, 42–43
Solar cells, 24–25
Solid-waste management, 198–199
Solution blending route, 12–13
Solution mixing, 172–173
Sonochemical preparation of hybrid polymer nanocomposites, 321
  case studies, 327–338
    layer-by-layer assembled nanocontainer, 335–338
    PANI/ZM nanocomposite, 329–331, 329*f*
    PMMA/$CaCO_3$ nanocomposite, 325*f*, 332
    poly(acrylamide)–kaolin composite hydrogel, 332–335
  future perspectives, 338
  properties of polymer nanocomposite, 322–327
Specular absorber for far field operation, 76–84
  angular performance, 83–84
  bandwidth-to-thickness ratio, 78–81
  fractional bandwidth, 81–83
Spinel ferrites, 88
Sporting goods, 43
Stain less steel (SS) counter electrodes, 24–25
Starch, 346
Stimuli-responsive nanocomposite-derived hydrogels, 360–361
Stober's process, 336
Stone–Wales defective graphene, 268
Strontium (Sr)-doped organic–inorganic 3D hybrid porous bone scaffolds, 350–351
Strontium ranelate (SRA), 243
Structure development and properties enhancement in conducting polymer composites, 13–19
  electrical properties, 14–15
  mechanical properties, 16–17
  thermal properties, 17–18
  thermoplastic–graphite composites, structure development in, 18–19
Structure oriented model, 6
Styrene–butadiene-rubber (SBR)-enhanced electrical conductivity, 14
Supercapacitors, 26–27
Superparamagnetic Fe_3O_4 NPs (SPIONs), 118, 121–123
Syndiotactic PS (sPS), 13
Synthetic hydroxyapatite, scaffolds "decorated" by, 352*f*
Synthetic polymer
  drug delivery with hybrid composites prepared from, 378–384

T
Telecommunication components, 43–44
Theranostic systems, 117–123
  polysaccharides and gold hybrid systems, 120–121
  polysaccharides and quantum dots hybrid systems, 121

Theranostic systems (*Continued*)
polysaccharides and SPIONs hybrid systems, 121–123
polysaccharides in, 118–120
Thermally reduced graphite oxide (TRGO), 16–17
Thermodynamic model, 6
Thermoelectric (TE) composites, conducting polymer-based, 169
Thermoplastic corn starch (TPS), 240
Thermoplastic resins, 37–38
Thermoplastic–graphite composites, structure development in, 18–19
Thermosets, 199
Thin noise-suppressing sheet materials, application of, 64–68
Thomson effect, 170
Three-dimensional interconnected graphene layers (3DGS), 16–17
Three-enzyme creatinine biosensor, 113
"Thru-reflect-line" pattern, 74–75
TiO_2 nanotubes (TNTs), 373
Tissue engineering, 123–134
applications, 347–373
biomimetic scaffolds, 356–357
bone scaffolds, 348–356
implant materials, 367–373
injectable scaffolds, 357–361
other scaffold applications, 362–367
scaffolds, 347–348
wound-healing materials, 361
applications of nanocellulose and nanochitin, 127–129
anochitin/chitosan-based scaffolds, 128–129
nanocellulose-based scaffolds, 128
applications of native cellulose and chitin/chitosan, 123–127
bone and cartilage, 123–126
skin, 126
biomimetic scaffolds for, 129–134
liquid-crystalline-based scaffolds, 132–134
Tissue-engineering applications of native cellulose and chitin/chitosan, 123–127
bone and cartilage, 123–126
skin, 126
Titanium dioxide–cellulose hybrid nanocomposite, 387
Titanium-based and magnesium-based FMLs, 37
Toxic tetrachlorodibenzo-*p*-dioxin (TCDD), 276
Transmission electron microscopy (TEM), 358–359
Transmitted wave absorbers, 92–95
mixed-type filler materials, 94–95
nonmagnetic filler materials, 93–94
Transpolyacetylene, 2
TX-100 wind turbine blade, 41*f*

U
Ultra-high-performance fiber-reinforced concrete (UHPFRC) slabs, 44–46
Ultrasound-assisted emulsion polymerization, 322
Ultrasound-assisted methods, 323, 334–335
Urea biosensors, 388–390
UUV propulsion, 296

V
Vapor grown CNFs (VGCNFs), 11
TEM picture of commercial VGCNF, 11*f*
Vector network analyzers (VNAs), 68–69, 70*f*
Visby class, 41–42

W
Water vapor permeability (WVP), 237–238
Water vapor transmission (WVTR), 237–238
Wiedemann–Franz law, 171
Wizard Genomic DNA Purification Kit, 410–412
Wood/recycled PP composites (WRPCs), 205–207
Wood–plastic composites (WPC), 201–202, 213–214
Wound-healing materials, 361

X
X-ray diffraction (XRD), 350–351

Y
Young's modulus, 370